PALEOIMAGERY

AF903846

PALEOIMAGERY

The Evolution of Dinosaurs in Art

Allen A. Debus *and*
Diane E. Debus

FOREWORD BY DONALD F. GLUT

McFarland & Company, Inc., Publishers
Jefferson, North Carolina, and London

The present work is a reprint of the illustrated case bound edition of Paleoimagery: The Evolution of Dinosaurs in Art, *first published in 2002 by McFarland.*

Also of Interest

Prehistoric Monsters: The Real and Imagined Creatures of the Past That We Love to Fear, by Allen D. Debus (McFarland, 2010)

Dinosaurs in Fantastic Fiction: A Thematic Survey, by Allen D. Debus (McFarland, 2006)

Library of Congress Cataloguing-in-Publication Data

Debus, Allen A.
Paleoimagery : the evolution of dinosaurs in art / Allen A. Debus and Diane E. Debus ; foreword by Donald F. Glut.
p. cm.
Includes bibliographical references and index.

ISBN 978-0-7864-6420-3
softcover : 50# alkaline paper ♾

1. Dinosaurs in art. 2. Paleoart—History.
I. Debus, Diane E. II. Title
QE861.4.D43 2011 567.9'022—dc21 2002007625

British Library cataloguing data are available

© 2002 Allen A. Debus and Diane E. Debus. All rights reserved

No part of this book may be reproduced or transmitted in any form or by any means, electronic or mechanical, including photocopying or recording, or by any information storage and retrieval system, without permission in writing from the publisher.

On the cover (top to bottom): *Triceratops and Tyrannosaurus*, Charles Knight (courtesy of the Field Museum of Natural History); *Orodromeus*, Douglas Henderson (reproduced with the permission of Mr. Henderson); *Painting of the K-T fireball*, Douglas Henderson (reproduced from *Asteroid Impact* [Dial, 2000], with the permission of Mr. Henderson)

Manufactured in the United States of America

McFarland & Company, Inc., Publishers
Box 611, Jefferson, North Carolina 28640
www.mcfarlandpub.com

To Juanita Ball:
Without her love, help and encouragement,
this book would not have happened.

To Allen G. and Brunilda Debus for their influence,
and to Kristen, Lisa and Jennifer for their patience and love,
and Vicki Wildman too!

Table of Contents

Foreword

I've been writing nonfiction books on various subjects for more than three decades. Many of these books have focused upon dinosaurs and other aspects of paleontology, topics in which I am keenly—even passionately—interested. One salient truth I have learned during these 30-plus years is that my most successful fac[illegible]th financially and critically, remain those which, at one time or another, I had wanted to *read* but which had not yet been written.

The key to a successful book is probably some combination of knowledge in a subject, passion for that subject, and the willingness to invest time and effort in writing about the subject. Together these three things produce the book that the author had wanted to read all along—and that others now will want to read as well.

This book, written by my good friends Al and Diane Debus, is a good example of a product in which the above factors all come into play. The subject is "paleoimagery," the definition of which will become clear as you read through the book's many chapters. The "Debi" (plural for Debus), the husband and wife team that squeezed all this terrific information between two covers, obviously know and love their subject, and they have done a remarkably thorough job (indeed, one of dinosaurian proportions) in bringing it, like a paleontologist assembling the component parts of a single massive fossilized skeleton, all together.

I've known Al and Diane, whom I originally met through correspondence, since the early 1980s. Al particularly shared my passion for the ancient world and its denizens. It was an interest in what—from the moment this book sees publication—may be for all time referred to as "prehistoria," a term that, perhaps not surprisingly, both Al and I independently concocted, although the two coinings were separated by some four and one half decades.

My own invention of this word was based, simply and honestly, upon childhood greed. As a young boy I was a big fan of the 1950s science-fiction television series *Space Patrol*. In 1954 that program launched a contest in which the grand prize was the ultimate kids' dream-come-true—a full-sized, truck-drawn rocket ship called the *Terra IV*. The ship actually "worked," doing just about everything but taking off, and the truck that came with it allowed it to travel through space, at least as long as that "space" was at street level. *Oh, how I wanted that rocket!* To win this most coveted of all the early TV premiums, all I had to do was name "Planet X," the enormous planet ruled by the evil Prince Baccarratti. As that planet was a prehistoric-type world complete with such threats as dinosaurs, cavemen and man-eating plants, I immediately came up with the name "Prehistoria." Unfortunately, this junior Space Patroller procrastinated to the extent that I never mailed out my contest submission. (I *still* want that spaceship.)

Al Debus had far loftier (and more Earth-bound) goals in mind when he came up with *his* "prehistoria." My recollection of Al's use of this word is fairly recent (even in the non-geological sense), within just the past couple of years. "Prehistoria" (*sensu* Debus) refers to the *gestalt* spanning the hundreds of millions of years preceding the dawn of recorded history—that sum total of fauna and flora and environment and climate and everything else that collectively make up an incomprehensively vast chronological stretch sometimes referred to as "deep time."

"Paleoimagery" gives substance to a world otherwise known only from the interpretation of fossil and geological evidence, much of this evidence quite fragmentary in nature. In short, most of what the public at large understands or believes to be true of these prehistoric life forms is *not* the result of studying their fossilized remains, but of experiencing the drawings, paintings and sculptures that constitute paleoimagery. And without prehistoria, there would be no paleoimagery.

Like Al and Diane, I have long been fascinated by the myriad ways extinct life forms have been depicted over the years, and by the artists—particularly the great Charles R. Knight—responsible for those depictions. I remember, as a child, spending hours poring through books about dinosaurs and other extinct life forms, noting how the reconstructions of those creatures varied, sometimes dramatically, from one picture to the next. Although I didn't know it at the time, such restorations could run the gamut from extreme accuracy to complete fancy, depending upon the completeness of the fossil materials upon which such depictions were based and how closely the artist based his or her work on those materials. Also they could have been influenced by current or "in vogue" scientific theories, not to mention, on occasion, factors as disparate as personal political agendas and religious beliefs.

Like Al in particular, I have long had a passion for life-sized figures of prehistoric animals, a kind of sub-genre of paleoimagery. I will *never* forget how, during a visit with relatives one fateful evening in 1954 or '55, somebody mentioned that there was a place called Dinosaur Park in Rapid City, South Dakota, where full-scale models (made by well-known local sculptor Emmett A. Sullivan, I would later learn) of those ancient animals were displayed atop a hill. I pestered my mother for months until she finally relented, agreeing—maybe after finding out that South Dakota's Black Hills had other things to offer tourists, e.g. Mount Rushmore and the famous "Passion Play"—that we would spend our 1955 summer vacation in part ogling those giant concrete figures. It was the first of many such parks I would see over the years and remains, in my grownup memories (despite the scientific inaccuracies of the figures), the one most cherished.

Yes, "paleoimagery"—and the "prehistoria" upon which it has been based—have both been passions of mine for many years. I am grateful that Al and Diane Debus feel the same way, and feel it so strongly that they have written this book—with scholarship and skill and passion. I thank them both for contributing this unique work to my library of "prehistoria"-based nonfiction, a book I have, for a long time, wanted to read.

Donald F. Glut
Burbank, California

Preface

You can never go back. That's what they say and, to our dismay, it's what we know from experience. Just as senior citizens dream of going back to recapture the glory days of youth, paleontology aficionados, like myself, savor the thought of going back ... way back, getting back to primeval periods when "prehistoria" roamed our planet. Dinosaurs and other prehistoric animals persist in human imagination because we *need* to see them "alive." They are creatures of our own making, monsters from our id. And, in a sense, they *are* us. *Paleoimagery* is one valuable means for restoring our favored prehistoria to life.

Not long ago, my wife and dedicated accomplice, Diane, asked me one of those hypotheticals that guys hate thinking about. "If you had the chance, what individual, living or dead, would you most care to meet?" Too fuzzy-minded at the time to answer, I let the question nestle in my subconscious.

Then, several days later, the answer came. What person? Only one. How intriguing to meet Charles Darwin, Albert Einstein, H. G. Wells, or even the last Neanderthal. Yet I knew who it would be: Charles R. Knight, the artist who, more than any other, shaped my visions of prehistory. You will read more about Knight and others of his calling in *Paleoimagery*. For this volume is all about time, travel, and time travel.

Twenty years ago I wrote a 1,000 word article, my first delving into paleontological terrain, entitled "Of Time Travelers & Mastodons." It was printed in *Crystal Cluster*, the news bulletin of the Des Plaines Valley Geological Survey. The article had been inspired by listening to a lecture concerning a reconstruction of America's noble Mastodon, now displayed at Wheaton College.

I recall fearing that anyone I knew might actually stumble across that issue of *Crystal Cluster*. If they read it, would they scoff at my penchant for "prehistoria"?

Since that first wave of temerity, Diane and I together have enjoyed the avocation of writing about and collecting just about anything having to do with prehistoric themes. However, no paleontological topic thrills us more than paleoimagery. We are a visually oriented species, after all, and beyond the evolutionary ideas inspired by or inherent in old bones, paleontologists' collective goal is to learn how fossil organisms, prehistoric vertebrates especially, appeared in life.

Inspirations principally fueling "prehistoria" include: (1) my father's scrapbooks pasted-in with newsclippings from Chicagoland's papers and other snippets of 1930s dinosaur memorabilia; (2) a "modern" scrapbook by Donald F. Glut, titled, what else, *The Dinosaur Scrapbook* (1980); (3) another classic—the one that lured me back into paleontology after completing a chemistry degree at Northwestern University—Adrian J. Desmond's *The Hot-Blooded Dinosaurs: A Revolution in Palaeontology* (1975); and (4) that riveting decade of dinosaur madness—the 1980s. During the 1980s, "prehistoria" and those who restored them from Time's tomb became media celebrities. Our perceptions of dinosaurs and other prehistoric animals "evolved" at an unprecedented, if not alarming, rate.

The 1980s and 1990s were post-college, post–Bakkerian formative years for *Paleoimagery*, as I regressed back to yesteryear's childhood fascinations, namely dinosaurs, fossils and paleontology. During these years we began writing about favorite animals of long ago, and the people and institutions responsible for restoring them to life.

Several years before Internet surfing became all the rage, dinosaur groupies and enthusiasts banded together, zealously united through their subscription to a handful of paleo-mags and dinosaur fanzines. On the heels of *Jurassic Park*'s record-setting release in 1993, the *Prehistoric Times* quickly rose to prominence, but counted among the ranks

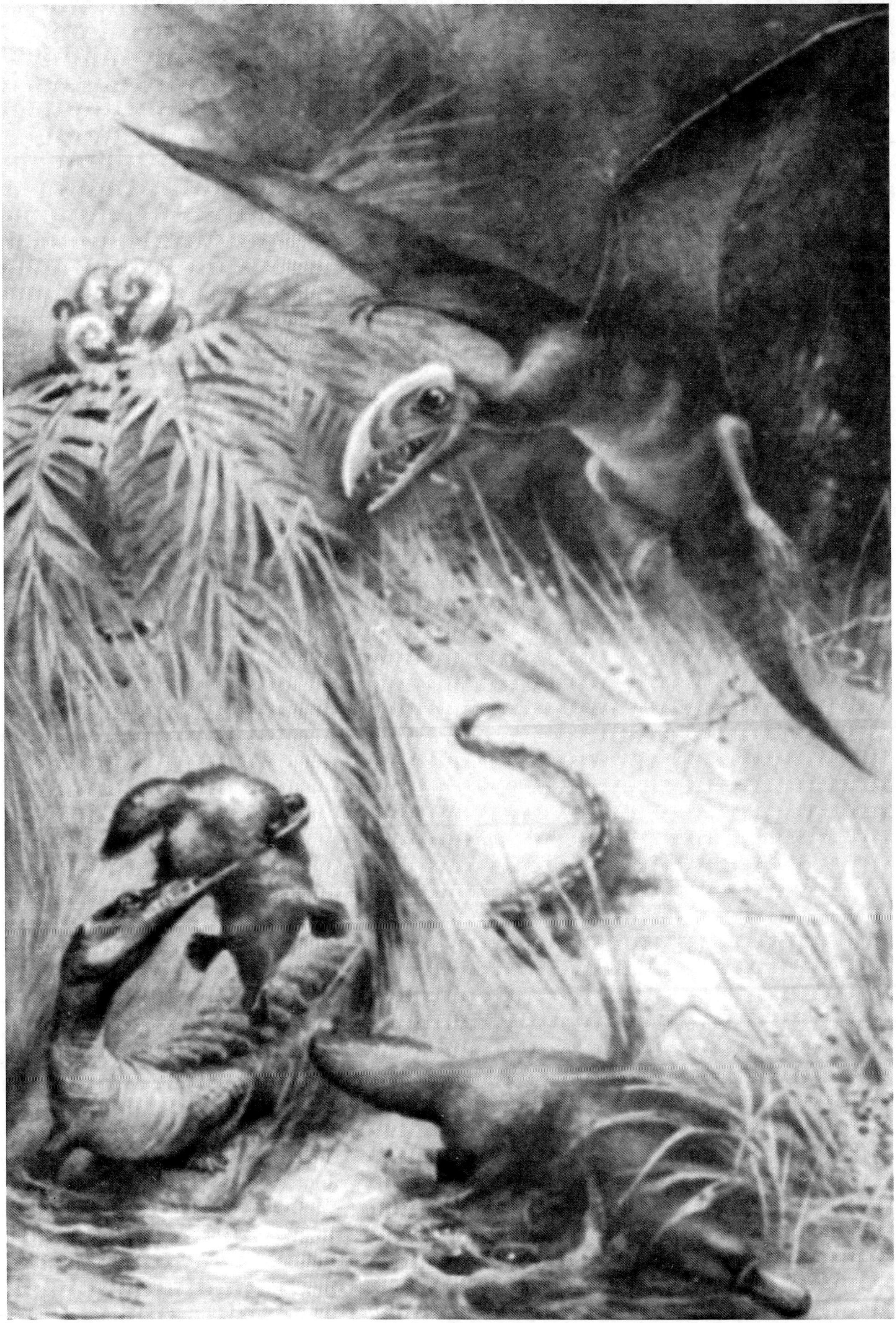

of publications were *Fossil News*, *The Mosasaur*, *Dinosaurus Magazine*, *Archosaurian Archive*, DinoPress, the Western Australian Museum's *Dinonews*, and news bulletins of both the U.S. and UK Dinosaur Societies.

Inspired by these worthy publications, I decided to present original material in another publication. So I joined with Gary Williams and John Lanzendorf to produce a new brainchild—*Dinosaur World*—conceived in the fall of 1996. Delving further into that dinosaur scrapbook universe than any paleomag had gone before, *Dinosaur World*'s mission has been to satisfy the continual cravings of dinosaur enthusiasts globally, and to probe all dimensions of the "Paleo-scene." Many of the chapters included in the present work were printed originally in these publications. Chapters 2, 10, 12, 13, 18 and 30 are presented under sole authorship. The rest were co-written with Diane.

So, what "prehistoria" lurk within *Paleoimagery*?

Paleoimagery addresses evolving ideas about prehistoric animal imagery in its varied manifestations (e.g. as science tool, cultural symbol, etc.). The recurrent theme in *Paleoimagery* is *visual paleontology*. Each chapter has anything and everything to do with prehistoric animal image-making and paleoart "imagetext."

Here you will encounter Chicago's regal *Tyrannosaurus rex*—her Highness, "Sue"; a genuine monster from the id—*Kronosaurus*, and a very special "inside-out" Mastodon. We will tickle you with dinosaur feathers! You will "visit" the Paleoartists' Hall of Fame and learn about Wisconsin's great dinosaur race. Prehistorians will be encountered at every bend along the way—giant pigs, plated and thorny lizards, mysterious giants of the Mesozoic. Throughout, we refer to our favorite prehistoric vertebrates—dinosaurs, aquatic reptiles, pterosaurs, and the great prehistoric mammals. Although modern paleontologists distinguish dinosaurs as a breed of organisms different from other prehistoric reptiles, during the 19th century, dinosaurs, marine reptiles, pterosaurs and prehistoric crocs such as teleosaurs, phytosaurs and giant turtles (chelonia) alike were often referred to as "saurians."

While there is a chronological flow to the chapters, from bare bones beginnings to the present Dinosaur Renaissance, it was not our intention to write a definitive history of the genre. *Paleoimagery* is an essay collection. It was also not our intention to write a definitive account of the evolution, classification and discovery of dinosaurs and prehistoric animals, or the entire history of vertebrate paleontology. Such books have already been written, expertly.

Why is this different from the scores of other dinosaur books already published? No books have treated the subject of paleoimagery in quite this vein before. There is a wealth of literature concerning dinosaur science and the history of paleontology, yet relatively little addressing paleoimagery. Read our new book, *Dinosaur Memories* (iUniverse, 2002) for additional insights on paleoimagery's nature and history.

In a book titled *Paleoimagery* one would hope to present myriad examples of paleoart and other imagery inspired by prehistoric animals. Because such an endeavor would quickly become cost-prohibitive, the authors have instead striven to place paleoart in historical perspective, considering how modern Dinosaur Renaissance artists have relied on their predecessors. For modern paleoart can be best understood in light of themes and landmark restorations created by the old masters. To incorporate other significant examples of paleoart not reproduced in this volume, the authors often refer to other books and sources cited in footnotes.

More than any other influence, the movie *Jurassic Park* introduced computers to the realm of paleoimagery. Thus, this short "history" of the genre comes at a pivotal time in the unfolding history of paleoart—when creative paleoartists find themselves increasingly reliant on software and cyber-images. I am fond of the old classic restorations, and revere the paleoart masters who created them long before the dawn of computer technology—using ink, pencil, canvas, paper and clay. The forthcoming restorations of our new millennium may outstrip those of yesteryear. But will they ever seem as magical or possess the same charm?

Each person's experience of paleoimagery must be unique because visual perceptions differ on an individual, psychological basis. We all see and experience life and art in our own unique ways. Therefore, one's journey through the world of paleoimagery is collective yet personal, shaped by individual experiences, moments, events and people. If not for each of the individuals mentioned below, our personal journey would have taken us to a different plateau, perhaps a lost one in which this book never would have appeared. Accordingly, the authors are especially indebted to the following people, for quite a number of reasons.

Opposite: **Early twentieth century restoration by C. Whymper illustrates fascination with prehistoric world's savagery while also blending in evolutionary ideas as Mesozoic reptiles (*Teleosaurus*—left, *Dimorphodon*—upper right) prepare to dine upon marsupial "duck-bill" snacks. (Figure from Henry Knipe's *Nebula to Man* [New York: J. M. Dent & Company, 1905].)**

With the exception of those most readily recalled names appearing near the very beginning of this compilation, the rest are not listed in any particular order. There is insufficient space to relate why they are individually named. We remember why though.

Allen G. and Brunilda L. Debus, Juanita Ball, George and Edna Debus, Kristen Debus, Lisa Debus, Rick and Michelle Debus, Karl Debus, Tanner Wray, Lynette Borders, Don Glut, Lynne Clos, Mike Fredericks, Rhoda Knight Kalt, Nicholas Hotton III, John and Ruth Lanzendorf, Gary and Glenda Williams, Richard Rush, Steve Brusatte, Brian Choo, John Long, Jack Arata, Ernie V. Nelson, Paul and Serge Dupuis, David Raup, William E. Swinton, Shannon S. Yeager, Zella Jonas Merritt, Edwin H. Colbert and Margaret M. Colbert, Donald Baird, Karol Sabath, Berislav Krzic, Riff Smith, Dan Petitpas, Jay Matternes, Copeland MacClintock, Jean Reynolds, Sara Debus, Jenny Debus, "B. C." Andrews, Mary Ann Gaydos, Diana Duckworth, Maurice Burton, William Stout, Douglas Henderson, Michael Skrepnick, Liz Moore, Vratislov Mazak, Krzysztof Kuchnio, Klaus Westphal, Chris Ott, Steve McCarthy, Gale Hruska, Kenneth Carpenter, Bob Morales, Tracy Ford, Paul Sereno, Ken Proffitt, Ely Kish, Tony Timpone, Jeff Liston, John Sibbick, Beverly Halstead, John Ostrom, Mike Henderson, David Thomas, Norbert Vavra, Ron Grey, James Gurney, Shane Foulkes, Cliff Green, David Weishampel, Ray Rimell, Walter Alvarez, Chuck Pillmore, Jack Bowman Bailey, Donald Prothero, Jim Gouldsberry, Stephen J. Gould, Wojtek Skarzynski, James Secord, Richard Rush, Eva Hochmanova, Kiki Halcumb, Dean Walker, Ellis Yochelson, Peter Crowther, Philip Dougherty, Ken James, Gerald M. Friedman, Steve Moshier, Thomas Kovacs, Daniel Searfoss, Thomas Carr, Richard Larmouth, Marilyn Gillespie, Brooks Britt, Bill Farrand, Michael Trcic, Pete von Sholly, Russell Hawley, John Rafert, Robert Bakker, John Klausmeyer, Robert Sanchez, John Fischner, Dan Alldredge, William J. T. Mitchell, William A. S. Sarjeant, Jeff Wells, Kim Hoffman (and "Chad"), Travis Hoffman, Peter Ward, John Bindon, Ted Daeschler, Peter Sheehan, Dougal Dixon, Rogelio Oran Fontaine, Gary Staab, Arthur B. Evans, Neil Gipson and Doug Sonnenschen, Debra Meier, R. George Corner, Brad Kosch, Joe Small, E. F. Allin, Christopher McGowan, Jeff Morris, Kevin Padian, Mark Hallett, Luis Rey, Tony Campagna, Franz Gruss, Elizabeth Lach, Bob Johnson, Diana Fucci, Russ Jacobson, Mike Evans, Frank Cozza, Bob Egan, Wen Huang, Mike Howgate, Dan LoRusso, Dennis Eberl, Robert M. Garrels, Matt Manit, John Lavas, Julie Rogers, Bob Nicholls, David Douglass, Charlie McGrady, David Peters, Tom Holtz, Jack Horner, Joseph M. Ramirez, George Olshevsky, Bill Vinikour, Barbara Slater and "Mark," Joe DeMarco, Greg Van Wie, Terry Hanson, Bruce Arthur Glazer ... and our furry "Masto-dogs"—Monty, Scruffer and Sarge, whose company I enjoyed while I typed articles for the paleo-mags.

Special thanks are in order to Don Glut for encouraging this endeavor, and to my Dad, Allen G., for seconding the motion. We hope you enjoy this.

In a nutshell then, *Paleoimagery* is about "time travelers and Mastodons." Please don't scoff! Now I'm back to where it all began. Maybe if we just take time to imagine, we really can go back after all!

Allen A. Debus
Hanover Park, Illinois
July 2002

CHAPTER 1

Dinosaurs and Evolution—Conflated Imagery

Once the last dinosaur expired, it required the intellect of Man to restore them as "living" yet virtual images. Curiously, however, dinosaurs have become enigmatic symbols of modern invention. While Paleontology's defining ideas (i.e. evolution, extinction and paleobiology) remain steadfast, laymen idolize paleontology for its "visual" nature. If not for "paleoart," paleontology would be far less popular today.

The dinosaurs' resurrection has been popularized through scientific restorations, cartoons, advertising logos, museum reconstructions and displays, dinosaur "theme" parks, animated sequences, and 3-dimensional toys and collectible figures. Without carefully worded explanations ("imagetext"),[1] the "meaning" of dinosaurs as cultivated through iconography can be easily distorted. (Conversely, the meaning and ideas of dinosaurs may be more precisely described in literature, whether of the sterile technical variety, a more popular style of science journalism, or, most cleverly, in fantasy and science fiction.) However, one thoughtfully prepared dinosaur graphic may seem everlasting, if not worth at least a thousand words, since we are a visually oriented species grown fearful of hard science.

Superficially, paleoart frees science from the shackles of abstruse morphological/anatomical studies, high falutin statistics and what must seem to many as geological "mumbo jumbo." Nearly any picture would seemingly be worth a thousand of *those* words, right? Few can resist glimpsing the "living" prehistoric zoo, or at least somebody's interpretation of it, even if the painstaking restorative effort involved isn't fully appreciated.

Those of you who have ever attended a lecture by Robert Bakker know how interrelated paleoart and paleontology can seem. The artistic Bakker lectures while referring to illustrations of prehistoric animals, boldly illustrated impromptu there on stage. Before giving his drawings away as prizes to audience participants who answer his questions correctly, Bakker will jovially draw our attention to them while offering amusing factoids about how the soft drink Dr. Pepper can keep one "regular" while on a dig, or how herds of "wind-breaking" Edaphosaurus fouled the Permian atmosphere. One might say Bakker's lectures have "artsy-fartsy" appeal.

Today it is virtually impossible to find a well attended natural history museum exhibit featuring dinosaurs and paleontology void of visual paleoimagery. It is relatively easy to spot the most popular paleontological books because they are richly fortified with artistic impressions of how prehistoric animals appeared in life. Fossils in stuffy glass cases may only illicit a yawn or two from Mom and Dad, but introduce, say, a lifelike sculptural restoration or a striking "scene from deep time," and the museum turnstiles will spin rapidly.

What is meant by the vague-sounding terms "paleoart and paleoartist," and what is their origin? Identifying the origin of the term "paleoart" is rather like considering the question of the origin of rock music's often used phrase "heavy metal." But, just as in the latter case (i.e. Steppenwolf's 1968 hit *Born to Be Wild,* which includes a key phrase—"...heavy metal thunder..."), we can acknowledge what I believe is the first published use of the term "paleoartist."

Mark Hallett used the word in an article contributed to Volume 1 of *Dinosaurs Past and Present* (1986). On December 23, 1999, Allen informed Mark by letter of his "discovery." Indeed, Mark had originated the term "paleoartist." Several weeks later, in a subsequent telephone discussion, Mark stated that "paleoartist" expresses a "neat and tidy way" of describing what he does, exemplifying his predisposition for unconsciously fashioning terms through abbreviation.[2]

Since Mark's first use of the term, "paleoart" has appeared in numerous popular publications, although, until recently, without formal definition.

Herein, whenever the terms "restoration" and "reconstruction" are used, we are referring to conventions introduced by William E. Swinton (1900–1994) in his book *Dinosaurs*. Here, a "restoration" is used for living appearances, and "reconstruction" for the completion of skeletons. Swinton states, "...it is obvious that if one is to make good restorations one must make, in fact or in theory, good reconstructions first."[3] During the 1950s, artist and author William Scheele, formerly a Director of the Cleveland Museum of Natural History, introduced another intermediate convention, the "skeletal restoration," in his popular prehistoric animals series of books for children. Paleoartists employing this illustrative technique enclose a skeletal reconstruction with a silhouette restoration of the fleshed-out body outline. As noted by paleoartist Dan Varner, "This has become an almost universally used method and has proven very effective. You can hardly open a dinosaur text without bumping into one."[4]

And although Mark Hallett should be credited for first using the term "paleoartist" in print, we are applying an original definition of the term "paleoart"—although not applied to that vintage cave art created by Cro-Magnon men and women. Therefore, "paleoartists" are (modern) artists who create original skeletal reconstructions and/or restorations of prehistoric animals, or restore fossil flora or invertebrates, using acceptable and recognized procedures. Paleoartists could themselves be scientists or individuals who are well informed about the relevant science concerning the organisms they seek to reconstruct and restore. Conversely, there are innumerable forms of "paleoimagery" not qualifying as "paleoart." The Society of Vertebrate Paleontology's recent definition for paleoart reads: "Paleo art [*sic*] is broadly defined as the scientific or naturalistic rendering of paleontological subject matter pertaining to vertebrate fossils."[5]

Funny how long it's taken to refine these working definitions, especially since the earliest examples of paleoart were created two centuries ago.

It's relatively easy to spot overall trends in the "evolution" of "paleoimagery" and paleoart.

One may discern the powerful meaning of dinosaurs by probing the bewildering array of graphical dino-devices. For in learning about our profound fascinations with dinosaurs (and their prehistoric ilk) we may better comprehend ourselves.

Here's how we see it.

For decades there was no recognized field for creating such imagery. In the 19th century, full-sized dinosaur statues erected at the Crystal Palace grounds at Sydenham, England, were utilized in a dramatic representation of the Earth's geological history. Shortly thereafter, Darwinian evolution galvanized the emerging science of paleontology. An 1860s effort to restore dinosaurs in New York City's Central Park proved unsuccessful, at least partially because the dinosaur image was conflated with objectionable aspects of Darwin's theory. In such cases, the "dinosaur theme" conveyed varied messages to different individuals, depending on their perspective.

In the twentieth century, dinosaurs became symbols of the "extinct." This was because they were regarded as evolutionary failures. According to one popular view, they were among the variety of prehistoric creatures which had evolved into obsolescence and therefore couldn't survive. The triumph of Technological Man was woven into these images, cartoons and museum exhibitions. Man was "superior" because, through evolutionary processes, we ascended to Creation's summit. In some unfathomable way, we seemed to have "conquered" our evolutionary competitors in dim prehistoric ages, symbolized by the bones of the most gigantic and fearsome dinosaurs mounted in museums. (Although, let's face it, collectively they outlived us by hundreds of millions of years.)

By the end of the nineteenth century, practitioners had invented nearly all of paleoart's major themes. However, it wasn't until about 1910 that use of paleoart restorations and reconstructions became recognized by a consensus of paleontologists as a suitable means for expressing scientific ideas.[6]

While much popular attention focuses on the current "Dinosaur Renaissance," a period characterized by numerous striking examples of paleoart and paleoimagery (i.e. representations of fossil organisms reconstructed or restored from fossil remains), it was during paleoart's earlier "Reformation" period that a consensus of paleontologists first confidently relied on the use of artistic, yet scientifically meaningful, forms of paleoart to convey theoretical ideas concerning prehistoric vertebrates. This Reformation period (circa 1900 to 1910) corresponds to the period in which Henry Fairfield Osborn (1857–1935) authorized artistic, scientifically sound displays for the public on paleontological themes at the American Museum of Natural History.

The development and origins of modern paleoart can be related to the late 19th century's onslaught of Jurassic and Cretaceous dinosaur, and Tertiary mammal, discoveries from America's western frontier. Whereas British paleoimagery had little influence in America, by the 1920s, and fueled by fossil discoveries (especially from the American west), a distinctively American paleoart genre had "invaded" Great Britain and the continent.

Thereafter, through the twentieth century the paleoimagery "phenomenon" manifested itself in myriad forms (movies, toys, games, advertisements, industrial prowess, socio-political ideologies and, yes, even religious themes). Dinosaurs would symbolize the war effort in World War II and our principal fuel source of the pre-nuclear age, petroleum. In the latter case, dinosaurs represented the source of fossilized organic matter being converted to hydrocarbons deep inside the earth. Paradoxically, dinosaur images would also represent both atheistic ideas and religious faith.

The manner in which we have viewed and interpreted paleoimagery has evolved too. While paleontologist Martin J. S. Rudwick has identified the "aquarium" perspective as having innovatively inspired many nineteenth century "scenes from deep time,"[7] most certainly the "fictional reality" of the television screen, our favored window for safely eyeing the prehistoric realm, has fortified the "credibility" of dinosaur images ever since the mid–twentieth century.

Paleoart has arguably enjoyed a rapid escalation to prominence as paleontology's most popular facet, especially during the past two decades. Even more popular than actual fossils? Well, just think how many people flocked to see *Jurassic Park* (1993), or tuned into *Walking with Dinosaurs* (1999), or shelled out hard-earned cash to see the new Disney *Dinosaur* (2000) animated film. How many of us purchase those all-too-tempting "scientifically accurate" model kits, either for our children or, greedily, for ourselves?

Now it seems we cannot evade prehistoric animals. They are literally everywhere, even in cyberspace. Lately, the universal appeal of paleoimagery has even prompted scholarly critique.

Politically correct Harvard paleontologist Stephen J. Gould bemoans the popularity of those prevalent Shopping Mall–invading animatronic dino-robots, "...featuring as their come-on, the very properties—mostly hideous growls and lurid colors—that leave no evidence in the fossil record and therefore remain a matter of pure conjecture to science."[8] Yet, in paleontology, isn't speculating half the fun? Apparently too much realism can be oppressive, however. At a popular shopping mall near Chicago, a *Utahraptor* animatronic robot was displayed with a victim's severed bloody head dangling from its mouth. So graphic was this that mothers worrying that their children would be terrified by the horrid apparition complained to the mall authorities. The severed head was replaced by a bale of hay! Vicious *Utahraptor* had transformed into a *herbivore*! Yes, highly *inaccurate*, but soothing.

Now the very practice of creating paleoimages has become fostered by the paleoart "phenomenon." In 1999, University of Chicago paleontologist Dr. Paul Sereno proposed that a "John J. Lanzendorf Paleoart Award" be presented annually by the Society of Vertebrate Paleontology to artists excelling in several categories—i.e. the best paleoart (not simply "dinosaur art," emphasizing strictly dinosaurs) created in two or three dimensional form, or as a technical illustration. Artists submitted over 250 entries for the 2000 inaugural contest, a total possibly exceeding the nineteenth century's entire modest output.[9]

Will John Lanzendorf's private collection someday inspire a Hall of Fame for paleoartists? Well, that might depend on who the "true" paleoartists really are.[10]

University of Chicago humanities professor and cultural historian William J. T. Mitchell has defined "paleoart" (in a "high brow" context) as "...art that engages the present and future of advanced industrial societies and reframes them in the temporal perspective of paleontology and geology."[11] Further, Mitchell extols the work of hard-to-decipher, postmodernistic paleo-abstractionists while diminishing the classical creations of those we would ordinarily regard as "true" paleoartists. Mitchell's unique perspective has been challenged by at least one paleoartist, Dr. Sereno himself, who stated, "To suggest that dinosaur images are not art, but rather scientific illustration or kitsch ... is myopic."[12]

Yet, even within the confines of dinosaur imagery and paleoart there are principles defining what we might regard as "artistic" renderings. Author and dinosaur expert Donald F. Glut claims, "True 'art' means more than just scientifically accurate pictures or models. It implies a certain feeling or spirit that transcends the lines or paints or shapes. It elicits an emotional response more profound than 'Gosh, wow!' or 'cool!' Sometimes it even tells a story, as in Knight's Field Museum murals."[13]

With so many degrees of freedom and expression, one is left with an impression that paleoart restorations are like a box of chocolates because you never know what to expect.

Perhaps paleontology's visual side most readily appeals to "escapists" longing to experience alternate realms of reality. We yearn for those lively, prehistoric settings

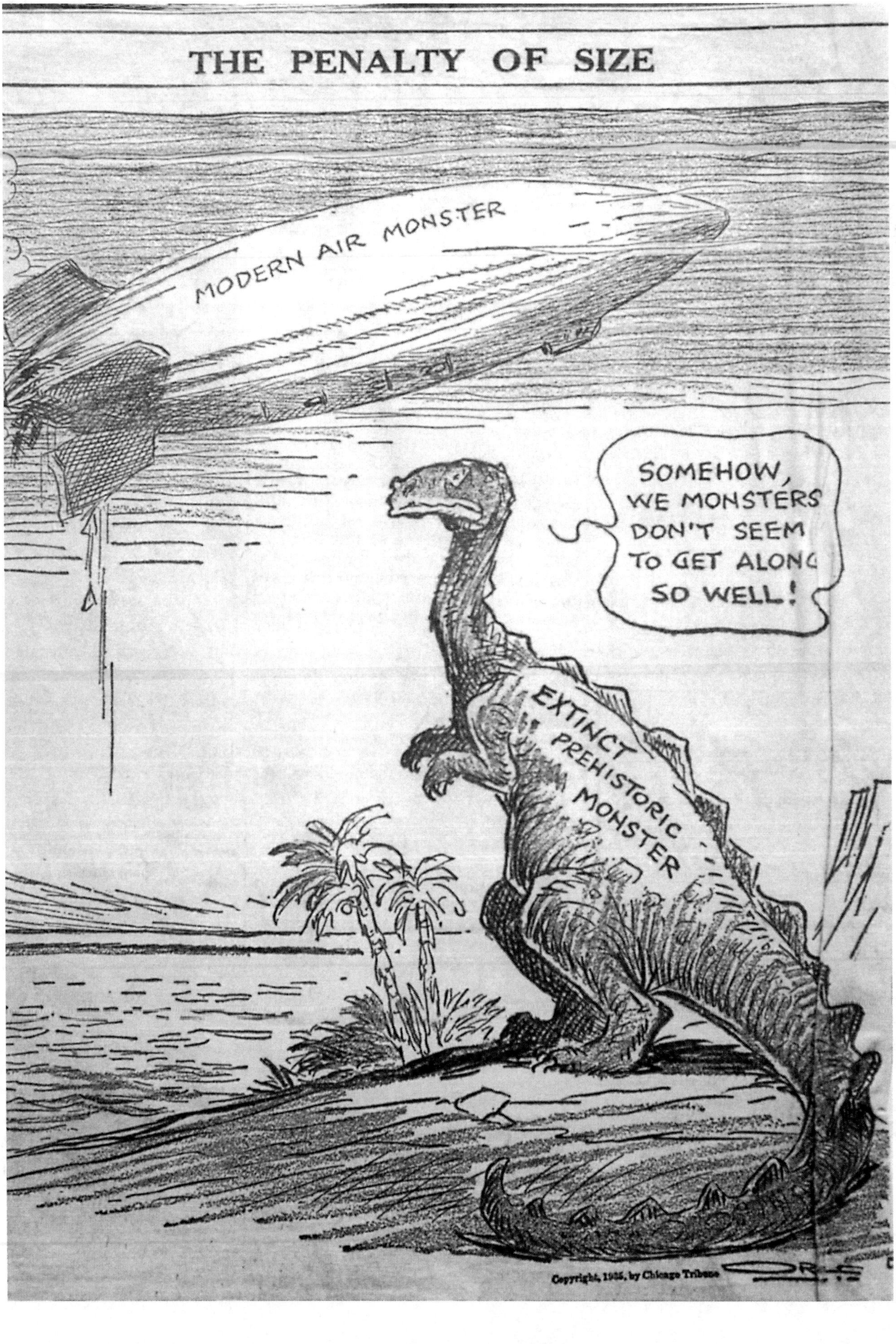
THE PENALTY OF SIZE
MODERN AIR MONSTER
SOMEHOW, WE MONSTERS DON'T SEEM TO GET ALONG SO WELL!
EXTINCT PREHISTORIC MONSTER
Copyright, 1935, by Chicago Tribune

"inhabited" by animals that most of us could scarcely perceive without the aid of skilled paleoartists.

Thankfully, "true" paleoartists deal in truth as opposed to willful abstraction. One premier paleoartist, Michael Skrepnick, echoed the opinions of scores of other bonafide paleoartists when he stated that, "As an artist, the thing which really fascinates me is, 'what did they really look like?' We have a tremendous amount of information concerning what they probably looked like, but it's that uncertainty of never knowing which really piques my curiosity. My artistic interpretations may seem quite reasonable, but was that really the animal? I wish I could see them for myself to satisfy my curiosity. I have a great deal of respect for these animals and I want to do them as much justice as I can. It may seem silly, but I wish I could do an artistic interpretation that they would feel 'proud of.' I would like them to be seen in their best light for what they were."[14]

Accuracy should be the goal, and that is the essence of paleoart's appeal. How close can our wonderful paleoartists come to authenticity? Science moves onward, new fossils are found and scientific interpretations change over time. Paleoartists perform their work conscientiously, magically. Nevertheless, every so often both the paleoart field and associated evolutionary sciences are plagued by embarrassments, such as the deceptive Piltdown Man or National Geographic's recent promotion of a self-created image, the faked fossil, "Archaeoraptor."

Isn't it self-evident that dinosaurs should champion evolutionary ideas? Perhaps until recently, dinosaurs haven't really been used that way.[15]

Dinosaur evolution wasn't diagrammatically reflected until 1933, when Richard Swann Lull (1867–1957) completed a chart showing the evolution of horned dinosaur lineages through time (focusing on skull bone morphology).[16] Based on recent analyses, the relationships of various genera represented in this diagram have been reassessed. However, this style of portraying evolutionary relationships—appearing in a scientific journal—particularly among ceratopsians, has proved enormously appealing since the time of Lull's innovation.

During the mid–twentieth century, however, there were few attempts to underscore prehistoric animal images with the science of evolution, or at least few that would leave lasting impressions. To a layman, perhaps, dinosaurs only seemed to make sense when viewed as some kind of weird "cult." What was the true meaning of dinosaurs? Just about anything, as long as there was a murky "prehistoric" connection. That made it possible to avoid associating dinosaurs with biological evolution, especially if the paleontologists didn't force the issue.

Perhaps this is another factor for why dinosaurs always seemed to be the least socially threatening—and hence most popular—group of prehistoric animals. Those who were disbelieving of evolution could still enjoy dinosaurs because they seemed only remotely associated with evolutionary principles! At another extreme, fossil hominids, and even the Mesozoic bird *Archaeopteryx*, were laden with evolutionary/theological "baggage."[17]

Remarkably, it has only been within the past decade or so that dinosaurs have come to popularly represent ideas in evolution more than the alternative causes they have been historically associated with. Partly this is because the scientific ideas of scientists now are reaching the public more than ever before through the outstanding popular contributions of Stephen J. Gould, Louis Jacobs, Michael Novacek and John Horner, to name a few. Instead of extinction, dinosaurs are beginning to symbolize evolutionary ideas.

Through an array of (cladistic) systematic graphical displays, which are even reproduced sometimes in media accounts, the traditional geological time variable is deemphasized. A cladist's concentration on form and skeletal features permits a more thorough and statistically rigorous analysis of phylogeny, acknowledging that many genera or species may forever be missing (in time) from the geological record. More recently, we note that dinosaurian evolutionary ideas have been popularly disseminated in the following arenas: 1) the evolution of modern birds, possibly from feathered dinosaur ancestors; 2) Dougal Dixon's fantasy portrayals of alternate evolutionary scenarios involving a post–Mesozoic world in which dinosaurs never succumbed to extinction; and (my favorite) 3) Tracy Ford's restorations of hypothetical evolutionary stegosaurs (transient between known species) published in an article by George Olshevsky. Much of this material may be speculative, but it is also instructive.[18]

Traditionally, restorations of prehistoric animals are far less likely to be found in technical journals than in popular accounts. However, detailed discussions of dinosaur evolution and phylogeny found in the scientific publications may be glossed over in popular books and articles. So it may be regarded as a minor triumph to find dinosaur reconstructions and restorations adorning the pages of the June 25, 1999, *Science* issue, dedicated to the theme of evolution.[19] Here, Paul Sereno's article "The Evolution of

Opposite: **Figure 1–1: In this editorial cartoon by Casey Orr, dinosaurs symbolize an extinction related to their outmoded size. (Copyright 1935, Chicago Tribune Company. All rights reserved. Used with permission.)**

Dinosaurs" firmly linked dinosaurs, through their imagery (as found in five detailed figures), with their phylogeny. Richard Swann Lull would have been pleased to see Sereno's publication. Let's use dinosaur images to promote the science of dinosaurs so that their imagery will seem less conflated.

It now seems that while dinosaurs can still symbolize a variety of prehistoric "themes," often through commercial exploitation, from this day forward they will also represent the ideas of organic evolution on a popular level! That's our opinion, but because of the conflated imagery we have become exposed to through the years, some of you may disagree.[20]

Chapter 2

Kronosaurus—An Imaginary Sea Monster That Got Away

Fossilized skulls stare sightlessly across the Museum halls. Widely gaping jaws and immense vertebrae symbolize a world remotely different from our own. However, to a child unaccustomed to the remoteness of deep geologic time, the primal memory is easily awakened. One of my fondest—though now faded—childhood memories concerns what to me is certainly the most spectacular sea monster that ever lived, the magnificent *Kronosaurus*, now exhibited at Harvard University.

My father, who was attending graduate school there in the late 1950s and early 1960s, would occasionally take my brother and I early on Saturday mornings to see the exhibits in the Agassiz Museum of Comparative Anatomy. Although it's been decades since we walked through the Museum halls together, I recall a Giant Squid model mounted overhead on the ceiling and a *Plateosaurus* skeleton as being the main attractions of the tour. But *Kronosaurus* always wound up as the awe-inspiring "grand finale." Is Harvard's *Kronosaurus* skeleton, which conjured a vivid restoration from my imagination, more imaginary than real?

Harvard's *Kronosaurus queenslandicus* skeleton is 42 feet long. No doubt, *Kronosaurus* was among the largest marine reptiles to have ever lived. No wonder its name means Titan (or Time) lizard. It was one of the most titanic of all time!

About one hundred and fifteen million years ago, during the Early Cretaceous, not far from what is now the sea coast 200 miles from North Queensland, Australia, *Kronosaurus* corpses came to rest on the bottom of a limey sea bed and were buried under a deposit of calcareous silt. Geologic ages passed steadily. The Australian continent rifted thousands of miles northward from its position at the time of *Kronosaurus*' death. The bones of Time's greatest sea monsters lithified in the fine-grained calcareous mud. Sediments became uplifted from the sea floor, weathered and eroded. One day, millions of years later, an Australian rancher noticed some splinters of fossil bone protruding from a limestone face....

What was *Kronosaurus*? First, it wasn't a dinosaur. Despite its general similarity in appearance to another Cretaceous sea monster, *Mosasaurus*, technically, *Kronosaurus* is classified among the marine pliosaurs. Pliosaurs, in turn, are related to the aquatic, long-necked reptiles of which the Jurassic genus *Plesiosaurus* is our most familiar example.

Long-necked, or "swan-necked," plesiosaurs of the Early Cretaceous period, such as the 46-foot-long *Elasmosaurus*, are conventionally thought to have evolved from swan-necked Jurassic ancestors, such as *Plesiosaurus*. However, in 1993 Robert Bakker suggested that both Cretaceous swan-necked plesiosaurs and Cretaceous pliosaurs evolved from Jurassic pliosaur stock.[1]

In other words, Jurassic swan-necked plesiosaur varieties vanished without leaving Cretaceous progeny. While Upper Jurassic plesiosaurs perished at the Jurassic/Cretaceous extinction boundary, pliosaurs survived the crisis. Then, in time, some Cretaceous descendants of Jurassic pliosaurs evolved swan-necks convergently on the plesiosaur design. Bakker's idea is supported on the basis of structures (i.e. heritage markers) noted in skulls of plesiosaurs and pliosaurs.

In contrast to Bakker, Ken Carpenter doubts whether the Upper Jurassic short-necked pliosaur lineage diverged into long-necked elasmosaurid varieties. Furthermore, according to Carpenter's detailed 1997 analysis, the plesiosaur lineage survived into the Cretaceous, giving rise to short-necked forms (i.e. the polycotylids) and long-necked elasmosaurids.[2]

Kronosaurus evolved from the Jurassic pliosaurs. *Pliosaurus*, a short-necked, long-headed Upper Jurassic evolutionary ancestor, is found in European deposits. Pliosaurs' heads grew to 10 feet in length and bore the trademark massive six foot jaw fitted with long, widely spaced teeth. (Teeth on the vertically opposing jaw were similar but grew into vacant spaces.) Front teeth reached one foot in length and were 2.5 inches in diameter at the base. Their strongly framed skulls were highly adapted for brief but vigorous encounters with struggling victims.

The short-necked, stubby-tailed *Kronosaurus* was among the few pliosaur genera surviving into the Early Cretaceous. (See Figure 2-1.) Like its evolutionary cousin the *Plesiosaurus, Kronosaurus* swam powerfully and swiftly through ancient seas. Each of *Kronosaurus*' oar-like paddles were formed from five toes comprising numerous joints that probably were embedded in a slick, tough skin. Like other plesiosaurs, pliosaurs and modern tortoises, the ivy league submariner propelled itself forward using oar-like paddle motions, alternating up and down strokes synchronously between front and rear limbs.

Older, Triassic ancestors of *Plesiosaurus* and the related pliosaurs are thought to have been semi-aquatic nothosaurs. Later, as a fully marine existence suited certain forms, webbed feet and legs evolved into strong-limbed paddles. The socketed teeth characteristic of pliosaurs were similar to those of the Triassic thecodonts.

Hungry kronosaurs may have relished more than simply fish and ammonoids. With huge jaws, great swimming efficiency and the hunting instincts of a great shark or killer whale, a voracious beast such as *Kronosaurus* would have been a formidable adversary for any other carnivorous sea monsters of the time. Bones belonging to *Elasmosaurus* and Mesozoic turtles have been identified in *Kronosaurus*' abdominal region.

Czech paleontologist Josef Augusta once vividly described the hunting habits of the pliosaurs: "These robustly built plesiosaurs were true hunters, able to pursue

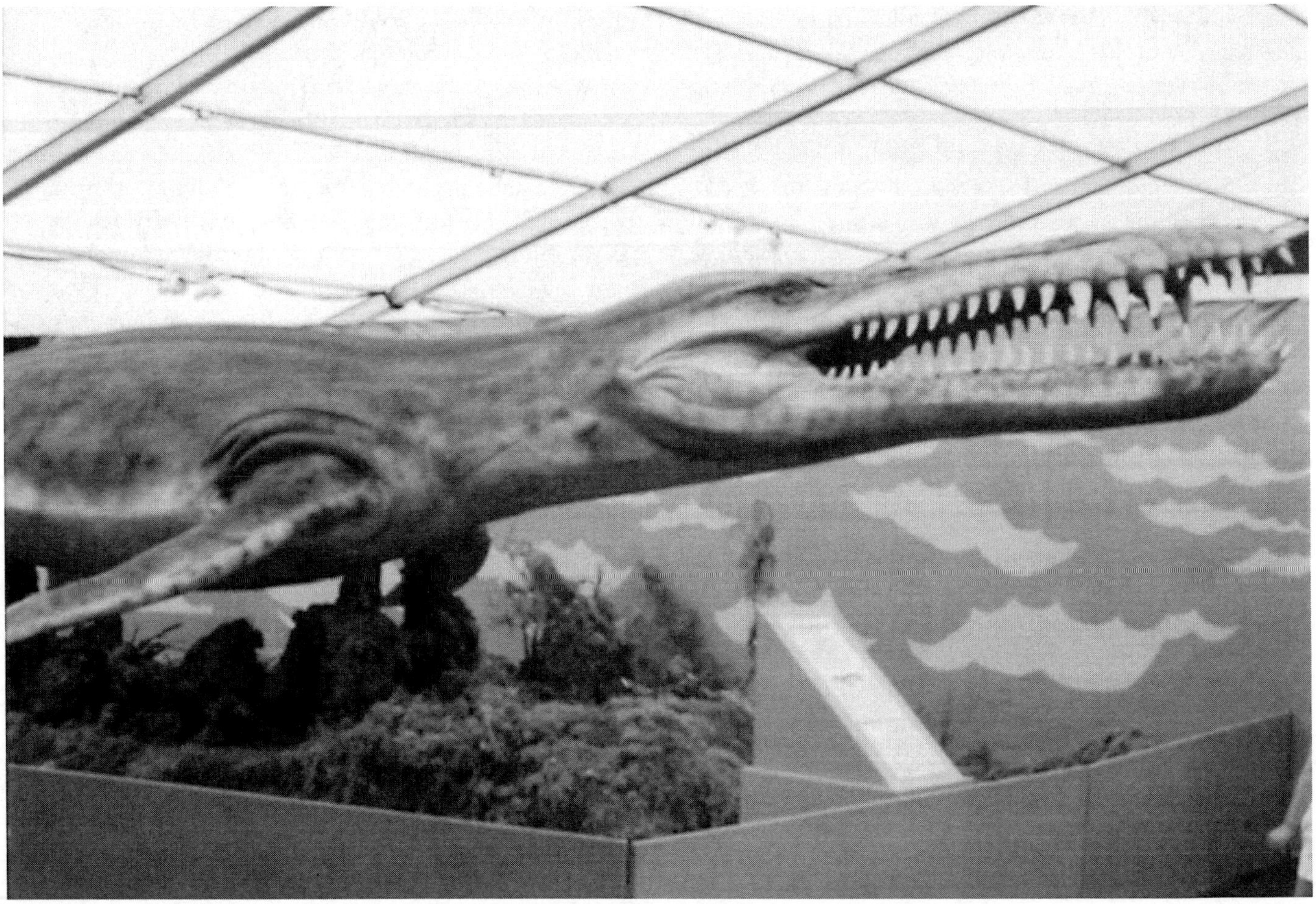

Figure 2-1: Dinamation's life-sized animatronic *Kronosaurus* robot, displayed in Chicago in the summer of 1990. (Photograph by Allen A. Debus.)

their prey for a long time.... Pliosaurs hunted somewhat differently from their long necked cousins. Having no long neck to help them, they had to jump on their prey suddenly. There was also a difference in the type of diet, since Pliosaurs because of their size hunted only large prey...."[3]

Bakker believes cosmopolitan, top marine predators such as *Kronosaurus* were less likely to succumb to extinction pressures than other aquatic marine reptiles, such as the Jurassic plesiosaurs, which were more dependent on localized ecological conditions. In the event of ecological adversity, *Kronosaurus* and other pliosaurs would swim long distances to less adversely impacted waters.

The mystery of pliosaurs' extinction may never be solved to the satisfaction of all, although surely the possibility of great climatic changes, triggered by volcanic outbursts and other terrestrial or cosmic phenomena, will be given due consideration. Bakker claims, "Only the most profound and prolonged sea crisis could eliminate the giant pliosaur guild ... a crisis like the apocalypse at the end of the Cretaceous."[4]

The first reported remains of *Kronosaurus*, a large anterior end of a mandible with six preserved, socketed teeth, were forwarded to the Queensland Museum from the Hughendon district of Australia by a Mr. A. Crombie in 1899. C. W. De Vis erroneously described the fossil as an *Ichthyosaurus*. Many years later, in 1924, Dr. Heber A. Longman (1880–1954), the Museum Director, realized that the socketed teeth could not have belonged to *Ichthyosaurus*. In fact, the relic seemed to be unique among the fossil vertebrates known at that time.

So Longman wrote, "Although tantalizingly incomplete, this fragment presents such definite characters that it is desireable to describe it, so far as the evidence permits and to give it a distinctive name."[5] In 1998 Dr. Ralph Molnar, then of the Queensland Museum, stated, "The original (or type) specimen of *Kronosaurus* is so poor that "masterly interpretation" ... is clearly an appropriate phrase. I must confess that nowadays no one would consider publishing on such scrappy material, and I think Longman only published it because it was the first and only specimen of a pliosaur from Australia ... at that time, that is."[6] Nevertheless, Longman reliably predicted that more *Kronosaurus* fossils would be found: "...It is ... believed that when other remains are forthcoming of *Kronosaurus* queeenslandicus that this course [i.e. of scientific description—authors note] will be found fully justified."[7]

Through newspaper accounts and popular lectures, Longman attempted to popularize the scientific study of prehistoric vertebrates in Australia. Unfortunately, however, during the Depression years the Queensland Museum ran on a strict financial budget. National interest in fossil collecting all but faded. What little collecting took place was conducted by foreigners. For instance, a number of important specimens were obtained from Australian soil for study by the British Museum of Natural History in London, England.

Such conditions hampered the success of the vertebrate paleontology program in Australia. Consequently, the greatest articulated skeleton of a *Kronosaurus* became identified by a foreigner, a young American geologist from Harvard University named William E. Schevill, in 1932. Australian paleontologists Harold Fletcher and Longman (of the Australian Museum and Queensland Museum, respectively) had both been invited to join Schevill's expedition. Lamentably, absence of adequate funding and lack of proper enthusiasm on the part of other Museum authorities prevented their participation.

Schevill was led to the site of an articulated (though incomplete) *Kronosaurus* skeleton preserved within a limestone exposure on a sheep ranch owned by a Mr. Thomas in North Queensland. A British expedition member (whom Schevill referred to as "the maniac" because of a rumor he'd heard concerning a murder the colleague had committed in Great Britain) was familiar with the use of explosives. Consequently, blocks of limestone bearing large sections of the *Kronosaurus* skeleton were "maniacally" blasted out with dynamite from the surrounding bedrock and eventually shipped to Harvard University. This rather crude excavation technique may have seriously eroded skeletal parts, haunting Harvard paleontologists years later. Two sites yielded partial remains of two pliosaurs that were shipped to Cambridge, Massachusetts.

The blocks had been packed in "end-of-sale bloody wool." This frightened the Director of the Museum of Comparative Zoology at Harvard, Thomas Barbour, who was fearful of the possibility of anthrax. Consequently, Schevill was ordered to wash each block with a disinfectant.

After *Kronosaurus*' bones were disinterred from its rocky tomb by Dr. T. E. White, an arduous task often delayed because of manpower shortages and budgeting constraints, it became evident that Longman's long-shot expectations founded on the single jaw fragment had been fulfilled. *Kronosaurus queenslandicus* was indeed a unique beast belonging to the Australian Mesozoic menagerie, and furthermore, as Longman had predicted, it was related to the plesiosaurs.

In 1939 the impressive nine-foot-long, toothy skull of *Kronosaurus* was mounted for exhibition. The remainder of the skeleton rested in the Agassiz Museum of Comparative Anatomy for 15 years. That is, until a man with a reputed passion for sea monster lore, Godfrey Lowell

Cabot, a Harvard graduate whose great grandfather actually claimed to have seen a living 40- to 60-foot-long sea monster off the coast of Gloucester, Massachusetts, in 1817, came to the aid of the partially assembled fossil. (Cabot's grandparents also wrote of a sea-monster they claimed to have seen in 1819. Such incidents fostered that ever-popular pastime of sea serpent watching.)

Cabot had often prodded vertebrate paleontologist Alfred Romer of Harvard for information concerning whether living sea monsters existed. When Romer mentioned that the remains of such a beast resided in a closet of the Zoological Museum, Cabot enthusiastically inquired as to the cost of complete reconstruction. "Romer, pulling a figure out of the musty air, replied, 'Oh about $10,000.' Cabot promptly sent a check. After all, *Kronosaurus*, though it has no direct descendants in the oceans of today, was itself a sea serpent of heroic proportions."[8]

The complete skeletal reconstruction quickly consumed Cabot's generous offering. Although erosion had claimed about one quarter of the skeleton, diligent extraction and preparation exposed all the remains within two years. Relying on their knowledge of other more complete plesiosaur fossils, under Romer's direction, Jim Jensen and Arnold Lewis supplied plaster cast "bones" to represent missing or damaged material.

Finally, the work was completed, and on June 10, 1958, the magnificent *Kronosaurus* skeletal reconstruction was unveiled. (Within the following year my father took me to behold this formidable reconstruction.) Cabot, who was then nearly 100 years old, was able to behold his magnificent sea monster!

Shortly thereafter, formal announcement of the exhibition began to create quite a stir in Australia. It seems that the jubilant Harvard crew had neglected to acknowledge the efforts of Australian paleontologists, notably that Longman had been *Kronosaurus*' original discoverer. Understandably, Longman was envious. Attempts had been made to rectify the situation even as early as 1939, when Longman arranged for his assistant, Tom Marshall, to go to Harvard to make casts of the *Kronosaurus* and assist with the preparatory work conducted by T. E. White. The opportunity was blocked by the onset of World War II. Longman never lived to see the *Kronosaurus* skeleton at Harvard.

If only Australian scientists had been able to make their casts, or if they had been able to retain the specimens found in 1931/1932 and, afterward, managed to finance their costly reconstruction, *Kronosaurus* would have remained in Australia. And by now, generations of Australian children would have gazed into the hollow eye orbits and gaping jaws of an enormous, bona fide sea monster, as I did 40 years ago. This is, partially, the fault of major institutions which reluctantly cannot finance the acquisition and preparation of large fossil vertebrates. According to Dr. Molnar, in 1988. "...this kind of thing is still going on here (unfortunately!)."[9]

Illegal seizure of significant fossils from locations around the world is a growing concern as the perceived monetary value of rare or unusual fossils skyrockets. We may expect stern policing of Canada's famed Burgess Shale fossils, fossils from U.S. public lands and those remarkable Chinese Liaoning vertebrate fossils. Australia has regulated "protected objects" since 1986, and recently, in May 1999, promulgated laws for permitting the export of significant specimens. Yet regulations are not foolproof. Sometimes they may even seem ambiguous. Sad tales, like those of "Sue's" entanglement by the legal system, will continue to make headlines. Today kronosaurs would doubtlessly be on Australia's "Control List."

Harvard's *Kronosaurus* was taken without license, although not illegally. Yet, while the *Kronosaurus* is our gain, it represents a loss to Australia's cultural heritage.

To a sensible adult, the bones of *Kronosaurus* now seem quiet and stately, peacefully at rest in a magnificent museum case. An illustration in a paleontology book satisfies idle curiosity as to what the immense animal once was. Yet, introduced to genuine 40-foot-long sea monsters (i.e. Giant Squids and *Kronosaurus*) in one of our most prestigious museums, imaginative children can conjure strange creatures from the ashes of time, haunting images unmatched by the most accomplished special effects paleoartists. "*Monsters from the Id!*"[10]

Paleoartists take pride in their imaginations. Although infrequently portrayed, *Kronosaurus*' form has captivated several prominent paleoartists. (See Figure 2-2.) Rudolph Zallinger's classic restoration, published in *The World We Live In* (1955), possessed a rather generous cervical region. Zallinger's restoration served as the design for a plastic toy figure, very popular in America during the late 1950s and 1960s.[11] However, Zallinger modified *Kronosaurus*' appearance in his painting, most likely after assembly of the Harvard skeleton, to conform with the mounted skeletal reconstruction. Zdenek Burian's spectacular pliosaur painting has a neck comparatively shorter than Zallinger's original *Kronosaurus* version.

Recent kronosaur/pliosaur restorations by Robert Bakker, however, sport prominently raised tail flukes, instead of serpent tails. The vertical tail fluke is based on fossilized skin impressions found in a Jurassic plesiosaur.[12]

But perhaps the most dramatic restoration of all is the life-sized animatronic marvel, created by Dinamation

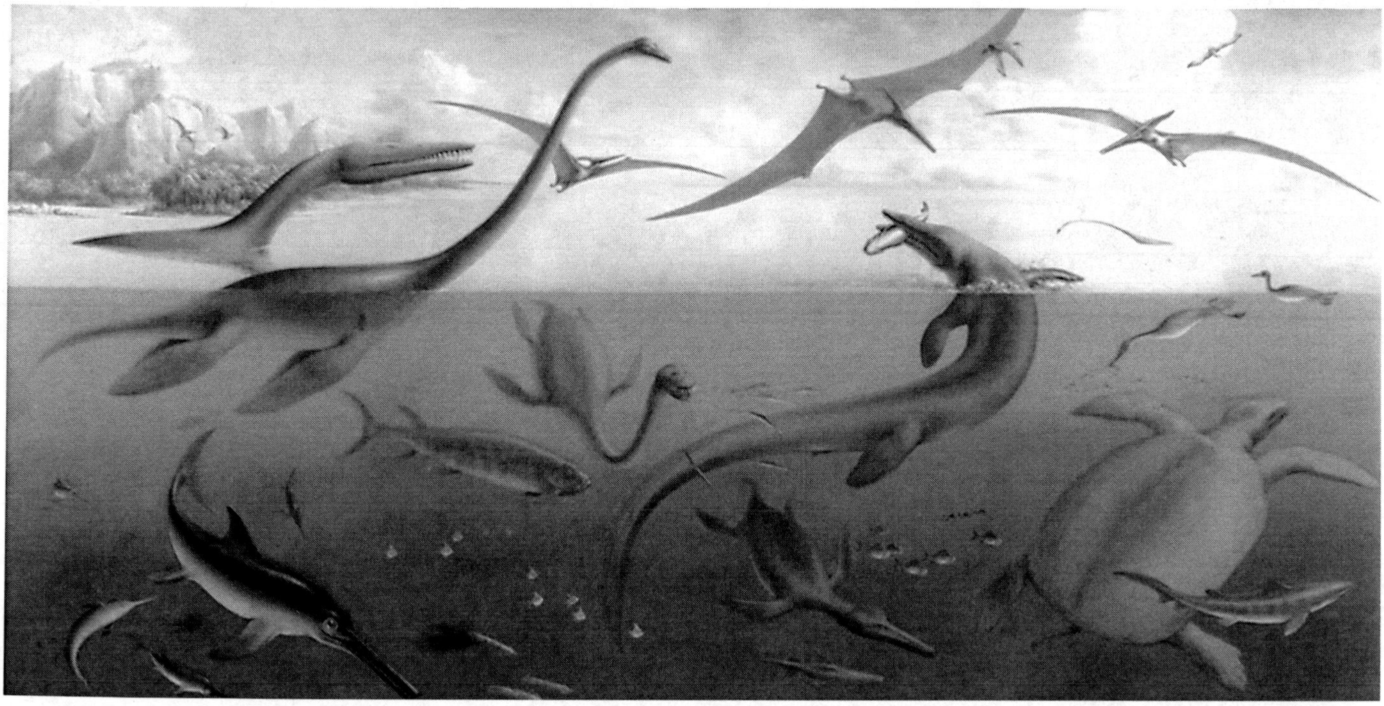

Figure 2-2: Zallinger's painting of aquatic Cretaceous reptiles. *Kronosaurus* is at upper left. This is the older version published in the 1955 *Life* production, *The World We Live In*. (Reproduced with permission. Courtesy of the Peabody Museum of Natural History, Yale University, New Haven, Connecticut.)

International Inc. in the late 1980s, faithful to the Harvard skeletal reconstruction in every detail.

Robert Bakker described a scene in his novel *Raptor Red* (1995) involving the *Kronosaurus* and a mating pair of *Utahraptors*. The title character, a female *Utahraptor* named "Raptor Red," learns of the fearsome beast while strolling along North America's western shoreline. Raptor Red spies their flippered forms lying secretively in the shallows, then teases the kronosaurs. Stray too close to the surf, Raptor Red discovers, and *Kronosaurus* will lunge furiously at prospective prey. She comprehends how easily unwary victims could be devoured. Eventually, Raptor Red applies her knowledge of kronosaurs to outwit a bullying predator, *Acrocanthosaurus*, which in turn is mauled and eaten by a *Kronosaurus*.

Recently, television viewers were treated to a shocking spectacle in the BBC production *Walking with Dinosaurs* (1999). Special effects imagineers animated an encounter between the European Upper Jurassic aquatic pliosaur *Liopleurodon* (imaginatively envisioned as a 60-foot-long brute) and a theropod dinosaur, *Eustreptospondylus*. In this restoration, *Liopleurodon* explodes from the surf, seizing the screaming dinosaur by the tail. *Eustreptospondylus* suffers a cruel death, dragged under and swallowed whole after being engulfed in *Liopleurodon*'s 10-foot-long, cavernous mouth.

What a grisly end—to be consumed by something frightfully unseen, a predator which should menace beneath the waves, not on the rocky shore. Few can relate to the horror. You may laugh, but I've very nearly suffered that experience.

Perhaps a year after Harvard's *Kronosaurus* skeleton forever captivated my soul, somewhere in the midwest, during an "amusement" park boat ride, suddenly and awfully, the placid pond waters began to froth and foam as a great water-spouting Leviathan surfaced alongside us.

I'm not very proud of how I handled myself that day. I was certain that we had invaded the territorial waters of the *Kronosaurus*, and that we were about to be savagely torn and devoured. I clawed, scratched and kicked my way down to the flooring of the boat, seeking sanctuary from, what turned out to be, a fiber glass "Moby Dick" rising on its mechanical assembly. While cowering miserably in the hull, I hollered louder than *King Kong*'s Fay Wray. Don't call *me* "Ishmael"!

Undoubtedly the immensity of the Harvard skeleton, not to mention the size of its teeth, especially when compared to my diminutive six-year-old frame, had left a deep impression. It required no painting by Knight or stop-motion horror sequence to unleash this appalling "Id monster." The nightmarish apparition, formed within, had brought about my undoing. (See Figures 2-3 and 2-4.)

My harrowing experience raises unanswerable questions. Our very existence may have stemmed from the

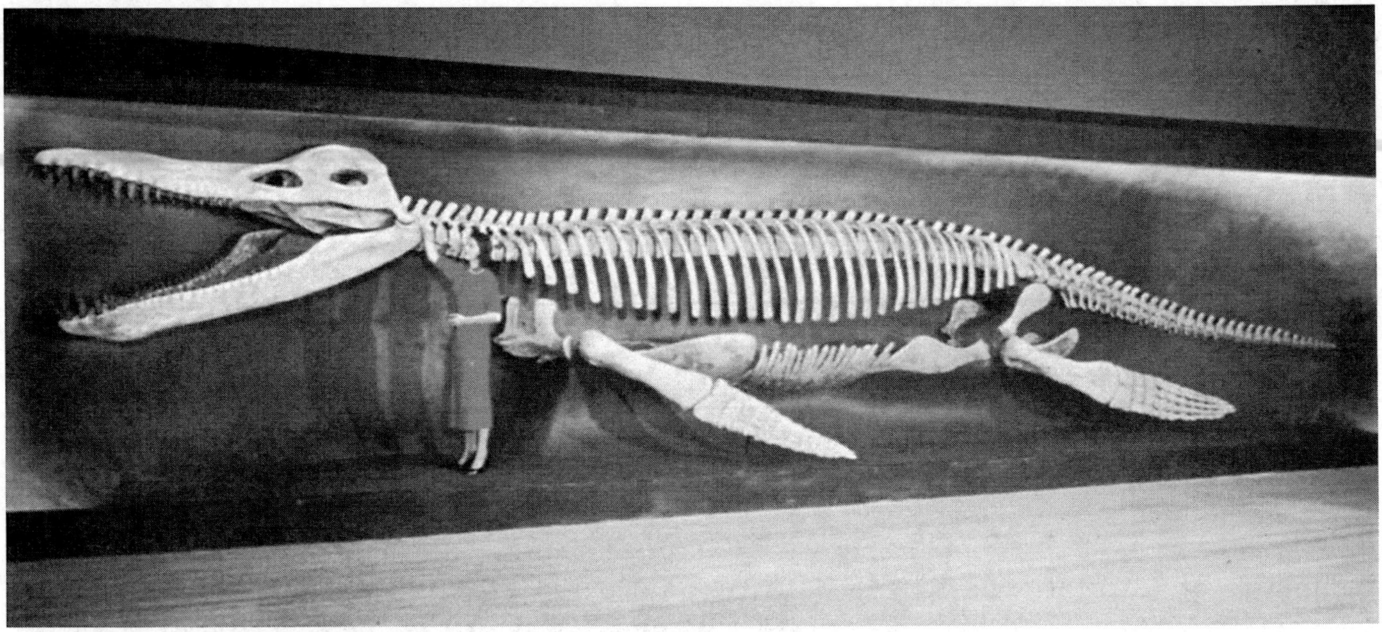

Figure 2-3: Harvard University's stately *Kronosaurus queenslandicus* remains an attraction at the Agassiz Museum of Comparative Anatomy in Cambridge, Massachusetts. (Reproduced with permission).

ability of the earliest mammals to hide from or cunningly outwit the Mesozoic overlords. Then they died and we emerged to "inherit" the earth. Yet, with such shadows lurking within, has our status truly altered?

Have the instinctive fears, desires and intuitive notions of our rodent-like ancestors become hard-wired in the human subconscious? Is our instinctive fear of reptiles, fanciful dragons, snakes and sea serpents a consequence of our mammalian heritage? Is child-like, often fear-driven fascination with our evolutionary predecessors, such as the dinosaurs, an uncanny inborn trait?[13] Certainly, what we vividly imagine can seem more "real" than stark reality.

It wasn't until three decades later when I finally appreciated the irony of *Kronosaurus*. You see, Romer's *Kronosaurus*, forever lurking in the depths of my lively imagination, was itself imaginary.

The "maniac" who blasted *Kronosaurus*' bones from bedrock may have done the specimen great disservice. For the eroded bones proved ever more challenging to reconstruct accurately. As mentioned previously, Romer resorted to plaster casting in order to fill in missing skeletal material. However, Romer's reliance on plaster became excessive. He added ten extra dorsal vertebrae to lengthen *Kronosaurus*' spinal column. And it seems as if that magnificent skull was also assembled erroneously. The undignified name, "Plasterosaurus" was conferred upon his creation.

So "my" *Kronosaurus*, now sadly christened "Plasterosaurus," has become reduced in stature from a monstrous 42 feet in length to a "mere" 27 feet. To date, no alterations have been made to the exhibited specimen. Meanwhile, the validity of Harvard's "Plasterosaurus" remains in doubt.

Investigations into the status of the genus *Kronosaurus* may lead to some important conclusions in the months ahead. Unprepared Australian *Kronosaurus* material, possibly representing two individual specimens, was collected in 1935 and in 1979. Some of this was collected by Dr. Molnar. One of these may even be more complete than the original Harvard specimen. (Another *Kronosaurus* block resides in the Harvard storerooms.) However, until perhaps recently, there has been little interest in studying these or preparing them for exhibition.

Because the Harvard material came from strata differing in age from deposits yielding Longman's type specimen (the *Kronosaurus* mandible described in 1924), Molnar is skeptical that Harvard's sea monster really is *Kronosaurus*.

Another *Kronosaurus* specimen recently surfaced in northern Columbia's Early Cretaceous deposits. The new fossils are thought to be approximately the same age as the Queensland deposits in which *Kronosaurus queenslandicus* was found. In his 1992 description, Oliver Hampe of the Frankfurt Museum named the new specimen *Kronosaurus boyacensis*.

The length of *Kronosaurus boyacensis*, comprising a well preserved skull and a fairly complete skeleton, minus the tail, has been estimated at 30 feet. Until the original *Kronosaurus* elements have been reexamined, however, it

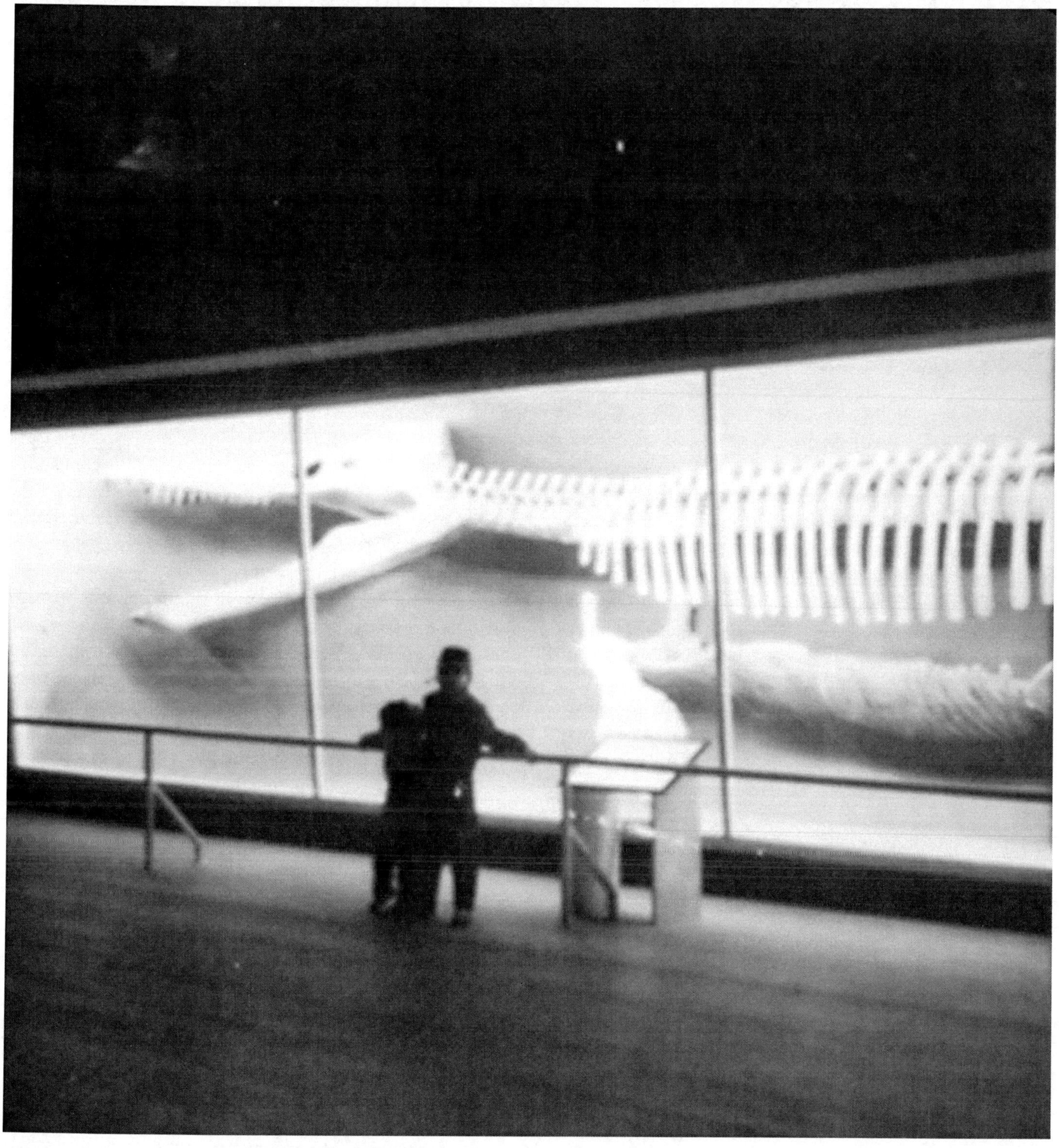

Figure 2-4: Allen A. Debus (right) bravely protecting a younger brother, Richard, from that menace of the Mesozoic seas—Harvard University's *Kronosaurus*. (Photograph by Allen G. Debus, circa 1960.)

will be difficult to confidently establish the relationship of newcomer *Kronosaurus boyacensis* to the originally described and reconstructed *Kronosaurus* specimens. Upon further inspection, a new generic category may be warranted for the Columbian material.

So that's the story of how *Kronosaurus* came to be the Mesozoic sea monster that "got away" from the land down under.

Just imagine!

CHAPTER 3

Grandmasters of the Paleoartists' Hall of Fame

We take for granted how freely the shapes and images of extinct animals come to mind whenever their names arise in idle conversation. Inundation of prehistoric animal images through the media, popular books, television documentaries and sci-fi movies lies at the heart of this fabulous phenomenon. So, understandably, we simply cannot appreciate how difficult it was for pioneering paleontologists and laymen to visualize "prehistoria" during the early nineteenth century, before the meaning of a "pre-historic" past was fully comprehended. Today, even grade school children can readily conjure quite accurate images of the most obscure dinosaur generic names faster than the two Grandmasters of Paleoart, Gideon Mantell and Georges Cuvier, could have ever produced a visual for the European dinosaur *Iguanodon*. Yet, someday perhaps, when the first spadeful of earth is taken from the Paleoartist's "Hall of Fame" museum foundation, an institution which assuredly someday must be established, the first two inductees will be two nineteenth century naturalists. Both are not typified in popular accounts as artists, per se. These are the French anatomist Georges Cuvier and British medical practitioner Gideon Mantell.

Since the early nineteenth century, the "sport" of fossil hunting has offered dedicated souls an avocational outlet and a means of financial stability. Early paleontologists took great interest in fossil species and published descriptions of new finds enthusiastically. But despite the scholarly intrigue focused on vertebrate fossils, there was remarkably little interest in restoring the old bones to life through illustrations, paintings or sculpture. Two of the earliest such depictions to become popularized were illustrations by paleontologist Henry De la Beche (1796–1855). Both illustrations were printed for distribution and sale in 1830.

De la Beche's scene, "*Durior Antiquior*," is regarded as the first reconstruction of a paleoecosystem, in this case showing the marine fauna and flora formerly existing during the British Liassic (Jurassic). His fanciful cartoon *Awful Changes* is a sarcastic stab taken at Charles Lyell's uniformitarian geology. *Awful Changes* features a well attired (post-historic) "Professor Ichthyosaurus" (i.e. Lyell caricatured) lecturing to a saurian audience millions of years hence about that "former," inferior race—*Homo*.[1]

Understandably, neither of De la Beche's illustrations were considered suitable for academic publications. The light tone of these illustrations is deceiving, however, because his illustrations were among the first to convey complex paleontologic ideas visually.

Certainly the two greatest early nineteenth century researchers of prehistoric vertebrates, Cuvier and Mantell, who tested the waters with early, serious attempts at scientific restorations, must have understood what little regard contemporary fossil specialists had for visual communication in scientific discourse.

CUVIER'S SCIENCE

Georges Cuvier (1769–1832), who dubbed himself a "new species of antiquarian," died a decade before Sir Richard Owen (1804–1892), often revered as the "British Cuvier," described the "Dinosauria." Cuvier is *the* pivotal character in pre–1850s vertebrate paleontology, and apparently also a key player in the development of geological systems. Prehistoric vertebrate fauna were essential to his perspectives on the recurrent nature of geological catastrophes. Had Cuvier not established a modern synthesis of extinct vertebrates, however, dinosaurs may not have even been comprehended by the mid–nineteenth century. Had Cuvier gained more fossil (theropod) evidence of the Honfleur "crocodiles," perhaps he and not Owen would have christened the Dinosauria. After all, the "Age of Reptiles" had been discovered by Cuvier. Were he alive today, what would Cuvier have contributed to the riddle of the Cretaceous-Tertiary (K-T) boundary mass extinctions?

Most popular works on dinosaurs reference Cuvier in historical, introductory chapters, but readily march on to British scientists William Buckland (1784–1856) and Gideon Mantell, both of whom Cuvier assisted in their early investigation of dinosaur fossils, followed by the disagreeable Owen. Such treatment yields false impressions that the origins of modern vertebrate paleontology and dinosaur science arose in England, when Cuvier deserves credit for erecting that field of research in France nearly three decades earlier.

Cuvier didn't discover or describe the first fossil vertebrates, but he was the first to explain their nature with sound reasoning. Intriguingly, Cuvier would elucidate the pattern of earth history through meticulous examination of fossil vertebrates (employing his "law" of anatomical correlation).[2] Cuvier's descriptions of early extinct mammals, written at a time when comprehension of their geological origin was in its infancy, are still regarded today for their accuracy.

Cuvier was meticulous in recording facts while carefully avoiding tendencies to speculate how these organisms came to inhabit the planet in their respective ages. Above all he was concerned that his observations should not become extrapolated into another "theory of the earth," as proposed by numerous colleagues, including the eminent Georges Leclerc Buffon (1707–1788).

In Cuvier's time, the earth sciences were in early developmental phases. Two "categories" of "geology" then existed, a nontheoretical, descriptive "Geognosy," in which facts and field observations were carefully recorded, and a more romantic, if not dramatic, "Geology," in which causal explanations charting the entire course of earth history were often loosely theorized, based on the most important or select observations.

Paleontology wasn't even a recognized field of study until Cuvier demonstrated how precise knowledge of prehistoric species testified to the nature of earth history in a "geognostic" fashion. Cuvier condemned the unbridled imaginations of the "geologists," claiming they often overlooked important facts. Unlike the "geognosists," it was the "geologists" who were prone to concocting fanciful "cosmic romances" of prehistory. Most importantly, Cuvier relied on vertebrate fossils in refuting the most popular eighteenth century theories of planetary development.[3]

Decades before Cuvier began his investigations of fossil elephants, Buffon (and others) had theorized that the earth had progressively cooled since an incandescent origin in gases expelled from the sun. This explained occurrences of fossil elephants and rhinoceros discovered in the northern hemisphere. As the earth cooled gradually, first at the poles and in temperate zones, such warm-loving species migrated to tropical regions.[4] (Recall that scientists [or "savants"] of Cuvier's time knew nothing of the Ice Age, astronomical Milankovitch cycles or radioactivity, which keeps the earth's deep interior hot.)

Through his examination of these fossil relics, Cuvier was led to a profound alternative interpretation. Because the fossil species were not the same as those living, and because even if such creatures were alive they would have already been witnessed by human explorations of the globe, the fossils must represent extinct life forms. Here was a radical and novel notion, a philosophy we now take for granted.

Cuvier had unexpectedly opened the barriers of time to a prehistoric period of uncertain duration, an age that ended with a bang, not a whimper. For "...only some sort of catastrophe" could have destroyed these animals in such complete stages of refrigeration as evident in some specimens of woolly mammoth and rhinoceros frozen in Siberian ice.[5]

While declining to speculate on the exact nature of any of the catastrophes, subsequent studies, particularly those conducted with colleague Alexandre Brongniart (1770–1847), led him to the belief that geognostic evidence of catastrophes represented "*irruptions of the sea,*" or repeated mass continental or more localized floodings that drowned terrestrial organisms of those lost ages.

Such a position, factored with his faith in the German school of "Geognosy," whose chief proponent was the "neptunist" Abraham Werner (1750–1817), and coupled with several unauthorized editions of English translations of his work by British geologists who sought to reconcile Genesis with Geology, has made some historians of science suspicious that Cuvier had a "Creationist," diluvial intent. (Interestingly, Cuvier also investigated human tales of genesis, many of which involve recorded catastrophes and deluges.) Based on his reading of Cuvier's original manuscripts, however, paleontologist and historian of science Martin J. S. Rudwick refutes such allegations.[6]

Cuvier was cautious about stating the nature of the "revolutions" that swept the globe in ancient times. The animals were "*covered in liquid*" in the places where they once thrived. They were not transported violently from one latitude to another. But the cold climate swing must have descended with sufficient rapidity, or else the carcasses of specimens frozen in ice would have quickly decomposed.[7]

Such evidence refuted Buffon's theory of a gradually cooling planet. But it would have been difficult to reconcile such a range of evidence into a new, self consistent "theory of the earth." And so Cuvier declined to do so throughout his career. In his view, there wasn't enough evidence on which to base a refined "theory of the earth" enlightened by geognostic facts.

Cuvier's perception of the scale of geological time evolved rapidly. In early papers he wrote of the *remote centuries* in which the more recently extinct creatures under his examination had lived. Thereafter, he spoke of *many centuries*, and then *thousands of centuries* as the probable age of the (Eocene) mammals found in the gypsum beds around Paris. On one hand, Cuvier was extending man's (biblically founded) scale of time. However, Cuvier departed considerably from the early "uniformitarian" geologists, such as James Hutton (1726–1797), who believed (geological) time had seemingly limitless boundaries.[8]

This rather restricted view of geological time's extent had bearing on two important areas, the tempo of planetary as well as biological changes. Based on the evidence, Cuvier held that, in some cases (but not all), extinct populations of animals must have experienced rapid turnover.

Cuvier's examination of Egyptian mummified remains (in particular a specimen of Ibis) indicated no substantial differences between features of the modern *Ibis*. Since the *Ibis* was of a fairly well established, ancient age, here was added information that species "transformation" did not occur. If no transformation took place over a shorter, yet extended, period of time, then there was no reason to expect "evolution" occurring over prolonged periods.

The revolutionary forces triggering planetary and faunal upheaval were unlike the processes presently shaping the surface of the planet. However, the causes were not supernatural. To Cuvier, the exact nature of these forces was of secondary importance and possibly indeterminate.

One of his greatest public "triumphs" was exposure of the "marsupial bones" (bones supporting the pouch in living marsupials) in a fossilized, opossum-like creature found in the gypsum beds of Paris. In the fossil slab, as originally preserved, the signature pelvic structure lay hidden under rocky matrix. Cuvier deduced from his detailed analysis of the exposed skeleton and anatomical principles that the little quadruped was really a marsupial. Then he proceeded to excavate the exposed pelvic bones and underlying matrix until the marsupial bones had been revealed, proving his hypothesis to a startled group of eyewitnesses.[9] Cuvier would also perform grandly in the "Homo diluvii testis" caper by demonstrating that this celebrated fossil, purported in 1726 to be that of a "man drowned in the Deluge," was really a species of Miocene salamander.[10] These are stunts that the fabled, fictional character Sherlock Holmes and his creator, Arthur Conan Doyle, would soundly applaud.

But the aim of these affairs was not simple showmanship. Once again, as in the case of the fossil opossum cited previously, Cuvier sought to demonstrate that facts must precede theoretical notions. Who would have surmised that a genus, now limited to Australia and the New World, was formerly native to Europe? It was a conclusion that falsified formerly held views of animal migration. Furthermore, in relying on his anatomical laws, Cuvier elevated the science of anatomy to the esteemed "geometrical" science of celestial mechanics and the emerging field of quantitative chemistry.

Cuvier estimated that the most recent catastrophe occurred about 10,000 years ago. Cuvier noted that no fossilized remains of humans had ever been found. However, Cuvier admitted in his *Preliminary Discourse* to his *Les Ossemens fossiles* (1812) that such findings did not prohibit the existence of ancient humans before that time. It remained possible that positions of land and sea had reversed. Traces of prehistoric, yet anatomically modern, men (the races eventually giving rise to modern humans) were now covered beneath the waves.[11] (Sometimes Cuvier's explanations can seem a bit "ad hoc.")

CUVIER'S ART

Cuvier applied his laws of anatomical correlation of parts to distinguish modern mammalian fauna from extinct fossil forms. Among others, Cuvier identified the *Megatherium*, *Mastodon*, Mammoth, *Anoplotherium* and *Palaeotherium*, the latter being a Tertiary "tapir." Cuvier also rendered the first reliable pictorial reconstructions of prehistoric vertebrate species.[12]

By the early 1800s Cuvier had distinguished the North American *Mastodon* from other varieties of fossil and living elephants. He reconstructed the genus in 1806, based on fossil bones shipped to Paris from Ohio. Cuvier's skeletal reconstruction of the *Mastodon*, depicted in a grassy setting, was the most "animated" ever done until that time.[13] Soon after, Cuvier reconstructed the *Paleotherium* and the *Anoplotherium* in lively poses. Then he took the next significant step toward restoring his *Anoplotherium.*

He sketched in muscles supporting the limbs, jaw, neck and abdominal areas, drew enclosing body outlines, and even inferred the appearance of nostrils, eyes and ears. (See Figure 3-1.) Martin J. S. Rudwick believes Cuvier's preliminary, undated reconstructions, replete with musculature, were made after 1812. Rudwick refers to these early studies in paleo-restoration as "...superb examples of his *[Cuvier's]* artistic skill."[14]

However, when it came time to publish an expanded, improved edition of his *Researches on Fossil Bones* (1821–24), Cuvier's groundbreaking reconstructions were omitted. Instead, simple restorations, body outlines of the *Paleotherium* and *Anoplotherium* (two species for each genus) drawn by Charles Laurillard, Cuvier's assistant, were selected to illustrate the text.

Apparently, Cuvier realized how conjectural were his reconstructions, as the inferred soft tissues had not been preserved for study. So, rather than risk criticism from colleagues who might bristle at the inclusion of speculative drawings in a scientific manuscript, instead Cuvier focused on his taxonomy, leaving Laurillard to take any blame for introducing uncertainty via the published reconstructed body outlines.[15]

Ironically, Laurillard's inanimate images proved quite popular, being reproduced in a number of textbooks and popular geology books throughout the nineteenth century. They were given greatest tribute in the hands of the "British Cuvier," paleontologist Richard Owen, and sculptor Benjamin Waterhouse Hawkins, who sculpted life sized paleotheres and anoplotheres as part of the geological menagerie first displayed in 1854 outside the Crystal Palace at Sydenham, England.

Meanwhile, Cuvier's brilliant designs rested in obscurity for decades. His *Anoplotherium* study was first published in Rudwick's book *Scenes from Deep Time: Early Pictorial Representations of the Prehistoric World* (1992). Although regarded as an early practitioner of paleoart, Laurillard is no "Hall of Famer." His restorations wouldn't have been possible without Cuvier's detailed reconstructions, several of which approximated lively restorations. Conversely, for his pioneering insights into the possibilities of what could be scientifically achieved, and for applying dashes of artistic license to scientifically sound skeletal reconstructions, Cuvier is recognized as the first Grandmaster paleoartist!

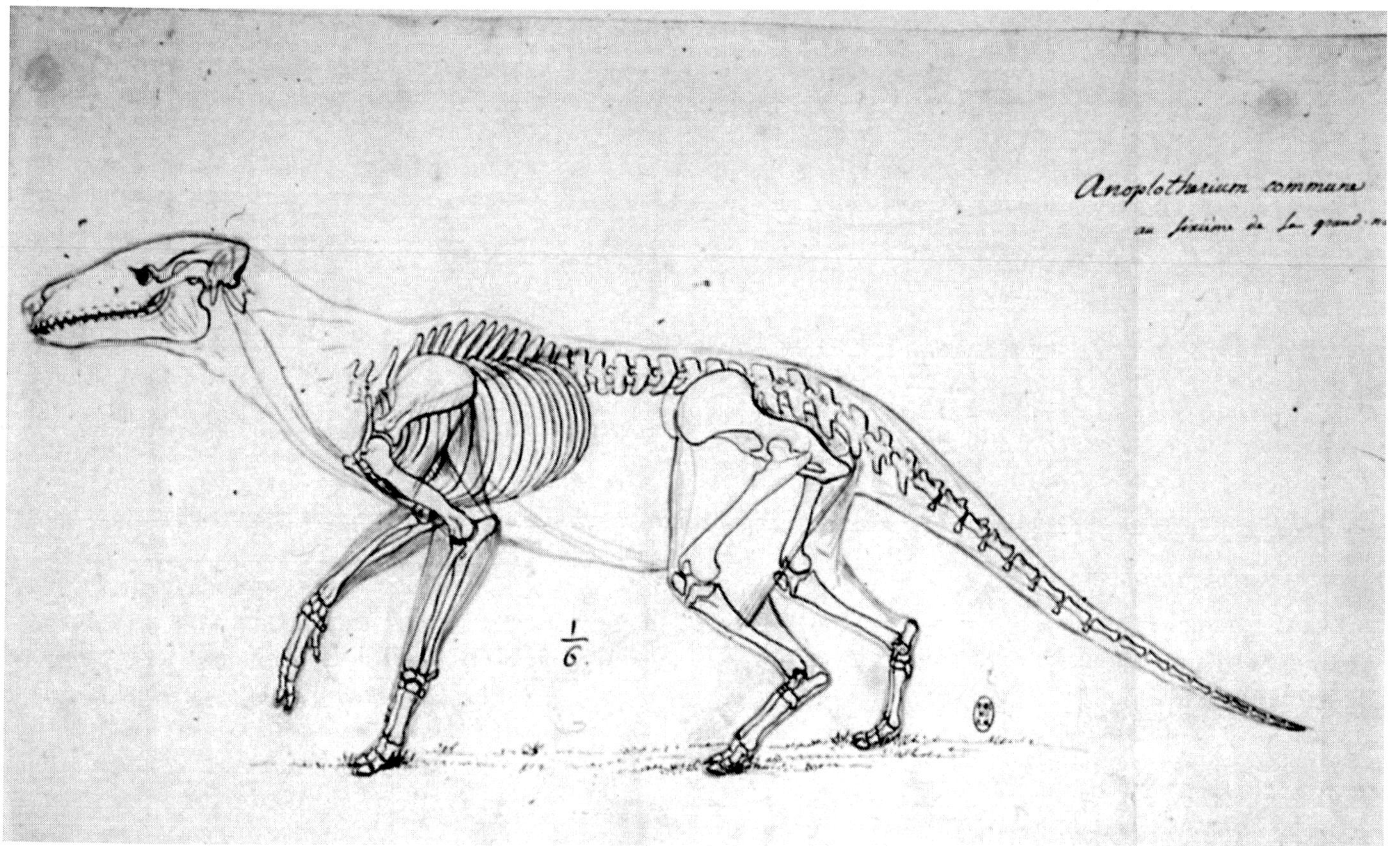

Figure 3-1: Cuvier's early restoration of the *Anoplotherium*, known from Parisian Tertiary deposits. (Copyright Bibliotheque Centrale du Museum National d'Histoire Naturelle 2000. Photograph ref. D10513. Reproduced with permission.)

MANTELL'S WONDERS OF PALEONTOLOGY

We take pride in our knowledge of popular dinosaur "myths." Yet so many of us know so little about the early history of paleontology or of how dinosaur science emerged. One often cited tale concerns the story of *Iguanodon*'s discovery. Supposedly, Gideon Mantell's wife, Mary Ann, happened to find a fossil tooth which, as Gideon astutely recognized, belonged to an extinct reptile. Supplemented by other related discoveries, the *Iguanodon* (Voila!) was described and eventually reconstructed (with that odd-looking "horn" on its nose). Sadly, this distortion may be the extent of what most dinophiles know about Gideon Mantell.

Gideon was one of the most prestigious paleontologists of his day. He wasn't just a fuddy duddy fossil collector who got lucky with a handful of dinosaur bones. In fact, Mantell was a more careful and accurate researcher than contemporaries William Buckland and Richard Owen.

Mantell was a patient, first-rate researcher of *Iguanodon* and other fossil saurians. He didn't jump to conclusions, and his hunches about the saurians he studied were startingly precocious. For instance, Mantell cautiously regarded the nature of the fossil tracks found in the Connecticut River Valley of Massachusetts, which had been described by Edward Hitchcock as avian in nature. Mantell suggested that the tracks would eventually be proven to be reptilian. In fact, these are now regarded as footprints made by bipedal theropod dinosaurs, such as *Coelophysis* and *Dilophosaurus*. Furthermore, Mantell's knowledge of stratigraphy, acquired through firsthand field experiences, allowed him to understand his saurians in a geological context, a skill Owen lacked.

Mantell eventually tangled with Owen, and a bitter "Marsh/Cope"–like rivalry ensued.[16] From the perspective of posterity, Mantell lost, as Owen's name has become more prominently associated with the pioneering days of dinosaur science.

In 1805, at the age of 15, Gideon entered into an apprenticeship under a surgeon and general medical practitioner. This led to formal instruction through anatomical lectures (completed in 1811) in London, and his first medical practice, beginning as a partnership in 1814. Based on surviving records, one gleans that Gideon was far more active and successful in his adult life with fossils, as opposed to his medical vocation.

Gideon's fascinations with fossils began at the age of 12 or 13 when he found an ammonite "curiosity" at a stream outcrop. Inspired by his "petrified serpent," as the 1797 edition of the *Encyclopaedia Britannica* referred to ammonites, by 1803 Gideon had become an avid fossil collector. He absorbed the richly illustrated *Organic Remains of a Former World* (1804, 1808, 1811), an early influential treatise on "oryctology" (paleontology), which led to a meeting with its prestigious author, Dr. James Parkinson (1755–1824). Around this time, Mantell began filling notebooks with his thoughts on the stratigraphical occurrences of Mesozoic fossils.

He expanded his network of personal contacts by joining societies, and in 1812 saw his first geological essay, *On the Extraneous Fossils Found in the Neighborhood of Lewes*, published in a popular journal. (Lewes is the town near the southern coast of England where Gideon then lived.) He corresponded with noted fossil collectors, exchanging specimens and communicating ideas on their origin and occurrence. One of these associates, Miss Etheldred Benett (1776–1845), one of the earliest female fossil collectors, facilitated the recognition of Gideon's "breakthrough" scientific study concerning a puzzling invertebrate fossil species (i.e. *Description of a Fossil Alcyonium*). This was published by the Linnaean Society in 1815. Gideon's life and career were gaining momentum, and soon he entered holy matrimony with 20-year-old Mary Ann Woodhouse in 1816.

Until their eventual breakup, Mary Ann assisted and collaborated with Gideon's fossil collecting and manuscript preparation. From Gideon's fossil illustrations Mary Ann engraved plates for his books. Nevertheless, when she left him on March 4, 1839, with their housekeeper, Gideon succumbed to fainting spells and neuralgic facial pains. Certainly the stress of being forced to sell the articles of his treasured Mantellian Museum to the British Museum when business failed to prosper after a move to Brighton, followed by yet another move to Clapham (precipitated by a need to secure his medical practice), would have strained any relationship. Mary Ann and Gideon never divorced. A son, Walter Mantell, became an important naturalist, instrumental in the discovery of New Zealand's ancient bird, the Moa, *Dinornis*.

Iguanodon was Mantell's dinosaur, really the first dinosaur to be described by any man with a sense of accuracy and diligence, although this description steadily evolved over a quarter of a century until Gideon's untimely death in 1852. According to Dennis Dean, Gideon didn't realize he had in his possession the bones of creatures later named as dinosaur genera, *Megalosaurus* and *Cetiosaurus*, half a decade before his interests turned to the *Iguanodon*.

By November 1821, Gideon had identified six teeth belonging to a new saurian, several of which had been found by Mary Ann in the Tilgate Forest. Difficulties in retracing the origin of the first *Iguanodon* tooth discovery stem from Gideon's varying recollections of the event. Mantell's biographer, Dennis Dean, has documented several versions of the story, recounted between 1825 to 1851.[17] According to the version celebrated by many modern writers, Mary Ann is cited as having discovered the first tooth in the spring of 1822. Mantell takes credit for the discovery in his other accounts. As we all know, memory fades in time and the mind can trick one's recollection of specific events, especially after one's spouse leaves unceremoniously.

Mantell was sufficiently talented to write in both a popular and scholarly vein. He wrote a number of books, all of which are important for either their early summarization of the regional geology and paleontology of southern England or their popular, edifying nature. Two of these, *The Fossils of the South Downs, or Illustrations of the Geology of Sussex* (1822), and *The Geology of the South-East of England* (1833), were even cited well over a century after their publication, in the fourth edition (1965) of *British Regional Geology: The Wealden District*, published by the Natural Environment Research Council-Institute Geological Sciences. In short, Mantell may be regarded as having been a "one man geological survey" of his time.

It is surprising how few copies of Gideon's early books were actually sold to subscribers. His first scholarly book, *South Downs* sold at a loss to Gideon, as only 152 copies were purchased. *Geology of Sussex* was really a separate book, although Gideon referred to it as a "volume II" to the former. Less than 50 copies of his "volume II" were purchased by subscribers. Mantell's biographer, Dean, considers this remarkable combination as "...the rarest and most historic dinosaur book in English."[18] Later books, such as the several editions of the two-volume set entitled *The Wonders of Geology* (1st ed. 1838), *Medals of Creation* (1844) and *Petrifactions and Their Teachings* (1851), popularized paleontology and his understanding of the "Age of Reptiles." Gideon would never complete a final book, concerning a reconstructive history of the Wealden beds.

Besides *Iguanodon*, eventually other fossil saurian specimens, the "Maidstone *Iguanodon*," the bones of the *Hylaeosaurus*, the "Swanage Crocodile" and others, fell into Gideon's hands. They were subsequently described, interpreted, sometimes illustrated, and exhibited. (Incidentally, in 1834, De la Beche caricatured Mantell in a cartoon, representing the Maidstone *Iguanodon* as an immense playful puppy.[19]) Gideon was meticulous in studying all of this material. Priority over the fossil reptiles became a significant issue for Owen once he came on the scene in the 1840s. For Owen usurped Mantell's position as the elucidator of British fossil reptiles with his own *Report on British Fossil Reptiles*, Parts 1 and 2 of which were read to the British Association on August 31, 1839, and August 2, 1841, respectively. In the published version of this paper (1842), Owen would refer to a newly created group of fossil saurians, classified by him as "Dinosauria."[20] Although Owen made many mistakes, he captiously noted several errors made by Mantell. Mantell, who described five fossil reptiles now regarded as dinosaurs, never endorsed Owen's classification.

Dean has stated, "Like other brilliant men, [Owen] had swelled too much to see the laborious foundations on which he stood. Though he refined and confused much about dinosaurs in his "Report," Owen discovered nothing.... In all justice, therefore, Gideon Mantell was both the primary discoverer and designer of dinosaurs."[21] While Owen may have been the "British Cuvier," Mantell was certainly the "Dean of Dinosaurs," ranking over Owen. (In June of 1824, Cuvier validated Mantell's original interpretations of the *Iguanodon* as an immense saurian. Cuvier's acceptance led to Mantell's formal description, published by the Royal Society in 1825.[22])

Tragically, Mantell fell from a horse-drawn carriage in October of 1841. He suffered a spinal injury, partially paralyzing him for an extended time and leaving him in great pain. Applying the contemporary medical knowledge, he took poison (i.e. opium and cyanide) to ease his suffering. His autopsy revealed a severe form of a condition known as curvature of the spine. It is remarkable how he was able to concentrate on his work at all during his final decade, which was also the first decade in which formalized dinosaurs had been identified.

Mantell's Paleoart

Mantell didn't sketch dramatic scenes of deep time's grandeur, or restore dinosaurs (as do today's "dinosaur artists") in dramatic, lively poses. But he skillfully illustrated fossils in his collection; the drawings were later engraved, often by Mary Ann, for publication. In 1832 Gideon's boyhood friend Warren Lee painted the *Iguanodon*'s rear leg, "...as restored by Gideon, who had calculated its musculature."[23] Gideon is also credited with what may be the earliest reconstruction of a dinosaur, a "horned" *Iguanodon* skeleton sketched by Mantell in 1832, left unpublished until 1968.[24] Dean has lauded Mantell's reconstruction as the "...earliest known scientific attempt to piece a dinosaur together."[25]

For Mantell, *Iguanodon*'s study occupied a lifetime's dedication. Mantell had commissioned two of the earliest scientific restorations of this genus in paleoecological settings. One of these, a restoration of a Wealden scene entitled, "*Reptiles Restored, the Remains of Which Are to be Found in a Fossil State in Tilgate Forest, Sussex*" was used in Gideon's popular lectures. The restoration was prepared by George Scharf (1788–1860) in 1833, and it recently adorned the dust jacket of Dean's book, *Gideon Mantell and the Discovery of Dinosaurs* (1999). Dean acknowledges Scharf's painting as the "...earliest known attempt to reconstruct the external appearance of dinosaurs and their environment."[26] Situated in this restoration are a plesiosaur, an armored dinosaur, *Hylaeosaurus*, and *Megalosaurus*, appearing as a strange monitor lizard. Unfortunately, a three-yard-long painting evidently derived from Scharf's surviving watercolor, pencil and ink design illustration has disappeared. (See Figure 3-2.)

Mantell also commissioned another artist, John Martin, to complete a painting titled "*The Country of the Iguanodon*," which appeared as an engraving, the frontispiece to several editions of *Wonders of Geology*, 1838 to 1857. Mantell was among the first paleontologists to incorporate restorations such as Scharf's and Martin's into his popular lectures on geology.

Perhaps another reason why Owen's fame as the principal researcher of dinosaurs spread in the latter nineteenth century was due to his involvement in the restoration of the three species he had described as "Dinosauria" in his 1842 report (*Hylaeosaurus, Iguanodon*, and *Megalosaurus*). As we all know, his visions are literally "chiseled in stone" on the Crystal Palace grounds at Sydenham, thanks to the early 1850s artistry of Benjamin Waterhouse Hawkins.

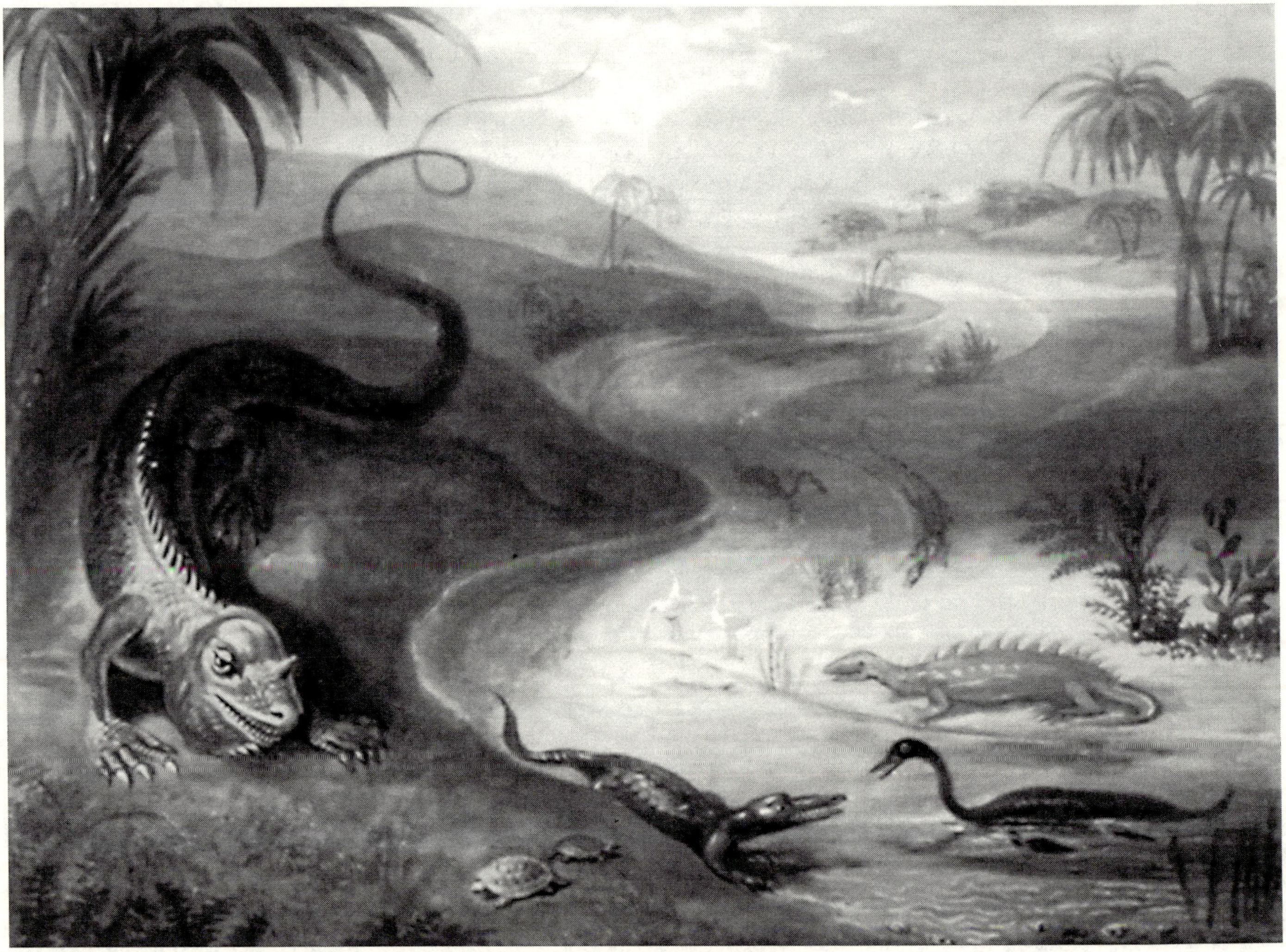

Figure 3-2: George Scharf's 1833 restoration of the British Weald-Tilgate Forest region, restored possibly for a book on saurians never completed by Gideon Mantell. Saurians chiefly visible are (left to right) *Iguanodon*—originally calculated from the remains to be 100 feet long, *Megalosaurus, Plesiosaurus* (in river) and *Hylaeosaurus*. Note wading birds at bend in river. Watercolor and pencil sketch. Megalosaurus actually lived during the middle Jurassic, not the Cretaceous. (Alexander Turnbull Library, National Library of New Zealand, Te Puna Matauranga o Aotearoa. Reference Number E-330-f-001. Reproduced with permission.)

Owen noted that Mantell's image of the *Iguanodon* was impossibly huge (exceeding 60 feet in length). Mantell, on the other hand, had at one time claimed he had "...assuredly underestimated than exaggerated its size, ... like Frankenstein, he was actually appalled at the being which rose from his meditations."[27] Owen had assigned to the *Iguanodon* ponderous forelimbs, while at one time Mantell had envisioned a nearly bipedal, (sloth-like) mammalian-looking creature.

Three months prior to his death, on August 10, 1852, an unhealthy Mantell was honored with a request from the Crystal Palace Company to "superintend" the erection of a "Geological Court" at Sydenham, a task which was declined because he preferred the idea of more scientifically oriented displays.[28] Owen now quickly rushed to restore the image (erroneously), in concrete, iron and cement, of two of Mantell's dinosaurs, *Iguanodon* and *Hylaeosaurus*, perhaps the final irony resulting from the Mantell-Owen feud.

Owen may have supervised Hawkins' wonderful sculptures at Sydenham, but it is Mantell, not Owen, who will someday merit induction into a Paleoartists' Hall of Fame—for devoting a lifetime to illustriously envisioning Britain's Wealden fauna as it once thrived! (See Figure 3-3.)

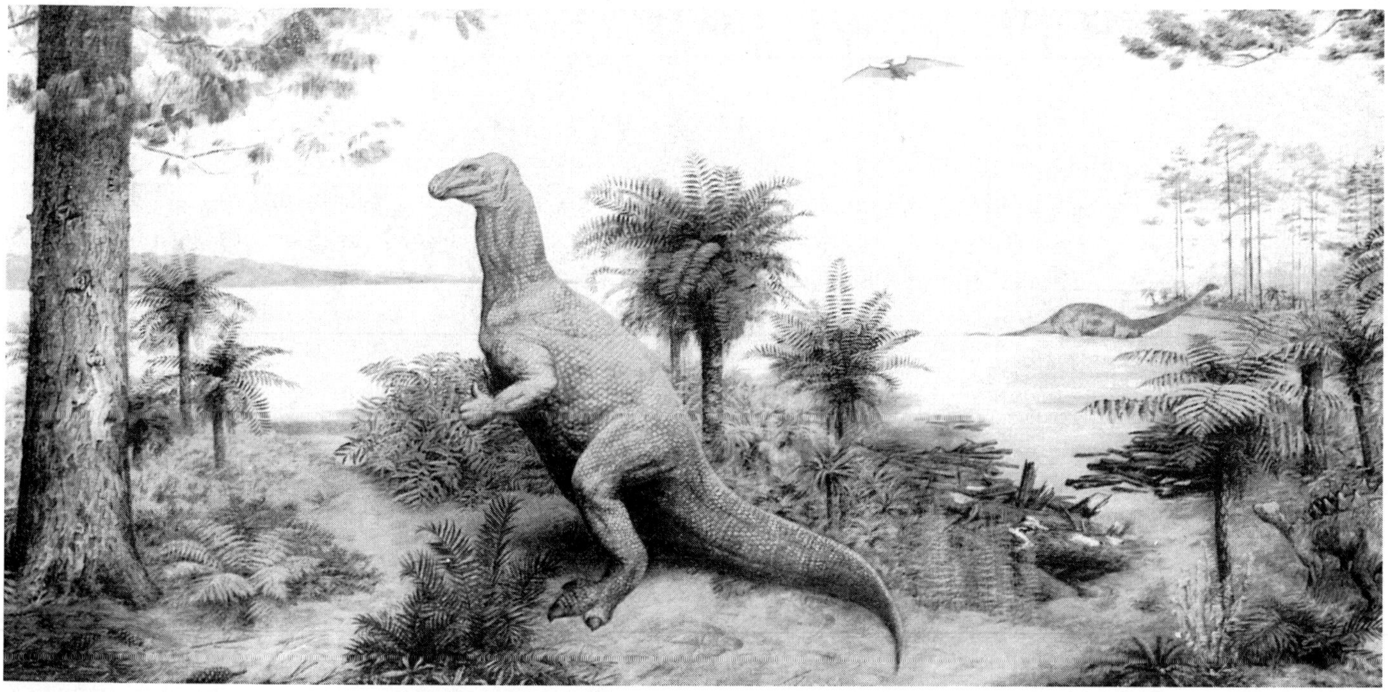

Figure 3-3: Idealized Landscape of the Wealden Lake (1935). E. Marsden Wilson's (1877–1965) impression of vegetation and reptiles of the early Cretaceous, including *Iguanodon, Polacanthus* and *Cetiosaurus*. While this view is more modernized compared to Scharf's older restoration, benefiting from further scientific discoveries and reinterpretation, by today' standards even Wilson's restoration seems dated. (Copyright The Natural History Museum—London. Photograph ref. T04324/A. Reproduced with permission.)

CHAPTER 4

John Martin and the Element of Paleoart Fantasy

The image of battling, ferocious prehistoric monsters strikes a familiar chord in us all. For many, such restorations form a principal attraction to paleontology. Through our decades of exposure to paleoimagery we can recall electrifying examples—Charles R. Knight's Field Museum "*T. rex* vs. *Triceratops*" mural; Zdenek Burian's battling aquatic reptiles; even *Jurassic Park's* climactic movie scene in which raptors engage the *T. rex* in a deadly struggle. We take for granted that popular paleoart emphasizes the theme of saurians battling in haunting, alien landscapes. However, few would recognize the unlikely origins of this particular genre.

"Paleoart" (a 1980s term) was a rare treat in the first half of the nineteenth century. Paleobiologists of the time generally avoided the practice of restoring fossil creatures because it seemed scientifically counterproductive, overly speculative. In this rather stifling culture, by the mid–1830s only a handful of restorations and skeletal reconstructions had been attempted. A theme of the earliest life scenes was to portray all the creatures (vertebrates *and* invertebrates) known as fossils from a locality together as a congregation of living organisms.

A particular attention to accuracy in detail can be discerned in the earliest examples, which were done by geologists also skilled at caricaturizing facets of paleontological ideology. There was minimal interaction among any members of the restored group, and although the vertebrates were depicted in likely poses, they were not very lively. Pains were taken to minimize speculation concerning scenery, life habits and appearance. Inert bones had been draped in flesh, and bodies were nestled in quaint paleo-landscapes.

A British artist, John Martin (1789–1854), introduced a distinctive "flavor" of paleo-fantasy in scenic restorations. Because even today the most popular restorations include the savage portrayals cast in the eerie, uncharted realm of prehistory, one could argue that the "Martinesque" style of paleoart has never waned. Who was John Martin and how did he become so influential in the field of paleontology?[1]

John Martin founded his reputation on painting (and engraving) cataclysmic scenes. He characteristically cast somber atmospheric skies over natural or civilized splendor, poised on the brink of catastrophe. Human figures were scaled in miniature against lofty mountainous horizons or mighty architectural displays. Martin's haunting imagery was well suited for his weighty themes: *The Deluge* (1834), *The Destruction of Pompeii and Herculaneum* (1822), *The Seventh Plague of Egypt* (1823), *Belshazzar's Feast* (1820), *The Fall of Babylon* (1831), *The Destruction of Pharoh's Host* (1830), *The Great Day of His Wrath* (1852). His specialty was in capturing the gravity of the moment from an omniscient, ethereal perspective.

Martin had the capacity to project his mind into the historical and antediluvian past to witness, in his mind's eye, the alien aspect of the earth's surface scarred by the mightiest supernatural or geological forces on record. In raising ancient cities from the ashes of time, Martin researched diligently. Contemporary archaeologists were opening up new windows into the past of Egypt and classical Greece and Rome, and he absorbed this information.

Martin's "Deluge" painting could have represented the first reconstruction of an antediluvian catastrophic scene involving prehistoric animals. After all, he offered a *natural* explanation for the event, and Baron Georges Cuvier (himself a French "catastrophist" geologist familiar with fossil fauna) may have visited Martin's studio while *The Deluge* was in progress. Apparently they agreed on the cause of the flood (i.e. it being a conjunction of the sun, moon and a comet). Yet we find no obvious trace of prehistoric mammals or other extinct fauna in this painting.[2]

By the time Martin visited the house and museum of British geologist Gideon Mantell (1790–1852), Martin's artistic accomplishments were widely known. Mantell, who had recently acquired the "Maidstone" *Iguanodon* specimen, recorded in his 1834 *Journal,* "Mr. Martin was deeply interested in the remains of the Iguanodon etc. I wish I could induce him to portray the country of the Iguanodon: no other pencil but his should attempt such a subject."[3]

Mantell believed that substantial book royalties would relieve his financial woes. Collaborating with an artist of Martin's stature would surely boost interest in Mantell's museum and popular books. So Martin completed *The Country of the Iguanodon* for Mantell, a scene which was later converted into mezzotint print (through a steel engraving process), and it appeared in editions of Mantell's *The Wonders of Geology* Volume 1. (This painting appeared as the frontispiece to Volume 1 in the first [1838], sixth [1848], and seventh [1857] editions.)

Martin produced a sparring throng of saurians that was printed from a steel engraving. (See Figure 4-1.) This scene, and one he would later complete for fossil collector and author Thomas Hawkins (1810–1889), could only be described as nightmarish, more intriguing for its somber atmospheric quality than the anatomical accuracy of depicted creatures.

In the "Description of the Frontispiece" (1848 and 1857 eds.), Mantell described Martin's restoration as a tropical setting "...peopled by colossal reptiles, and traversed by a river which is seen to empty itself into the sea in the distance."[4] Mantell interpreted plant fossil remains from Tilgate Forest in the Isle of Wight and at Portland for Martin, who introduced a verdant splendor of coniferous trees, arborescent ferns and zamias.

Yet, as before, the scene is dominated in the foreground by thrashing, bloodthirsty saurians. Fighting for its life, a herbivorous *Iguanodon* grapples with *Megalosaurus* and a massive crocodile. In the distance, a *Hylaeosaurus*

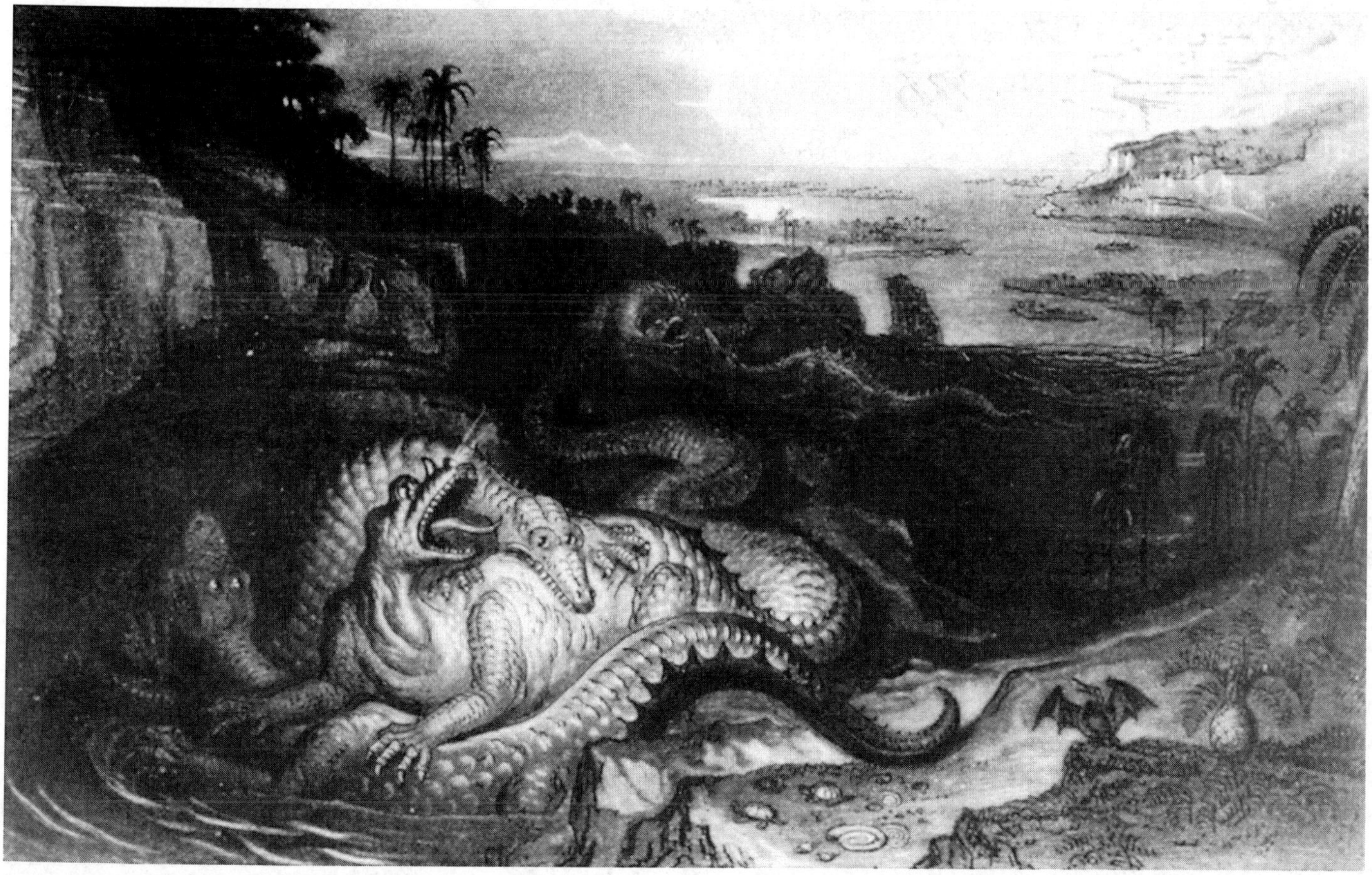

Figure 4-1: John Martin's *Country of the Iguanodon* (1838), showing battling Iguanodon dinosaurs (foreground), and armored *Hylaeosaurus* fighting *Iguanodon* in distance.

and *Iguanodon* prepare for battle. It is not known why Martin did not also restore the warm-blooded insectivore "spalacotheres" or avian creatures that Mantell believed also existed in the "country," based on the available evidence.

Mantell believed the "country of the Iguanodon" was irrigated by streams and rivers. The bones of fossil vertebrates were broken and scattered, indicating that they must have been transported to the deltaic Weald, thought to be as mighty as the modern Mississippi River. The carcasses of the two dinosaurs had floated hundreds or perhaps thousands of miles prior to deposition. Mantell delighted in the fact that Martin had "rescued from the oblivion of countless ages and placed before us in all the hues of nature, its appalling dragon-forms, its forests of palm trees and tree-ferns, and all the luxurious vegetation of a tropical clime."[5] For Martin, the collaboration fortified his (irreverential) belief that humans could not have coexisted with the great Secondary reptiles, as there was no fossil record of Man in Secondary rocks. Illustratively and narratively, the artist/scientist team dramatised the "wonders of geology" by projecting a rare, violent vision of prehistory.

Thomas Hawkins (unrelated to paleoartist Benjamin Waterhouse Hawkins) has been described as an eccentric poet-geologist who passionately collected Liassic (Jurassic) fossils. While preparing a second monograph on his collection, *Book of the Great Sea Dragons* (1840), Hawkins came across the scene Martin had painted for Mantell. Cognizant that Martin could greatly contribute to the success of his publications, Hawkins commissioned Martin to complete a study titled *The Sea-Dragons as They Lived.* (See Figure 4-2.) Martin depicted a ferocious encounter between ghoulish-looking ichthyosaurs and plesiosaurs along a moonlit, wave-swept shoreline. The accuracy of the creatures is suspect, yet the cataclysmic tone of his final contribution to paleoart is distinctively "Martinesque."[6]

According to Martin J. S. Rudwick, Martin produced another restoration that was printed in George Richardson's text *Geology for Beginners* (1842).[7] Aquatic reptiles and

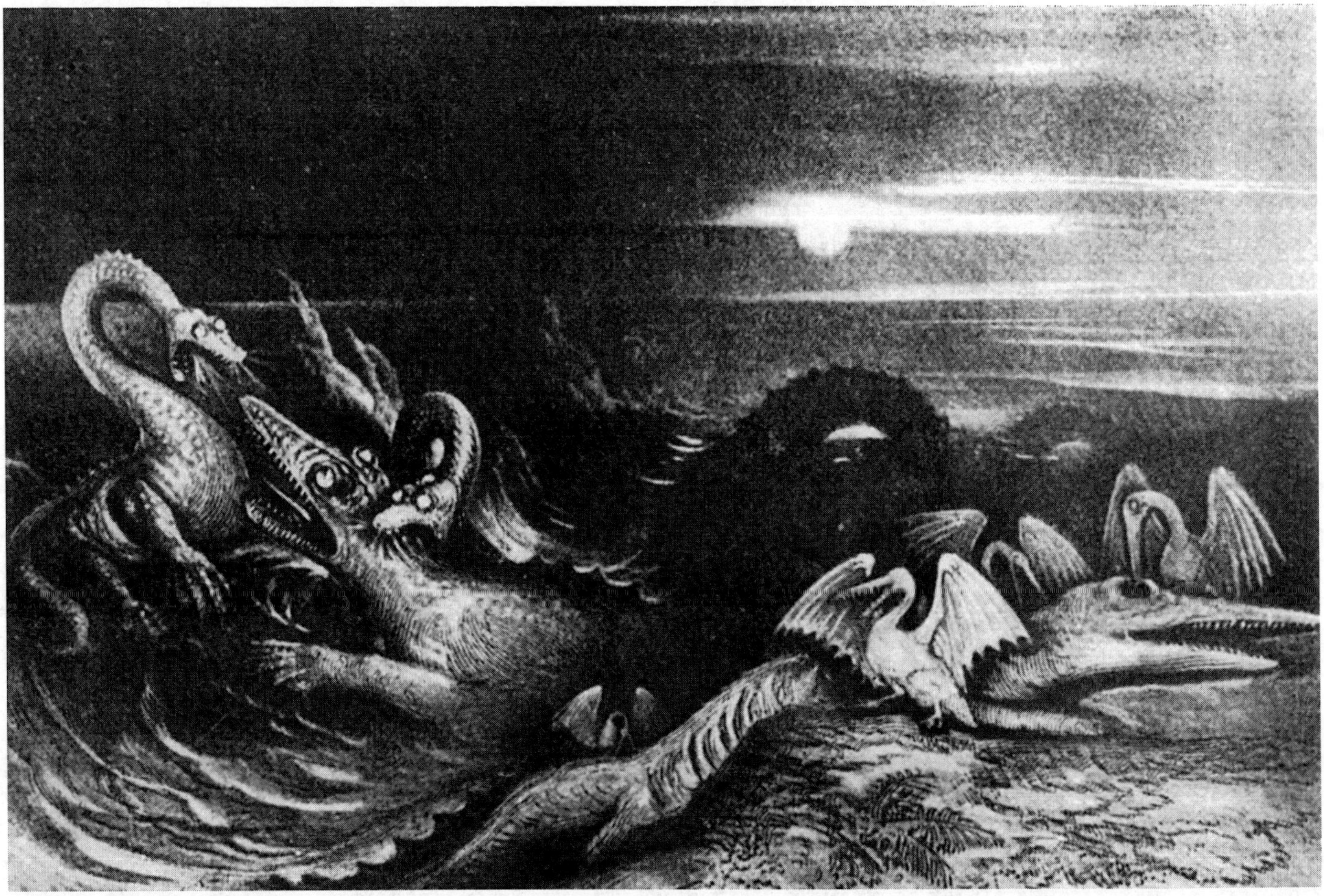

Figure 4-2: John Martin's restoration of a hellish primeval age, the British Liassic, showing sea dragons (*Ichthyosaurus* and *Plesiosaurus*) "as they lived." Presumably there were more tranquil moments as well in prehistory.

dinosaurs convincingly menace one another while a squawking pterodactyl (which appeared previously in Martin's two other restorations) surveys the unfolding chaos. Yet it is the least inspiring of Martin's three prehistoric scenes. (It is possible Martin created this restoration during his period of association with Mantell, and that it may have even appeared in an edition of Mantell's book.)

With the resolve of a gifted "Renaissance Man," Martin promoted original engineering designs for improving water quality. Epidemics caused by unsanitary conditions, as represented by pollution of the Thames river, and a reading of Mary Shelley's science fiction novel *The Last Man* (1826), describing the extinction of mankind caused by a plague-like illness, inspired one of Martin's final masterpieces, *The Last Man* (1849).

Throughout Martin's career, as he was both branded a charlatan by detractors and lauded for magnificent visions, his apocalyptic paintings were thought by some to be products of a delusional mind. In the decades following his death (1856), Martin's influential artistry went largely unacknowledged.[8]

Dinosaurs and paleoart represented only a brief artistic excursion for Martin, as relatively little was known of prehistory in the first half of the nineteenth Century. Curiously, given his penchant for recreating cataclysmic moments on canvas, Martin never painted catastrophes in geological prehistory (neglecting his Deluge paintings). However, had the asteroid impact model of mass extinctions been known to his contemporaries, wouldn't Martin have seized the opportunity to paint the event expertly? One can only dream of how the contributions of "Mad Martin" (as referred to by critics during his heyday) would have impacted art in the era of "science gone mad."

CHAPTER 5

Henry Ward's Last Great Siberian Mammoth

Undoubtedly the most astounding way in which to witness a prehistoric animal would be to bring it back, alive, refreshed from the vault of Time. Someday scientists may raise Mammoths from icy graves using the latest advances in biogenetic engineering. But then they will have Mammoth calves to raise in an alien landscape, an environment perhaps unsuited for the extinct. Few realize, however, that the most effectively recreated Mammoth of all time bedazzled mankind only a few generations ago—before being itself razed. This was the mightiest "prehistorian" ever sprung from the fossil cabinets of Museum-builder Henry A. Ward (1834–1906).

More rugged than the hills where he collected fossils—that was Henry A. Ward. Ward logged more miles in his quest for fossils and geological specimens than any other nineteenth century naturalist. His adventures off the beaten path make Darwin's voyage on the Beagle seem like a mere waltz in the park. He instituted the first American company dedicated to supplying a host of venerable universities and other institutions with specimens and other materials for disseminating public knowledge concerning natural and geological history. Although Ward was among America's earliest influential, yet unheralded, paleontological promoters, he is an unfamiliar character.

Driven by a wanderlust uncommon even to nomads, Ward was among the hardiest of old-time fossil collectors. Enduring extreme physical discomfort and peril on his journeys, Ward was a resolute ramblin' "hard guy." His unbridled passion for collecting zoological, geological and paleontological specimens rivaled the mania of Cope and Marsh. But whereas Cope and Marsh preferred to hide and horde their respective "quarries," so as to quietly but quickly establish priority of authorship, Ward, who published infrequently, sought zealously to educate the public with his procured items. This led to the establishment of his highly reputed company in 1862, Ward's Natural Science Establishment, specializing in sales of fossil cast specimens.

While not realizing it, many of you may have seen paleontological exhibits originally purchased from Henry A. Ward displayed in various museums around the country—say, a *Megatherium* skeleton here, a Giant Squid model or *Glyptodon* mount there. Although Ward also provided meteorite samples, and specimens and models broadly representing the zoological kingdoms, perhaps his greatest fame stemmed from the prehistoric vertebrates that were popularized through his "cabinet" displays at an impressionable period in American history.

Ward's collecting pursuits were established precociously. For in 1837, at the age of three, he is said to have snuck out through the front door of his Rochester home to seize a rock from his front lawn, a volcanic prize containing glacial striations. In time, the little boy who liked to run away (once to find his father who had fled the family to take on a new life way out west in Chicago) became a modern "Ulysses" in his unceasing quest to obtain fossils and specimens the world over.

Ward started formal schooling rather late in life, at the age of ten, yet immediately enrolled in high school classes, which is probably why he found it difficult to perform in mathematics and "engineering" classes. Yet he mastered foreign languages and had a profound enthusiasm for natural history. An early influence was his first formal teacher, Dr. Chester Dewey of Rochester High School, the first instructor who could answer any of Ward's questions concerning his cherished fossils. Through his readings, young Henry also became intrigued by the tensions between the clergy, who preached that the Creation took place in six biblical days, and "ungodly" geologists who claimed the earth was considerably older. Given that Ward's mother was highly religious, and that a common objective for college training back then was to prepare for the ministry, Ward must have been keenly sensitized to these discussions.

Ward repeatedly attempted to complete formal courses of instruction at various educational centers, including the Jardin des Plantes, the Paris School of Mines and Harvard's Lawrence Scientific School, yet kept dropping out both for financial as well as academic reasons. He always excelled at geology but found it hard to concentrate on other classes, especially when there was an entire world of "petrifactions" waiting to be found outside the classroom doors. Typically he would hike many miles, sometimes through inclement weather, on fossil hunting expeditions or to meet noted geologists, such as James Hall and Louis Agassiz. Eventually, however, in recognition of his remarkable accomplishments, in 1862 Williams College awarded him an A.M degree, and in 1896 he received an L.L.D. honorary degree from the University of Rochester, where he had held a professorship.

Ward was by no means an "arm chair" naturalist. For example, after visiting his father in Chicago, he hiked over to see the Galena Lead Mines (a four hour highway trek these days often taken by the authors on the way to a favored camping spot). Smitten by the collecting bug, Ward and his comrade, Charlie Wadsworth, traveled to Nova Scotia, the Spanish Peninsula, Europe and the Middle East (Egypt, Nubia and Palestine) in search of fossils and rocks. Trips such as these were bankrolled by Charlie's father, General Wadsworth, who encouraged the making of a noteworthy geological cabinet. No stranger to seasickness, during his lifetime Ward would cross the Atlantic 52 times and the Pacific Ocean 28 times.

Ward became well known to many of the leading public officials and scientific figures of his time. In England Ward befriended Sir Roderick Murchison (1792–1871), and David Livingstone (1813–1873). In France he attended a lecture by paleontologist Alcide D'Orbigny (1802–1857). In America he hobnobbed with Phineas T. Barnum (1810–1891), inventor Cyril McCormick and, his most unusual "specimen," William "Buffalo Bill" Cody (1846–1917). And, yes, he cheated death on more than one occasion.

Along the Cameroon coast, after had succumbed to a contagion, a sea captain left him for dead, marooned on an island. There a kindly native washerwoman nursed him to health. In Arabia he and Charlie twice encountered peril in the desert when they strayed into the midst of a civil war. In the American west, aboard a stagecoach, he outsmarted gold bandits, and at another time, after a shotgun was thrust into his hands to thwart a surly yet unsuccessful rescuing party, he became "deputized" by a band of justice-seeking vigilantes out to hang a criminal named Slade. His life would make a great television documentary!

In later years, the 5' 8", 175 pound Ward was characterized by one of his protégés, William Hornaday, thus: "If one would examine him analytically, it would be found that internally he is composed of rawhide, whale bone, and asbestos; for surely no human being could for 45 years so successfully withstand the bad cooks, bad food and bad drinks that have necessarily been encountered by anyone who has, so recklessly of self, traveled all over creation."[1]

Writing in 1948, Ward's biographer, Roswell Ward, a grandson of Henry, stated that fossils at that time "...had the same relation to the scientific world that cosmic rays or plutonium have at the present time. They were the center of speculation; the pathway to new knowledge...."[2] But fossils also offered more than knowledge; they were a means of survival.

Several circumstances led to Ward's conviction that selling natural history cabinets was a worthy, if not noble cause. First, while studying in France, Ward began selling fossils to private collectors and students—at substantial profit. So, needy of tuition and travel expense funds, fossils became a natural focus of his life. Second, the Paris Exposition of 1855, featuring many scientific and commercial achievements of the day, deeply impressed Ward, who became infected by the "exposition bug." Ward wondered why America shouldn't have its own expositions, a philosophy leading to the Centennial Exposition in Philadelphia and the earliest World's Fairs held on western soil.

In Europe, vertebrate fossils, especially, were guarded more carefully than even the crown jewels. Few people had access to them. Ward's "mission," however, was to facilitate natural history education in America through dissemination of fossil reproductions and the creation of natural history cabinets and geological museums. In America Ward would make these prehistoric giants accessible to everyone. Most significantly, he would foster the images of prehistoric vertebrates in America through replication of articulated or composited fossil skeletons.

By 1864 Ward had acquired molds from which plaster casts could be formed of magnificent fossil specimens, such as the 18-foot-long *Megatherium*, a specimen excavated from a South American Pampean deposit between the years 1831 to 1838. The original skeleton was jointly owned by the British and Hunterian Museums. Ward had also molded a *Glyptodon* skeleton. The latter was originally found in 1846 at Montevideo, but later presented to French officials. Incidentally, because the *Megatherium* bones were found in close proximity to a *Glyptodon* carapace, naturalists were "naturally" confused. Deceived by the evidence, the great beast became thought of as a "mammoth armadillo." Eventually, scientists demonstrated that the two were distinct genera, and that *Megatherium* was really a large genus of sloth, not an armored form.[3] (See Figure 5-1.)

The practice of replicating fossil skeletons through plaster casting had already been refined in England, and it is possible that while in London Ward had met artisans trained in the studio of Sir Francis Legatt Chantrey (1781–1841). By 1836 Chantrey had already molded the British Museum's *Megatherium*, and it is likely that here Ward may have learned the art of fossil preparation, plaster molding and casting. Ward may have gained expertise in fossil replication techniques from British specialists.

With such wonderful acquisitions, Ward reveled in the thought that he could now "...offer to American savants and institutions faithful copies of almost every noted fossil in the noted European cabinets.... These two larger pieces (i.e. Megatherium & Glyptodon) will form fine centers of attraction in my large museum—just such objects as any wealthy friend of a college would (or should) like to give it."[4]

Casts of the Ice Age brutes proved to be reliable attention grabbers. Ward popularized his star player in a late 1860s pamphlet titled *Notice of the Megatherium cuvieri, the Giant Fossilized Sloth of South America*.[5]

Ward's 1866 catalog of available specimens, *Catalogue of Casts of Fossils from the Principal Museums of Europe and America*, bearing a gold imprint of a quadrupedally posed *Megatherium* skeleton on the cover, is a pleasure to behold today. While *Megatherium* was apparently intended to be displayed quadrupedally, as indicated in a Catalog illustration, from contemporary photos and illustrations, museum curators also mounted Ward's *Megatherium* in upright position, poised as if tearing down a tree.[6] (See Figure 5-2.)

Neglecting the mineralogical specimens and models, there were approximately 1,200 specimens offered for sale. And check out these prices! Imagine purchasing a complete 23-foot-long *Plesiosaurus* skeleton on a slab for only $150, painted no less! A complete skull of the *Deinotherium*, thought then to be a marsupial, sold for $80.

Ward's prize *Megatherium*—complete skeleton—sold for a mere $250 (some assembly required!). *Megatherium*'s tree was sold separately, and there was an additional (optional) labor charge for mounting and painting the bones. Furthermore, an iron railing, adorned at the top of each post by the bronzed statues of extinct and extant edentate animals, could be furnished for another $175. Total charges for materials, shipping, labor and the ornate railing could come to $800. What a treat for Christmas morning—an assembled Ward's *Megatherium* towering over the star on

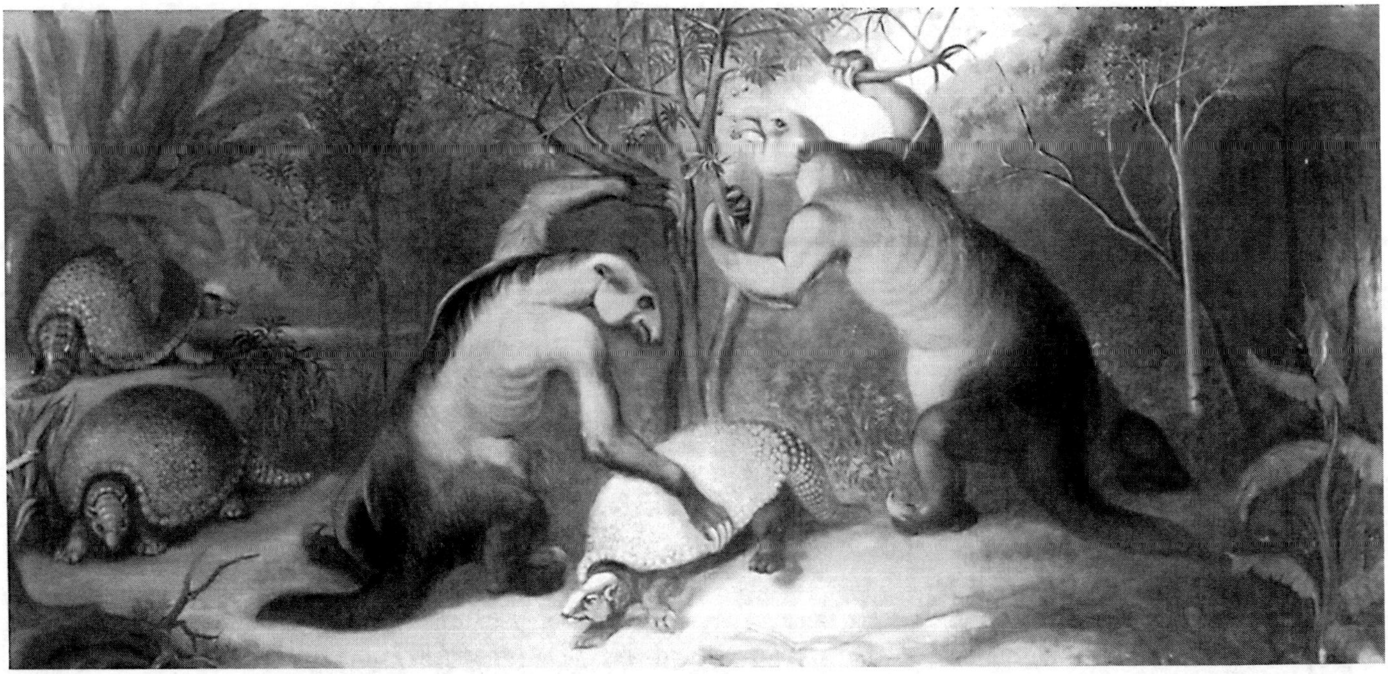

Figure 5-1: Benjamin Waterhouse Hawkins' contemporary vision of Patagonian Pleistocene edentates—*Megatherium* and *Glyptodon*. Circa 1877. (Copyright 1980, Trustees of Princeton University. Used with permission.)

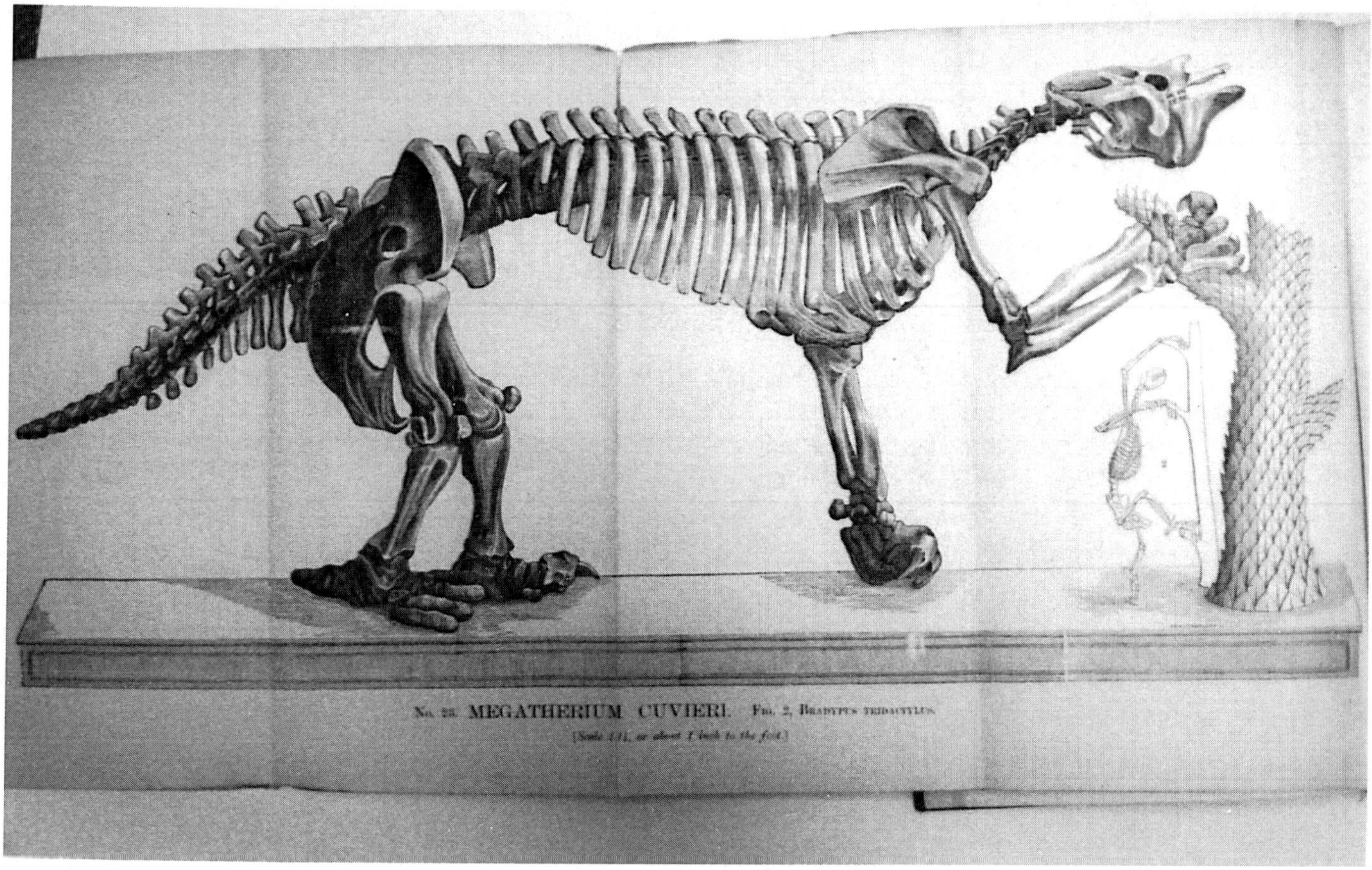

Figure 5-2: Reconstruction of Ward's *Megatherium* fossil cast, as appearing in Ward's 1866 *Catalog*. (Authors' collection.)

top of your decorated tree! Then again, maybe not such a good idea for the creationists of that time

Just about any fossil known to science back then was available through Ward's Catalog, both vertebrate and invertebrate—from a cast of the Neanderthal skull found in 1857 to ammonoids of every shape and variety. Ward had them all. Even today some museums exhibit pieces purchased from Ward in their collections.

Despite the fabulous array of fossils offered for sale, as we know, one doesn't become rich doing paleontology. And Ward, whose business was nearly always in the red, was among the earliest to make that disappointing discovery. Yet, Ward, having attended some of the greatest shows then on earth (in Europe), was himself a master showman.

Ward's experiences in helping to debunk the notorious Cardiff Giant underscored his commitment to public education. For one thing, the large human-shaped figure, sculpted in gypsum, heightened interest in fossils.[7] But the sensational hoax, perpetrated by the imaginative rogue George Hull, also raised questions over why people could become so easily duped. Were schools and colleges ill equipped to teach students about the true nature of fossils? Orders for Ward's fossils skyrocketed!

Circumstances led to the first of Ward's grandest exhibitions—a display at an 1875 industrial exposition held in Chicago. Amidst many splendid geological and mineralogical specimens, and "...hair-raising drawings of pre-glacial landscapes and prehistoric monsters...," Ward's plaster *Megatherium* stole the show![8]

As Roswell Ward noted, "Many a school boy and many a grownup went away from the Ward cabinet with a much clearer idea of the mysterious and alarming theories that the evolutionists were propounding. After all, there was nothing so sinister in this deep inquiry into the works of the Creator Himself. And the awe-inspiring monsters, the neatly labelled mineralogical specimens ... furnished ammunition for many a school boy's exercise, many a school master's lecture, and many a sermon."[9] A large proportion of the collection exhibited in 1875 was sold to the Chicago Academy of Sciences.

Eventually, because ground sloth fossils are rather common, museum institutions acquired genuine bones to display. So Ward's *Megatherium* casts gradually fell into disfavor. Thirty-five years ago an attempt was made to identify how many of them were still around. Six institutions divulged that Ward *Megatherium* skeletons still graced their exhibit areas. One instance of demolition was documented too. A sledgehammer was taken to the mounted bones, and the plaster fragments were discarded through an upper story window.[10]

Another odd "tale" surrounds the Ward *Megatherium* formerly displayed at the Colorado College in Colorado Springs. In 1977 Dr. William Fischer, Professor of Geology, recalled giving tours to students through the museum halls, offering lessons in comparative anatomy. Fischer stated, "On the last occasion of such a trip I discovered the class convulsing in laughter as I stood with my back to the sloth—the center of attention was the product of student ingenuity whereby the previous night the three foot long armored tail of the glyptodon had been securely wired in the pelvic region of the sloth, thus creating a Guinness Book of Records entry for phallic length."[11]

This particular rascal of a sloth is now owned by the Denver Museum of Nature and Science.

Ward's "signature" *Megatherium* was stunning. However, in those pre-animatronic/Dinamation days, perhaps the most publicized of Ward's paleontological displays became the life-sized restoration of the Siberian Mammoth that made its way through many U.S. cities during the late nineteenth century.

Ward had visited Moscow in 1857, becoming fascinated with the remains of a Mammoth, splendidly preserved, frozen in Siberian river mud. The "cool" fossil wasn't available for sale. However, little did Ward realize how great a sensation *his* own Siberian Mammoth would eventually create across the Atlantic.

Then, twenty years later, in Stuttgart, during one of his fossil roundup ventures, Ward came across a sculpted marvel—a life-sized restoration of a Mammoth created under advisement of naturalist and preparator Dr. L. Martin of the Royal Institute of Stuttgart. Martin had been inspired by the bones of a Mammoth skeleton held in the Royal Institute collection. Martin's observations of a Mammoth corpse displayed in St. Petersburg, found frozen along the Lena River in 1799, led him to recreate its massive frame in a lifelike pose.[12]

Martin's Siberian Mammoth, prepared under direction of German geologist Dr. Oscar Fraas, Director of the Royal Museum's paleontology section, had been constructed of wood, wire, cloth and papier-mâché. Their Mammoth was modelled after the largest bone samples then existing in the Royal Museum collection, explaining why the resulting restoration was so huge. The Mammoth's hair was made of Indian palm fiber dyed black with a reddish tinge, after hair coloration noted on the Lena River specimen. Its curved, six-foot-long tusks were fashioned in wood, and the trunk is said to have been made of cardboard. The magnificent restoration was 26 feet long (from tip of the tail to extremity of tusk curvature) and 16.5 feet high. It was a huge mammoth and it seemed alive. Ward dearly wanted the colossus![13]

Although mounted plaster skeletons were certainly impressive, Martin's Mammoth provided a striking "living" image of prehistory. As Roswell Ward recounted 70 years later, "It was all very well to exhibit fossil bones and fossil skeletons, but they did not carry conviction. Even the skeleton of the Megatherium, rearing up fourteen feet and forming the center of attraction of most of Ward's early cabinets, was after all, just a skeleton. Here was the real thing, a gigantic facsimile of the mammoth just as it had ranged the Siberian steppes centuries before. Ward had attempted to give some impression of the monsters of the past through drawings and paintings as well as fossil skeletons, but the average person did not regard these as much more than artists' fancies and geologists' dreams. In this case, however, it was as if he had been permitted to penetrate into the past and actually captured a mammoth and bring it before the public."[14] This would be the reaction to the earliest, authentic-looking, life-sized prehistoric animal restoration to invade America's public places.

With a little haggling over the price, the Mammoth was transferred to Ward's Natural Science Establishment for less than $3,000. Later, after a three week period, the giant had been carefully disassembled and crated for shipping.

The Mammoth received much fanfare as it rafted up the Erie Canal to Rochester in June 1877 from New York City. Reporters carried news of the grand procession into Ward's facility on College Avenue—a caravan of 14 horse-drawn wagons, each weighted with an enormous box. Although initially considering a sale to Princeton University, Ward came to his senses, realizing the appeal of his latest "find." "...But he fortunately discovered, before it was buried in a museum, that his new exhibit was just what he needed to appeal to the general public—something to convince the average citizen that such monsters could have existed. No argument or skepticism could stand up before the overwhelming bulk of the mammoth."[15] Ward, evidently, had a showman's good sense, allowing him to recognize the uniqueness and broad appeal of his Mammoth at a time when the American public was relatively unfamiliar with scientific restorations of prehistoric animals and other forms of "paleoimagery."

Yet, Martin's Mammoth wasn't the first life-sized sculptural restoration of a prehistoric animal. Benjamin Waterhouse Hawkins had already sculpted an "antediluvian" menagerie for the Crystal Palace exhibition at Sydenham over two decades earlier. In fact, Hawkins had even sculpted a *Megatherium*, standing in a pose similar to that of Ward's skeleton, before exhausting available funding. From surviving design sketches (circa 1869), it is apparent that Hawkins was planning to sculpt a Mammoth for New York City's ill-fated "Palaeozoic Museum" too. But Hawkins' Mammoth never materialized, so Ward's Siberian Mammoth may have been the first life-sized Mammoth sculpture ever completed.

Following its reconstruction at Rochester, Professor Alexander Winchell paid a visit to the "antediluvian." Astounded by the imposing sight, Winchell reported to the *New York Tribune* in August of 1878: "As the visitor enters the door of the building ... a dark mountain of flesh rises before him.... The monster's brow rises like some old granite dome, weather-beaten and darkened by the lapse of geologic ages. Two winding streams of ivory descend like glaciers from the base of the dome.... Behind expands and uprises the mountain mass ... this majestic form stands awaiting our wonder and adoration."[16]

Two weeks later, after observing the Siberian Mammoth at Chicago's Exposition building, Professor W. D. Gunning wrote, "Such restorations as those of Professor Ward give to the bygone creations a new life, a life in the minds of men.... You, of Chicago, have the proud scientific honor of first presenting the colossal mammoth to the eyes of the American nation."[17]

Some confusion lingers regarding how many cast replicas were made of the Siberian Mammoth and whatever happened to each. Winchell suggests that the first assembled specimen exhibited in America was built "...after the model of the original but with several minor improvements. The hair is rendered incombustible by steeping in a solution of tungstate of soda."[18]

The "original" was first unveiled to the public on "Mammoth Day" in 1877 at a charity event held in Rochester. Thereafter, promoters contracted an appearance of Ward's Mammoth at the Pittsburgh Industrial Exposition in 1879, for which Ward received $1,000. Upon sight of the apparition on display in Pittsburgh, poet Frank Cowan was moved to verse. In August of 1879 he penned a ballad entitled *The Last of the Mammoths*, concerning the Mammoths' last stand versus a prehistoric human hunter.

Demand for the "hairy" spectacle soon escalated, and soon it was whisked to expositions held in Louisville, Milwaukee, Cincinnati, New Orleans, the west coast and a host of other destinations.

In spite of the toll such commuting must have inflicted upon the fragile restoration, the Mammoth would hold up for a decade, until its untimely demise in a Louisville storage building. A leaky roof exposed the treasure to irreparable water damage in 1887 (or at least prior to 1889), and soon Ward found himself adding newer, improved paleontological displays, such as the "polyzoopticon," to make up for his "late lamented mammoth."

The polyzoopticon consisted of five compartments constructed on a circular stage, rotating to show life-sized dioramas of diverse wildlife habitats, ranging from Arctic to equatorial desert zones. Also, Ward commissioned a Mrs. Mathews to paint a Noah's Ark–like procession of living and extinct creatures. At the time, Mathews' composition was thought to be the largest painting in America, having dimensions of 18 feet high by 22 feet long.

Fortunately, as previously done with his *Megatherium* and *Glyptodon* cast skeletons, Ward managed to produce several replicas of the Siberian Mammoth, evidently cast in plaster. At least one of them survived from Victorian times into the atomic age.

One of the first Mammoth casts was installed, in 1877, at the Lewis Brooks Museum at the University of Virginia in Charlottesville. Brooks Museum was named after a wealthy Rochester entrepreneur, Lewis Brooks, who had helped support the creation of the University of Rochester's geological cabinet. In the case of the University of Virginia, Brooks had taken on a new cause in the aftermath of the Civil War—the reconstruction of the South. Brooks desired to remain anonymous until after the museum's grand opening.

In April of 1876, Brooks had generously donated $50,000 to be used in the construction of a natural history museum on the University of Virginia grounds, exhibiting casts and specimens to be supplied by Ward's Natural Science Establishment. Ward opportunistically launched an expedition overseas to scrounge up more valuable specimens. When he returned, Mammoth in tow, museum construction was well underway, being nearly completed by June 25, 1877.

Tragically, the 70-year-old philanthropist died on August 10, 1877, before the museum's grand opening. Only then did the University learn of their mysterious benefactor. Shortly thereafter, University officials named the museum in Brooks' honor.

Along with Ward's starring *Megatherium* and *Glyptodon* skeletal attractions, the Siberian Mammoth would become the favored attraction among all the 25,000 other paleontological displays at the Brooks Museum.

Although the museum would not officially open until June 27, 1878, visitors first had the opportunity to greet the mammoth after its assembly on December 12, 1877. An article in the January 9, 1878, *Jeffersonian* referred to the new museum—graced by its central attractions, a Mammoth and a "dinosaur" (the latter term most likely an erroneous reference to Megatherium)—as the nation's most "remarkable" collection. An 1878 article in *Frank Leslie's Illustrated Magazine* described the splendor therein:

"Here is a full procession of life through the ages, starting with the earliest animals of the Silurian and leading along in the series and down through time to the advent of man, who while the most modern of all, has lent his remains to the series in a cast of the famous man of Guadaloupe. In the central area are standing on broad pedestals a few colossal forms—the great Siberian mammoth and the glyptodon of South America."[19]

The aforementioned *Jeffersonian* article enthusiastically noted: "The most conspicuous and striking object in the Lewis Brooks Museum is the mammoth...."[20]

Attendance proved so encouraging that it soon became necessary to hire a janitor to maintain the premises. A true sign of success! (See Figure 5-4.)

The Pleistocene giants would remain prime attractions until late 1948, when a Geology Department office expansion resulted in their demise. An article in the November 30, 1948, *Daily Progress* reported that, "Two life size replicas of the gigantic fauna of prehistoric ages were disposed of yesterday by the University of Virginia Brooks Museum to make room for the present generation of expanding student body. And, in the opinion of the head of the Corcoran and Rogers School of Geology, there will be no wailing or gnashing of teeth.... It took a corps of Buildings and Grounds workmen most of yesterday to reduce the mammoth and dinosaur to rubble. The job proved what has been suspected, the dinosaur was a plaster of paris cast, and the mammoth was made of plaster and hollow inside. The hair of the mammoth, it was revealed today, was pampas grass from Argentina."[21]

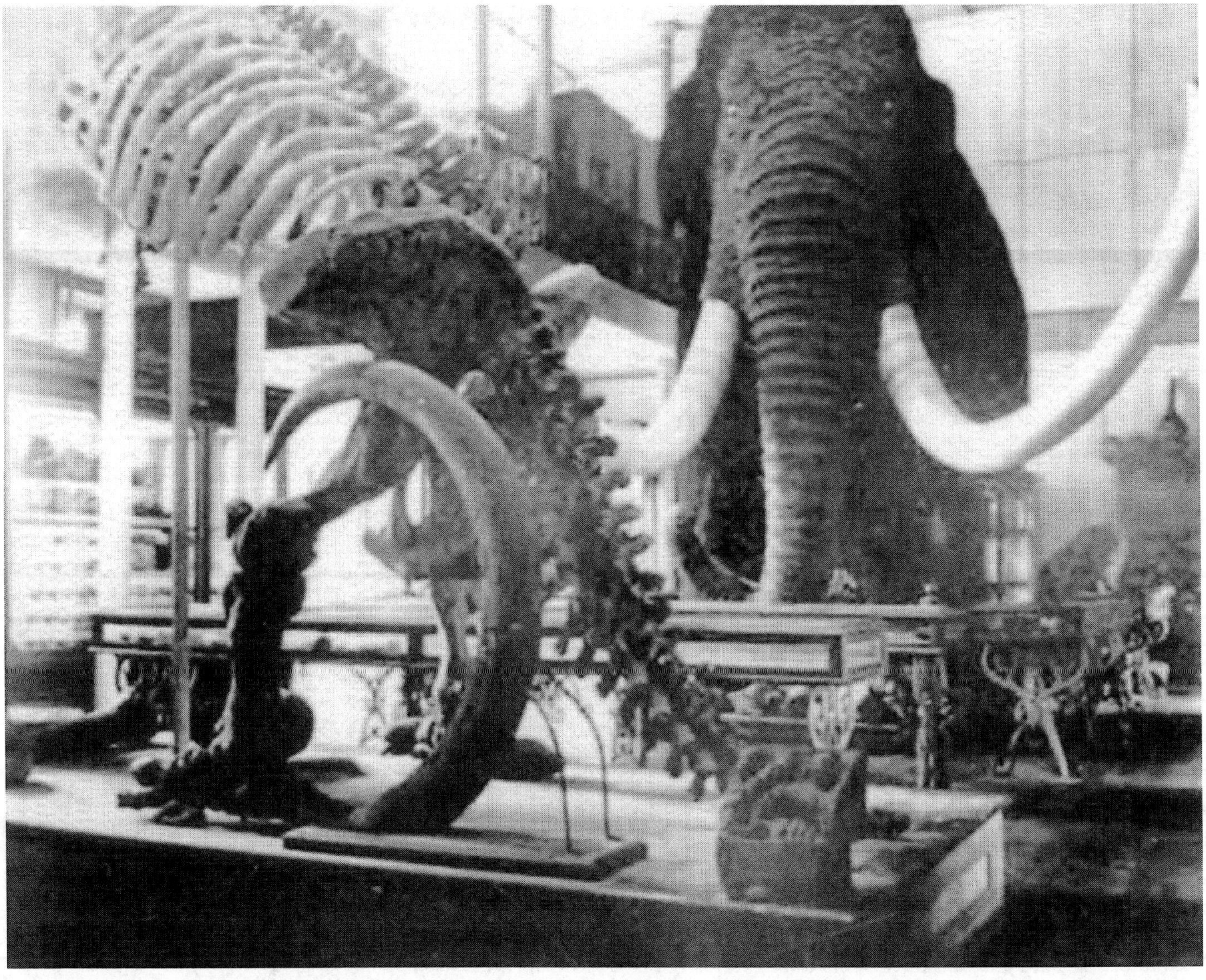

Figure 5-4: Rare photograph showing interior of the Brooks Geological Museum, circa 1880, with Ward's *Megatherium* fossil cast partly visible in front of Ward's Siberian Mammoth restoration.

Archaeologist Jeff Hantman, an ardent researcher of the history of the Brooks Museum, informed the authors that, "The mammoth was here until the late '40s. Many UVA alumni and visitors recall seeing it.... It is not uncommon for me to be sitting on the front steps of what is now Brooks Hall (now home to the Anthropology Dept.) and have an older visitor cautiously ask if 'this is the building that used to have the elephant in it?' That's how many remember it."[22] The building still stands, exterior architecture girded by a frieze of "paleolithic monsters," sans the Ice Age menagerie formerly occupying its interior.

Was this the last of the Siberian Mammoths supplied by Ward, to "inhabit" America?

No. Other casts were made as well. Roswell Ward's biography of Henry Ward attributes to the Lewis Brooks Museum a photo showing a dapper-looking Henry gazing fondly at a displayed Mammoth surrounded by a series of large mural panels showing restored scenes from geological history.[23] (Hantman, however, believes the caption is incorrect; this photo wasn't taken inside the Brooks Museum but at some other institution.)

Following construction of a museum building around 1890, another Siberian Mammoth became a central display at the California Academy of Sciences. This particular specimen, a Ward's *Megatherium* and a *Glyptodon* cast, seem to have been destroyed in the San Francisco earthquake of April 18, 1906 (about two months prior to Ward's death). According to Jean DeMouthe of the California Academy of Sciences, the tusks of their replica, salvaged from the rubble, were genuine mammoth ivory.[24] (See Figure 5-5.)

However, perhaps the most famous Siberian Mammoth replica of all was that exhibited during Chicago's Field-Columbian Exposition. The "Chicago World's Fair" opened to the public on May 1, 1893, closing on October 30, 1893. Six hundred and sixty six acres of Jackson Park lakefront property were converted into a "dream city" occupied by 150 buildings constructed of plaster of paris and jute fibers, a finished formulation resembling white marble. An average throng of 172,712 visitors attended the Exposition daily.[25]

Installation of the geological exhibits was supervised by Oliver C. Farrington, a Yale geologist who later would establish the Field Columbian Museum's Department of Geology. Ward's mighty Mammoth stood prominently in the Anthropological Building adjacent to a number of other striking paleontological displays (including Ward's 22-foot-long *Plesiosaurus* cast skeleton, and cast skeletons of the *Megatherium*, an Irish Elk, a wingless moa "...and other evolutionary forms of bird, beast and fishes....").[26] Counted among the bestiary were Benjamin Waterhouse Hawkins' *Ichthyosaurus* sculptural restoration, a plaster *Hadrosaurus* skeleton, Ward's *Glyptodon* and *Mastodon* skeletons, and a "*Dinoceras*" (uintathere) reconstructed at Yale.

Through the latter 1890s the Columbian Museum of Chicago, as it was named following the World's Fair, made its 1893 debut in the Chicago World's Fair's former Palace of Fine Arts building, situated in Jackson Park. The Siberian Mammoth graced the main ground floor exhibit hall of this Jackson Park building. In adjacent hallways the previously mentioned skeletons were on view, as well as a life restoration of the gigantic prehistoric tortoise *Colosochelys*. There is little record indicating how long the Mammoth remained on display at the Field Museum. But by 1921, when the Field Museum of Natural History (as it was renamed in 1906) moved to its current Grant Park location, the Mammoth seems to have mysteriously vanished among the 321 railroad cars of shipped inventory.

Perhaps Field Museum officials perceived that as their paleontological expeditions began bearing fruit (in the form of genuine fossil vertebrates instead of casts), there was no longer need to exhibit, or sufficient display room for some of the older, antiquated displays. A caption describing a photo of the Siberian Mammoth reproduced in a 1982 issue of the *Field Museum of Natural History Bulletin* bears the disappointing news, "...the Indian elephants and the mammoth were disposed of before the museum moved to its present building in Grant Park in 1921."[27] Please say it ain't so!

The Field Museum still retains Ward's several displays included in the original collection purchased by Marshall Field for $95,000.

Today, in the Field Museum's *Life Over Time* display, one may spy an impressive *Megatherium* skeleton, nearly identical in size, posture and appearance to Ward's casts exhibited in London, the former Brooks Hall Museum and the former California Academy of Sciences' museum building. Although this skeleton is said to have been composited from real bones, the Field's *Megatherium* resembles the specimen formerly greeting Chicago World's Fair patrons over a century ago.[28]

Intriguingly, Yale artist James H. Emerton modelled a life-sized, 40-foot-long papier-mâché Giant Squid for Ward in 1883. Marshall Field purchased a replica for the Columbian Museum. Although once suspended above the floor facing the Siberian Mammoth in its former Palace of Fine Arts Building home, the Giant Squid can still be seen in the Field Museum, anchored from the ceiling only a few yards from the newly acquired *Tyrannosaurus* skeleton "Sue."

If Squid (and other displays) made their way to the Field Museum, then is there any hope that other Field

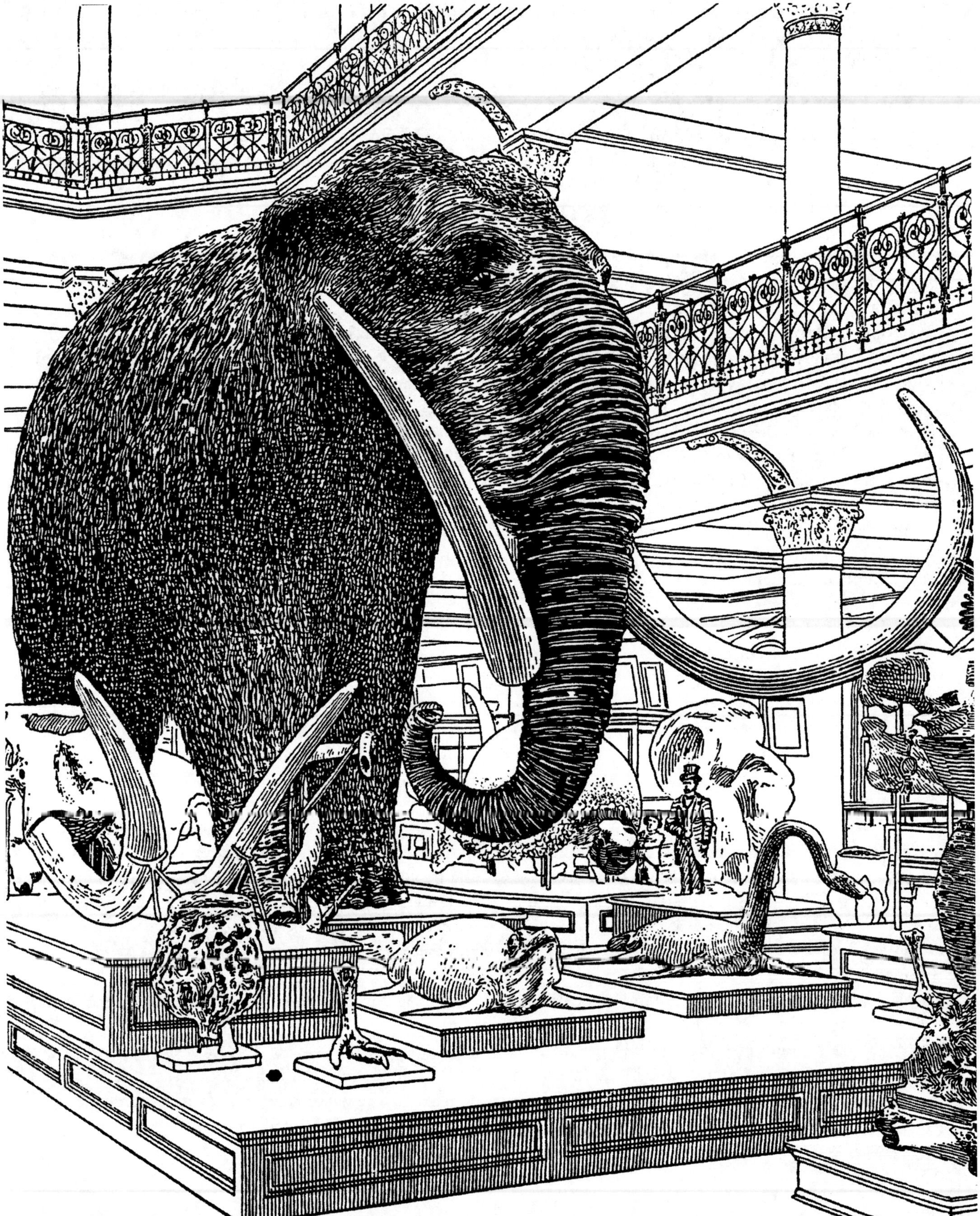

Columbian "prehistorians" are held securely in storage in the bowels of Chicago's Field Museum? Could that 1982 *Bulletin* caption be incorrect? Might the Siberian Mammoth reappear after so many decades?

Even if the Field Columbian Siberian Mammoth met its untimely demise in Chicago, hope lingers that the few other institutions that once displayed them retain their Mammoth replicas to the present day. Because Ward's company records were destroyed in 1930, it would be a "mammoth" undertaking to track down their whereabouts.[29]

Will a Ward's Siberian Mammoth surface somewhere before the next frozen mammoth specimen is found in Alaska or Siberia? Will plaster and dyed palm fiber triumph over DNA strands? The search continues. The race is on!

Opposite: **Figure 5-5: Ward's Siberian mammoth displayed at the California Academy of Sciences, circa 1892. (From the *California Illustrated Magazine*, 1892/93 vol.)**

Chapter 6

Benjamin Waterhouse Hawkins— A Modern Pygmalion

A decade before the Civil War raged in America, and a handful of years before publication of Charles Darwin's *Origin of Species* (1859), a very talented artist was first employed in the creation of lifelike restorations of prehistoric animals—the most realistic the world had ever witnessed until that time. Benjamin Waterhouse Hawkins (1809–1889) is often overshadowed by the luminary scientists he worked with. In England he assisted Sir Richard Owen (1804–1892), a brilliant anatomist who was disliked by many colleagues, and Thomas Henry Huxley (1825–1895). Huxley gained fame as Darwin's "bulldog" defender of the theory of evolution through natural selection. In the United States Hawkins teamed with paleontologists Joseph Leidy (1823–1891), Edward Drinker Cope (1840–1897) and Arnold Guyot (1807–1884). Unfortunately, history has been unkind to Hawkins. This master paleoartist of his day "came before his time." Had he begun his prehistoric animal restorations, especially the dinosaurs, only 50 years later, at a time when such creatures had been more fully understood scientifically, Hawkins' restorations would be far more celebrated today.

Of his major prehistoric visions, only sculptures at Sydenham, England; fifteen of his original seventeen mural paintings at Princeton University's Guyot Hall; and two damaged *Paleotherium* casts, discovered in 1978 during renovation of the Crystal Palace grounds near London, have survived to this time. Although today one might chuckle at Hawkins' rudimentary attempts to portray vertebrate "antediluvians," Hawkins' achievements represent an early influential effort to present the science of vertebrate paleontology for public edification. Considering the relative scarcity of scientific information available to him, and the adversity he struggled against, Hawkins' efforts were heroic.[1]

Hawkins obtained his formal education from St. Aloysuis College in England. He learned how to sculpt under artist William Behnes. Between 1827 and 1852, Hawkins became a student of natural history, focusing his interests on the subjects of zoology and geology. Hawkins also made extensive zoological observations of living animals between 1842 and 1847 for the Duke of Derby in Knowlsey Park. Such experiences undoubtedly honed his interests in vertebrate anatomy. During his career, Hawkins published five books on this subject, one of which was coauthored by Huxley.[2]

It was in 1852 that Hawkins began his first major undertaking as a fossil vertebrate restorer under the auspices of the eminent Sir Richard Owen, an established authority on the subject of "saurian" paleontology. With Owen, at Crystal Palace Hawkins would forge his name forever in the annals of vertebrate and, particularly, dinosaur paleontology.

In 1852, following the reestablishment of Joseph Paxton's Crystal Palace in its new location in Sydenham for the cultural and scientific edification of the public, Prince Albert suggested that the new park grounds be adorned with life-sized models of antediluvian denizens for the purpose of *revivifying the ancient world*. Hawkins, who had served as assistant superintendent during London's Great Exhibition of 1851, was appointed to the task. Although Hawkins' original aim was to restore large extinct mammals, Owen's dinosaur monographs inspired him to concentrate on Mesozoic saurians instead.

To the Victorian public's delight, Owen and Hawkins strove for more lifelike appearances in their recreations than could be portrayed by simply reassembling the fossil bones into a probable "best fit" configuration. Assembling bones in itself was a tricky task, since fossil skeletons, especially dinosaur specimens available to Owen, were often poorly preserved. But by "clothing" them with "flesh," scales and claws, thereby breathing the essence of life into their creations, Owen and Hawkins stirred the imaginations of an adoring public like never before.

Their three dinosaurs, *Megalosaurus*, *Hylaeosaurus* and *Iguanodon* (replete with nasal horn—actually a misplaced thumb claw), were restored as immense rhinocerine lizards. Though incorrect, these peculiar designs proved influential for decades. (See Figures 6-1 and 6-2.) It is said that Owen's decidedly inaccurate dinosaurs served as a visual expression countering the prevalent theory of Lamarckian evolution of organisms throughout geologic time.[3] Such ideology countered Owen's (and possibly Hawkins') mystical ideas on evolution. So Owen enlisted dinosaurs in his crusade to defend paleontology from the progressive evolutionists, who claimed species naturally transmuted into more advanced forms. If dinosaurs were, as Hawkins constructed them, grandiose Mesozoic "overlords," this would imply that there had been no progressive evolution.

Owen's participation secured a sense of accuracy in the resulting restorations. Never mind that no living animals seemed comparable to dinosaurs, or that their remains were highly incomplete. There were few restorations available at the time, and of these, none could be judged scientifically accurate. Assuredly, it must have seemed, no individual other than the "inventor" of the "Dinosauria," Owen himself, could have been better suited to serve as Hawkins' mentor.[4]

First, Hawkins visited fossil collections to better comprehend Britain's prehistoric menagerie. His proposed sketches were approved by Owen. Next, Hawkins proceeded with small-scale clay miniatures, also prepared for Owen's approval. Then Hawkins sculpted life-sized clay models, and molds were taken from these. An iron armature was built, of sufficient strength to support several tons of concrete. Final casts were produced after concrete was poured into the molds to form the dinosaurs' outer "skin." Because the models have been painted numerous times

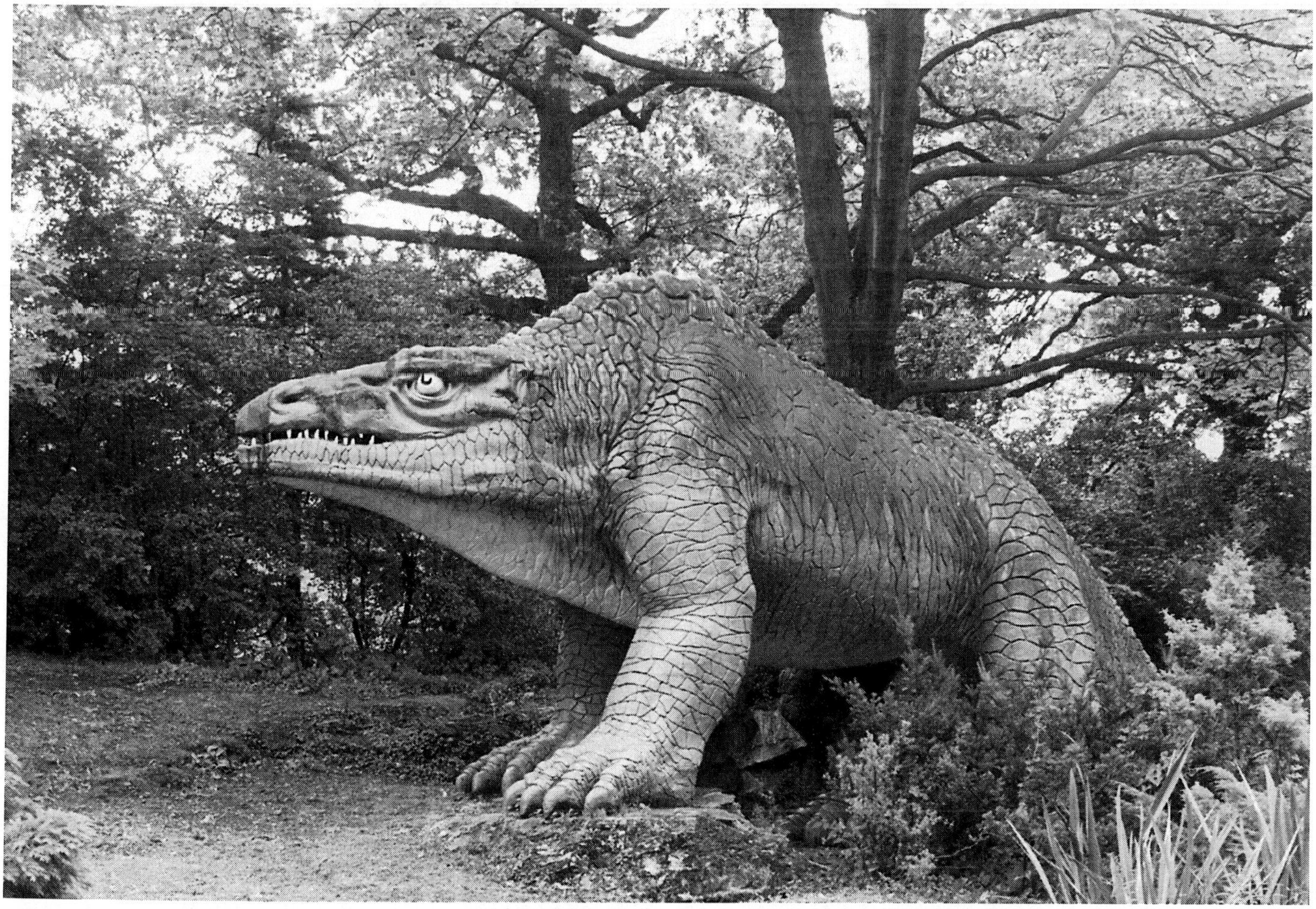

Figure 6-1: Hawkins' Crystal Palace *Megalosaurus* sculpture. (Photograph by Karl Debus, 1985.)

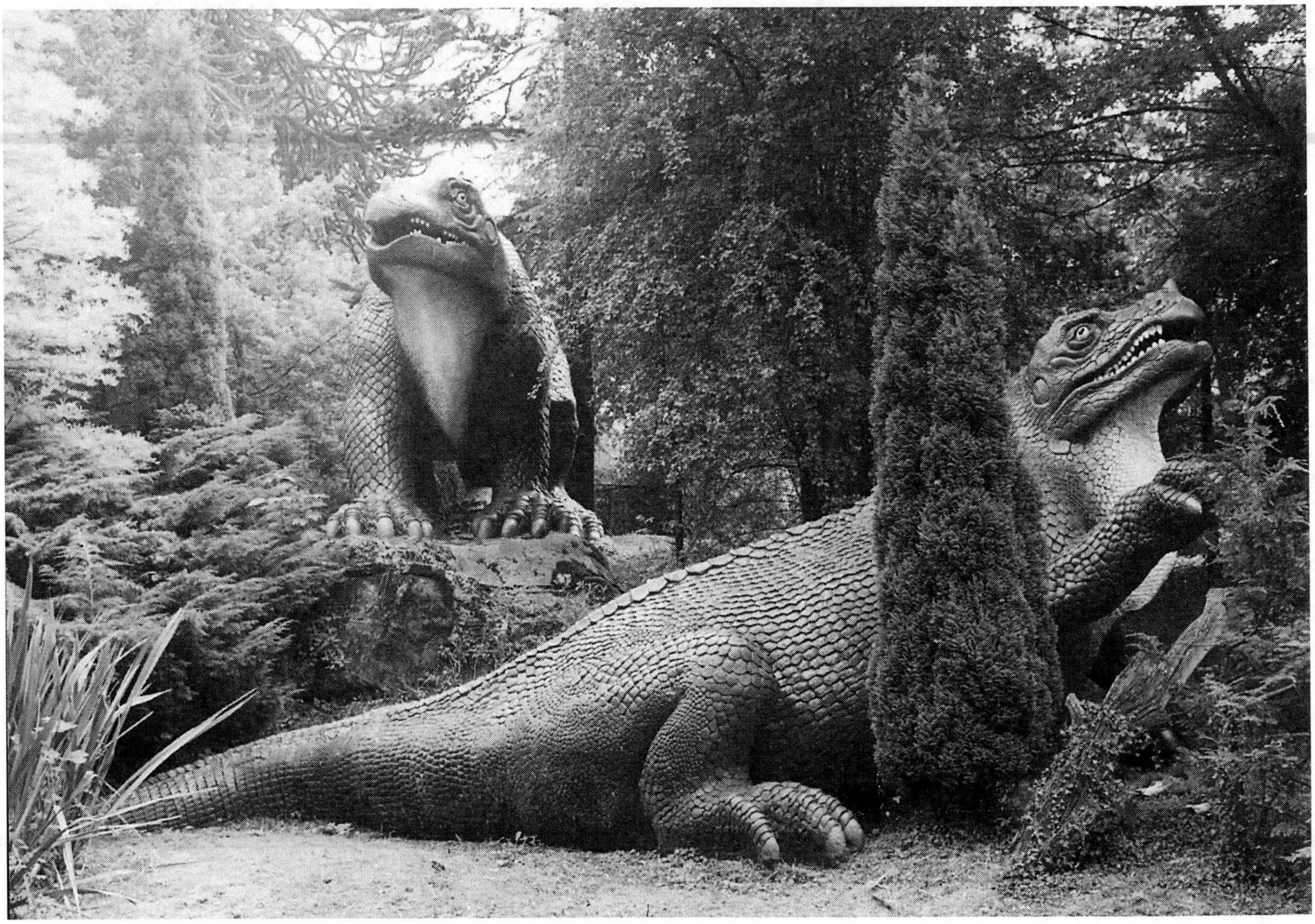

Figure 6-2: Hawkins' Crystal Palace *Iguanodon* sculpture. (Photograph by Karl Debus, 1985.)

since 1854, it is challenging today to determine what colors they had on the Exhibition's opening day. Hawkins referred to his labors as summoning "...from the abyss of time and from the depths of the earth, those vast forms and gigantic beasts which the Almighty Creator designed with fitness to inhabit and precede us in possession of this part of the earth called Great Britain."[5]

On December 31, 1853, the project's near completion was celebrated by staging a festive dinner held for 21 scholars, who dined within the hollow *Iguanodon* mold. Hawkins was heralded as the "modern Pygmalion." The merry occasion was even noted in the *Quarterly Review*: "Saurians and Pterodactyles all! Dream ye ever, in your ancient festivities, of a race to come, dwelling above your tombs ... dining on your ghosts, called from the deep by their sorcerers."[6] (Interestingly, Hawkins' father had, in 1802, attended a banquet hosted by Charles Wilson Peale inside the skeleton of a mounted *Mastodon*.[7])

The park, with its island of restored *prehistoria*, was officially opened on June 10, 1854, to a throng of 40,000 visitors. Thirty-five models were strategically placed on three specially constructed islands, surrounded by an artificial six acre lake. (Only 29 of these remain on display.) Vegetation was added to create an eerie, primeval atmosphere. The rock formations on which the animals stood represented the deposits in which their fossils had been found. It was intended to have the lake waters periodically rise and fall at regular intervals, suggesting seashore tidal action, although this plan didn't reach fruition. Apparently, even then, bringing the age of dinosaurs back to life proved easier than finding a competent and inventive plumber!

The geological islands were intended as a "life through time" display, the most magnificent of its kind ever created, and the first "real" *Jurassic Park*. Although reptilian fauna of the Great Secondary Period (i.e. Mesozoic era) were emphasized, Paleozoic animals and Tertiary and Pleistocene mammals also inhabited the grounds. There were four genera of Mesozoic aquatic reptiles—*Teleosaurus*, *Mosasaurus*, *Plesiosaurus* and *Ichthyosaurus*. Of these, the *Ichthyosaurus* would seem most peculiar today. (See Chapter 34 for more on the history of ichthyosaur restorations and reconstructions.) In the case of the aquatic *Mosasaurus*, only the skull was available for study. So Hawkins cast only its head, which was placed in the lake, suspended at the water line. Bat-winged, dragon-like pterodactyls presided over the primeval setting from a cliff face. Many of the restorations at Sydenham may seem incongruous today, but at the time they were sensational!

Buoyed by his sculptural successes, Hawkins turned to the canvas, depicting a scene of marine Liassic animals for an 1858 edition of William Buckland's *Geology and Mineralogy, Bridgewater Treatise* volume. Also, in 1860 Hawkins prepared a wall chart, titled *Struggles of Life Among the British Animals in Primaeval Times*, showing glacial age mammals figured in the prehuman, "nature red in tooth and claw" vision fostered by Darwin's *Origin of Species*. Then, in 1862, Hawkins was commissioned to produce six scenes of prehistoric vertebrates designed for classroom educational purposes. The wall chart and scenes were sold by British and fossil proprietor James Tennant (1808–1881). With his work apparently done in Great Britain, Hawkins was off to America.[8]

In 1868 Hawkins' presence in Philadelphia became known to the Commissioners of the Central Park in New York City. Miniature replicas of several of Hawkins' Crystal Palace dinosaurs had been advertised for sale in an 1866 Ward's Catalog, which promoted Hawkins' good name overseas. His reputation as a great sculptor of free-standing extinct vertebrates soon made him the obvious target for another grand undertaking. Andrew H. Green, Comptroller of the Park, requested that Hawkins "...undertake the resuscitation of a group of animals of the former periods of the American continent" for public display.[9] New York's "Palaeozoic Museum" in Central Park was to be enclosed in an iron-framed glass building, having an arched roof similar to its prototype at Sydenham. (See Figure 6-3.)

For several months, Hawkins devoted his time to the study of North American fossil vertebrates, viewing fossil specimens in several American museums. He was chiefly aided by vertebrate paleontologists Edward D. Cope and Joseph Leidy at the Philadelphia Academy of Natural Sciences. Leidy and Cope particularly inspired Hawkins' interest in two American dinosaurs discovered in New Jersey's Cretaceous deposits, *Hadrosaurus* and *Laelaps* (now *Dryptosaurus*). Even with two of America's premier paleontologists serving as consultants, as he had before with Owen, Hawkins struggled for accuracy in his work.

Leidy had excavated the bones of *Hadrosaurus* over a decade earlier. Many large bones had been unearthed, with lozenge shaped teeth indicating the dinosaur's close relationship both to modern iguanas and a herbivorous dinosaur, *Iguanodon*. However, no skull had been found and there was no collarbone or pelvis. The bones had been left in an unassembled state through the Civil War. With no skull available, but aware of the anatomical similarities between the living iguana and the known portions of *Hadrosaurus'* skeletal remains, Hawkins cast an oversized iguana skull with jaws sufficiently large to accommodate the discovered teeth. Human shaped collarbones and pelvic bones were also constructed, scaled to fit the analogous missing skeletal structures in the *Hadrosaurus* reconstruction. Despite Cope's antagonism for Hawkins, the team persevered.[10]

By 1868, a skeletal reconstruction of Leidy's *Hadrosaurus* was placed on exhibit at the Academy. From surviving sketches, it is apparent that a plaster cast of the *Hadrosaurus* skeleton and a life-sized clay sculptural restoration had also been completed in New York. *Laelaps*, however, never progressed far beyond the armature stage of reconstruction.[11]

Unfortunately, *Hadrosaurus* turned out to be the most enduring of his Palaeozoic Museum labors. For, in the spring of 1871, all of Hawkins' in-progress work was demolished by henchmen of a prominent local politician, "Boss Tweed."

At Sydenham, Hawkins unified natural history and geology through paleoart restoration. Hawkins had similar intentions for his New York project. Although Hawkins' views on evolution were most likely allied with Owen's, in America Hawkins had associated with the prominent Cope, a noted "evolutionist." While Hawkins, who believed in a course of evolution guided by the Creator, opposed Darwin's views, the "struggle for survival" theme evident in his "Palaeozoic Museum" figures could have become conflated with Darwin's blasphemous theory of evolution through natural selection of the fittest. These circumstances, and the fact that Hawkins was a foreigner, made the as yet unprofitable "Palaeozoic Museum" an easy political target for those who feared the museum project was draining municipal funding. Hawkins' venture had already cost $300,000, far more than the Crystal Palace expenditure, and to make matters worse, at a public meeting held at the New York Lyceum of Natural History in March 1871, he had antagonized politicians who carried the financial burden. Hawkins had expressed concern over the association between Tweed and the planned museum. Was Tweed's disposal of Hawkins' magnificent creations merely an act of retaliation? Quite possibly.[12]

Shortly thereafter, Tweed's fraudulent activities were exposed. Tweed descended into exile but later, after extradition, died in prison. While justice was served, Hawkins had been irrevocably victimized.

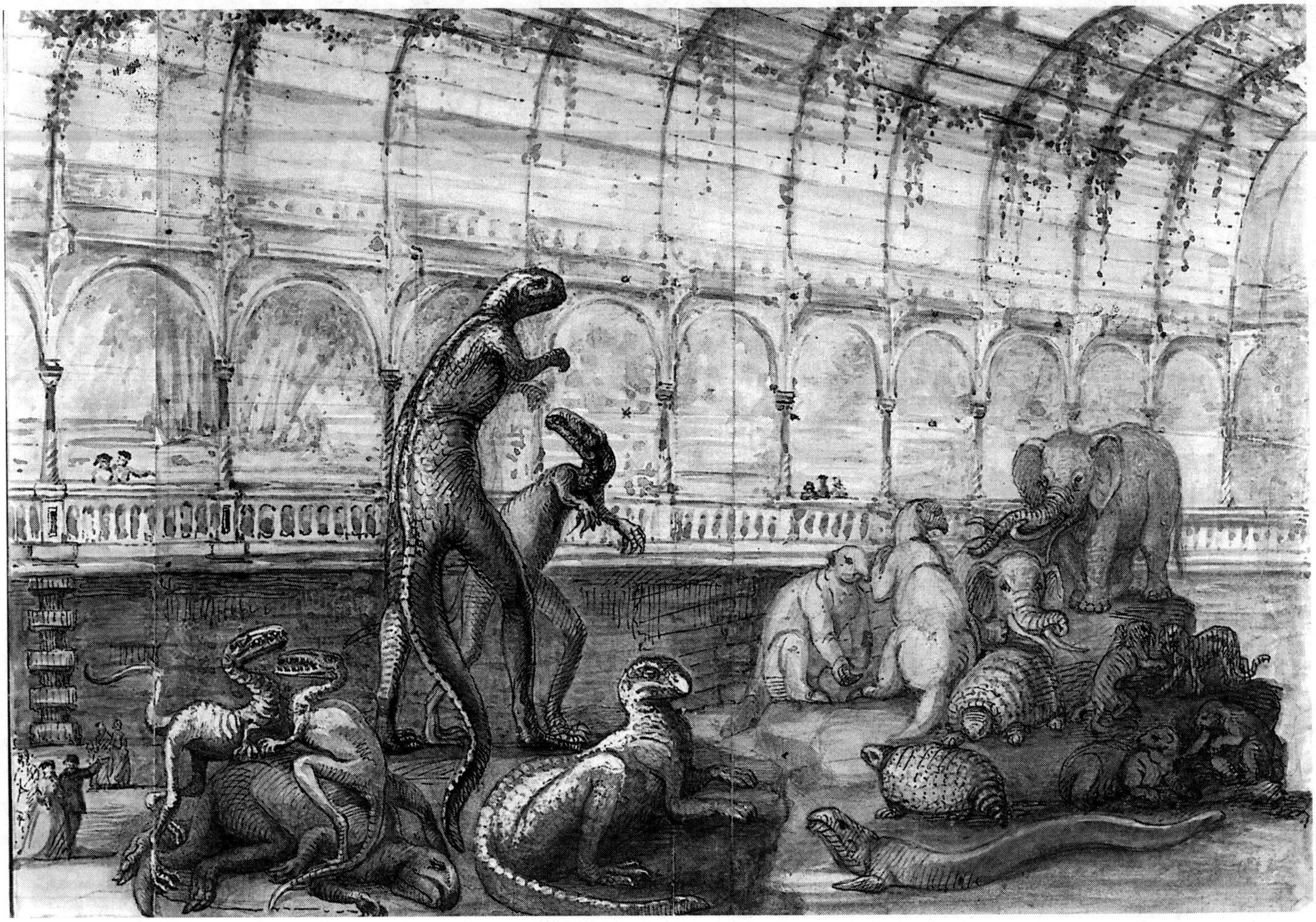

Figure 6-3: Hawkins' design, circa 1868, showing prehistoric diorama intended for Central Park's "Palaeozoic Museum," showing battling *Hadrosaurus* and *Laelaps* towering over other antediluvians, including *Glyptodon, Megatherium, Megalonyx*(?), *Mastodon, Smilodon/Machairodus, Mosasaurus/Basilosaurus*(?). (Reproduced with permission from the Ewell Sale Stewart Library—The Academy of Natural Sciences, Philadelphia.)

Several plaster casts of Hawkins' *Hadrosaurus*, referred to as "plasterotheres,"[13] survived the Palaeozoic Museum debacle. These were displayed at Princeton University and the U.S. National Museum (Smithsonian). The National Museum specimen was subsequently transferred to Chicago for the Field Columbian Exposition of 1893. This later was displayed in Chicago's Field-Columbian Museum when it opened in 1894, but disassembled by 1905. (One wonders whether these bones still remain in museum storage.[14]) The Princeton skeleton was carelessly exposed to a series of "accidents" perpetrated by the museum curator, leading to its demise in 1909.[15] Another cast made for the 1876 Philadelphia Centenary was shipped to Edinburgh's Royal Museum of Scotland in 1879, where it became the first dinosaur skeleton mounted in Europe. During World War II it was dismantled, the ironwork being needed for the war scrap drive. The plaster bones were ultimately discarded in 1928.

Although Hawkins also proposed a life-sized sculptural restoration for the Centenary event, this was initially declined. Othniel Marsh had refused to allow Hawkins to study the bones of his vertebrate fossil collection at Yale University. Marsh evidently feared that, due to the incomplete nature of the remains available for study, many mistakes would result if Hawkins were to attempt restorations. Hawkins' proposed model would create false, lingering impressions of how these fossil animals supposedly appeared in life. Hawkins made a plaster *Hadrosaurus* sculpture anyway, but this was eventually allowed to slowly disintegrate outside on the Smithsonian grounds.[16] His proposal to create a scaled down version of the "Palaeozoic Museum" exhibit for the Smithsonian was never approved.[17] (See Figure 6-4.)

Down on his luck, and embittered by his experiences in America, Hawkins became an academic recluse at Princeton. In 1874 Princeton University rewarded him for his life

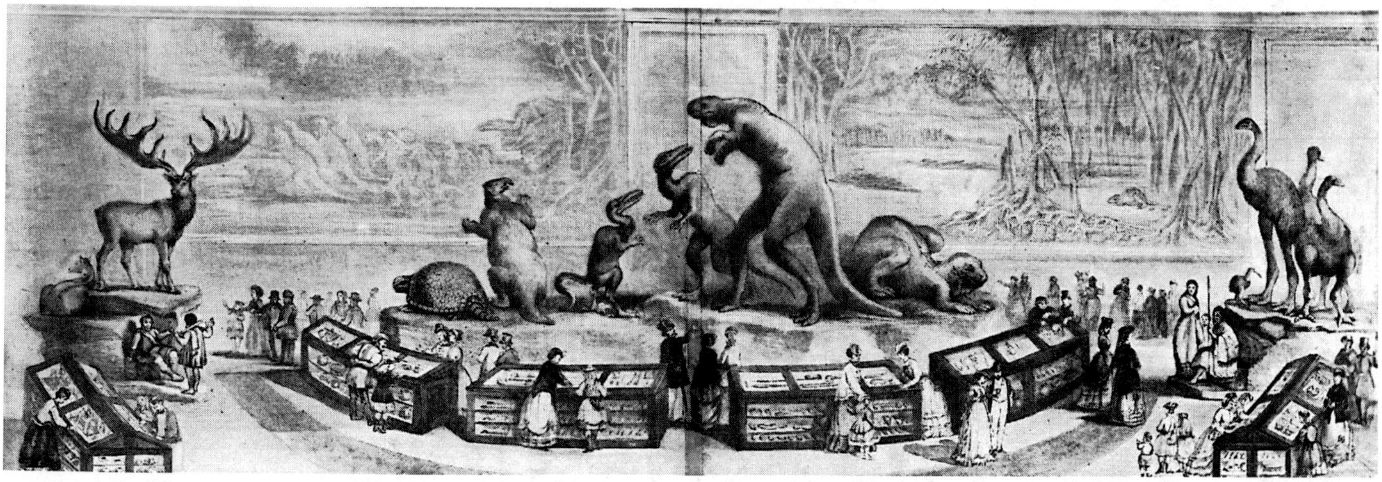

Figure 6-4: Hawkins' unapproved design for the Smithsonian's prehistoric display, circa 1870, featuring (left to right) Irish Elk (male and female), *Glyptodon, Megatherium, Laelaps* (pair), *Hadrosaurus* (pair, one reclining) and *Moa* trio. (Reproduced with permission from the Ewell Sale Stewart Library—The Academy of Natural Sciences, Philadelphia.)

achievements with an Honorary Doctor of Science degree. Here, working under Arnold Guyot, he completed a series of 17 paintings depicting the Earth's geological/paleontological history, showing 17 scenes from the "Age of Invertebrates" through Recent time.

By the latter nineteenth century, a host of new discoveries revealed that the Sydenham dinosaurs were hopelessly misconceived and ill proportioned, prompting Othniel Marsh to unfairly lambaste their appearance during an 1895 British Association meeting.[18] Today, however, the Sydenham models' historical significance has become fully appreciated. A three phase plan is underway to restore the deteriorating statues, many of which are cracking because the underlying iron has rusted and bulged, cracking the exterior. The curation effort is sponsored by the Crystal Palace Foundation and the London Borough of Bromley. (The *Hylaeosaurus*' head, which fell from the body, has already been recast in fiberglass, and a missing pterodactyle bat wing has been recovered from the lake.) In the final reconstructive phase, artisans will attempt to create, as fiberglass replicas, the remainder of the total of 22 Tertiary mammals which Hawkins never received funding to produce for the original display.[19]

Hawkins should be recognized as the first of a long illustrious line of artist/popularizers in the field of vertebrate paleontology. His methods were state-of-the-art, and he relied on information that was scientifically current. Time stands still for Hawkins' menagerie. Even though the Crystal Palace burned down in 1936, his signature concrete "monsters" still reside in their primeval setting. At Princeton, 15 separate "windows" to antediluvian lands, more imagined than real and presaging the Field Museum murals of Charles R. Knight by half a century, frame the extent of Hawkins' extraordinary "prehistoric" visions.

Someday, rightfully, Hawkins' Palaeozoic Museum sculptures will be recreated in fiberglass to adorn New York's Central Park, and to commemorate the setting and circumstances of the greatest museum that never was![20]

CHAPTER 7

A Mythic Place in Time—Spirit in the Knight

Quite unwittingly, Charles R. Knight (1874–1953) set a wave in motion over a century ago at the American Museum of Natural History. Ask any dinosaur aficionado from the "baby boomer" epoch about the origin of his (or her) interest in "dinosaurology," and most will mention their love for Charles R. Knight's paintings of prehistory.

Despite the fact that he was blind in one eye, Charles Knight possessed an extraordinary vision—an ability to project himself into prehistory.[1] His sensitivity for comprehending animal expressions was unbounded temporally. Knight was the consummate wildlife artist, employing keen understanding of the expression and character of modern mammals, reptiles, birds and amphibians during an illustrious career spanning nearly six decades.

Knight firmly believed that the key to success lay in understanding his subjects, literally inside and out. While developing skills as a commercial artist, Knight would often sketch zoo animals. He also sketched the carcasses of various animals in the taxidermy department of the American Museum of Natural History. Through such studies Knight reached the very souls of the creatures he animated on canvas and in sculpture.

Knight was among the first to specialize in the field of restoring extinct animals, combining a rare medley of scientific and artistic energies and insights. In 1894 he restored the pig-like *Elotherium* for Dr. Jacob Wortman of the American Museum who was deeply impressed with the result. Commercial assignments and museum commissions swiftly followed, and his reputation escalated.

Knight began providing illustrations of prehistoric animals for magazine publications, notably *McClure's Magazine* and *The Century Magazine*. In one famous 1897 issue of *The Century Magazine*, eight of Knight's most famous restorations of prehistoric reptiles appeared.[2] Used to illustrate articles by journalist William Ballou and paleontologist Henry Fairfield Osborn, these restorations were a product of Knight's abbreviated association with the legendary vertebrate paleontologist Edward D. Cope, who died some weeks after sharing his knowledge of dinosaurs and Permian finback reptiles with Knight.[3]

Knight learned the value of sculpting prehistoric animals prior to painting them under direction of the artistically inclined Cope. Knight sculpted some of his most famous dinosaur sculptures under Cope's tutorial—the doubtful horned genus *Agathaumas, Hadrosaurus* and a pair of leaping *Laelaps* (now *Dryptosaurus*). I am aware of 20 different miniature prehistoric animal sculptures Knight created for artistic study, scientific modeling or purposes of distribution.

Following an extended Eurpoean trip to Paris, Antwerp, Amsterdam and London, Knight returned to New York City where his long association with Osborn and W. D. Matthew blossomed. By this time, he was recognized by professional paleontologists and other animal and wildlife painters as the leading authority in the field of fossil vertebrate restoration. In his quest to restore extinct creatures, Knight found it essential to first study the fossilized bones and consult with scientific authorities.

Knight's artistry had a profound influence on several generations of fellow artists, scientists, movie producers, scholars and young paleo-enthusiasts. Recently, however, Knight's name surfaced in a debate over the physiology of dinosaurs. One paleontologist held the assertiveness of Osborn, and the influence of his colleagues (paleontologists

Barnum Brown and William K. Gregory, and artist Charles Knight), responsible for a perceived state of "paleontologic lethargy" during the 1920s and 1930s.[4] More recently, a prominent paleo-artist has regarded Knight's "every foot on the ground" rule (a "symptom" of cold-blooded metabolism) as a product of incorrect scientific reasoning.[5]

Knight's influence certainly cannot be overstated. However, it would be unfair to implicate Knight for causing "leaden patterns" in the field of dinosaur paleontology. In fact, rather ironically, dramatic restorations (some of doubtful accuracy) created by modern Dinosaur Renaissance paleoartists must be highly influential in popularizing paleontologists' visions of "dancing dinosaurs" and the theory of "warm blooded dinosaurs." Restorations, both past and present, have championed theoretical ideas.

In some of Knight's early restorations, dinosaurs were painted in lively poses, or in poses commanding a warm-blooded metabolism. Knight also astutely noticed the dinosaur-bird relationship, using *Archaeopteryx* as an evolutionary link in 1946, over two decades before the issue became delicate.[6] It can only be stated that Knight's artistry mirrored the state of contemporary science, as is largely the case today with newly created dinosaur art. (See Figure 7-1.)

His most famous mural, a misty Upper Cretaceous scene showing the mighty *Triceratops* squaring off against North America's all time (literally) king of carnivores, *T. rex*, has influenced many artists through the years.[7] Conceptions about the life appearances of the dinosaurs he restored have altered in the 70 years since Knight created his

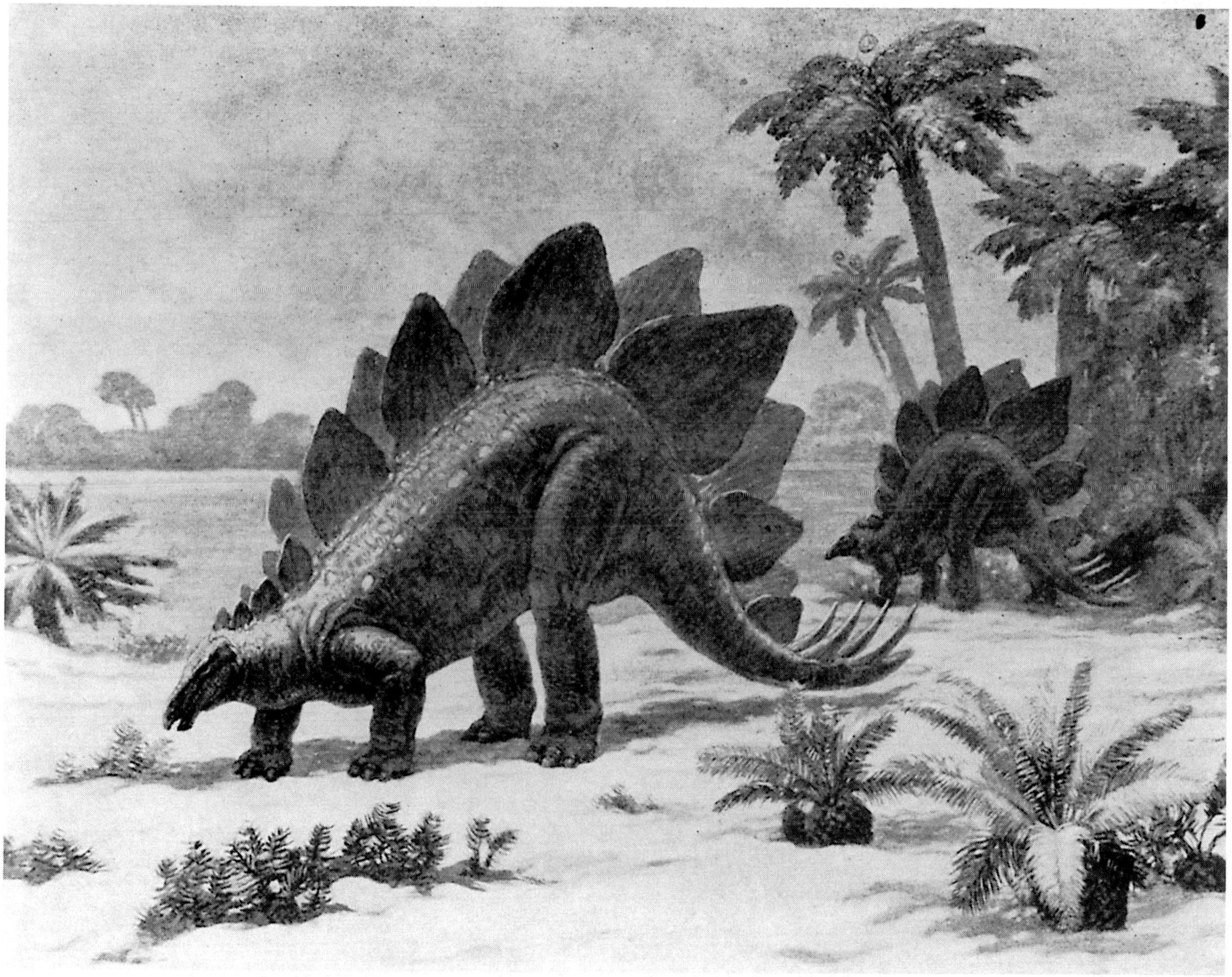

Figure 7-1: Charles Knight's influential *Stegosaurus* restoration remains on display at Chicago's Field Museum of Natural History, reflecting contemporary views concerning osteoderm arrangement as a staggered double row of plates. (Negative #320611, courtesy Department of Library Services, American Museum of Natural History. Reproduced with permission.)

greatest masterpiece for the Field Museum of Natural History. However, no dinosaur painting has proved so completely captivating as this wonderful composition, portraying an episode in the life of two immortal creatures! (See Figure 7-2.)

Dinosaurs and their prehistoric brethren have attained cult status, and yet little is generally known about the personable side of the man who fueled this inspiration. Nor are dinosaur fans generally aware of Knight's great love for the American Museum of Natural History, the institution that inspired most of his magnificent artwork. Knight restored numerous fossil animals for our visual and imaginative senses. So, in kind, let us pay tribute to Knight by restoring some visage of his character.

We marvel at his extraordinary efforts, but our imaginative senses strain for the beauty and wonder Knight beheld in capturing the essence of Life so convincingly. Knight was an absolute perfectionist, seldom satisfied with the results so painstakingly achieved. It is difficult for us to imagine an artist so skilled ever suffering from the pangs of artist's remorse! Yet he would require the opinions of others—peers, colleagues and family—before convincing himself that the work was truly done. But if he judged that Life had not yet been breathed into his subject, then there was more to do.

When he especially enjoyed a particular subject or design, he would recreate the scene in detail several times, with slight variations, before selecting an optimal depiction. Accuracy was a chief concern to him in all stages of design. Energies would be focused on the heart of the matter until a design seemed as convincingly correct as was humanly possible.[8] For example, his set of small preliminary design paintings for the Field Museum's world famous 28 murals, portraying the Earth's geological evolution, are so exceptional that they merit pride of place at the entrance to Princeton University's geology museum in Guyot Hall! In recent years, many of his practice sketches, notebooks and painted compositions (representing early stages in the design of many famous murals) have been sold to private collectors.

Knight, or "Toppy" (as known to intimates), really was a down-to-earth, stout-hearted sort of person. Knight was characteristically self-effacing, extremely outgoing, and lacked any sense of class consciousness. As the title of his unfinished autobiography, *Autobiography of an Artist* (circa 1940?), conveys, Knight's life truly embodied the spirit of what being an artist is all about.

Knight expressed a "fraternal emotion" while in the presence of cave dwelling art made by ancient Europeans, yet had an apprehensive perception of Mankind's tendencies. Writing in *Prehistoric Man: The Great Adventurer* (1949) following the chilling mortality of a second world war, he stated, "Too much brain may be even more dangerous than too little, and an intelligence run wild,

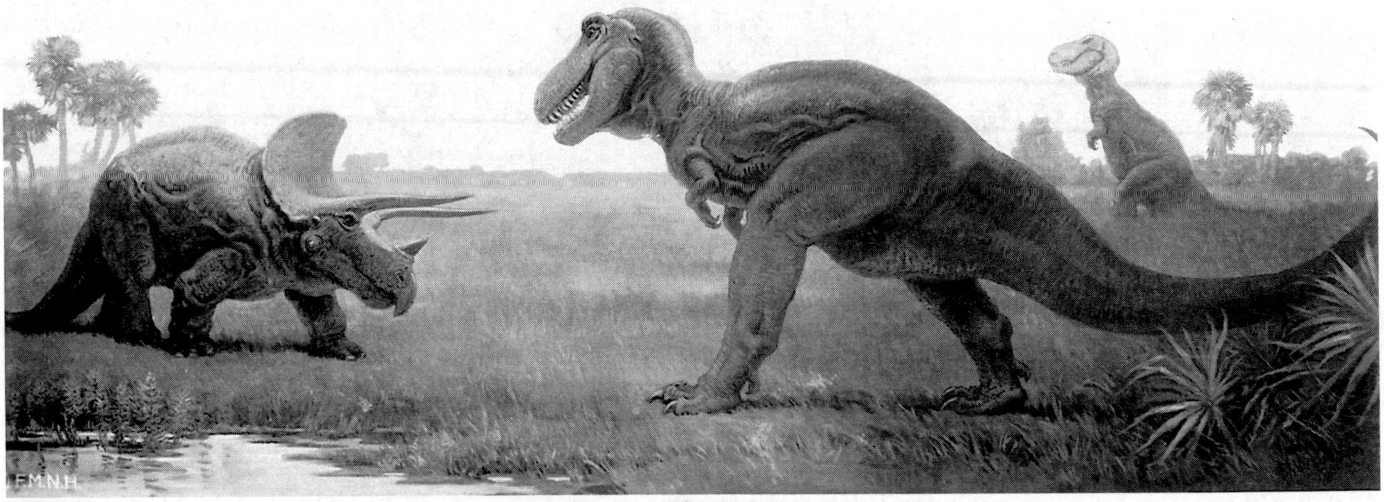

Figure 7-2: Perhaps *the* most admired dinosaur restoration of all time—Charles Knight's *Tyrannosaurus vs. Triceratops* remains on display in the Field Museum's *Life Through Time* display. (Negative #59442, courtesy of the Field Museum of Natural History.)

without the possibility of curbing its exuberance, has recently brought us to a terrible and ominous impasse in our struggle for supremacy."[9]

Knight was also a talented author. He wrote four treasured books, including *Before the Dawn of History* (1935), *Life Through the Ages* (1946), *Animal Drawing: Anatomy and Action for Artists* (1947) and *Prehistoric Man: The Great Adventurer* (1949). His lengthy illustrated article "Parade of Life Through the Ages" (*National Geographic Magazine*, February 1942, pp. 141–184) is written in the classical "life through time" style.

Revered by many admirers, paleontologists, artists and writers, Knight left about 800 drawings and 150 oil paintings in the wake of an illustriously productive career. In the mid–1980s a traveling exhibit, "*Dinosaurs, Mammoths, and Cavemen*," memorialized his talents.[10] This wonderful exhibit, sponsored by the Los Angeles County Museum and developed by Sylvia Czerkas, toured the United States, reaching Chicago in late summer of 1985. There, at Chicago's Academy of Natural Sciences, visitors could view many of Knight's original paintings, notebooks and bronze sculptures. On display was a sketch of the famous *Tyrannosaurus* vs. *Triceratops* scene, a preliminary exercise on brown paper done before he attempted the Field Museum mural. A *Styracosaurus* grinned from behind a picture frame years after its first publication in 1942. Prehistoric men lived again, hunting reindeer and prehistoric bison, or were themselves menaced by other mammalian predators.

The magnitude of his brilliance still shines in the American Museum, where several of his most beautiful murals, long archived and flaking into deterioration, were recently restored at great expense. The ghosts of extinct creatures revivified by Knight still haunt museum corridors graced by his art displays. The legacy of the Man who saw such beauty and wonder in Nature still burns!

CHAPTER 8

Paleoart—Search Image and Theory

...every reconstruction of an extinct life-form—whether a scientific artist's drawing, a full-scale museum replica, or a computer-animated dinosaur in a Hollywood movie—is a form of paleontological fiction.

(Adrienne Mayors, *The First Fossil Hunters: Paleontology in Greek and Roman Times*, 2000, p. 244)

...Unfortunately, since only a few bones, teeth, and fragments of these animals have been discovered, it has been necessary to trust to imagination for their greater part of the work of their restoration, but we shall gradually have less and less occasion to have recourse to its aid; for as new discoveries are made, less difficulty will be found in divining approximately or ascertaining accurately the physiognomy of these strange beings.

(Franz Xaver Ungers, *The Primitive World in Its Different Periods of Formation*, 1851, as translated by Martin J. S. Rudwick in *Scenes from Deep Time: Early Pictorial Representations of the Prehistoric World*, 1992, p. 119)

When it comes to paleoart, does a fine line exist between truth and imagination? Or does a generous "grey zone" lie in between?

Through classical folklorist Adrienne Mayor's eyes, how fascinating is the startling array of "restorations" conceived through human interpretation of fossilized bones! Fanciful griffins, dragons, the mythical "Cyclops," celebrated gods, mythohistorical heroes and even mythological monsters of antiquity have all resulted from ancients' examination of fossilized bones. Such interpretations may be attributed to the human mind, a hardwired system producing what is referred to by Mayor in *The First Fossil Hunters* as an "anthropomorphizing search image." This happens because, as the Greek philosopher, Xenophanes claimed in the sixth century B.C., human intellect reaches for "what matters most ... humanlike models."[1] The human mind, the brain's right, intuitive hemisphere especially, interprets visually, what it "sees" in terms of what it "knows." (See Figure 8-1.)

According to Dr. Jack Horner, the ancient Scythian nomads who apparently "restored" the fossilized bones of *Protoceratops* as living Griffins three thousand years ago had a different "conceptual toolbox" than do the paleontologists and paleoartists of today. Horner claims, "If the nomads had viewed the natural world [in Linnaean terms] they would have felt compelled to place Protoceratops in one category or another, bird or mammal. They would not have been so open-minded about what their eyes told them—that the creatures they found in the Gobi Desert possessed features from both categories and so belonged to both, or neither."[2] (See Figure 8-2.)

In *The Last Dinosaur Book: The Life and Times of a Cultural Icon* (1998), William J. T. Mitchell discusses the union of human intuition with scientific interpretation in our understanding of dinosaur images: "The dinosaur image is not, then, simply a popularization of scientific understanding, a mere vehicle for reporting and representing scientific knowledge. The visual image is also a key element in the process of scientific thinking and discovery as such, not just as a descriptive afterthought or after image, but as a constitutive element, a speculative, theoretical construction. The moment of scientific intuition is often a vivid insight, a daring projection of a visual or spatial model, and not merely a summing up of empirical data. This is especially true for paleontologists, who must turn fragments into wholes, skeletal structures into living, breathing, flesh-and-blood animals. Perhaps more than any of the other historical sciences, paleontology requires the aid of visual

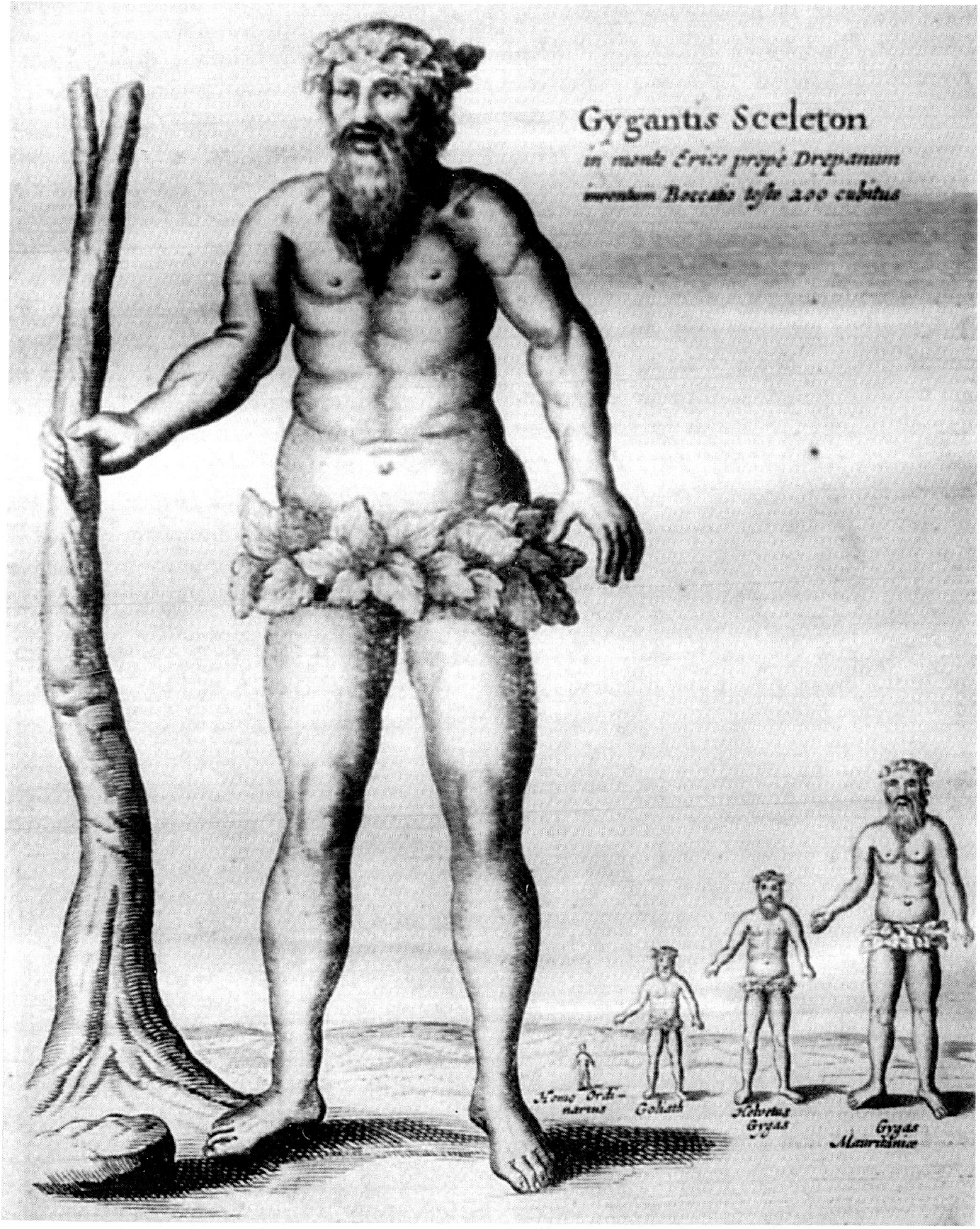

Figure 8-1: Illustration showing antediluvian giants "restored" from fossil bones, as perceived by the German scholar Athanasius Kircher in the 1665 edition of *Mundus Subterraneus*. A normal human stands to left of largest giant, and the biblical Goliath stands in the middle. Homer's Polyphemus is second from right. (Photograph by Allen A. Debus, courtesy of Allen G. Debus.)

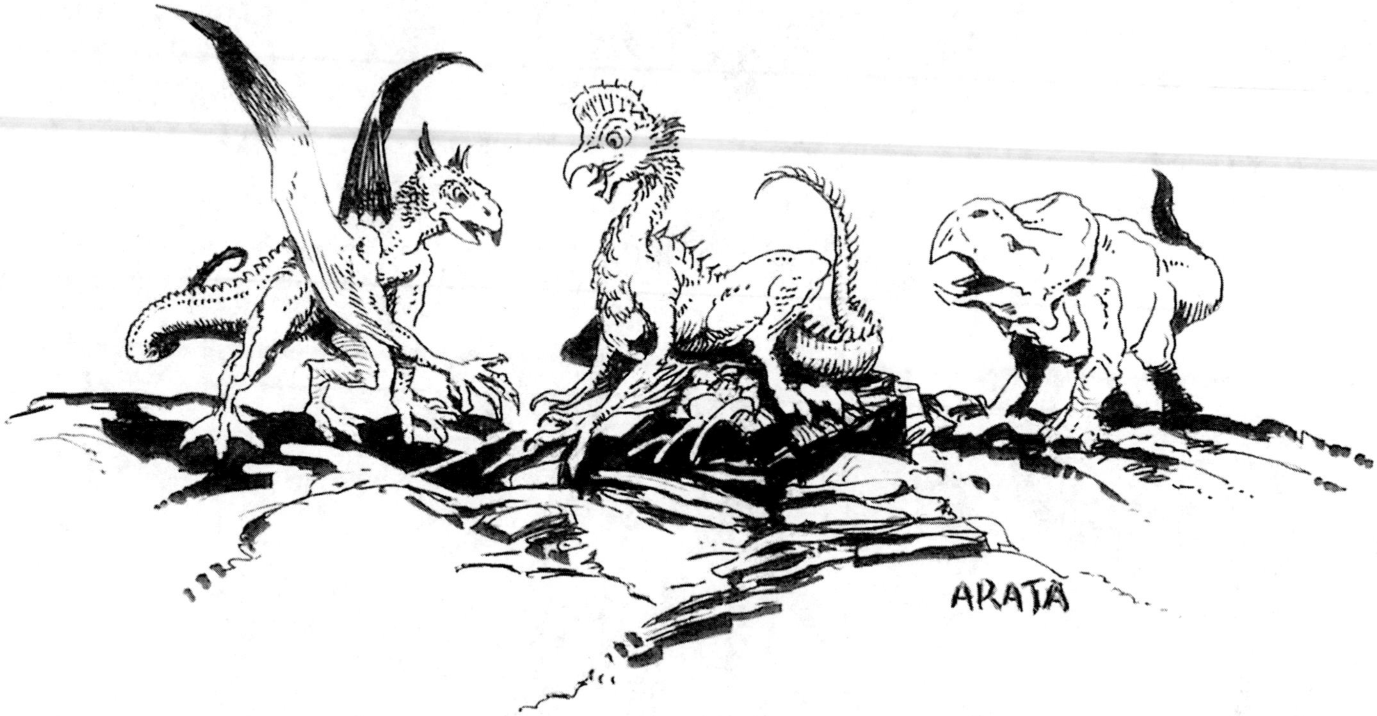

Figure 8-2: Here Jack Arata illustrates how evidence of fossils can be interpreted by humans in widely contrasting ways. Skeletal remains and fossil eggs found in Mongolia's Gobi Desert have been attributed to winged griffins (left) by the ancients, and by early twentieth century paleontologists as the bones and nesting sites of a dinosaur—*Protoceratops* (right). Many of these nests and fossil eggs have recently been attributed to a presumably feathered, brooding oviraptorosaur dinosaur. (Reproduced with permission.)

artists and illustrators to depict in concrete, visible form what the scientist conceives in the imagination."[3]

Life image conceptualizations of what we would regard as prehistoric animals, then, are conjured from ancient neural pathways, hardwiring used by ancient observers, nineteenth century pre–Darwinian natural historians and modern degreed scientists alike. Had they comprehended prehistory and the immensity of geologic time, it is tempting to wonder whether Plato or Socrates could have "published" the first scientifically valid description of a prehistoric vertebrate thousands of years before Cuvier.

Yet, as Mitchell maintains, beyond the psychological factors influencing "theoretical" visual perceptions of prehistoric fauna, sociopolitical and cultural factors may also be at large. Such factors must be carefully considered in light of the emerging American paleoart "genre."

At the outset, American science had a diminished reputation relative to the more esteemed state of British and European science. Americans were known for providing facts and observations, especially about the vast, largely unexplored western territories. While contemporary American scientists were sometimes lauded for technological inventions, by the 1870s they had not established themselves in the pure and theoretical realm.[4]

In fact, American scientists were careful to "watchdog" the work of (American) colleagues, so that "embarrassments" didn't happen. For it was important to avoid publishing "quackery" that would later prove damaging to the developing reputation of American science. Thus, the arena of theory and speculation seemed risky territory indeed! Best to stick with the facts and details only, as did Joseph Leidy, an American scientist highly lauded by his British and European scientific "superiors" throughout his long, illustrious career.

However, a risk-free strategy to American science was not without its own risks. For example, while British "theoretician" Thomas Henry Huxley praised Othniel Marsh's theoretical insights into equine evolution, an anonymous, apparently British critic condemned one of Leidy's monographs, *Cretaceous Reptiles of the United States* (1865), stating that it wasn't of the caliber of Cope's recent, more daring, theoretically oriented publications.[5] A fine line indeed!

Certainly prehistoric vertebrate reconstruction and restorations fell into that uneasy realm of scientific speculation and "theory," territory generally avoided by Joseph Leidy. Leidy was adept at *scientific* description, yet he avoided the theoretical realm in his science and art.

However, it is possible to project theoretical ideas

through artistic reconstructions and restorations, convincingly and confidently, provided caution is taken to assure their accuracy. Caution is needed because, according to Kevin Padian, "The starting point for the unfamiliar is the familiar ... what an artist believes he sees in an exotic subject merely reflects his own experience.... Theory colors perception, and belief is often stronger than observation."[6]

Armed with twenty-first century hindsight, it would seem most appropriate and logical to create valid life restorations of fossil vertebrates by proceeding through a series of conceptual design stages.

First, the prerequisite to preparation of restorations is adequate condition of the fossils. Understandably, the accuracy of life restorations ultimately hinges on the quality of fossil remains available for study. Naturally, modern paleontologists would doubt the authenticity of any restoration founded on inadequate fossil remains.

After excavation, preparation and scientific study of the bones, if the fossils are of sufficient nature, a two dimensional (skeletal) reconstruction can be prepared in either a static or lively pose. (More dramatic poses may reflect deeper theoretical insights.)

Next, the illustrated reconstruction can be fabricated as a three dimensional miniature model of the skeleton, then subsequently scaled into a life-size reconstruction made of either original bone or bone casts. (Missing skeletal material can be inferred and modeled, although in the case of poorly known species, reconstruction increases the likelihood of introducing inaccuracies.)

Life restorations can now be attempted with greatest confidence. Whether to attempt sculptural or illustrative designs first entirely depends on the paleoartist's skills and objectives. However, some artists will attempt both three and two dimensional restorations. For example, Charles Knight painted restorations while referring to his own miniature sculptures, while Benjamin Waterhouse Hawkins scaled full-sized prehistoric animal sculptures from his miniature sculptures. (See Figure 8-3.)

Ultimately, after scientifically comprehending the prehistoric species and devising a conceptual model of its life habits and paleoecology, a theoretically daring scene from deep time may be prepared, illustratively incorporating the species in a naturalistic setting.

While not all modern paleoartists perform all these steps in the order stated, it is evident that, historically, scientists' confidence in the creations of contemporary paleoartists suffered because either their paleoart was founded on incomplete fossil material or because the artistic approaches were not judged to be of sufficient rigor.

Let us briefly consider how the significance of critical factors affected the perceived "acceptability" of several early modern paleoart examples.

The critical factors are "luck—completeness of specimens," "theoretical correctness," "restorative methods and techniques," and "artistic execution":

1. First there is the "luck" factor. The completeness of specimens has a direct bearing on the credibility of restorations. "Sue," the *T. rex*, is over 90 percent complete skeletally. *Agathaumas*, the horned dinosaur, is over 90 percent incomplete. (See Figure 8-4.) So, in the hands of knowledgeable paleontologists and skilled paleoartists, restorations of the former have greater potential to be "accurate" portrayals.
2. Confidence in the accuracy of any particular restoration, as perceived by contemporary paleontologists, will also be judged by the "correctness" of the underlying theoretical ideas embodied in the restoration. Here, cultural biases can creep in.
3. The method and techniques used by paleoartists to produce a restoration will be scrutinized by paleontologists. Paleoart used for museum displays must be scientifically founded and meticulously prepared so as to offer a genuine perspective on how the extinct species appeared in life. This work involves careful measuring and examination of bones, and may entail intermediate design stages to perfect the restoration. Clearly, the objective is to "get it right," and great pains may be taken to make it so. Following or approximating the meticulous "ideal" strategy for restoring fossil vertebrates outlined above can create a sense of accuracy in the final result.
4. Finally, how convincing is the overall result? This is a collaborative function of the scientists' vision and paleoartists' abilities. Even if their conceptual model was accurate, if something was poorly perceived or inexpertly rendered along the way, the sense of accuracy will be suspect.

Now let us survey five episodes in the evolutionary development of modern paleoart, while considering the critical factors:

1. In their collaborative restorations of prehistoric animals prepared for the Crystal Palace exhibition at Sydenham, England, during the early 1850s, Sir Richard Owen and Benjamin Waterhouse Hawkins employed theoretical concepts, even though,

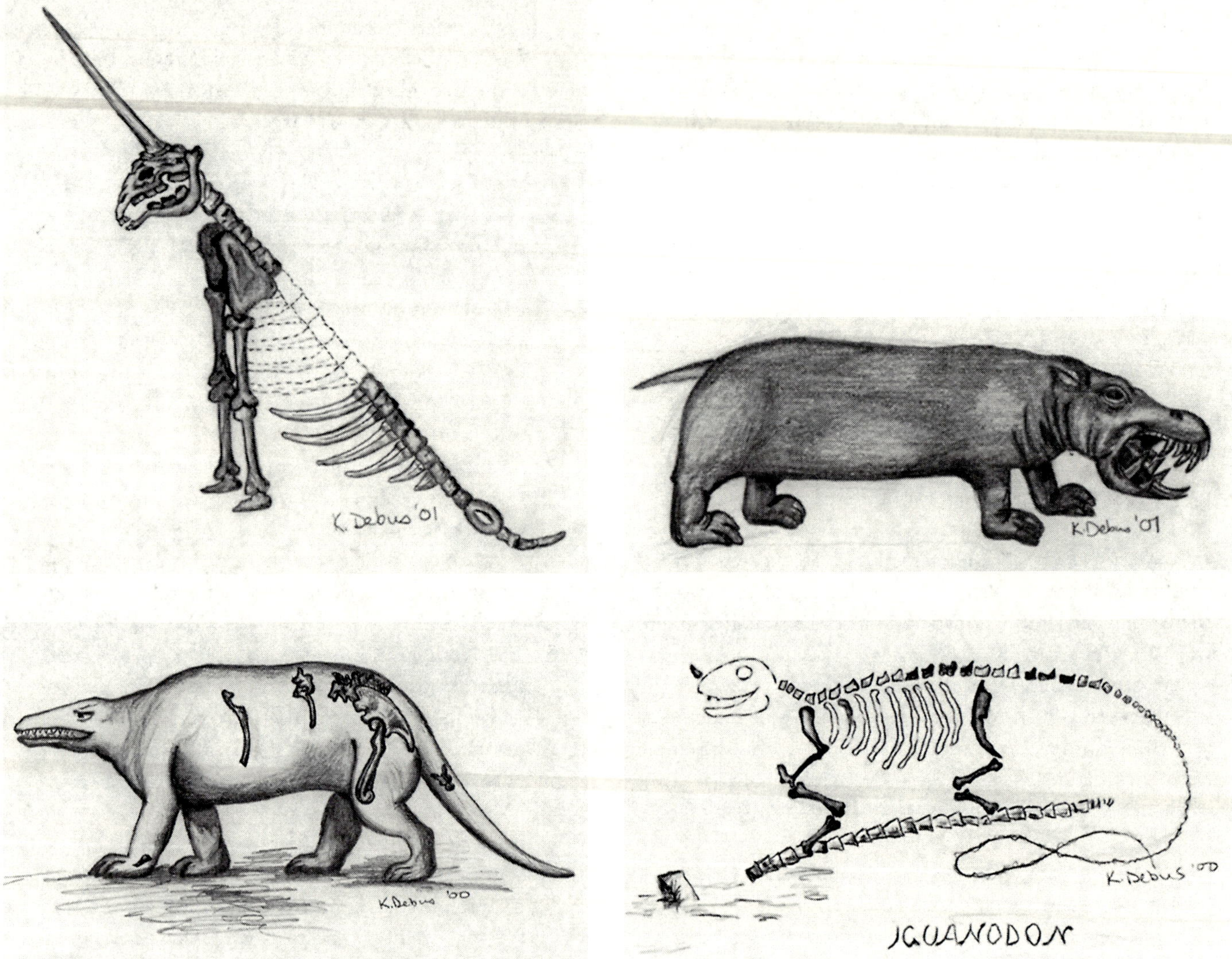

Figure 8-3: Four antiquated views of how early scholars viewed fossil bones. Clockwise beginning at top left: (a) Skeleton of unicorn as reconstructed from mammoth bones by physicist Otto von Guericke (1602–1686) and later reproduced in Gottfried Wilhelm Leibniz's (1646–1716) book *Protogaea* (1749); (b) A "behemoth," really a hippopotamus, but in 1682 associated with bones of the Siberian Mammoth—based on engraving reproduced in Paul Semonin's *American Monster*, New York: New York University Press, 2000, p. 73; (c) Gideon Mantell's early 1830s vision of how his dinosaur, the *Iguanodon*, appeared are reflected in this diagram. See Chapter 3; (d) During the nineteenth century the middle Jurassic dinosaur *Megalosaurus* was restored fancifully by several artists on the basis of scanty remains. This drawing illustrates how incompletely known the genus is, both then and now. After figure, circa 1850, appearing in David Norman's *Dinosaur!* 1991, New York: Prentice Hall, p. 56. Figure by Kristen L. Debus, 2000/2001.

notably in the case of their dinosaur statues, they lacked the prerequisite fossil material.

2. By the late 1860s and early 1870s, when Hawkins, Leidy and Edward D. Cope pooled their knowledge in considering how Leidy's *Hadrosaurus foulkii* could be accurately restored, the methods and techniques leading to scientifically valid restorations had been somewhat refined. Leidy speculated that *Hadrosauru*, "...may have been in the habit of browsing, sustaining itself, kangaroo-like, in an erect position on its back extremities and tail."[7] Leidy envisioned *Hadrosaurus* as "transitional" between quadrupedal reptiles and avian bipeds, a creature matching Huxley's theoretical (evolutionary) expectations. Accordingly, Hawkins ambitiously prepared skeletal reconstructions either during or prior to restoring prehistoric animals for New York's ill-fated "Palaeozoic Museum." Lacking the pelvic bones, however, Hawkins refreshingly sculpted a gigantic human-shaped pelvis for his *Hadrosaurus* skeletal recon-

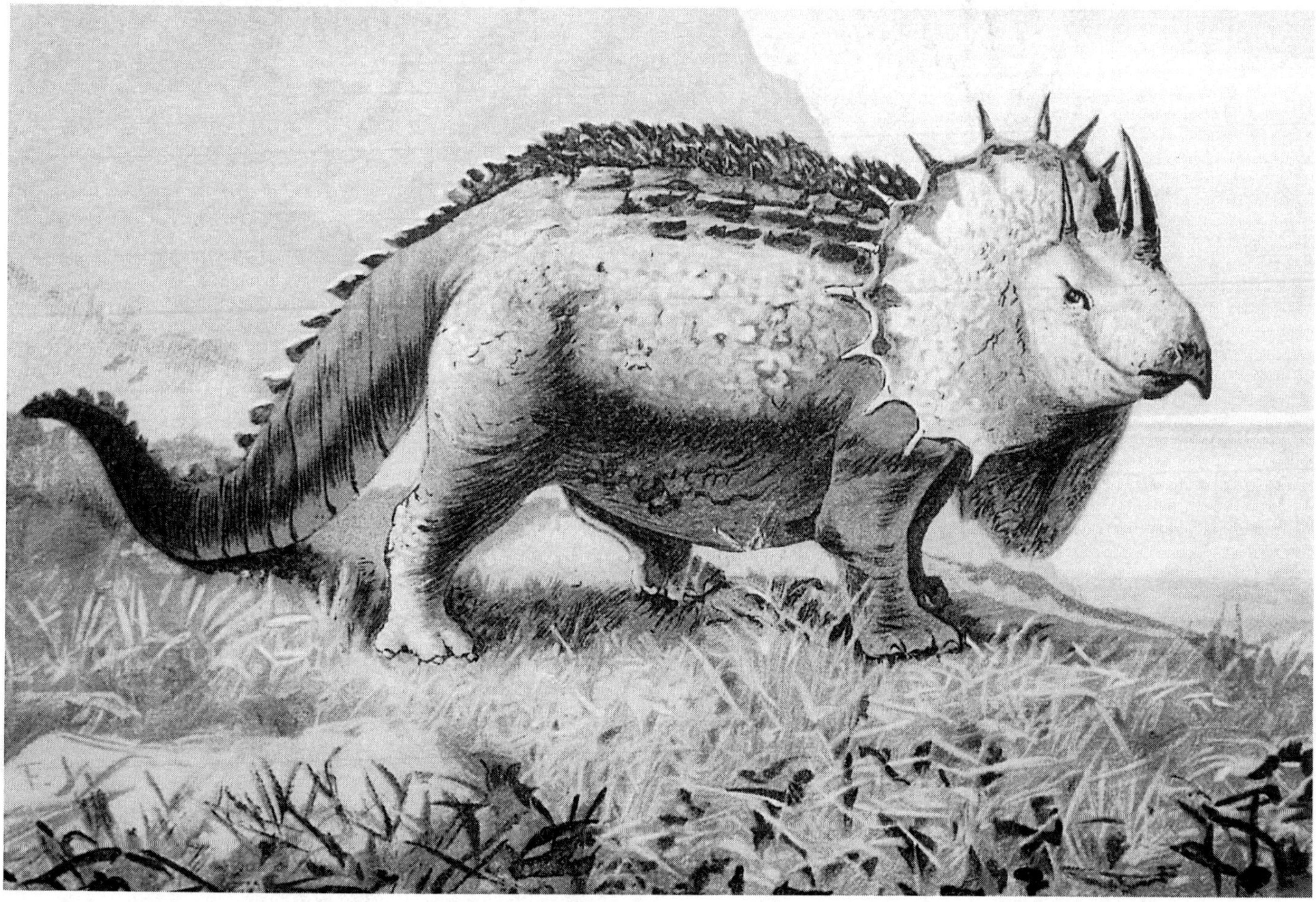

Figure 8-4: F. Long's illustration, circa 1905, of the poorly known horned dinosaur *Agathaumas*, inspired by Charles Knight's influential yet fanciful 1897 restoration. This collectible card was originally released by the Kakao Company of Hamburg-Wandsbek, Germany, as a promotional item. Compare the dorsal armor with that shown in Figure 26-5. (Photograph by Allen A. Debus, courtesy of Allen G. Debus.)

struction. In view of the creature's bipedal stature and the makeshift pelvis, in his perception of Leidy's *Hadrosaurus*, evidently Hawkins (like his ancient predecessors) may have been afflicted by an anthropomorphizing search image.[8] (See Figure 8-5.)

3. At first, Othniel Marsh would not condone the practice of scientifically reconstructing prehistoric vertebrates. Later, by the early 1880s, he fostered two and three dimensional reconstructions. Yet, although he had abundant fossil material at his fingertips for confidently attempting life restorations, Marsh remained critical of paleoartists' life restorations.
4. Marsh's archrival, Cope, on the other hand, invested energies in restorations without compunction. Unfortunately, two of the prehistoric reptiles selected by Cope for restoration (*Laelaps and Agathaumas*) lacked sufficient fossil material for confidently establishing their life images. These may be regarded as examples of "fine paleoart," yet not "fine paleontology."
5. In fact, it wasn't until the very late nineteenth century, when Henry Fairfield Osborn recognized the benefits of preparing scientifically founded paleoart (e.g. dinosaur skeletons reconstructed in lively poses and brilliantly rendered paintings of prehistoric vertebrates) for display in New York's American Museum of Natural History, that the practice of paleoart preparation had achieved a state of scientific "sensibility." Osborn's paleoart "team" developed, refined and used the idealized scenario, outlined previously, to greater degree than any predecessors. Within reason (i.e. within constraints of the four aforementioned factors), paleoart was finally deemed "acceptable." Thereafter, during the early years of the twentieth century, American paleontologists confidently utilized paleoart to convey scientific ideas.[9]

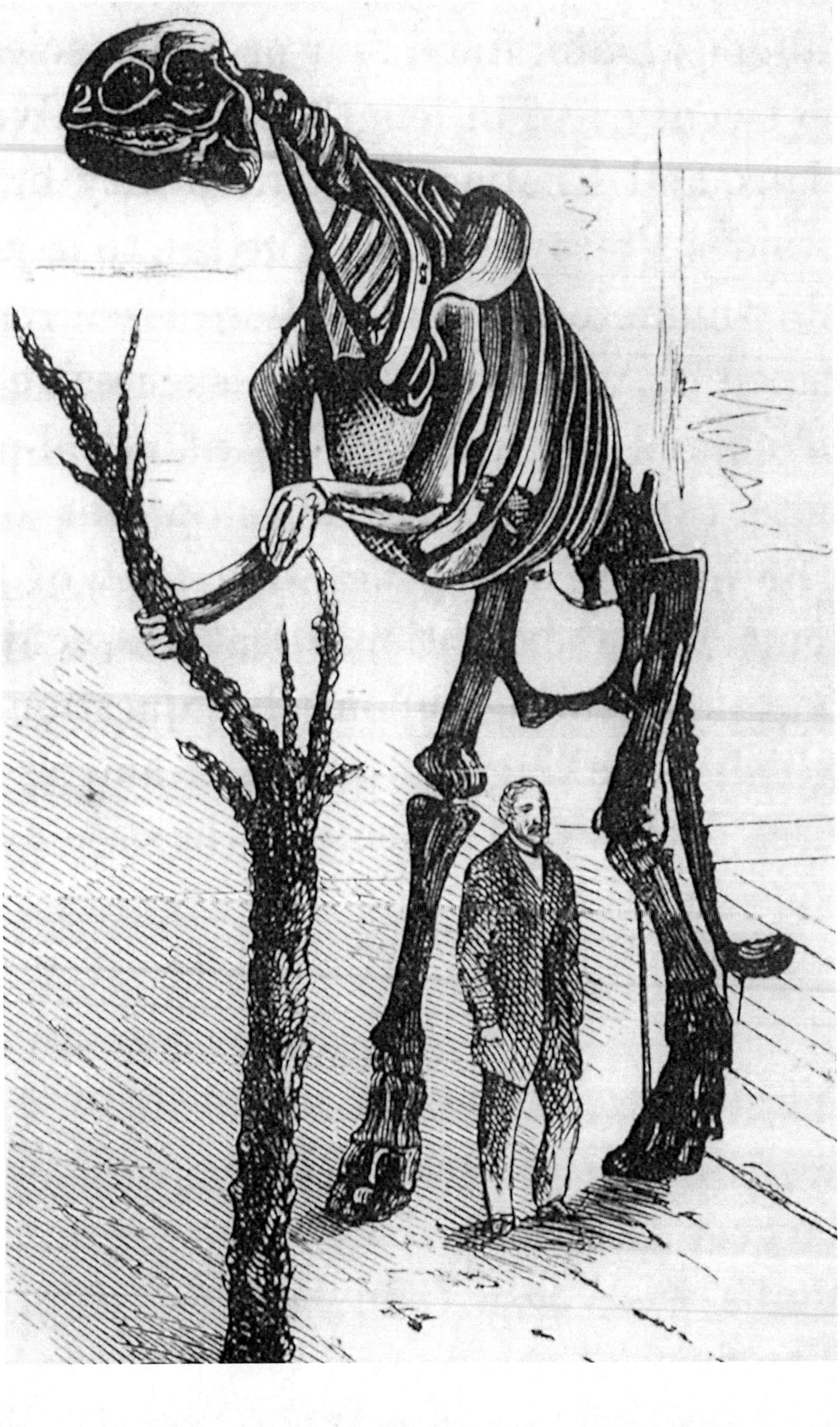
2

Theoretical ideas in paleontology, as exemplified through a longstanding tradition of paleoart, have taken a convoluted path from murky beginnings. Historian of science Thomas Kuhn has noted how science often doesn't follow what would be regarded, retrospectively, as a logical progression.[10] Artistic portrayals of creatures represented by fossils, in turn, cannot be expected to have improved or progressed "logically," illuminating truth and wisdom incrementally through the centuries as fossil evidence steadily accumulated. This is why ancients could apply their acquired natural knowledge to the consideration of (fossilized) bones and "see" beings which today are known to be entirely unnatural. This is why, in what may be the earliest recorded case of an attempted three dimensional "restoration," during the reign of the Roman emperor Tiberius (42 B.C.–A.D. 37) a woolly mammoth's tooth became the basis for a heroic giant's head, sculpted by an ancient artist named Pulcher. This is why the remains of mammoths could be viewed by Medieval and Renaissance scholars as monstrous hippos and unicorns.[11]

In modern times, the artistic Leidy generally refrained from using paleoart to exemplify theoretical ideas. Leonard Warren believes that, "...Leidy had brought to life a great menagerie of extinct beasts, but he never made drawings or sketches of how these ancient animals might have appeared. Such an activity would have drawn too much attention to himself, from both the scientific community and the lay public."[12]

In contrast to Leidy, Osborn, however, became highly influential in promoting the use of paleoart for conveying scientific ideas about prehistoric vertebrates.

Modern paleoart is usually a resultant, a visual whole documenting how scientists interpret the fossil organisms in question. As such, paleoart may often be viewed as an end result of tests, inquiries, observations, determinations and, yes, speculations. Ordinarily, paleoart isn't utilized experimentally, as a direct means for testing theoretical notions.

Yet, recently a skeletal cast reconstruction of the right forelimb bones of the dinosaur *Torosaurus* was used to test hypotheses concerning posture and movement in ceratopsian dinosaurs.[13] And cyberspace dinosaurs have been designed to test how sauropod dinosaurs may have held and moved their necks and tails.[14] Consequently, paleontologists are richly rewarded with visual data from experimental models.

Seeing is believing! Humans can observe fossils yet "see" feathery-winged griffins or visualize various kinds of dinosaurs, with or without feathers and wings. In classical times, individuals may have believed that the mythological monsters of old, represented by the lithified bones turning up in farmers' fields, were struck down with lightning bolts hurled by Zeus from the heavens. In fact, we now perceive bolides bolting from the heavens instead. Paleoartist David Krentz claims, "[Dinosaurs] are real life monsters and real animals, a link between fantasy and science. We'll never know what they really looked like, and I'm sure our representations of them today are ultra conservative."[15]

What passes for scientific knowledge can be triggered by flash (visual) insights. Paleoart is inherently theoretical. Fossils fool devilishly!

It is a fine line indeed!

Opposite: Figure 8-5: Benjamin Waterhouse Hawkins standing adjacent to a reconstruction of his mounted *Hadrosaurus* at the Philadelphia Academy of Sciences. From D. Alexander Winchell's *Sketches of Creation*, 1874. Hawkins also sculpted a prostrate hadrosaur, reflecting its 'transitional' nature.

Chapter 9

Verne's Paleontological Journey

Every picture tells a story, but the best tales inspire unforgettable visions. For many paleo-philes, the name Jules Verne (1828–1905) kindles fond memories of James Mason acting in 1950s Hollywood movie productions, one being *Journey to the Center of the Earth* (1959, Twentieth Century–Fox).[1] Verne, author of the 1864 French novel that inspired the movie, and author of numerous outrageous adventure stories ("*Les Voyages Extraordinaires*"), many demonstrating precocious inventions, is often revered as the "inventor of modern science fiction."

Journey was Verne's second novel and one of the earliest paleontologically-founded science fiction tales ever written. Unsurprisingly, however, much of Verne's story, reflecting contemporary paleontological themes, was not faithfully projected on screen. Accordingly, paleo-philes may question the geological and, in particular, paleontological inspirations for Verne's *Journey*. Should Verne, who was not a paleontologist, be counted among the early evolutionists? We suggest that Verne's chambered, honeycombed subterranean realm was a geological artifice, framed for illustrating paleontological opinions and speculations.

Although the first English edition of *Journey* appeared in 1871, Verne did not read or write English. The favored English edition of *Journey*, however, is a 1988 translation of Verne's first illustrated 1867 edition, by William Butcher.[2]

In the 1988 translation, Professor Otto Lidenbrock and his 19-year-old nephew, Axel (played by James Mason and singer Pat Boone, respectively, in the 1959 movie) launch their travels from Germany. As in the movie, they join their servant, Hans, in Iceland before departing for the summit of a volcano, Mt. Sneffels, the subject of an encrypted note written by alchemist Arne Saknussem centuries earlier.

Verne admitted no female participants on his journey, nor was there an evil-hearted descendant of Arne Saknussem to reckon with once they reached the bowels of our planet. Hollywood used these characters to create comedy and tension, but especially to have conservatively-minded moviegoers shaking their heads in disbelief at Arlene Dahl's (Greta Goetaborg) portrayal of a woman so obviously "liberated" before her time.

Because we have become so accustomed to the movie, however, most of us don't understand where the idea of a cavernous terrestrial interior originally sprang from. Certainly Hollywood gave no clues as to Verne's scientific inspiration (unless you believed Arne Saknussem really existed).

Suspense builds with translation of the runic cryptogram:

"Descend into the crater of Yocul of Sneffels, which the shade of Scartaris caresses, before the kalends of July, bold traveler, and you will reach the center of the earth. I did it. [signed] Arne Saknussem"[3]

But where did Verne get the idea for his geological epic? What was the inspiration? What mad science did Verne employ in fabricating his center of the earth "universe"? Our belief in the "reality" of the 1864 *Journey* is reinforced by Verne weaving scientific debate into the plot. The sources for his tale may be categorized as "scientific" and "literary."

Verne simply borrowed the idea of a hollow earth from fellow writers, notably Edgar Allan Poe, who had already written three eerie tales concerning maelstroms and holes at the earth's poles. Poe, in turn, was influenced by American War of 1812 hero and (pseudo-) scientist Dr. John Cleves Symmes, who believed the earth was hollow in between the spaces of a series of concentric shells comprising the earth's framework.

Symmes even fictionalized his ideas in an 1820 novel, *Symzonia: A Voyage of Discovery*.[4] Other writers, Baron Ludwig

Holberg (1741), Alexandre Dumas (1853) and George Sand (1864) among them, wrote tales concerning exploration of the earth's interior which are thought to have influenced Verne's writing of *Journey*.[5] Pierre Boitard (1861) wrote of a mystical journey into the past where prehistoric animals are encountered, including a plesiosaur and a fossil man.[6] One fiction writer sued Verne for plagiarism, without prevailing in court.

We suspect a scientific influence may have been Athanasius Kircher, who published a Latin tome, *Mundus Subterraneus* (1665). Here Kircher theorized a "central fire" at the earth's center and described fossil bones as the remnants of antediluvian, subterranean creatures. His book contained a lavishly prepared figure showing how waters entering the earth are heated from within and rise from underground ocean-sized caverns to the crust where they emerge as mountain streams.[7]

It doesn't take too much imagination to segue from Kircher's seventeenth century views to Verne's nineteenth century science fiction. In fact, curiously, Kircher was born in Germany and died in Italy, the circuit taken by the explorers in *Journey*. There were other, more direct, inspirations for the geological elements of *Journey*. For instance, Charles Sainte-Clark Deville (1814–1876) discussed his observations of the 1858 eruption of Stromboli with Verne.

By the mid–1860s, informed readers would have realized that explorers descending into the shaft of an active volcano eventually would encounter molten rock. So Verne deftly played sleight of hand, dusting off some antiquated, or downright fanciful, theories of the earth's interior to allow his explorers safe passage to his prehistoric setting at the "Central Sea."

British chemist Sir Humphry Davy (1778–1829), the inventor of "laughing gas," had claimed that reaction of water with metals under the earth's surface caused evolution of gases and, hence, eruption from volcanic chimneys. By relying on Davy's by then old and erroneous theory, Verne could sidestep the problem of the earth's temperature gradient and allow Professor Lidenbrock, Alex and Hans to penetrate 87 miles below the earth in relative comfort. While Lidenbrock believes Davy's theory is correct, Alex, who believes in the theory of "central heat," remains continually concerned for his safety.

Of considerable importance to Verne's understanding of paleontology were two French publications of 1863. Guillaume Louis Figuier's *The Earth Before the Deluge* was magnificently illustrated by popular French artist Edouard Riou, who would later illustrate the seventh (1867) edition of *Journey*. Although written in the "Darwinian" period, Figuier avoided discussion of organic evolution. As in the arrangement of Hawkins' display at Sydenham, chapters in Figuier's *The Earth Before the Deluge* have a "life through time" organization.

Another significant publication of 1863 was the first French edition of Charles Darwin's landmark book *On the Origin of Species by Means of Natural Selection or the Preservation of Favoured Races in the Struggle for Life*. Controversy over Darwin's theory, public exhibitions of outstanding paleontological displays (e.g. the fabulous Crystal Palace "dinosaurs" first placed on display outside London in 1854, and the Paris Exhibition of 1867), and news of dramatic fossil discoveries (e.g. *Archaeopteryx*—1860/1861 and Neandertal Man—1856) all created quite a sensation in the years and months leading to the first and seventh editions of Verne's *Journey*.

It is likely that Verne may have either visited Sydenham to see the Crystal Palace "dinosaurs" or must have known about them from reports published in French newspapers. Verne may have visited Sydenham when he traveled to Scotland and London during the summer of 1859.[8] As many paleontological passages were added to the seventh (1867) edition of *Journey*, it is apparent Verne's sensitivity to the science of prehistory only heightened during the mid–1860s.

In Verne's subterranean universe, the drama involves the explorers' encounters with prehistoric organisms, with much activity near the Central Sea. (See Figure 9-1.) *Journey's* prehistoric passages are organized as a "life through time" presentation of the history of organic life. There is broad movement from the "primordial soil" to the appearance of fossil man. Before revealing his reconstructions (either real or imagined), Verne shows us fossils. Ever the science popularizer, Verne describes the geological age of each fossil and "living" creature encountered by the protagonists.

Readers of *The Earth Before the Deluge* would have been positively influenced by artistic restorations sprinkled generously through the text, showing splendidly illustrated panoramic scenes of prehistory. These were the handiwork of artist Edouard Riou, who eventually was enlisted to illustrate Verne's *Journey* and three of Verne's other most beloved novels, including *Twenty Thousand Leagues Under the Sea* (1870—French edition; English edition—1873).

Although, due to his collaboration with Figuier, Riou has recently won fame from dinosaur enthusiasts as an early "paleoartist," prehistoric themes were not his specialty. Riou was a popular artist of the 1850s through 1870s, contributing artwork to a great number of French magazine publications and books. He was a landscape painter and commemorative illustrator. Of his style, Edmondo Marcucci has stated, "...Riou's drawings are rich with light, and

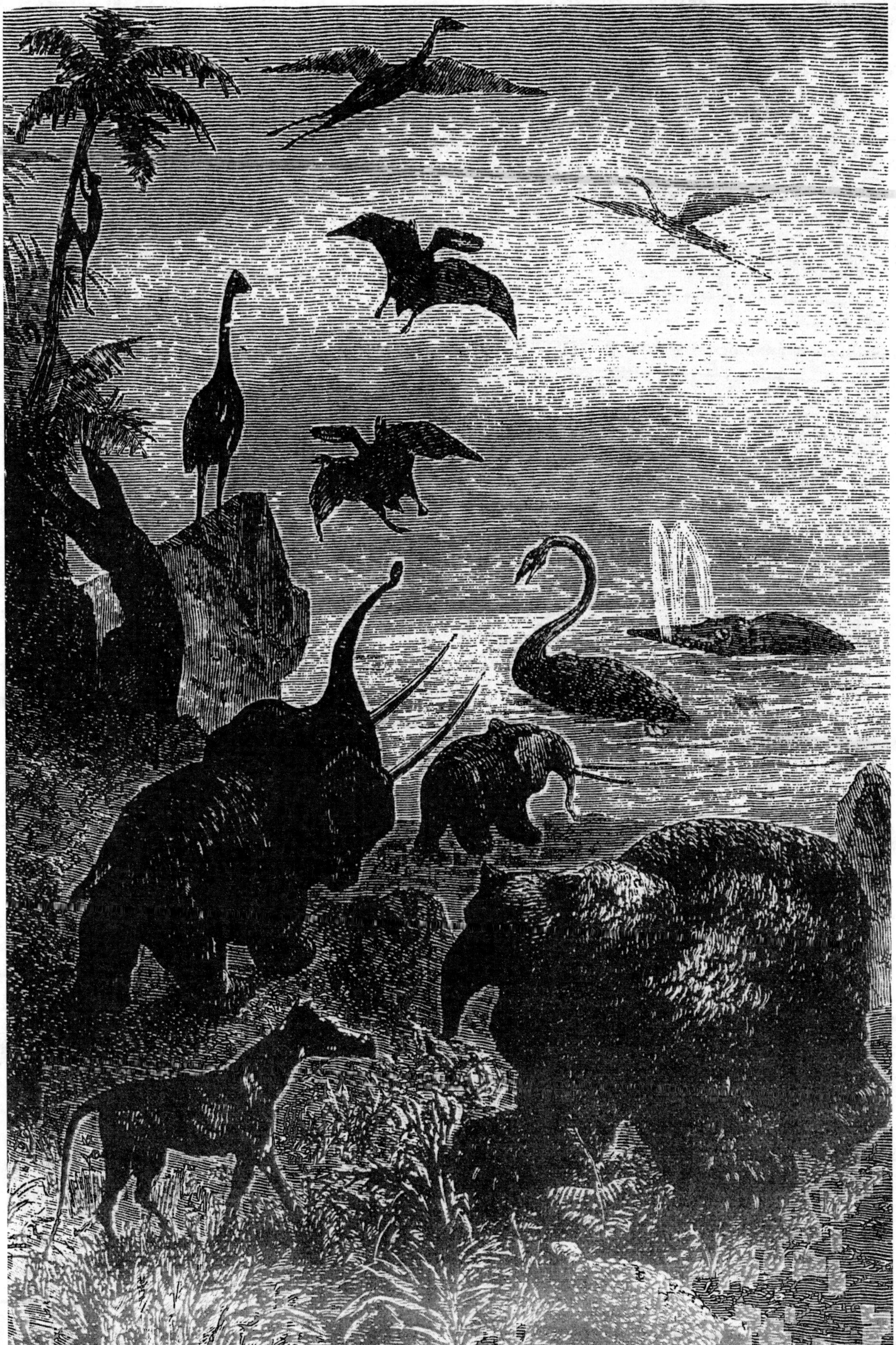

the traits of his characters have a vigorous expression. Riou succeeds in adapting himself artistically to the realistic reproduction of the many fantastic locales of the Vernian fictional geography ... the shadows inside the earth's crust, the deserted and expansive beaches, and the many bodies of water and their movement. Everything is both ordered and evocative in Riou's work—his style might be called 'romantic realism.'"[9]

For Verne's 1867 illustrated edition of *Journey*, Riou supplied fifty-six illustrations (reproduced through a wood engraving process), all of them moody and effectively projecting the stark surroundings of the outer world as the explorers prepare for their descent, as well as the sublimity of the earth's interior. Paleophiles would most enjoy Riou's visions of "prehistoric" settings, fossil bonebeds and animal restorations. The prehistoric animals, both in *Journey* and *The Earth Before the Deluge*, are to some degree patterned after Hawkins' "antediluvians." Riou's illustrations are presented shortly before the scene is narratively described, offering encouragement for the readers to "carry on."

In *Journey's* prehistoric passages, Verne sets the table with fossils early on before revealing his reconstructions. At this stage he is building the credibility of his subterranean universe, founded on Davy's theoretical ideas. Then, recreated animals are confronted, either directly "in-the-flesh," or purely imagined. There are no dinosaurs per se, even though Verne must surely have known of them by that time. Remember this was a "free-for-all period," with many "prehistoria" jockeying for position in public imagination. Prehistoric mammalian and aquatic reptilian forms (especially) were more familiar, and so they simply outstripped the dinosaurs in popularity. In fact, 140 years ago the idea of a "dinosaur" would have seemed vastly more obscure than in the present dinosaur-saturated world. Yet the telling of these encounters in *Journey* is brilliantly done; readers' encounters with Michael Crichton's *Jurassic Park* (1990) "raptors" seem bland by comparison.

Along their journey the party first identifies a Silurian trilobite, then a fossil fish, followed by a coal age deposit. Later, anticipating the Hollywood "B movie" tradition of the next century, bones of "Tertiary" mammals (e.g. *Deinotherium, Mastodon, Megatherium*) are found, stirring debate over whether a surficial deposit of bones simply fell through a chasm to be redeposited below, or whether they actually lived and died on the shore of the Central Sea. Alex fears such monsters may be lurking about in the nearby Giant Mushroom forest (which is said to be of Tertiary age). Near the end of the novel there is a climactic scene where a fossil man of Quaternary age is found. (Along the way, Professor Lidenbrock is figuratively transformed into a frightening antediluvian man, and, depressively, Verne suggests the eventual fossilization of Alex's skeleton.)

In the novel, fossils foreshadow the dramatic appearance of prehistoric animals. Unfortunately, in the 1959 movie there were few suggestions of prehistoric life, other than the larger-than-life finbacked *Dimetrodons* that James Mason and comrades encounter as they are about to launch their raft upon the Central Sea. Verne couldn't have written about this famous non-dinosaurian, however, because Edward D. Cope didn't describe the species of *Dimetrodon* until the 1870s.

So, while on board their raft (made of fossil wood that should sink, not of mushroom stalks as in the movie), Hans hooks a prehistoric fish (*Pterichthys*) in the waters of the Central Sea. Here is the first "proof" of the existence of prehistoric animals at the earth's "center" (which, incidentally, is not even close to the true center of the earth). Then, in a dream-like state, Alex experiences an "epiphany," envisioning the entire geological evolution of the globe and transformation of life from recent to primordial times. Verne offers vivid glimpses of many prehistoric animals, especially those known to fellow countryman and noted paleontologist Baron Georges Cuvier.

"My dream then goes ahead of the appearance of the animate beings. The mammals disappear, then the birds, then the reptiles of the Secondary Period, and finally the fish, the crustaceans, the molluscs and the articulata. The zoophytes of the Transition Period themselves return to nothingness. The whole of the world's life is summed up in myself, and mine is the only heart that beats in this depopulated world!"[10]

This marvelous description forms the nucleus of Verne's novel. Yet artist Riou provides only a single scene showing a grouping of the antediluvian fauna. A few years earlier he had done an entire sequence of scenes to capture similar ideas for Figuier. While Alex's description of prehistory may appear "uniformitarian," it is not "evolutionist." In fact, it borrows more from Figuier than from Darwin. From passages such as this, it becomes clear that Verne did not accept Darwinian evolution.

A titanic sea struggle (waged between an *Ichthyosaurus* and *Plesiosaurus*) forms the next striking scene in the novel. The crew grows alarmed when Hans notices bite marks on

Figure 9-1: Edouard Riou's vision of the Central Sea in *Journey*. Visible are a variety of Tertiary and Mesozoic vertebrates, including pterodactyles, *Megatherium, Plesiosaurus, Ichthyosaurus* and Mastodonts. (Figure reproduced from 1871 edition.)

a crowbar used to sound the sea depths. Alex worries, "Is my dream of last night about to become reality?" You betcha! Verne expertly builds the tension, casting science factoids into the storyline concerning the fossil aquatic saurians they are about to confront. If we are already fearful at the thought of mere museum skeletons, then how petrified will we be in the actual presence of such frightful monsters? Then we are treated to one of the first epic battles ever to appear in science fiction literature, staged between two colossal prehistoric monsters. If ever the phrase "We're going to need a bigger boat" was warranted, this was that time!

The classic chapter (titled Terrific Saurian Combat in the 1871 translation) could not have been better written by any of today's famed writers. Unfortunately, movie viewers were deprived of this dramatic scene. The raft rises from the waves on the back of the plesiosaur, signalling a monumental struggle to the death between the mighty *Icthyosaurus* and the *Plesiosaurus*. "These animals attacked one another with inconceivable fury. Such a combat was never seen before by mortal eyes, and to us who did see it, it appeared more like the phantasmagoric creation of a dream than anything else."[11] Had Victorians the means of making feature films, this scene alone might have inspired a fantasy movie, "Plesio-zilla versus Fish-Lizard!" (See Figure 9-2.)

Hours later, the monsters disappear beneath the waves. The mortally wounded plesiosaur resurfaces shortly after. Verne leaves us spellbound. "As for the Ichthyosaurus, has it gone down to rest in its mighty cavern, or will he reappear on the surface of the sea?" Riou's illustration of the combatants, tossing the tiny raft to and fro in the sea spray, is reminiscent of his similar scene for *The Earth Before the Deluge.* (Figure 9-3.)

Why did Verne select the *Ichthyosaurus* and *Plesiosaurus* for his most dramatic scene instead of dinosaurs? Perhaps because to many contemporary readers, these rather well known, enormous aquatic reptiles seemed to be more "terrible lizards" than the lesser known dinosaurs. Also, perhaps, as Peter Benchley proved to us with his

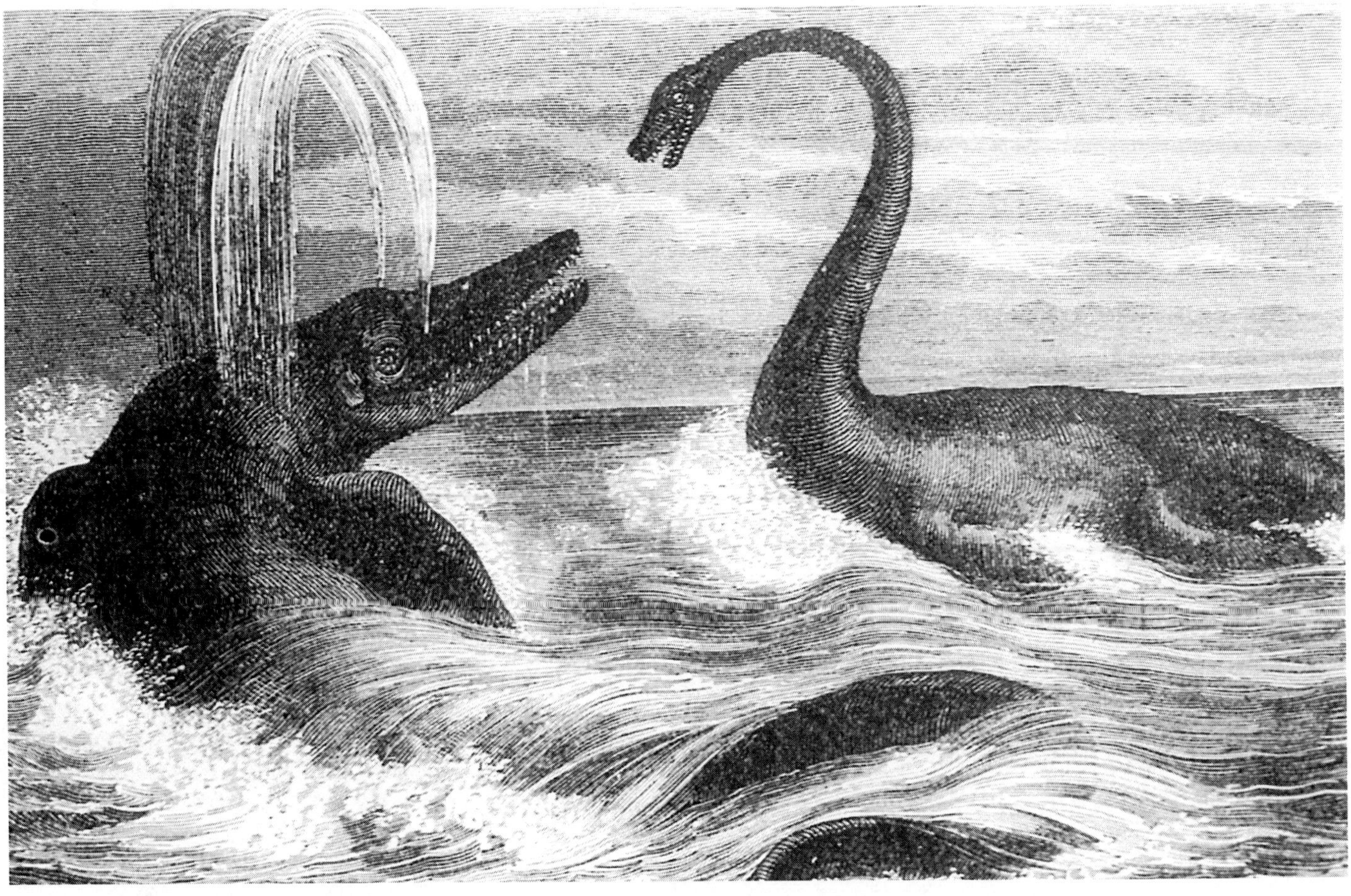

Above: Figure 9-3: Riou's less fanciful vision of battling sea saurians—*Ichthyosaurus* and *Plesiosaurus*—as depicted for Louis Figuier's *Earth Before the Deluge*. (Authors' collection.) _Opposite:_ Figure 9-2: In this scene Riou captured the fury of Verne's scripted titanic sea battle waged between *Ichthyosaurus* and *Plesiosaurus* in the Central Sea. (Figure reproduced from 1871 edition.)

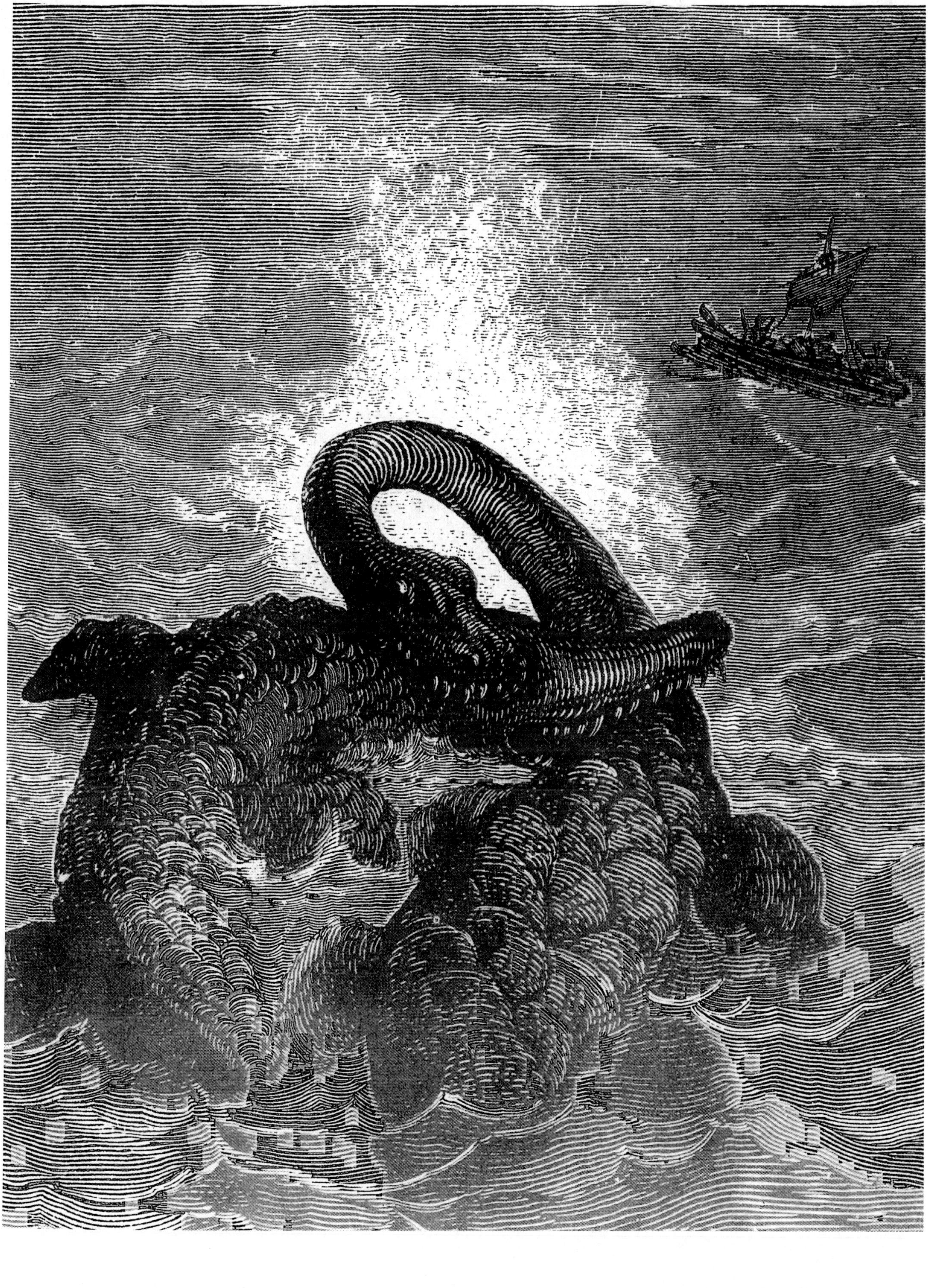

novel *Jaws* (1975), people are simply scared of the water and what lurks therein. In *Journey*, Verne, like Benchley a century later, capitalized on our innate fears. In earlier chapters, Verne's explorers had already contended with the fear of darkness. Here they worried over what hideous monsters thrived below the surface of the Central Sea's murky waters.

Furthermore, published illustrations of plesiosaurs interacting with icthyosaurs had been available since 1830. Henry De La Beche's cartoon, "*Durior antiquior*" (1830), for instance, in which a battling plesiosaur and ichthyosaur were depicted, was redrawn for a volume in Francois Jules Pictet de la Rive's (French) reference series (1844 to 1846) titled *Elementary Treatise on Palaeontology* (*Traite Elementaire de Paleontologie*). And Emile Le Puillon de Boblaye incorporated Johann Baptist Pfitzer's illustration showing an ichthyosaur/plesiosaur battle—obviously derived from De La Beche's "*Durior antiquior*"—for installments of an 1834 book, *Illustrated Dictionary of Natural History* (*Dictionnaire Pittoresque d'Historoire Naturelle*). Also around this time, an idiosyncratic British fossil collector, Thomas Hawkins, wrote a book, *Memoirs of Ichthyosauri and Plesiosauri* (1834), including a restoration of these animals drawn by John Samuelson Templeton.[12]

Did Verne remember such scenes from childhood reading? Did the thought of a plesiosaur-ichthyosaur battle haunt him from childhood through his adult life, much as Charles Knight's famous Field Museum painting of the *Triceratops* and *Tyrannosaurus* captivated one of us (the youthful Allen)? Vivid scenes by talented artists such as Riou and others who illustrated combating sea saurians were bound to leave an impression. Verne scholars may question the inspiration for this chapter. But there is no mystery. Verne was simply employing the most accurately known, most popular and most terrifying prehistoric monsters of his time as literary devices.

Also in this chapter, Verne revealingly refers to even the greatest of modern reptiles as "feeble reductions" of plesiosaurs and ichthyosaurs. This reference to "reduction" of reptiles reflects a Victorian age anti-evolutionary principle formerly used to falsify the (non–Darwinian) evolutionists' belief in a stately "ladder of progress."

Scenes involving prehistoric "man" follow. First, buried amidst the jumbled bones of Pliocene species, they find a human fossil. This scene was inspired by the Moulon-Quignon hoax played half a century before the more famous Piltdown Man prank. In 1863 French paleontologist Boucher de Crevecoeur Perthes thought he had found the remains of an "antediluvian" fossil human. But other authorities proved the fossil was a fake.[13] So, in the 1867 edition of the novel, Verne presents a deranged Professor Lidenbrock comically defending the nature of this fossil (Quaternary) man without ever mentioning the word "evolution."

Next the party notices a 12-foot-tall humanoid shepherding *Mastodons* through the Tertiary forest. They flee from the apparition in terror, and Verne even suggests the experience may have been imagined. (Riou's accompanying shadowy, sketchy-looking illustration underscores this suggestion.) By openly questioning the sanity of the explorers at key points in the novel and suggesting absurd extrapolations of evolutionary science, and in siding with the skeptics, Verne bares his religious beliefs.

A final prehistoric encounter, first appearing in the 1871 English edition, was not written by Verne but by an unidentified translator. As the explorers travel aboard their raft along a torrent into a volcanic shaft, created after explosives are detonated so they can continue following the marked trail of Arne Saknussem, Alex has another fitful dream. This time, however, the creatures which haunt him are imaginary.

Alex spies a "shark-crocodile," which challenges the enormous "Ape Gigans," a 14-foot-high, hairy, mammoth-toothed monster. After a nightmarish battle, the Ape emerges victorious. Too bad the name of this creative writer has been lost to posterity, because although he may have tampered with the flow of Verne's novel, he may have invented the original "King Kong!"[14] Riou would have surely produced an unforgettable illustration of this scene had Verne ever written it.

Jules Verne's *Journey* has fascinated many science fiction fans for many years, even though it has never been regarded as a paleontologically oriented "package." This may be the fault of Hollywood producers who chose to omit all of the novel's original paleontological elements from the 1959 film adaptation. Verne was clearly fascinated by paleontology, although not in a Darwinian sense. Somewhat ironically then, given *Journey's* mid–nineteenth century competition from writers of similarly themed fantasy tales, only *Journey to the Center of the Earth*'s popularity has survived, evidently through a process of (natural) selection.

CHAPTER 10

Time for Dinosaurs

At a young, impressionable age, Josef Augusta's majestic account of Life's origin and history, *Prehistoric Animals* (1956), had me nearly believing the impossible! Featured were the superb paintings and illustrations of Czechoslovakian artist Zdenek Burian. The photographic quality of the restored animals nearly convinces the casual observer that "prehistoric" creatures still roam today, even though such a conclusion contradicts sensibility. Years ago, I did not appreciate how the Augusta/Burian collaboration represented one of Paleontology's earliest traditions, the artistic portrayal of a series of scenes representing successive periods of earth history. Subconsciously, this early experience did forever meld the image of dinosaurs with my perception of Geologic Time.

Paleontologists' impressions of prehistory are mirrored in artists' restorations of extinct animals and the landscapes in which they lived. Ever since it became understood that fossils represented the organic remains of extinct organisms, their existence and interpretation have become closely linked to Man's concept of geological time. Traditionally, the "visual language" of dinosaurs and other prehistoric animals fosters our popular perception of geological time. However, this association is often taken for granted today.

Because dinosaurs often appear as favored subjects in a series of scenes designed to show "progression" of Life's history through geologic time, dinosaurs can conjure both a spirit of mystery and connote an eerie sense of "deep time." Dinosaurs evolved through 150 million years of time before disappearing from the fossil record 65 million years ago. Try to imagine how long these intervals really are, as I first tried to 40 years ago, and one recoils in apprehension.

Largely because they were so poorly understood during the first half of the nineteenth Century, dinosaurs were rarely emphasized in early efforts to depict the earth's geological history. Therefore, an association of dinosaurs with Man's understanding of Geologic Time may be related to the most startling examples of this artistic genre created after this period, when more complete specimens were discovered, especially in North America.

The earliest diagrams designed to represent geological time were void of life, that is, "living" prehistoric animal imagery. An early example was a tabular display prepared in 1821 for the idiosyncratic British geologist William Buckland, who once allowed his pet bear to attend a party at Oxford University dressed in student attire. Despite his predilections for keeping the strangest of pets, Buckland was quite the innovator when it came to considerations of how the idea of geologic time could be expressed diagrammatically. Rock types were classified according to age, and the oldest rock was placed at the bottom of the sequence, a convention still followed today.[1]

Another example was the "idealized column," in which rock strata were illustrated on the page in a column adjacent to a description of stratigraphic age and physical characteristics of the deposits. Georges Cuvier's 1822 stratigraphic section of the Paris Basin is regarded as a classic example.[2]

Geologists also represented geological time using horizontally extended charts showing idealized cross sections through the earth's crust. William Buckland had suggested in 1831 that a series of illustrations showing the earth's landscape in prehistoric ages would be precedent setting.[3] Several years later, in 1837, Buckland proudly presented Thomas Webster's idealized geological cross section in his contribution, "Geology and Mineralogy Considered with Reference to Natural Theology," to *The Bridgewater Treatises: On the Power, Wisdom and Goodness of God as Manifested in the Creation*.[4] Along the top of his diagram, Webster restored successively arranged faunal and floral groupings in order to represent fossils found in the indicated strata. This colorful engraving, an early form of Geologic Time Table, may perhaps be regarded as the earliest effort to associate a dinosaur image, that of an *Iguanodon*, with a drawing intended to represent geological time.

During the next 14 years it became stylish to present the succession of geologic "ages" pictorially. Geological periods were represented either as a series of isolated scenes showing discrete events, "windows" into the past, or in panoramic perspective encompassing broader temporal range. However, contemporary opinions varied considerably over the extent of time thought to have elapsed between represented strata. At first, dinosaurs only made token appearances.

For instance, in 1841, one year before Richard Owen published the name "Dinosauria," John Whichelo sketched four remarkable geological epochs for Joshua Trimmer's *Practical Geology*. Whichelo's scenes were vertically formatted in order to represent stratigraphical succession. In the Cretaceous scene, a crudely drawn, double-horned *Iguanodon* strolls along the beach.[5] Then, in 1849, a more continuously arranged scene, suggesting Time's flow from the Silurian period through the Tertiary, was created by John Emslie. Emslie's Mesozoic scene contains only one animal that appears "dinosaurian."[6] Next, Josef Kuwasseg crafted a series of 14 beautiful drawings for the 1851 edition of Franz Unger's *The Primitive World in Its Different Periods of Formation*. Only one dinosaur scene was included, because Unger, a botanist, placed emphasis on the flora (not the fauna) of "former worlds."[7] (See Figure 10-1.)

Faunal components of the "antediluvian world" were showcased on the Crystal Palace grounds at Sydenham, where in 1854 a number of full-sized sculptures were placed on exhibition. These were the handiwork of sculptor Benjamin Waterhouse Hawkins, completed under advisement of British paleontologist Sir Richard Owen. (See Figures 6-1, 6-2, and 33-1.) At Sydenham, a "continuous"

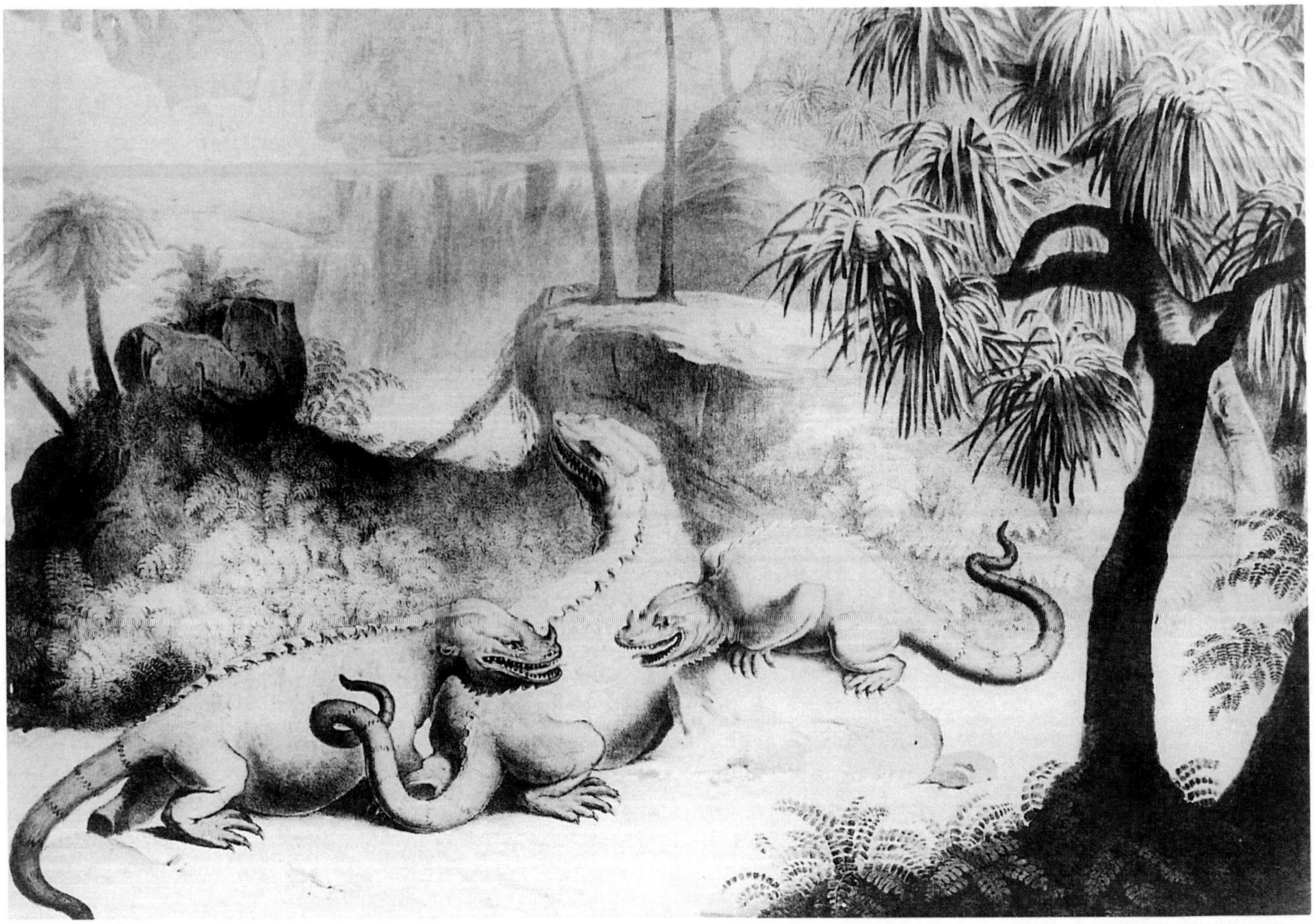

Figure 10-1: Josef Kuwasseg's *Iguanodon*, from Unger's *Primitive World* (1851), one of several restorations organized as an innovative series of 14 drawings illustrating the earth's geological history. (Library of Congress.)

three-dimensional section of geological time has been projected across three islands "inhabited" by creatures of the Late Paleozoic through the recent Ice Age. Due to financial constraints, nearly a dozen of the originally planned Tertiary animals were never completed. Consequently, for the first time ever, saurians, the dinosaurs (*Iguanodon* and *Megalosaurus* especially), commanded top billing in a life through time display! Versions of Hawkins' Crystal Palace dinosaurs also appeared in 1860s editions of Louis Figuier's *The Earth Before the Deluge*. However, in Edouard Riou's complete sequence of 25 restorations showing how the earth's surface had been populated by successions of extinct life forms, dinosaurs only appeared in two scenes.

During the 1870s, Hawkins created a series of 16 scenes ranging from the Silurian period, *Age of Invertebrates*, to the Pleistocene for Princeton University. Although many scenes emphasized ancient European settings, Hawkins' exposure to the emerging North American Geological Sciences now permitted him to add at least one scene that was unambiguously "American," titled *Cretaceous Period in New Jersey* (1877). To execute this scene, Hawkins relied on contemporary knowledge concerning the newly discovered dinosaurs *Hadrosaurus* and *Laelaps* (now *Dryptosaurus*). Although it was fashionable in America to view dinosaurs as monstrous bounding kangaroos, Hawkins' erect, bipedal New Jersey dinosaurs are not leaping. (See Figure 10-2.) In an accompanying scene, *Jurassic Period in England*, featuring the British dinosaurs *Iguanodon*, *Hylaeosaurus* and *Megalosaurus*, Hawkins reverted to his quadrupedal Crystal Palace designs, all restored from scanty fossil material. These paintings embody the evolution of opinions concerning dinosaurian locomotion.

Hawkins' Princeton paintings are strikingly similar in style and content to those created by another artist, Charles R. Knight, 50 years later for the Field Museum of Natural History in Chicago. Their murals cover similar themes and periods of geologic time, but, possibly, with deeply contrasting messages. In 1868, physicist Lord Kelvin, claimed life originated only 100 million years ago. This estimate, founded in geophysical experimentation, seriously restricted the intervals of time which could be envisioned between Hawkins' scenes.[8] However, during the years Knight labored on his Field Museum designs, more accurate radiometric dating measurements revealed that the earth's age was considerably older than Kelvin formerly believed. Whereas Hawkins rejected Darwinian evolution, the essence of geological and biological evolution occurring through considerably longer ages was captured in Knight's murals.

By the 1920s, Knight had become the prominent authority in the art of scientific restoration. In emphasizing North America's primeval frontier, his subjects were often cast in an eerie, supernal light, or revealed in lush, misty settings. Through his artistry, Knight placed the Mesozoic overlords firmly and popularly within the broader context of geological history, thereby contributing to the dinosaurs' early popularity and commercial success. In particular, the Field Museum's series of 28 murals, including five popularly regarded dinosaur scenes, have left deep impressions on several generations of paleontology aficionados. In his most famous mural, a blood thirsty *T. rex* squares off against the mighty *Triceratops*, illustrating contemporary paleontologists' views on the cruel struggle for existence in the upper Cretaceous. (See Figure 7-2.) Knight fortified his views on the evolution of life in a book further popularizing the "life through time" theme, *Before the Dawn of History* (1935), and an article as well, *Parade of Life Through the Ages*, published in the February 1942 issue of *The National Geographic Magazine*.

The world's first major indoor display of animatronic prehistoric animals, Messmore and Damon's *World a Million Years Ago*, debuted at Chicago's Century of Progress Exposition in 1933. While a number of life-sized statues of prehistoric animals had been built during the 80 year period since Hawkins and Owen conspired at Sydenham, George Harold Messmore and Joseph Damon offered the first life-sized life through time assemblage ever to appear in America. Naturally, dinosaurs, "*Brontosaurus*," *Tyrannosaurus*, *Stegosaurus* and *Triceratops* included, were a featured part of the program. Visitors entered a globe shaped rotunda and ascended on a conveyor belt walkway through major periods of earth history, along the way encountering "life-like ancient monsters that move and roar." In 1934 the revamped display was billed as "*Down the Lost River*." While not all attendees may have readily appreciated the fact that fauna featured in *The World a Million Years Ago* rotunda didn't all live contemporaneously, "one million years ago," Messmore and Damon's dramatic images of Permian reptiles, dinosaurs, prehistoric mammals and even cavemen unified our impressions of geological time's "flow."

Perhaps *the* most celebrated panoramic life through time painting is Rudolph F. Zallinger's *Age of Reptiles* mural (110 feet long and 16 feet high) completed for Yale's Peabody Museum between the years 1942 and 1947. In *The Age of Reptiles*, time "flows" from the Devonian to the upper Cretaceous, tracing evolutionary history in a horizontal sense.

Zallinger's painting dispels misconceptions that all dinosaurs co-habitated *in time*. However, the mural conveys an impression that the restored fauna inhabited the *same* continental region. Actually, fossils used as basis for

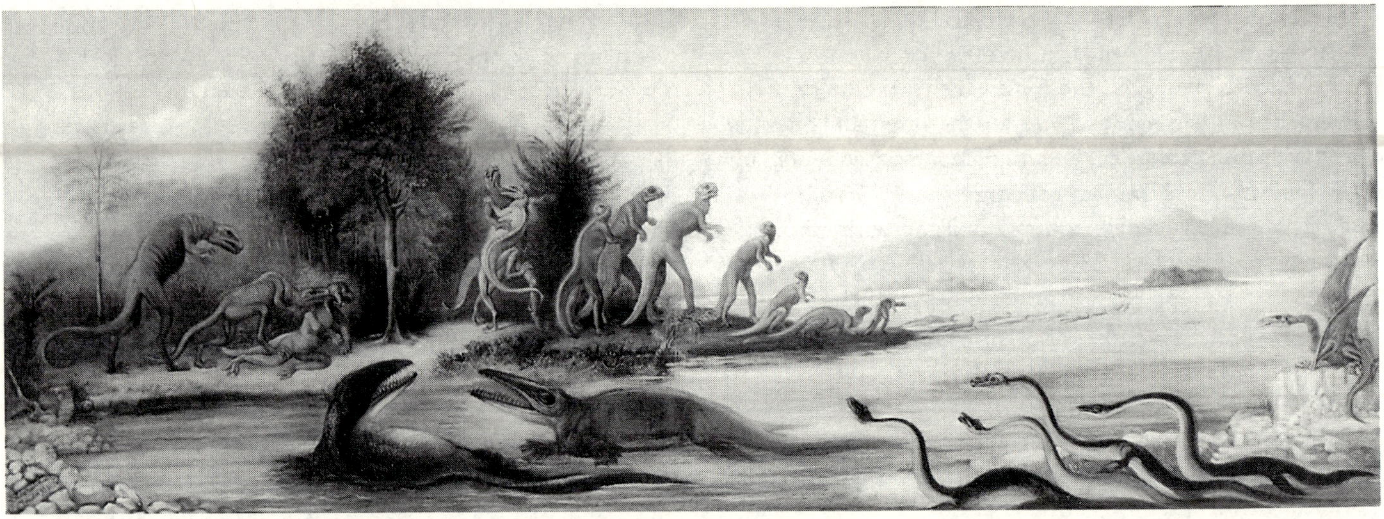

Figure 10-2: Benjamin Waterhouse Hawkins' 1877 restoration for Princeton Univesity showing the *Cretaceous Life in New Jersey*. On land, carnivorous *Laelaps* (left) attacks *Hadrosaurus*, while a herd of hadrosaurs swims to safety. Also, two *Mosasaurus maximus*, four long-necked elasmosaurs and a gratuitous pterodactyl view the carnage. According to Donald Baird, this scene was "based on the life-sized habitat for the Palaeozoic Museum—1869." (Copyright 1980—Trustees of Princeton University. Used with permission.)

restoration have come from a broad range of localities. However, the most popular "subscenes" are of North American Jurassic and Cretaceous dinosaur fauna. Prominent figures include a swamp-dwelling brontosaur and a *T. rex*, ruling imperiously in flowering splendor amidst attentive herbivores. Zallinger's brushstrokes carry us through over 300 million years of time, with strategically planted trees separating the three Mesozoic periods. Erupting volcanoes sound a death knell for the Age of Reptiles.

By the 1950s, dinosaurs could be regarded as veteran "old timers" of the motion picture industry. Yet, in nearly every case, fantasy dinosaurs were cast as vengeful Hollywood monsters hellbent on destruction or included for their horrific value when menacing puny humans. That is until Czech director Karel Zeman created a wonderful film, *Ceste do Proveku* (1954), that includes dinosaurs among a host of prehistoric animals. The prehistoric menagerie were animated as "puppets" using stop-motion animation technology. Reproduction rights were gained for an American release when the movie was retitled *Journey to the Beginning of Time* (1960). Zeman's tale sends four boys down a river of time, from modernity, past the Ice Age and to the primordial sea at the "beginning of time." Along the way, the boys encounter many expertly recreated prehistoric animals, some of which are thought to have been influenced by the paintings of master Czech paleoartist Zdenek Burian. As expected, scenes of dinosaurs spied along the Mesozoic "leg" of the river proved most captivating to audiences.

Compared to other movies in the fantasy dinosaur film genre, Zeman's dinosaurs in *Journey* seem pastoral, shown simply living their lives as they did in *their* world and time, not as they would in ours. The animated "living" images of dinosaurs appear even more lively than Messmore and Damon's creations. Yet the prehistoric animals may almost be regarded as elaborate props used to underscore the film's principal idea—the flow of geological time along the metaphorical river. Rather than being stricken with fear by the prehistoric monsters lurking about, the boys have a "healthy fascination" for the pageant of prehistoric life unfolding with every bend in the river.[9]

In America and Great Britain, where the film was broken down and played in five minute–long segments, *Journey to the Beginning of Time* became a favored feature item of many children's afternoon programs. During the 1960s, a generation of baby boomers, young impressionable "dinophiles" in the making, avidly awaited the next daily installment of *Journey*. Over the course of several weeks, we would complete the journey down the proverbial river of time with the boys, all the while marveling at the grandeur of geological time and Life's procession through the ages.

***Opposite:* Figure 10-3: Portion of wall mural linking dinosaurs, evolutionary procession and geological time on display at the Exhibit Museum of Natural History of the University of Michigan, Ann Arbor. This mural, *Life Through Geological Time*, was painted by artists Robert M. and Jane Mengel in 1950. (Photograph by Allen A. Debus, 1998. Courtesy of the University of Michigan.)**

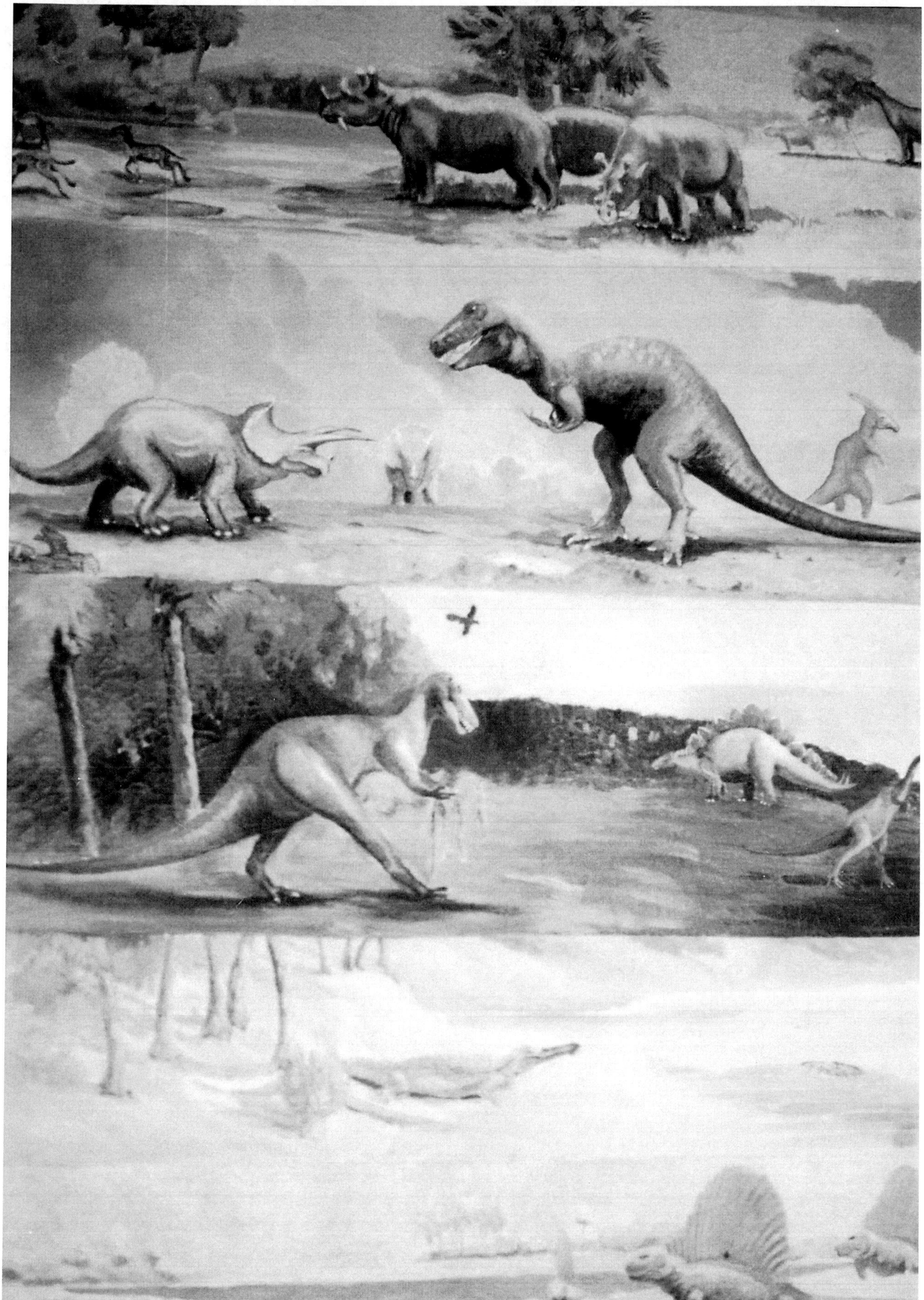

Scenes from deep time, especially "dinosaur art," have long fostered a romantic sense of mystery and awe. Following Hawkins' and Owen's early 1850s successes at Sydenham, dinosaurs became centralized figures in depictions of life through time. During the past quarter century, instead of framing dinosaurs within a broader evolutionary vista, paleontologists and their associated artists have focused on the *biology* and *evolution* of dinosaurs. Today, dinosaur images are rarely placed in the context of Life's broader history. Usually only an "age of dinosaurs" is showcased.

Now, especially due to Robert Bakker's artistic renditions of sleek and fast looking dinosaurs, we have generally become accustomed to the idea of warm-blooded dinosaurs leading vigorous, socially complex lives, fleeing from predators and rearing young while periodically dodging those accursed bombarding comets and asteroids. Nevertheless, through artistic tradition in the depiction of Life's history, one interesting and genuinely taken-for-granted paleoart association is the degree to which the image of dinosaurs can evoke feelings of Time's vastness![10] (See Figure 10-3.)

CHAPTER 11

Plated Puzzle

Few dinosaurs have suffered more indignities than the plated Upper Jurassic dinosaur *Stegosaurus*, pejoratively labeled as "ugly," "stupid" and "cold-blooded" by Man. Some have even irreverently suggested that the main reserve of *Stegosaurus'* brains resided in his rear rather than inside his skull. However, *Stegosaurus* has done his best to snub our inhospitable welcome to Modern Times by confounding our best efforts to accurately restore his remarkable frame.

Although known from two nearly complete and several partial specimens, including two juveniles, no dinosaur has been scientifically described and pictorially restored by artists in such an intriguing variety of ways. By 1947, *Stegosaurus'* curious assortment of bony plates and fearsome defensive tail weaponry had attracted 23 artists to complete life restorations, resulting in 38 noteworthy interpretations.[1]

Some of these restorations vary so considerably that they have often been the subject of special papers or chapters in paleontology books. For instance, paleontologists Charles W. Gilmore, Othenio Abel and Edwin H. Colbert, and paleoartist Stephen Czerkas, devoted special attention to historically "evolving" *Stegosaurus* restorations in their research.[2] Only recently, with the discovery of an especially complete specimen during the summer of 1992, has the mystery of how *Stegosaurus* may have really appeared possibly been resolved.

Because Othniel C. Marsh first restored *Stegosaurus* in an 1891 skeletal reconstruction with a single row of plates, most early life restorations were of the single plated variety. These had been done by artists Arthur Lakes, Carl L. Dahlgren, Joseph Smit and Charles R. Knight. Later, by 1905, only three published life restorations reflected Marsh's 1887 and 1896 *written* descriptions, in which the cervical area was thought to have been protected by a paired, double row of plates. So Marsh's written descriptions apparently contradicted his drawing.[3] It is now believed that the second pair of cervical plates may have been erased from the original 1891 drawing prior to publication, "for reasons unknown." Another possibility is that the illustrated plates were paired, in which case they would look like a single row.

Carl Dahlgren's restoration was published only five months before the first printed edition of Henry Neville Hutchinson's *Extinct Monsters* (September, 1892).[4] (See Figures 11-1 and 11-2.) Hutchinson included a *Stegosaurus* restoration by wildlife artist Joseph Smit. When the second edition of *Extinct Monsters* became available in 1893, Smit had revised his *Stegosaurus* restoration quite extensively. It is apparent that, based on "some remarks of Mr. Lydekker," both the front and hind limbs of Smit's revision have a more crocodilian, sprawling stance, and the four pairs of tail spikes are more needle-like in appearance.[5] Changes were also made to the plate geometry. Smit's resulting, often reproduced 1893 restoration retained the single row of plates but exhibited a "sprawling alligator-like stance, sadly out of keeping with the anatomical evidence of the skeleton," as stated by Colbert 70 years later.[6]

In 1897 Charles R. Knight was commissioned to produce a stegosaur restoration for *McClure's* magazine.[7] Knight depicted a stegosaur possessing a single row of plates. Marsh's great scientific rival, Edward Drinker Cope, had also described stegosaur material and even named one genus, the *Hypsirophus*, later attributed to *Stegosaurus*. Then, complicating matters further, journalist William H. Ballou suggested in 1897 that perhaps Marsh's species, *S. ungulatus*, had a single row of plates, whereas a species attributed to Cope, *S. latus*, had the double row of plates.[8] Charles R. Knight's 1897 restoration of single-rowed *Stegosaurus ungulatus*, similar to his *McClure's* design

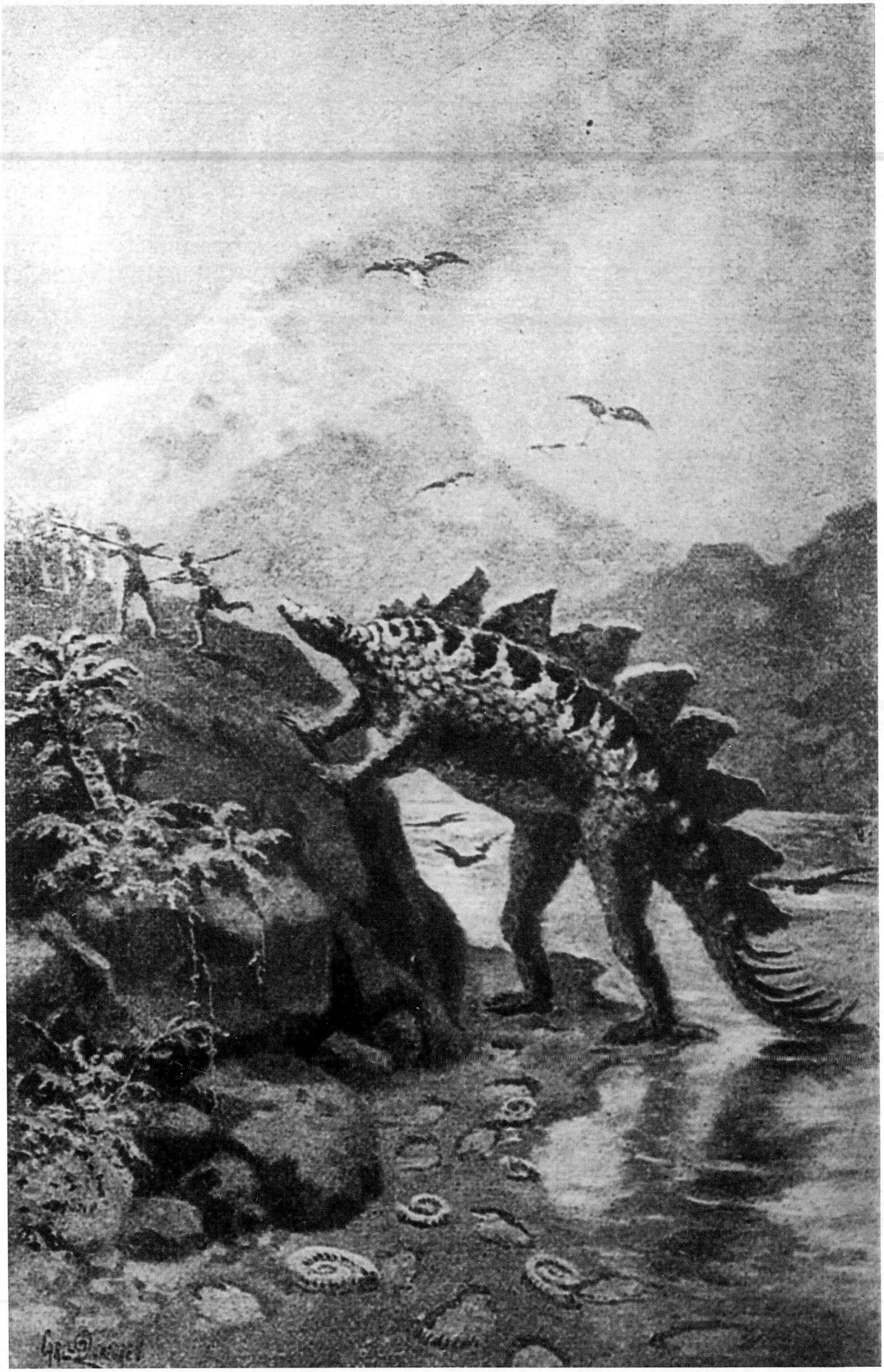

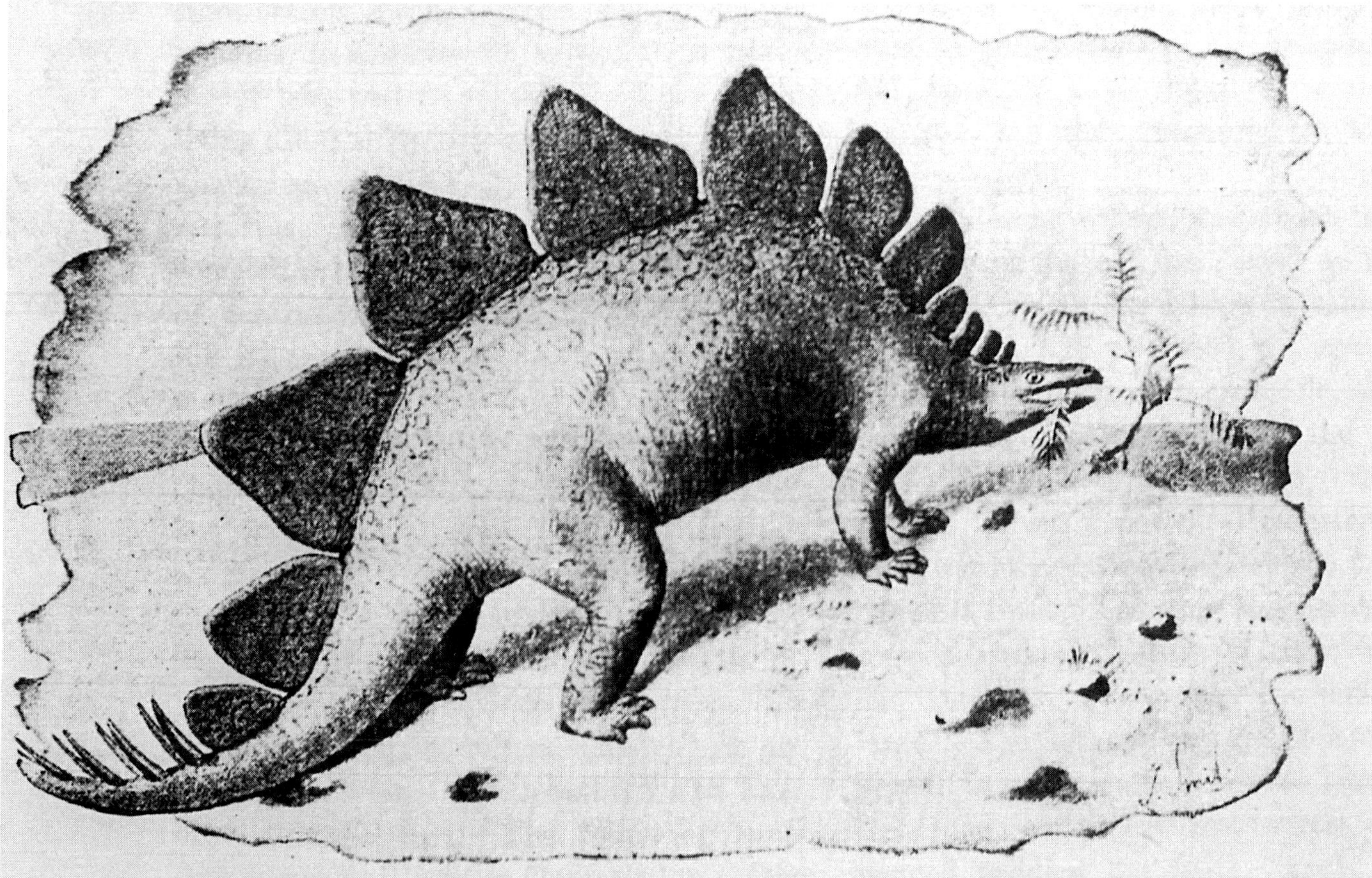

A GIGANTIC ARMORED DINOSAUR. LENGTH, ABOUT THIRTY FEET.

Figure 11-2: Joseph Smit's 1892 *Stegosaurus* restoration. (Courtesy of Allen G. Debus.)

accompanied Ballou's *Century Magazine* article "Strange Creatures of the Past." (See Figure 11-3.)

Knight's *Century Magazine* design was copied a few years later, in reversed pose, to illustrate E. Ray Lankester's *Extinct Animals* (1905).[10] (See Figure 11-4.) This was roasted by paleontologist, Edwin H. Colbert in 1961. "Why a man of [Lankester's] ... insight should approve of such a weird effort is hard to understand. The bird-like beak, the swollen body, the five pairs of spines on the tail have nothing to commend them."[11] By the time of Lankester's publication, it is true that Knight had completed two other much improved restorations that were considerably more accurate than his attempts of 1897 that inspired Lankester's similar version.

Sir Arthur Conan Doyle referred to "Lankester's" stegosaur restoration in *The Lost World* (1912), affirming Knight's influential artistry, even then. "'This is an excellent monograph by my gifted friend, Ray Lankester!' Said [Professor Challenger] 'There is an illustration here which would interest you. Ah ... here it is.... Probable appearance in life of the Jurassic Dinosaur Stegosaurus. The hind leg alone is twice as tall as a full-grown man. Well, what do you make of that?'"[12] Early restorations of *Stegosaurus* also inspired author (and inventor of Tarzan) Edgar Rice Burroughs to include a passage describing the dramatic charge of an oversized 40-foot-long stegosaur in his novel *Out of Time's Abyss* (1918). Burroughs' description seems remarkably familiar to a beloved stop-motion animated scene of another stegosaur death charge, created by Knight admirers Marcel Delgado and Willis O'Brien, fifteen years later for RKO's *King Kong* (1933).[13]

In 1899 another restoration was completed under advisement of another Knight, geologist W. C. Knight.[14] This represents the most serious departure from conventional thinking. Historically, paleontologists have cited this mutant variety of *Stegosaurus*, with its turtle-like carapace of

***Opposite:* Figure 11-1: Carl Dahlgren's restoration of "*Hypsirophus*," now *Stegosaurus*. From the *California Illustrated Magazine*, 1892. Spear-wielding savages reflect the popular view of American prehistory's "savage" nature. In early restorations, stegosaurs were often depicted with 8 to 10 tail spikes.**

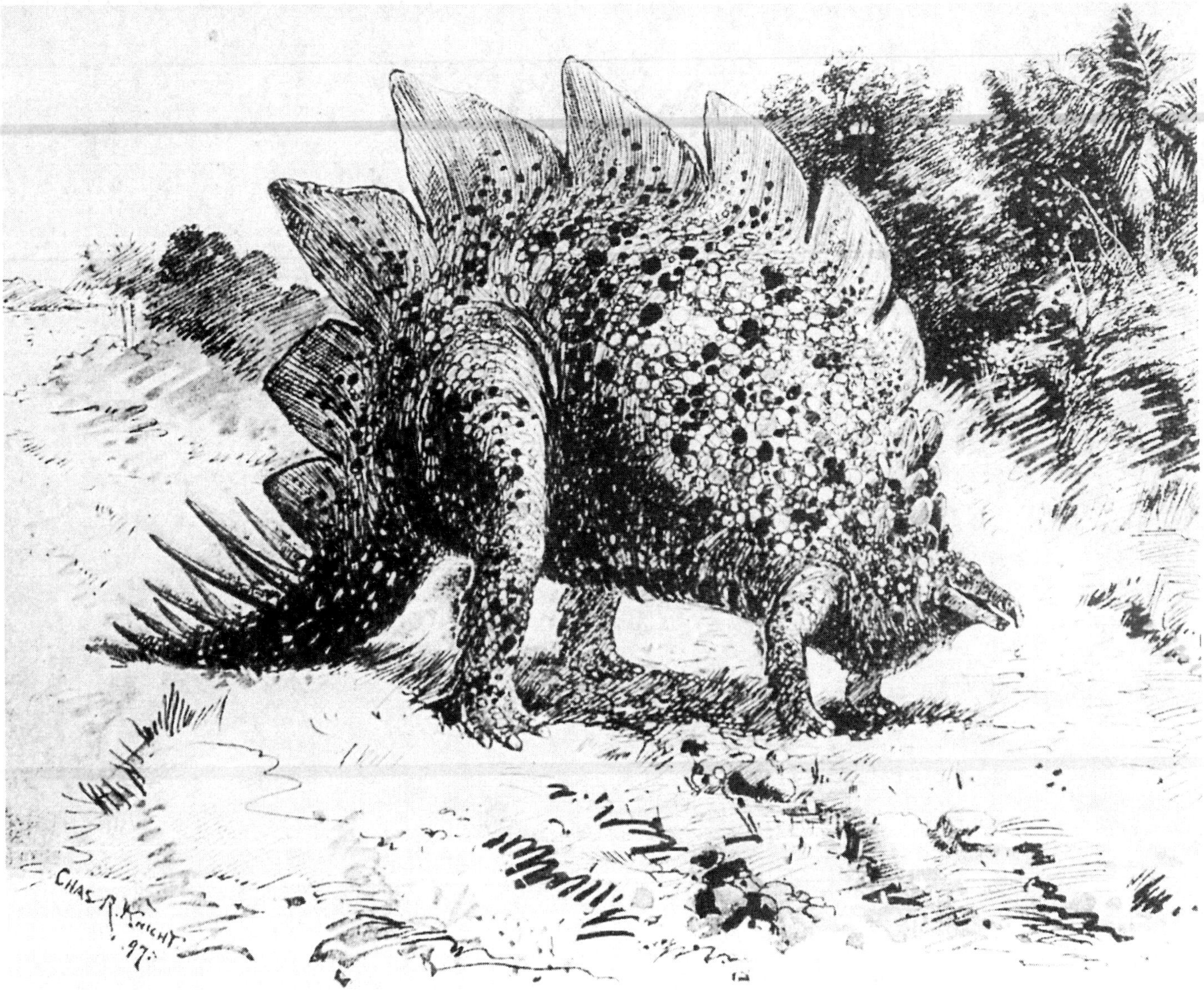

Figure 11-3: Charles Knight's 1897 *Stegosaurus* restoration. (From *Century Magazine*—authors' collection.)

plates and armadillo-like arrangement of spikes, created by artist Frank Bond (under W. C. Knight's direction) as the most fanciful restoration ever made of this dinosaur. Artistically, it was a nice try, but with no cigar awarded for anatomical accuracy! In fact, Marsh himself had toyed with the possibilities of a recumbent plate geometry,[15] an aquatic habitat and tripodal posture.

By 1901, further study of a fossilized specimen (discovered in 1885–86 and retained in the Smithsonian's collection) in which the proper plate orientation was thought to have been preserved opened the possibility that *Stegosaurus* had *two* rows of plated armor. But were these arranged in an alternating or paired configuration? A century long debate transpired, with consensus opinion generally favoring the alternating arrangement after the late 1920s.

The first restoration ever to unequivocally represent the paired, double-row hypothesis was Charles R. Knight's 1901 restoration done under the direction of paleontologist Frederic A. Lucas.[16] But Lucas also considered the "alternate" possibility that *Stegosaurus'* double row of plates alternated, and in the same year commissioned another artist, G. E. Roberts, to design the first restoration illustrating this view.[17] In 1903 Lucas also served as Knight's advisor for a miniature sculpture furnished with a staggered double row of plates. Knight's sculpture was the inspiration for the first life-sized sculptural restoration of a *Stegosaurus*, first seen by the public at the St. Louis World's Fair Exposition in 1904.[18]

Joseph Smit completed yet a third *Stegosaurus* restoration, published in Henry R. Knipe's epic poem of galactic evolutionary history, *Nebula to Man* (1905). In arranging the plates into two paired rows, Smit was evidently

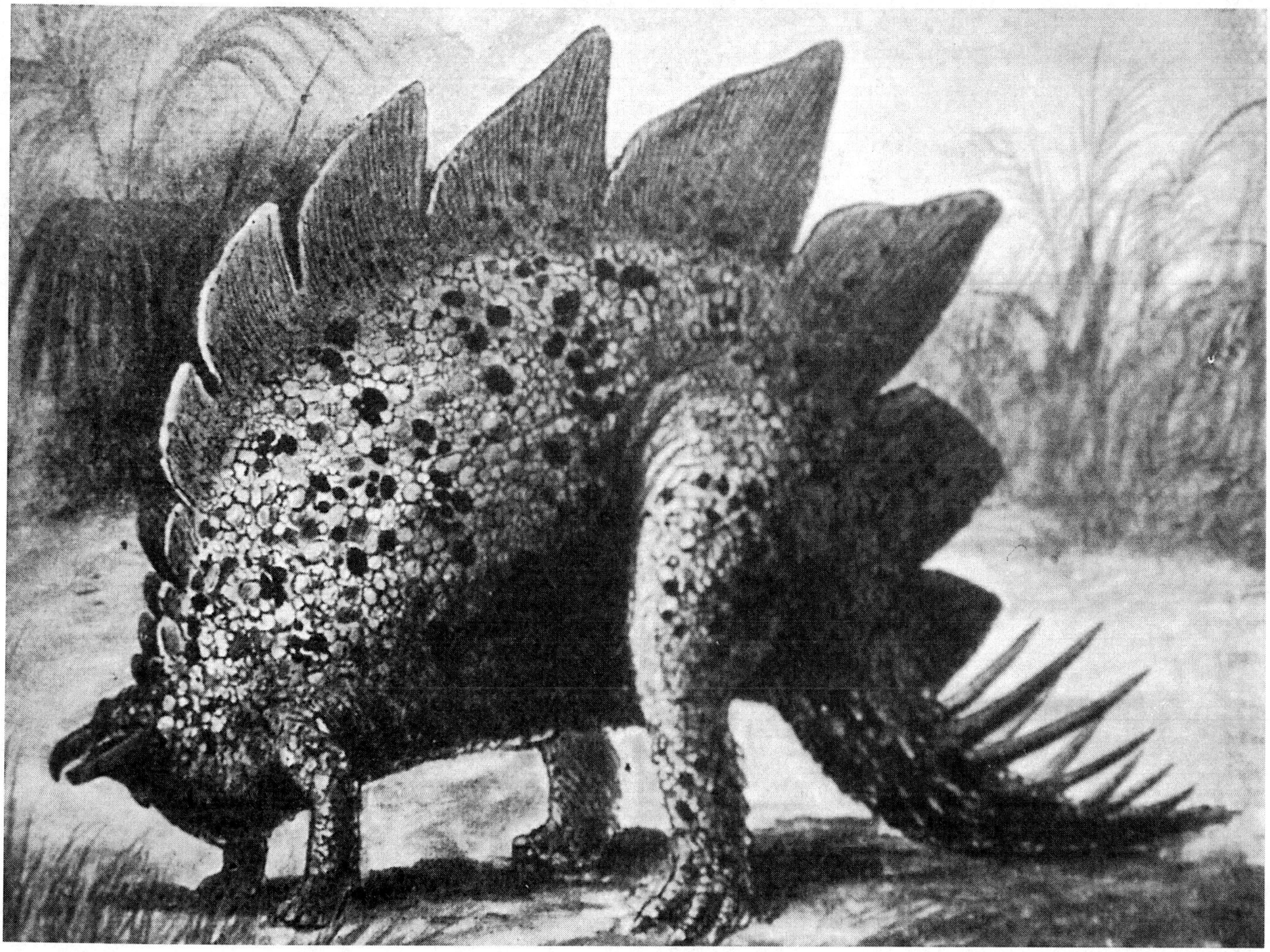

Figure 11-4: E. Ray Lankester's 1905 *Stegosaurus* restoration. (Courtesy of Allen G. Debus.)

inspired by Knight's 1901 restoration. However, Smit's dinosaur lacks any plates along the greater part of the tail.

Incidentally, Knipe's book is a genuine curiosity. It is an epic poem tracing from the origin of the Milky Way Galaxy, through millions of years of evolution, to the dawn of Man. Many dinosaur enthusiasts may be familiar with the famous poem, titled *The Dinosaur*, concerning the supposed two brains *Stegosaurus* possessed. This poem, written by B. L. Taylor, was published in a 1912 issue of the *Chicago Tribune*. Few may realize, however, that Knipe's epic was published seven years before Taylor's. Thus, Knipe may be credited with the first piece (20 lines) of published stegosaur poetry.[19] Alice B. Woodward, whose restorations were criticized by Hutchinson for their baggy looking skin, completed an uninfluential, double-rowed *Stegosaurus* restoration for another Knipe tome, titled *Evolution in the Past* (1912).[20]

The image of Knight's 1901 restoration popularly persisted through the 1920s. Both Hubert V. Zwickle, under advisement of paleobiologist Othenio Abel (1908), and Danish ornithologist Gerhard Heilmann (1916) restored *Stegosaurus* with a paired double row of plates.[21] Restorations of *Stegosaurus* for museum displays (e.g. Richard Swann Lull, 1910, and Josef Pallenberg, 1910) and major movie productions (*King Kong*, 1933, and *The Lost World*, 1925) also evidently followed suit.[22]

Visitors to the Chicago World's Fair (1933–34) would have seen a *Stegosaurus* with a staggered double row of plates at Sinclair Refining Company's dinosaur display. Only a few doors down they would have witnessed Messmore and Damon's animatronic *Stegosaurus*, bearing a staggered double row of plates.[23] Evidently, during the decade after 1925, popular views on *Stegosaurus*' appearance were shifting rapidly.

By 1915, Charles Gilmore challenged the paired double row perception popularized by Lucas and Knight with an original sculpture favoring the staggered, or alternating, plate geometry.[24] Both Charles R. Knight and Othenio Abel furnished restorations reflecting this alternate opinion. Abel's *Stegosaurus* restoration, in which body scutes have been added, was completed in 1920.[25]

The paired double row form finally met its arch rival in the alternating double row version completed by Charles R. Knight in 1927. (See Figure 7-1.) This new restoration was included in his famous series of 28 murals for the Field Museum of Natural History in Chicago, tracing the evolutionary history of the earth . Twenty years later, when artist Rudolph F. Zallinger finished his famous 110-foot-long and 16 foot high mural *The Age of Reptiles* for Yale University's Peabody Museum (which included an impressive *Stegosaurus* with plates in the by then conventional staggered, double row pattern), it was clear that the alternating double-rowed view had prevailed!

For over 40 years the staggered, alternating view represented conventional understanding. That century old Smithsonian specimen, thought to have preserved the life orientation of plates, had certainly inspired quite a bundle of varying interpretations!

During the late 1980s and 1990s Stephen Czerkas, brimming "...with confidence and caution...," introduced three striking restorations of stegosaurs (one life-sized and two miniature sculptures).[26] Departing from both Lull's paired double row and Gilmore's staggered double row restorations, Czerkas reverted to Marsh's "single row," century-old hypothesis of 1896. Czerkas had taken what may be regarded as a "middle ground" in his reinterpretation of the evidence. (See Figure 11-5.)

"...I realized ... in 1986 ... that the 17 plates ... were all there probably were, and ... all the plates fit perfectly along the length of the vertebrae in a single row. The only plates that really alternated to the left or right were those over the neck and back which had grown longer than their base. Those of the tail did not alternate so much, but were nearly vertical following one right after the other. I believed that there was an alternating pattern, although it did not represent two rows, but one."[27]

At least two stegosaur skeletons mounted in North American (Utah) museums sport the single row arrangement.[28] (See Figure 11-6.) Nevertheless, Czerkas' effort to

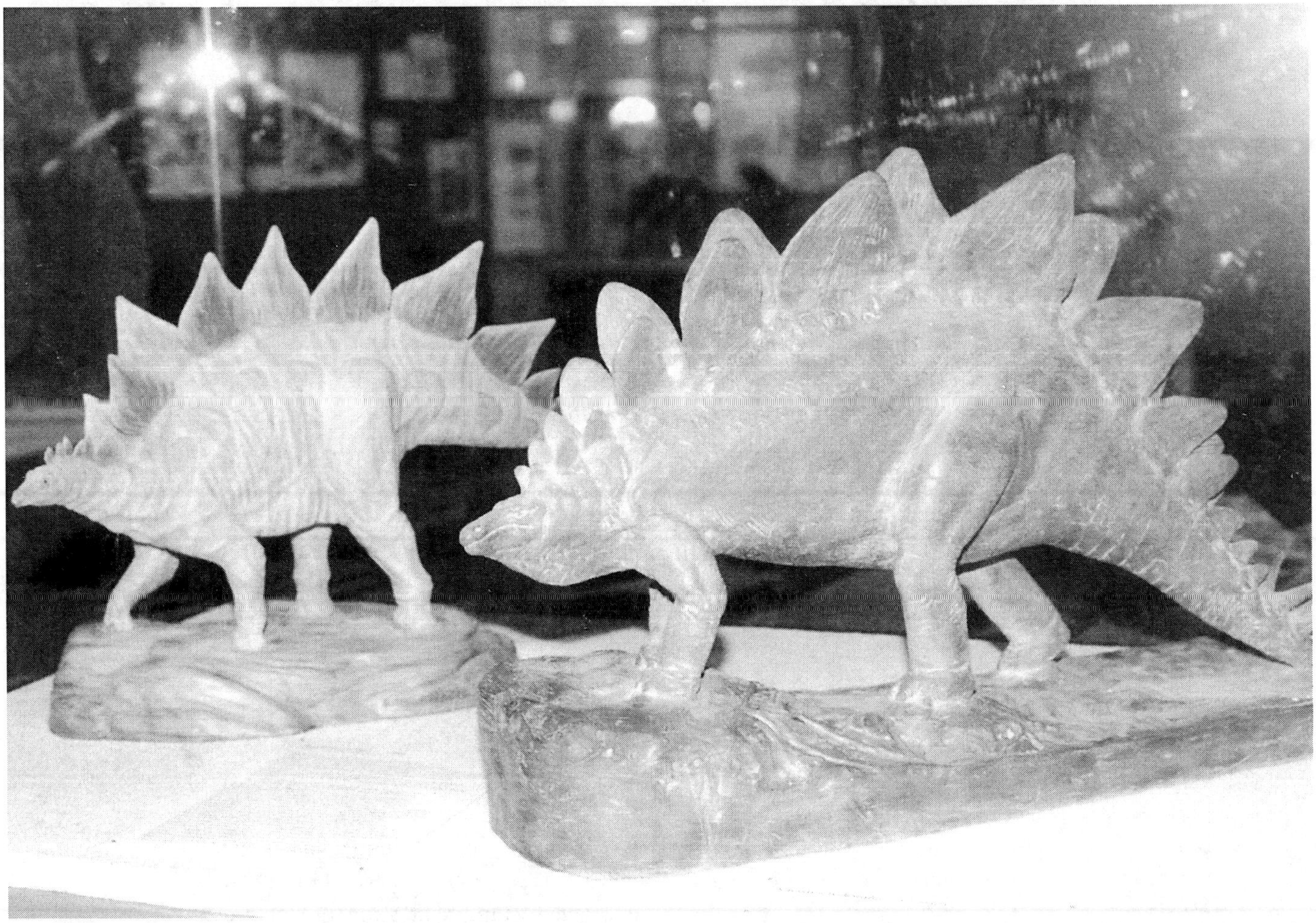

Figure 11-5: At left, Stephen Czerkas' *Stegosaurus* restoration, adjacent to Charles Knight's 1899 American Museum model. (Photograph by Allen A. Debus, 1990—courtesy of Field Museum of Natural History.)

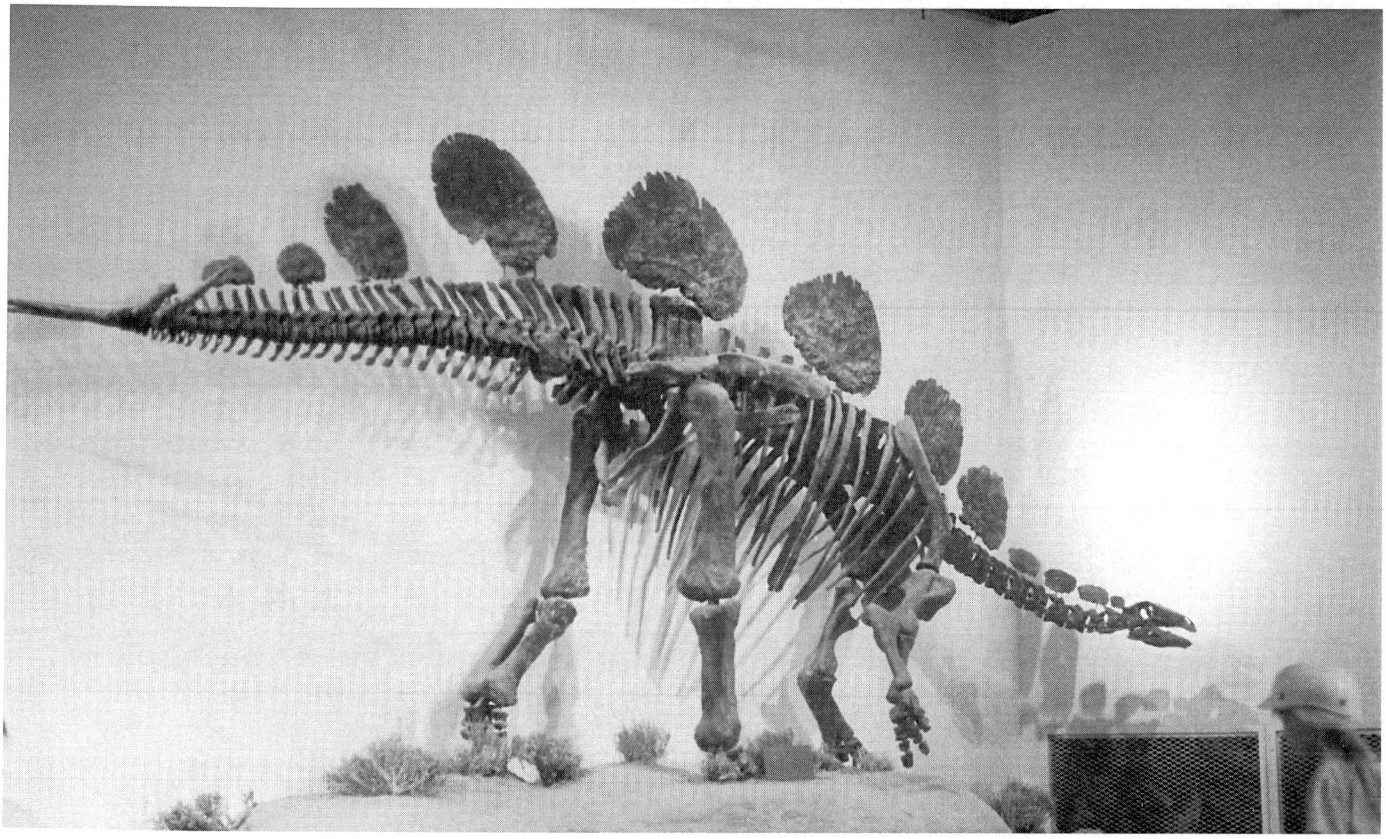

Figure 11-6: Photograph of a stegosaur skeleton sporting a single row of plates mounted in Museum of Prehistoric Life, Utah. (Photograph by Allen A. Debus, 2000, courtesy of Museum of Prehistoric Life.)

reintroduce the single row hypothesis in the case of *S. stenops*, based on a reinterpretation of the old specimen, has proven unpersuasive to a majority of dinosaur paleontologists.

Initial reaction to Bryan Small's 1992 discovery of a nearly complete *Stegosaurus stenops* from Canon City, Colorado, was that Gilmore was correct after all. Plates in the new specimen were preserved in life position, not in a single row but alternating from one side to the other. Furthermore, another stegosaur, found 14 years earlier, in 1978, by Donna McKowen, revealed two dermal plates preserved in a staggered, alternating orientation.[29] These discoveries prompted Kenneth Carpenter to conclude that *Stegosaurus* had armor plates arranged in two closely set alternating rows.

"...My return to the 'traditional view' of plate arrangement is based upon the discovery of two very important specimens of Stegosaurus. Both specimens have rock separating the plates of one side from those of the other. If the plates had been in a single alternating row, then they should have lain against each other. Even taking into account the compaction of the original sediments into rock, there is no getting away from the fact that the plate rows had to have been separated to allow the sediment to get between them."[30]

One may ask, however, that if *Stegosaurus* did have a double *alternating* row of plates, then why are its evolutionary cousins, the Asian *Tuojiangosaurus* or *Huayangosaurus*, and the African genus *Kentrosaurus*, reconstructed possessing a double row of *paired* plates? Carpenter is currently studying a new stegosaurid genus, found in 1985 near Buffalo, Wyoming, dating from the Upper Jurassic–Morrison formation. Tentatively, he believes it is more closely related to the European Jurassic genus *Dacentrurus* than either species of North American *Stegosaurus*. His studies may someday offer a solution to the evolutionary plate patterns evident in stegosaurids.[31]

The thriving "evolutionary" history of *Stegosaurus* restorations continues to be "revolutionary." A speculative restoration by John Sibbick illustrated how muscular attachments embedded in the skin-covered plates may have been capable of "wagging" the plates to some degree.[32] An incongruous restoration by John C. McGloughlin shows a *Stegosaurus*, with staggered alternating rows of plates, curled up submissively with defensive armor dangling uselessly above the ground surface, while a nasty predator comes in for what must be a relatively easy kill.[33] Through his imaginative restorations during the mid–1980s, Robert Bakker introduced the idea of stegosaurs rising into tripodal position with front feet raised off the ground.[34]

Yet there are other possible functions for those mysterious plates besides armored protection. As James Farlow has asserted, stegosaurid plates may have also served thermoregulatory purposes, much like the sail of the Permian pelycosaur *Dimetrodon*.[35] Ken Carpenter speculates that the plates, which in life were covered in thick skin, could have flushed deep colors that would have proved exciting to prospective partners during mating rituals. Noting that two North American *Stegosaurus* species (*S. armatus* and *S. stenops*) had differently sized plates (*S. armatus*' were smaller), Carpenter claims that having, respectively, more smaller sized plates versus fewer plates of larger dimensions aided stegosaur individuals in distinguishing members of the opposite sex *and* correct species. Also, large plates would appear threatening to unwary predators and "sexy" to wooed females.[36]

Paleoartist Luis Rey has seized the opportunity to depict two stegosaurs mating in "porcupine-like" fashion, the natural aftermath of becoming seriously turned on by all that flashing plate coloration.[37] Copulating stegosaurids—ouch! Robert Bakker has humorously noted that the spikes of certain genera, (i.e. *Kentrosaurus*) would have made "...sexual intercourse even more dangerous and fraught with problems."[38] Others have referred to stegosaurid sex as downright "kinky."[39]

The life positioning of *Stegosaurus*' tail spikes raises thorny questions. Bryan Small's *Stegosaurus stenops* "road kill" specimen indicates a sideways spike orientation, the position originally suggested by Heilmann in a 1920s restoration.[40] An incomplete specimen discovered in the 1980s, apparently possessing a fifth tail spike, tentatively suggests that some *Stegosaurus* individuals may have had more than the "traditional" two pairs of tail spikes.[41]

Credit goes to paleoartist Tracy Ford for clarifying how the tail spikes of *Stegosaurus* were really oriented in life. Those spikes had been traditionally mounted in a vertical "V" shape. Ford, however, noticed "...a large gap ... between the base of the tail spikes and caudal vertebrae.... Either there was cartilage or perhaps "fat" between the base of the spikes or the spikes were mounted wrong."[42] Ford's suspicions were honored by Carpenter, and now it is accepted that tail spikes were positioned jutting outward horizontally from the tail. (See Figure 11-7.) The first series of "corrected" tail spikes appeared in an article by George Olshevsky titled *The Origin and Evolution of the Stegosaurs* (1993).[43] Ford accurately reconstructed an entire evolutionary "lineage" of stegosaurids.

Ever since the 1980s it has become fashionable for paleoartists to restore *Stegosaurus* with its spiky tail extremity suspended in the air, suggesting "fitness" and a vigorous metabolism, as opposed to being lethargically dragged behind on the ground. Tracy's intuition has fortified that view and also led to an intriguing conclusion about how stegosaurs carried their tails aloft.

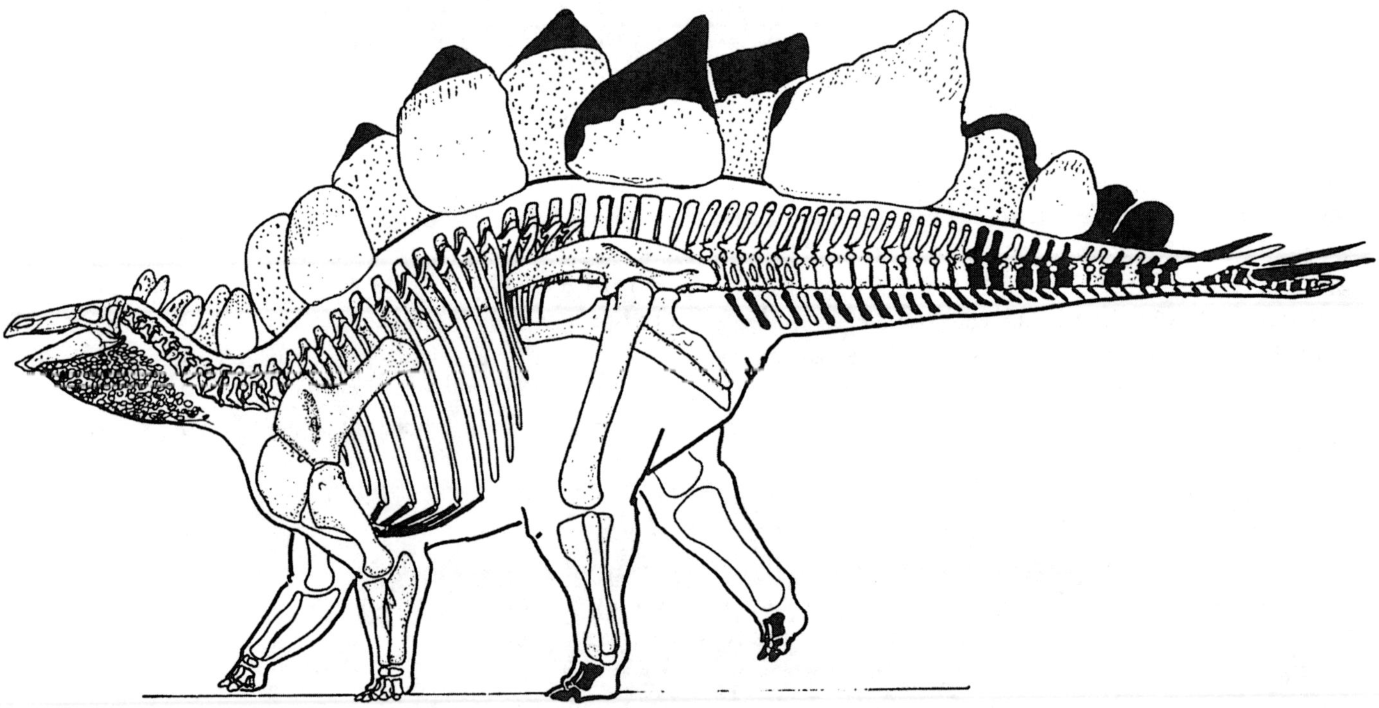

Figure 11-7: Tracy Ford's skeletal reconstruction of *Stegosaurus*, the most accurate to date. This drawing shows the correct number of tail spikes—four. (Reproduced with Tracy Ford's permission.)

In many of the old restorations, stop-motion animated and cartoon sequences, and sculptures, stegosaur tails appear "snaky," capable of whipping those spikes around into an attacker's belly. However, using illustrations, Ford has demonstrated that stegosaur tails weren't that flexible. In fact, plates on the tail would have restricted horizontal motion. A *Stegosaurus* tail would have been held rigidly by overlapping tail plates. Tracy claims, "...A tail without the plates would have been very mobile, but the plates would lock the tail in place.... It couldn't move horizontally or vertically, since it would displace the plates out of the skin.... The stegosaur genera with smaller plates would have a much more mobile use of the tail and be able to defend itself in a manner the public is more used to seeing [in] the media."[44]

Surprisingly, the evidence of a juvenile *Stegosaurus* has failed to inspire many restorations of this genus. Although differing in species (i.e. *S. stenops* and *S. ungulatus*), a composite mount was made from the combined remains. A cast of this composite juvenile may be seen at Denver's Museum of Science and Nature.

Artists much prefer to depict *Stegosaurus* battling theropod dinosaurs, notably traditional foes *Ceratosaurus, Allosaurus* or even (erroneously) *Tyrannosaurus*.[45] How many of you recall the image of Spielberg's cuddly *Stegosaurus* baby in one of *The Lost World*'s (1996) pivotal scenes, restored through the wizardry of computer graphic animation? However, Spielberg's restoration of a stegosaur "family" wasn't the first. For, in 1942 American Museum of Natural History artist George Geselschap painted the first grouping of stegosaurs, including what appears to be a *Stegosaurus* juvenile standing in the foreground. (See Figures 11-8 and 31-8.) In the early 1990s Bob Morales sculpted a miniature diorama combining the maternal and combat themes, featuring a stegosaur parent defending its offspring from a hungry *Ceratosaurus*, the latter evidently inspired by a 1986 illustration by Robert Bakker.[46]

With respect to limb posture, was *Stegosaurus* bowlegged to some degree, or did it stand upright on all fours or even on its hind limbs to forage? Emily Giffin has devised new techniques for researching dinosaurian posture. Her work on the spinal column of *Stegosaurus* led her to conclude that, "...the animal put its weight between its forelimbs, giving it a somewhat bowlegged, or less dignified, posture."[47] This ungainly image counters recent impressions of *Stegosaurus* standing proudly upright like an elephant, browsing off the high branches of trees.

Stephen Czerkas' most recent stegosaurid faces are "classical," having turtle-like, beaked jaws lacking cheeks.[48] Conversely, stegosaurs shown in recent animation, such as the Jurassic Park movie sequel *The Lost World* (1996) and *Walking with Dinosaurs* (1999), all have cow-like cheeks. Far

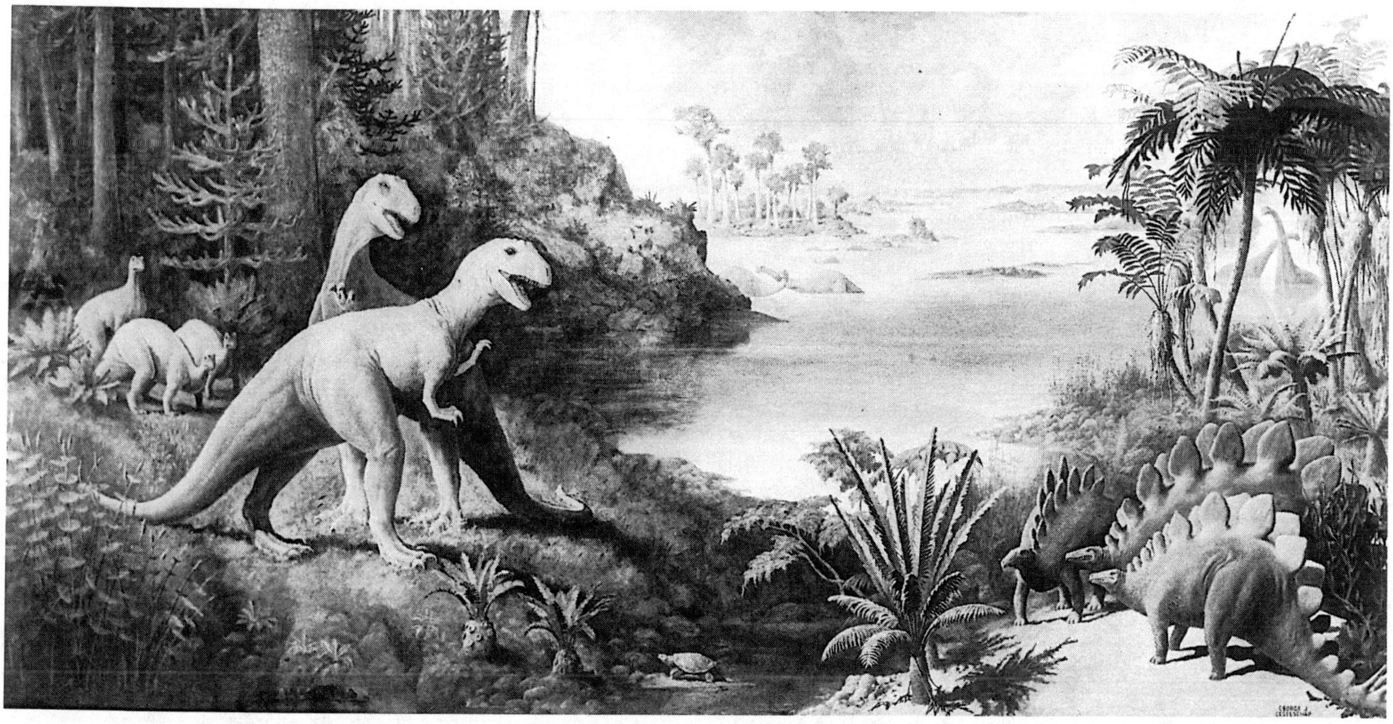

Figure 11-8: A scene in Wyoming during Upper Jurassic times—painting by George J. Geselschap (1942). Dinosaurs (left to right): *Camptosaurus, Allosaurus, "Brontosaurus," Stegosaurus*. (Reproduced with permission of the American Museum of Natural History—Department of Library Services. Negative #322674-A.)

less debatable is evidence of a relatively complete specimen excavated in 1997 revealing that *Stegosaurus* had bony armor scutes protecting its throat region.

Stegosaur restorations evolved considerably during the 1990s. Would stegosaur restorations of a century from today seem as "alien" to us as Joseph Smit's and Frank Bond's century-old restorations now appear? Possibly within the next two centuries, if ever their DNA can be recovered from an amber specimen or a hunk of bone, living stegosaurs can be reincarnated. Until then, however, unless we attempt to "exorcise" his avenging spirit by recanting some of the criticism and ridicule undeservedly directed toward *Stegosaurus*, many of his secrets may remain concealed forever.

CHAPTER 12

America's Paleoart Reformation

When in the history of fossil vertebrate restoration did it become fashionable for scientists and their associated artists, or scientists who were themselves artists, to confidently wield restorations as scientific, interpretative "tools"? Many would be surprised to learn that the modern origins of this form of "paleoimagery," as used to confidently project scientific ideas, projected from America's Western frontier.[1] During the early twentieth century paleontologists increasingly tackled *biological* questions using fossils. While nineteenth century paleontologists focused on classification and description of fossils, "new wave" twentieth century paleontologists increasingly regarded fossil species as formerly *living* organisms. In America, talented artists collaborated closely with museum scientists in resurrecting the living appearances of extinct fauna and flora from the vault of time.

For decades there was no recognized field for creating "prehistoric imagery." By the end of the nineteenth century, with two notable exceptions (the paleoimagery of mass extinctions phenomena and the paleoimagery of biological reproduction—e.g. "dinosaur sex"), practitioners had invented all of "paleoart's" major themes.[2] However, it wasn't until about 1910 that use of paleoart restorations and reconstructions became recognized by a consensus of paleontologists as a suitable means for expressing scientific ideas.

A century before the current "Dinosaur Renaissance," visionary American scientists and artists monumentally shifted our comprehension of prehistoric animals (especially dinosaurs) during an earlier "Reformation" period by allowing us to "see" them "accurately" conveyed for the first time as artistic interpretations.[3] In consequence to the under appreciated or neglected efforts of Edward D. Cope, Henry F. Osborn, Frederic A. Lucas (1852–1929), William Berryman Scott (1858–1947), Richard Swann Lull (1867–1957), William J. Holland (1848–1932), and Samuel Wendell Williston (1851–1918), the practice of restoring fossil vertebrates finally became "respectable," even though many of these early restorations were later shown to be inaccurate.

Why did it happen, and what does this illustrative transformation signify? Let us survey benchmarks leading to paleontologists' established confidence in the creation and use of restorations as scientific tools for comprehending prehistoric vertebrates.

FIRST GLIMMERINGS

Two centuries ago Cuvier drew the first "life" restorations of fossil vertebrates at a time when savants and scientists frowned upon the union of speculation and hard science (i.e. the latter accomplished through scientific observation). Prehistoric animal restorations were viewed as less than factual, fanciful notions to be avoided by keen observers of natural history. "Flesh" restorations were usually reserved for popular consumption. Prehistoric animal skeletons were rarely reconstructed three dimensionally as museum displays, even in Great Britain where several of the earliest discoveries of "antediluvians" had been made.

An Austrian botanist, Franz Xaver Unger (1800–1870), had commissioned a landscape painter, Josef Kuwasseg (1799–1859), to complete a series of scenes of the Earth's geological past for his 1851 publication *The Primitive World in Its Different Periods of Formation*. At first reluctant to incorporate artistic scenes because they would be too speculative, Unger overcame his hesitancy after seeing Kuwasseg's preliminary results, which were splendid. Then Unger was no longer fearful of "...wandering from the actual domain of science, into the realm of imagination." *Primitive World* made only a minor impact on the paleontological community, however, as it is likely that less than 100 copies were printed altogether.[4] While Unger's expertise in paleobotany authenticated Kuwasseg's illustrations, understandably, because this was a book about fossil flora there were few prehistoric animals incorporated into the landscapes. (See Figure 10-1) Those that were included were illustrative props, often "witnessed" in the distance. Unger's masterpiece offered little to those interested in perfecting techniques for prehistoric vertebrate restoration.

In America, however, art and vertebrate paleontology were at least symbolically wedded from the beginning. For our first prehistoric vertebrate, a composited Mastodon skeleton was excavated and later mounted for public exhibition in 1801 by artist Charles Wilson Peale (1741–1827). Through its noble Mastodon, America helped unveil the "prehistoric world" to Cuvier.[5]

Between 1850 and 1875, paleontological restorations became tools for public edification in the service of several scientists, including the "British Cuvier," Sir Richard Owen. For instance, the famed Crystal Palace "dinosaur" statues first exhibited at Sydenham in 1854 took on an ideal significance, manifesting what can be referred to as the "single bone theory."

Owen and his artist, Benjamin Waterhouse Hawkins, had restored dinosaurs and other genera from relatively few remains. It would certainly have been sensible to reconstruct their skeletons in intermediate stages prior to attempting the full-sized restorations. But erected (incomplete) skeletons housed inside a building would have seemed far less inviting to the public than "revivified" monsters prominently displayed outside on their own antediluvian island. Furthermore, both Great Britain's Prince Albert, who presided over the grand opening, and the glory of the imperial British Empire would have been served to a lesser degree by mere skeletons. So, at Sydenham, Owen cleverly wielded the "laws" of paleontology in a program referred to by Hawkins as "visual education."[6]

Men such as Baron Georges Cuvier had believed that paleontology could be elevated to an "exact" science, given understanding of the laws of anatomy. Cuvier had, in fact, elucidated a "law of the correlation of anatomical parts," predicating that an extinct species could be reconstructed from a single bone![7] Rather dramatically, in the late 1840s Owen had reconstructed the Moa from New Zealand from a single fragment of bone.[8] So, given a few partial skeletons, by the early 1850s it certainly seemed possible to recreate the living form of ancient England's fauna.

Meanwhile, American paleontologists were cutting their teeth on the bones of North American fossil genera. For example, an 1869 *American Naturalist* article by Edward D. Cope incorporated one of the earliest scenes, drawn under Cope's direction, showing prehistoric (Cretaceous) life.[9] The illustration featured restorations of "*Laelaps*" (now *Dryptosaurus aquilunguis*), *Mosasaurus* and the *Elasmosaurus*. Cope's artistic visions, very apparent in this restoration, were used to convey his scientific ideas concerning the fossil reptiles of New Jersey.

Three years later, Archibald M. Willard (1836–1918), renowned for his portrayal of *The Spirit of '76* (1876), created the earliest known "scene from deep time" ever painted by an American artist. This 1872 painting, recently discovered in Chicago by Donald F. Glut, was evidently inspired by the work of Hawkins, Edouard Riou (1833–1900), and British artist John Martin (1789–1854).[10] Yet, although he symbolically spirited the flag forward through his famous 1876 painting, for his 1872 masterpiece Willard relied on European, not American, fossil genera and imagery. (See Figure 12-1.) Nevertheless, in the efforts of Cope and Willard we see the first glimmering of the development of American paleoart, a mere presage of what was to come. (Another rare, early example is John A. Ryder's restoration of the South American glyptodont, *Panochthus*, published in an 1878 issue of *Popular Science Monthly*.[11])

In a letter dated January 12, 1873, Cope produced the earliest restoration of the uintathere "Dawn Emperor" *Eobasileus*, viewed incorrectly as a proboscidean.[12] In the same year, a more interactive scene appeared in Cope's popular article for *The Penn Monthly*. The similarly styled published restoration of a grouping of uintatheres, "*Loxolophodon*" (i.e. *Eobasileus*), was drawn by Edwin Sheppard under Cope's direction.[13] The imaginative Cope thought uintatheres had elephant-like trunks. Othniel Marsh disagreed with Cope's opinions.[14] Although not a polished artist himself, the artistically inclined Cope would later become influential in producing the imagery of the extinct.

Despite Cope's pioneering efforts in America, by 1875, prehistoric animal restorations had become standardized. This is exemplified by (both) the figurative and literal overseas passage of two artists, Edouard Riou and Hawkins.

The popular French artist Riou is best known to paleophiles for his prehistoric landscape scenes originally published in 1860s editions of Guillaume Louis Figuier's (1819–1894) popular book, *The Earth Before the Deluge*.[15] He also did illustrations of prehistoric animals for an 1867 edition of Jules Verne's popular novel *Journey to the Center of the Earth*.[16] Three of his *Earth Before the Deluge* scenes would be reproduced in a book published in the United States, authored by D. Alexander Winchell (1824–1891), Director of the State Geological Survey of Michigan,

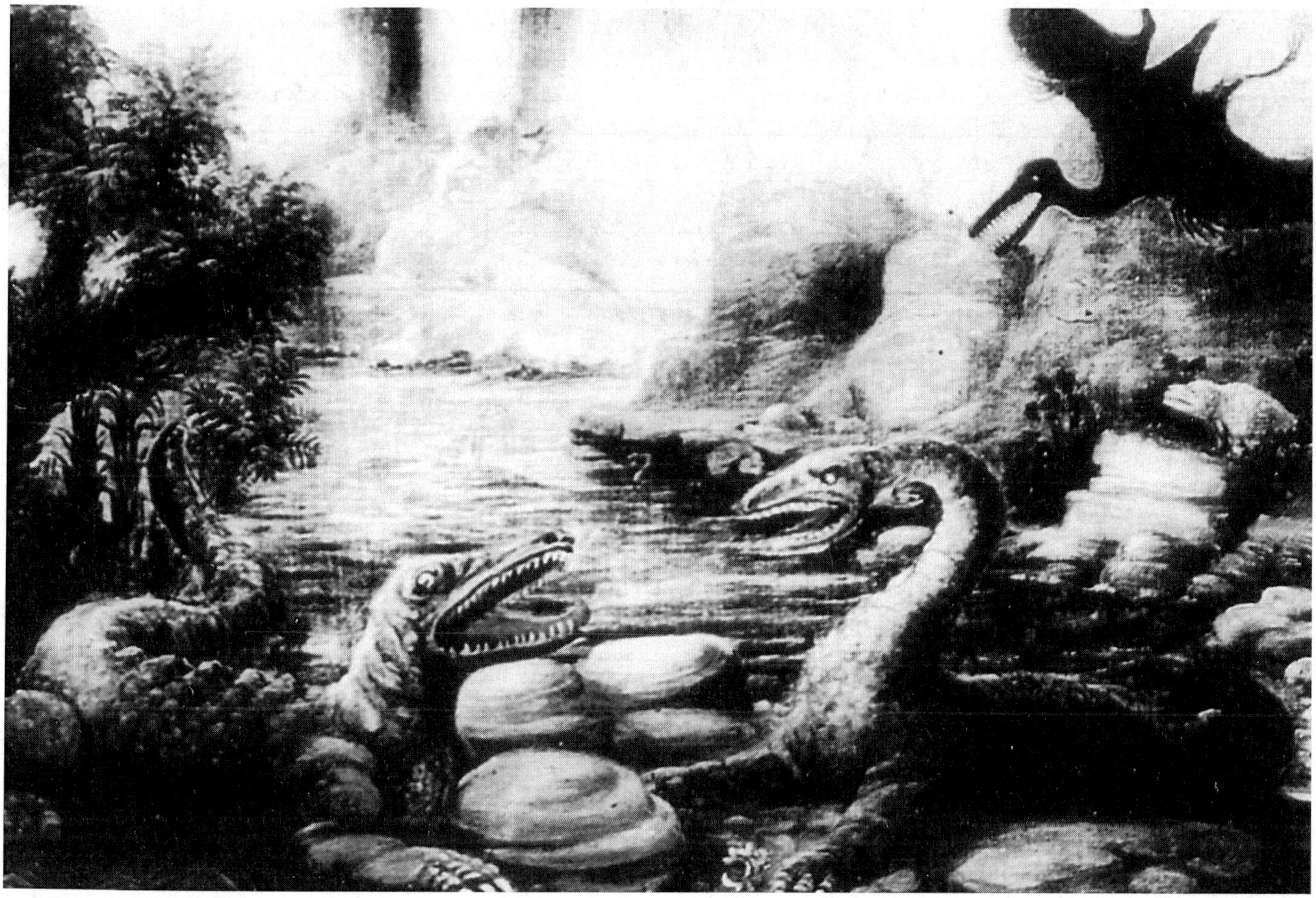

Figure 12-1: Archibald Willard's 1872 painting of prehistory. (Reproduced with permission of Donald F. Glut.)

entitled (in abbreviated form) *Sketches of Creation* (1874), and intended for students as well as the educated public.[17]

The appearance of Riou's work in an American publication remains significant in that paleontologist Martin J. S. Rudwick has stated, "The sequence of pictures that Riou drew for Figuier established the genre of scenes from deep time, not only for the scientists whose research had provided the materials for the reconstructions, but also for a vast literate public spread around the Western world ... Figuier's book illustrated the long history of life on earth in greater and more specific detail than ever before.... Riou's illustrations ... set the pictorial precedents and established the visual language, for all later scenes from deep time, not least those in our modern museums and television programs."[18]

Although Riou had become transplanted directly into an American geological text, his work may be most closely associated with the scientific understanding of European and British geologists. Riou's scenes were not of prehistoric North America, the "flavor" of his scenes is distinctively "un–American."

In *Sketches of Creation*, Winchell offered views on the restoration of prehistoric animals: "It is almost incredible that information so exact can be drawn from the few scattered fragments which have been brought to light; but such is the unity and persistence of plan which runs through the different classes of the animal kingdom, that a single tooth, whether of a living or extinct species, will often suffice to enable the expert to disclose all the zoological relationships of the animal to which it belonged...."[19] In other words, Winchell remains reverent to the "single bone theory" of Cuvier and Owen.

A striking, gold-embossed image of Benjamin Waterhouse Hawkins' American hadrosaur skeleton (actually known from highly incomplete fossils), as reconstructed for New York's ill-fated "Palaeozoic Museum" and several other institutions, appears on the cover of Winchell's *Sketches of Creation*. Unlike what had been done before for his Crystal Palace concrete statues, in the case of several sculptural restorations planned for New York's "Palaeozoic Museum" under direction of Leidy and Cope, Hawkins had begun by reconstructing skeletons.

All the "Palaeozoic Museum" models were destroyed in early 1871.[20] However, a mold of Hawkins' *Hadrosaurus* skeleton survived that was later used in America's first case of dinosaur "cloning," performed well over a century before novelist Michael Crichton wrote *Jurassic Park* (1990). Four replicas were made, apparently. One was erected and then destroyed in New York in 1871, a second was made for Princeton in 1874, and by 1875 a third was made for the Smithsonian which was later transferred to the Field Columbian Museum by 1894. Another hadrosaur "clone" was made for the Philadelphia Exposition of 1876.[21]

While Winchell's book was making progress to eventual publication, Hawkins was at Princeton, busily painting a series of scenes showing the succession of North America's geological ages. In 1875, with the support of Cope and Arnold Guyot (1807–1884), who both admired Hawkins' artistry, Hawkins generously offered to cast a restoration of his *Hadrosaurus* in plaster (from molds which survived the debacle of 1871), for the 1876 Declaration of Independence Centennial exhibition program in Philadelphia.

In November 1875, Othniel Marsh's opinion in the matter of Hawkins' proposal was sought by Spencer F. Baird (1823–1887) of the Smithsonian Institution. Marsh proclaimed a month later, in December 1875, "In the few cases where the materials exist for a restoration of the skeleton alone, these materials have not yet been worked out with sufficient care to make such a restoration perfectly satisfactory, and to go beyond this would in my judgement almost certainly end in serious mistakes. Where the skeleton, etc., is only partially known, the danger of error is of course much greater, and I think it would be very unwise to attempt restoration, as error in a case of this kind is very difficult to eradicate from the public mind.... A few years hence we shall certainly have the material for some good restorations of our wonderfully extinct animals, but the time is not yet."[22]

A hadrosaur flesh restoration seems to have been built anyway for the Philadelphia Exposition. This cast was commissioned by the Smithsonian. Afterward, the plaster cast was delivered to the Smithsonian where it was used to decorate the grounds. Unfortunately, because it was susceptible to weathering, it had crumbled by 1893.[23] Meanwhile, the original *Hadrosaurus* was maintained at Philadelphia's Academy of Natural Sciences until the mid 1980s, when the museum's paleontology displays became renovated. Then, Hawkin's hadrosaur was remounted in conformance with modern interpretations.

While Marsh saw merit in attempting two dimensional skeletal reconstructions of some of the better known prehistoric animals, he evidently had a low opinion of restorative (in-the-flesh) efforts. In 1895, at a British Association meeting held in Ipswich, Marsh lambasted Hawkins' Crystal Palace dinosaur statues.[24] Other writers and scientists, such as Henry Neville Hutchinson, author of a popular book, *Extinct Monsters: A Popular Account of Some of the Larger Forms of Ancient Animal Life* (1892, 1st ed. and 1893, 2nd ed.), and geologist Charles Schuchert (1858–1942) and Clara LeVene (Marsh's biographers), had similar opinions concerning Hawkins' dinosaur artistry.[25]

By 1877, American paleontologists were benefiting from their fossil bone collections unearthed from the western states and territories, fossil bone treasure far richer than could have ever been dreamed of by mid–nineteenth century British paleontologists. In contrast to Owen's practice of occasionally throwing caution to the wind and resorting to the "single bone theory," Cope recognized the value of both reconstructing and restoring fossil species, sometimes conceptually "composited" from the remains of other related fossil forms.[26]

For instance, years later, in 1897, artist Charles R. Knight collaborated with Cope to produce a sculptural restoration of the poorly known dinosaur, "*Laelaps*" (i.e. *Dryptosaurus*).[27] This rendition was fortified with knowledge of the analogous genus *Ceratosaurus* (described by Othniel Marsh in 1884). Although inaccurate even by contemporary standards, it was a striking example of state-of-the-art paleontological "Fine Art." Doubtless, Marsh wouldn't have approved, although in designing the first skeletal reconstruction ("*Brontosaurus*"), Marsh had erred in combining the bones of *Apatosaurus* with Cope's sauropod, *Camarasaurus*!

Hawkins' brilliant visions of prehistoric life had fared well in America. They had not caught on either in museums, or as commercial commodities. Hawkins returned to England, and prophetically, one of his clones did as well. For the 1876 Philadelphia Exposition cast was ousted and delivered to Edinburgh in 1879, where it became the first dinosaur skeleton of any kind mounted in the United Kingdom.[28] The tables had been turned. The British Hawkins, America's first great "paleoartist," had been thwarted and "used." The British "dinosaur invasion" had ended. The American invasion had begun!

Meanwhile, a phenomenon had been set in motion in New York, literally across the street from where Hawkins' "Palaeozoic Museum" would have stood. Certainly, restorations of American "prehistorians" were being attempted and published, although by the mid–1890s these appeared only in popular accounts. Life restorations still weren't considered quite "respectable" enough to be useful to scientists as a visual form of scientific discourse. Even reconstructions had limited utility! Indeed, one prominent museum official believed that, "fossil skeletons were not for the understanding of the general public, but that the bones should be left inarticulated in museum drawers or on shelves for the edification of paleontologists alone!"[29] One wonders whether this was done at least partially to restrict Cope's access to Marsh's specimens.

In 1896 Schuchert had been instructed that a grouping of fossilized skeletons being prepared for assembly at the American Museum of Natural History was "...not Fine Paleontology, but Fine Art."[30]

From Popular Paleoimagery to "Paleoart"

While the erection of skeletal displays (a distinctively American tradition) could be viewed as more "art" than pure "science," by the 1890s American scientists had dispelled the notion of in the "single bone theory."

Although the term "paleoart," as used in its modern sense, didn't appear in print until 1986, certainly by the mid–1890s, paleoart was proliferating in America. Although most of the themes (and styles) noted in modern fossil vertebrate iconology had been tried by the mid–1890s, techniques and means for preparing restorations thought to be "accurate" by contemporary standards were only beginning to be developed.[31] The limits to what qualified as valid artistic interpretations were vaguely defined.

From contemporary publications we can comprehend the reaction to American paleoart as it emerged in the latter half of the nineteenth century.

The use of two dimensional reconstructions for describing fossil vertebrates scientifically had originated on the eastern side of the Atlantic. This form of paleoart was reserved typically for the more completely known specimens. For example, by the early 1800s, Cuvier had reconstructed skeletons of the *Paleotherium, Anoplotherium* and America's Mastodon.[32] And there are only a handful of skeletal reconstructions in Owen's book *A History of British Fossil Mammals and Birds* (1846), all representing well known genera, such as the Irish Elk.[33]

Years later, in 1882, James W. Hulke (1830–1895) incorporated a skeletal reconstruction of the Wealden dinosaur *Hypsilophodon* in his monograph for the Geological Society of London.[34] In the following year (1883), Louis Dollo (1857–1931) published a figure of his *Iguanodon* reconstruction.[35] Neither Hulke's nor Dollo's reconstructions are considered correct today, although in their time they were regarded as scientifically accurate.

Significant advances were made in our understanding of that time-honored genus, *Ichthyosaurus* as well. Between 1890 and 1891, the artistically-gifted preparator Bernard Hauff (1866–1950) noted carbonized tissue on the periphery of ichthyosaur remains discovered in Holzmaden, Germany. He realized that ichthyosaurs possessed a vertical shark-like tail rudder (e.g. rather than horizontal serpent-like tail).[36]

Previously, as reflected in the Crystal Palace restorations of Hawkins and Owen, the down-turned tail end was considered to be a "post-mortem" fracture. The carbonized coating also indicated that ichthyosaurs had dorsal fins in life, much like the dolphins of today. Even though Hawkins'-style ichthyosaur restorations continued to be exhibited well into the 1890s, such investigations led to the earliest accurate restorations of ichthyosaurs, notably by Richard Lydekker (1849–1915) in 1892, Joseph Smit (1836–1929) in 1893 and Charles Knight in 1902.[37]

By 1900, restorations had really started to "soar." For example, to augment his authoritative account of pterosaur paleobiology, *Dragons of the Air* (1901), British paleontologist Harry Govier Seeley (1839–1909) included many illustrations and restorations well executed by Miss E. B. Seeley.[38]

However, these early advances in the paleoart of Great Britain and Europe would soon be quickly surpassed by American workers. In the late 1880s through 1900s, analysis of bones belonging to uintatheres and chalicotheres (including the North American genus *Moropus*) would shatter confidence in Cuvier's "law" of the anatomical correlation of parts, meaning that application of the "single bone theory" was risky indeed! In America, confidence would become established in artists' abilities to accurately restore living images of prehistoric animals.

As suspected, life restorations of fossil vertebrates are encountered even less frequently than skeletal reconstructions in late nineteenth century literature. Dramatic portrayals generally only appeared in popular books and articles, even if these restorations had been "respectably" prepared under direction of knowledgeable paleontologists or museum scientists. Such interpretations were evidently regarded as too speculative to be of scientific value in "specialist literature." Of the restorations, Martin J. S. Rudwick has claimed, "...in expert discourse, abstract or schematic representations have come to be regarded as more useful, for conveying scientists' interpretations ... than the "realistic" scenes that appeal so widely to the general public."[39]

For example, although Richard Owen is famous today for his collaboration with Benjamin Waterhouse Hawkins on the Crystal Palace restorations, in his scientific writings intended for paleontologists' consumption, Owen remained fairly conservative. In his 1846 *Fossil Mammals and Birds* book, for instance, only two conventional life restorations appear, and these are mere body outlines of the *Paleotherium* and *Anoplotherium*, inspired decades earlier by Cuvier.[40]

Another instance concerns Charles R. Knight's 1897 painted and sculpted restorations of Edward D. Cope's "*Laelaps*" and *Agathaumas*. Although Cope shared firsthand details of the dinosaurian skeletal anatomy with Knight, so little information was available (and still is) that Knight's beautifully rendered *Agathaumas* and "*Laelaps*" can only be regarded as fanciful creations.

During the mid–1880s through the mid–1890s, popular books and magazine articles were illustrated with (both) reconstructions and restorations. Interesting books of this period, incorporating life restorations of America's prehistoric animals as figures, include two American publications, *Life History of our Planet*, (1879 and 1886 eds.) by William Gunning and illustrated by Mary Gunning, and Arnold Guyot's *Creation, or the Biblical Cosmogony in the Light of Science* (1887), and a London publication by the Reverend Henry Neville Hutchinson, *Extinct Monsters: A Popular Account of Some of the Larger Forms of Ancient Animal Life* (1st ed., 1892).[41]

In particular, Hutchinson's book was attractively illustrated by Joseph Smit, a wildlife artist best known for his renderings of modern birds. In the context of describing the history of life on earth, American fossil vertebrates were discussed along with prehistoric animals from other continents. This book provided a popular, illustrated account of America's extraordinary prehistoric vertebrate fauna, such as the *Brontops*, *Stegosaurus*, "*Brontosaurus*" and *Triceratops*. Hutchinson included skeletal reconstructions to complement Smit's life restorations. Although Hutchinson steered around the topic of Darwinian evolution, *Extinct Monsters* remains a remarkable achievement.

Popular articles of the period addressing the subject of America's prehistoric animals also incorporated restorations. James Erwin Culver's article, "Some Extinct Giants," appearing in *The California Illustrated Magazine* (April 1892, published five months prior to publication of Hutchinson's book), introduced America's dinosaurs. In fact, this may be the first popular article addressing the subject of America's dinosaurs to include restorations. The illustrations are quaint, however, as Carl Dahlgren's restorations of the "*Hypsirophus*" (i.e. *Stegosaurus)* and sauropod *Amphicoelias* are posed facing primitive humans wielding spears! (Just exactly what did some people think went on in the "old" west anyway?!) Another drawing shows a mounted "*Hypsirophus*" skeleton with people milling around the display, even though no stegosaur skeleton had been exhibited by that time. An illustrated reconstruction of Marsh's uintathere *Dinoceras* was awkwardly mislabeled as the *Agathaumas*.[42]

Another article, written by writer William Ballou (1857–1937), *The Serpentlike Sea Saurians*, appeared in the June 1898 issue of *Popular Science Monthly*. It featured a dramatic restoration of "The Great Cretaceous Ocean," a Niobraran scene illustrated by J. Carter Beard. A quick glance reveals that Nature could never have been redder in tooth and claw![43] (See Figure 12-2.)

Osborn's September 1896 entry for *The Century Magazine* may represent the public's first modern and accurate introduction to prehistoric North American mammalian fauna. For this article Charles Knight was enlisted to do restorations featuring nine mammalian genera, all of them striking, each the result of his tutelage under American Museum scientist Dr. Jacob Wortman (1856–1926). Wortman had conducted the American Museum's fossil collecting expeditions to the Rocky Mountain region during the 1890s. So he was highly familiar with the paleontology of Tertiary mammals, and consequently was able to offer expert opinion to Mr. Knight. Knight's paintings for Osborn's 1896 *Century Magazine* article were first exhibited at the American Museum in New York in October of 1896.[44]

Osborn's often neglected 1896 article offers clues as to the incentive for Knight's restorative efforts. For, after discussing the limitations of prehistoric restoration and favorably judging the accuracy of Knight's portrayals, Osborn claimed, "...The complete restoration of the glories of that earlier era is the dream and ambition of the fossil-hunter."[45] In other words, it is the restoration of the beast which matters most, although we understand Osborn was deeply interested in the phylogeny and evolutionary tendencies exhibited in the fossil fauna as well.

In 1897, Knight's paintings of prehistoric saurians, such as the "*Laelaps*," *Hadrosaurus* and *Agathaumas*, prepared under Cope's direction would grace the pages of two articles (one written by Ballou and the other by Osborn) of another often cited *Century Magazine* issue.[46]

Few realize just how many paleontological three-dimensional restorations graced the exhibit halls of American museums at this time. One recently rediscovered illustration showing exhibits in the old halls of the California Academy of Sciences, circa 1892, reveals a number of oddities. First, we spy a full-sized mammoth sculpture by Ward's Scientific, replete with hair, being a "...perfect restoration from the hair and skeleton..." of a mammoth found frozen in Siberia.[47] Leading the parade of displayed "prehistoria" were two casts of Benjamin Waterhouse Hawkins' "pre–Adamite" antediluvians—the *Ichthyosaurus*

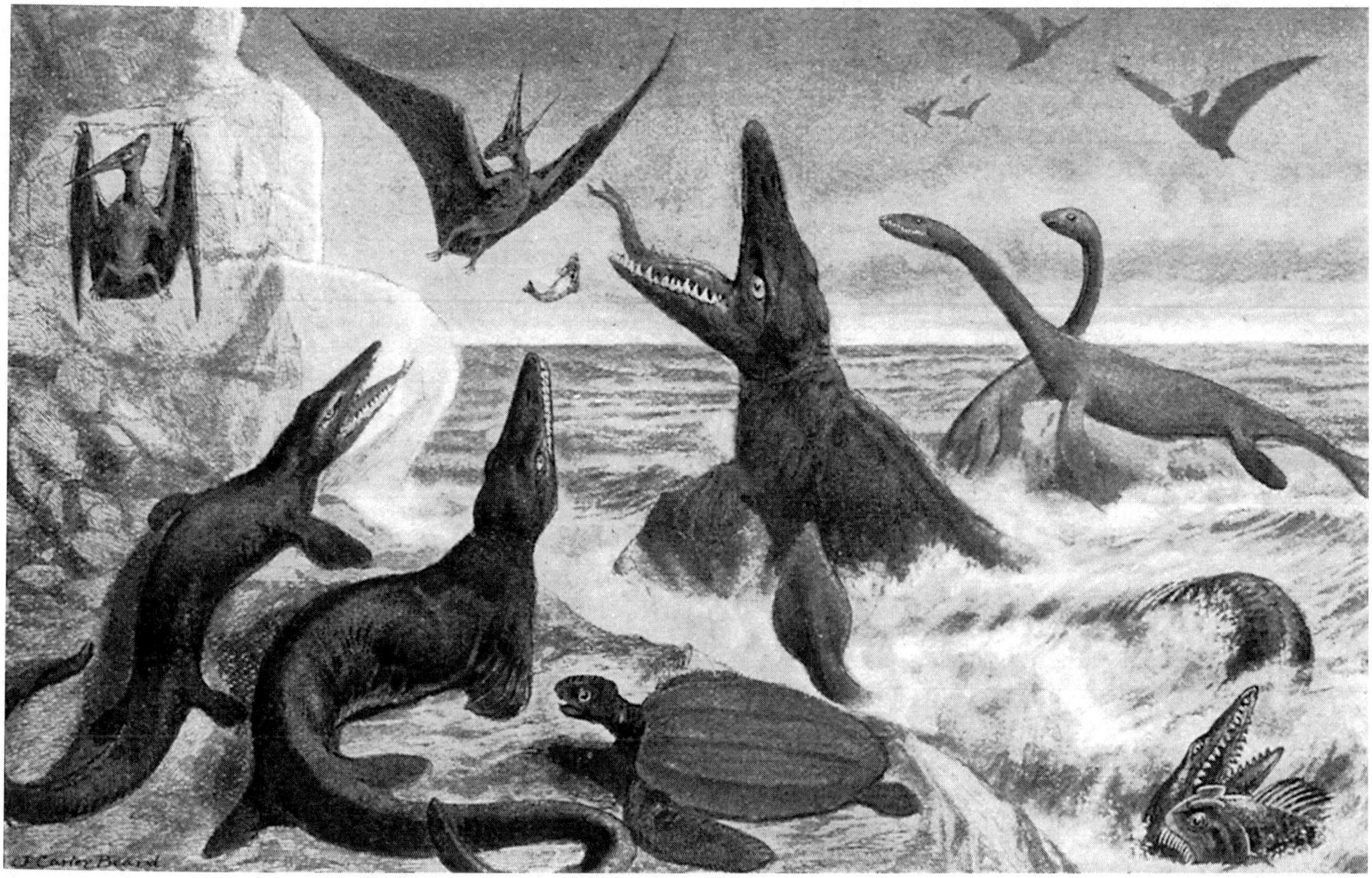

Figure 12-2: J. Carter Beard's 1898 restoration showing fearsome reptiles of western America's *Great Cretaceous Ocean*. Depicted in this idealized assemblage are three mosasaurs—(left to right) *Platecarpus, Tylosaurus*, and *Mosasaurus* winged pterodactyles *Pteranodon* and *Ornithostoma*; a marine turtle—*Protostega*; the "bulldog fish" *Portheus*; and two long-necked plesiosaurs—*Polycotylus*.

and the *Plesiosaurus*. These wonderful exhibits perished, ironically perhaps, in the San Francisco earthquake of 1906.[48]

Henry A. Ward (1834–1906), of Ward's Scientific Establishment, situated in Rochester, New York, pioneered the business of supplying museums with woolly mammoth replicas and casts of prehistoric animal skeletons (such as the cast *Megatherium* and *Glyptodon* exhibited at the California Academy of Sciences circa 1892).[49] By the early 1890s, as did many museum curators of the time, Osborn relied on Ward for the purchase of display specimens. But Osborn wanted to establish a self-reliant, centralized department under his authority at the American Museum. So, accordingly, the first American Museum expeditions were launched in 1891.[50]

Osborn had a profound fascination with artists' potential for producing accurate scientific restorations and for understanding the limitations of the restoration process. For instance, in a 1904 illustrated *Century Magazine* article, "Fossil Wonders of the West," Osborn prophetically attempted to place matters in context: "These restorations are the 'working hypotheses' of our science; they express the present state of our knowledge, and, being subject to modification by future discoveries, are liable to constant change."[51]

In order to place Osborn's contributions to an emerging American Paleoart properly in perspective, one must contrast Osborn's efforts with those of Othniel Marsh (who equated the term "restoration" with "skeletal reconstruction" in his publications).

Although Marsh had "pooh-poohed" Hawkins' restorative efforts, Marsh had no qualms over reconstructing fossil vertebrates—first on paper, and later three-dimensionally—provided there were sufficient fossil remains for an accurate reconstruction. So, between 1880 and 1895, Marsh championed paleoart reconstructions by including figures of the skeletons of several prehistoric animals, all illustrated under his direction for his published scientific descriptions of the respective genera.[52]

Perhaps unsurprisingly, the first form of paleoart to become "blessed" by a consensus of paleontologists as being of sufficient scientific standard were skeletons reconstructed in life poses, first two-dimensionally, and, later, increasingly prepared and mounted for museum display. Othniel Marsh had popularized the method of two dimensional reconstructions. Eventually, the practice of mounting skeletons "accurately" in life poses for museum display would also become refined in America. Nevertheless, as shall be explained, it was Henry Fairfield Osborn, not Marsh, or even Hawkins, who principally influenced the "Reformation" period (1895–1910).[53]

On paper, Marsh and associated artists reconstructed skeletons of the Mesozoic birds, *Ichthyornis* and *Hesperornis* (1880), the uintathere "*Dinoceras*" (1881; now known to be a junior synonym of *Uintatherium*), "*Brontosaurus*" (1883), *Triceratops* and *Stegosaurus* (1891), *Claosaurus* and *Ceratosaurus* (1892), *Anchisaurus* (1893) and *Camptosaurus* (1894).[54] In 1895, for comparison he also published reconstructions of four European dinosaur genera, *Compsognathus*, *Iguanodon*, *Scelidosaurus* and *Hypsilophodon*. Next, these were issued, together with the American dinosaur illustrations, on a single sheet.[55] In Belgium, during the early 1880s, Dollo had reassembled the bones of a single genus, *Iguanodon*. Marsh, however, was investigating a pantheon of bountiful North American fossil forms.

Marsh labored to study and reconstruct the four European genera, eventually encompassing them within his illustrated poster. Unsurprisingly, however, Cope's dinosaurs and Hawkins' hadrosaur were not included in Marsh's visual compilation. Nevertheless, the illustrated dinosaur reconstructions greatly extended both scientists'—and the public's—familiarity with North America's spectacular fossil genera. Marsh's acclaimed efforts prompted British paleontologist Richard Lydekker to republish several of the reconstructions in a *Nature* article, although two of them were given names created by Cope for similar fossil remains. Marsh dutifully chided Lydekker for the error.[56]

Marsh also authorized the three dimensional reconstruction of a complete skeleton, even though he did not permit the fossil bone reconstruction to be assembled for display purposes. In the days before invention of plastics and resin, Marsh relied on papier-mâché casting techniques. So, as an experiment, Marsh authorized the molding of the bones of his uintathere, *Dinoceras*, such that a complete skeleton could be cast and assembled in papier-mâché.[57] (See Figure 12-3.)

As Charles Schuchert and Clara LeVene have stated, "At that time, no paleontologist would have committed

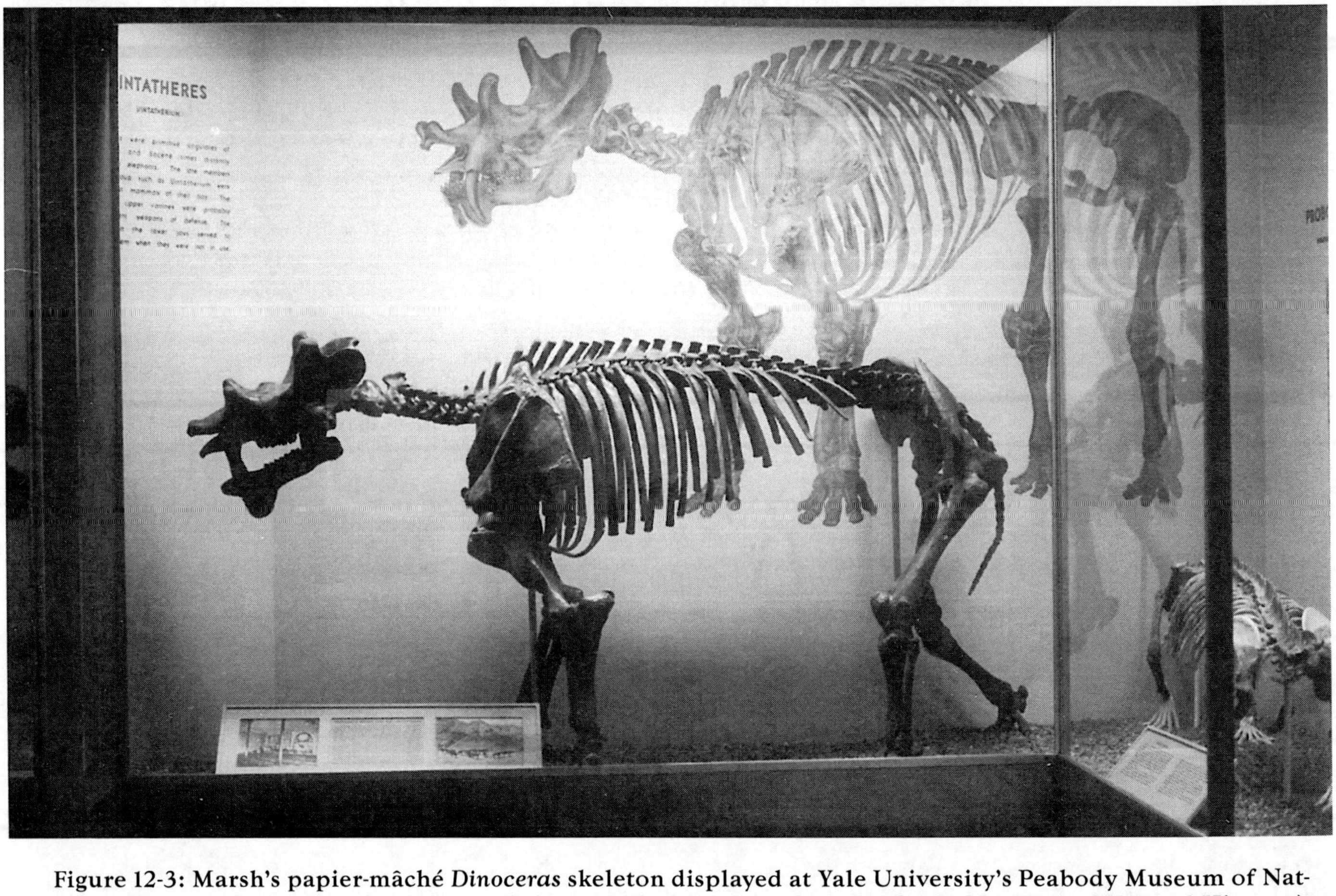

Figure 12-3: Marsh's papier-mâché *Dinoceras* skeleton displayed at Yale University's Peabody Museum of Natural History. (Photograph by Allen A. Debus, 2000. Courtesy of the Peabody Museum of Natural History.)

the sacrilege of showing the bones themselves mounted in a lifelike attitude, or even of modeling the missing parts in plaster. No!"[58]

Marsh's modeler, J. H. Emerton (1847–1930), completed the *Dinoceras* casts between Jan. 1885 and August 1888. Interestingly, the source of the paper for the papier-mâché is said to have been American "...greenbacks redeemed by the United States Treasury. Truly, millions of greenback promises must have gone into these fossil replicas!"[59] The resulting papier-mâché *Dinoceras* model weighed only 40 pounds. One of these models was exhibited at Chicago's Field-Columbian Museum in 1894.

Incidentally, a decade later, a life-size *Stegosaurus* papier-mâché model was created for the St. Louis World's Fair Exposition of 1904. This was based on a small sculpture made in 1903 by Charles Knight under Frederic A. Lucas' direction. By 1914 this model, also widely rumored to be composed of American money, was placed on exhibition at the U.S. National Museum (Smithsonian) following the Exposition.[60]

In 1940 Schuchert and LeVene lauded Marsh's contributions to the development of early American paleoart. But in praising Marsh, Cope and Philadelphia's Joseph Leidy were unfairly criticized for not promoting paleoart to the same extent as Marsh.[61] This conflation merits correction. For while Marsh established the use of skeletal reconstructions as scientific tools to a degree further than any other paleontologist, Marsh also denounced the depiction of life restorations. To his credit, however, Marsh did uncharacteristically include a life restoration of a pterosaur species, *Rhamphorhynchus phylluris*, in his 1882 *American Journal of Science* memoir.[62]

Cope, on the other hand, promoted the use of life restorations for scientific discourse and public edification.[63] Leidy, who was an exceptional scientific illustrator in his own right, consulted Hawkins and Cope in the reconstruction and restoration of *Hadrosaurus foulkii.* At the American Museum of Natural History, Cope's protégé, Osborn, would build on the foundation set by the pioneers, Cope and Marsh.

Paleoart Established

By 1910, a real slice of Americana had taken hold in Europe. As settlers and pioneers infiltrated the west, transplanting indigenous Native American populations, America's dinosaurs traveled east, eventually spilling into the Atlantic and overseas, where they were greeted enthusiastically, overtaking the indigenous fossil fauna. Benjamin Waterhouse Hawkins' sculptures of American prehistoric fauna were literally a "bust." So it was left to German sculptor Joseph Pallenberg (1882–1946) to get things "right" for the first time. Yet this feat was accomplished not in America but on foreign soil, in Stellingen, near Hamburg, Germany.[64]

I can think of no grander tribute to America's triumph in comprehending the "prehistoric" than Pallenberg dedicating years of his life to the task of accurately sculpting several of North America's dinosaurs at life scale. Pallenberg's life-size statues are the closest realization to what contemporary American scientists had been striving for, symbolically linking scientific method, art, American ingenuity and the western frontier with the latest European paleontological advances.

Pallenberg's grouping "lives on" savagely, America's dinosaur invasion incarnate! (See Figure 12-4.) A pair of bathing adult horned dinosaurs coax their offspring into cool waters, a *Ceratosaurus,* with intended victim's tail spikes protruding from a gruesome wound, tears flesh from stegosaur hide. During the 1910s, intermingled with Louis Dollo's vision of the Bernissart *Iguanodon* and other saurians native to the European Mesozoic landscape, visitors to Hagenbeck's "Tiergarten," Zoological Gardens, would have also spied restorations of Marsh's *Diplodocus* and a sunning *Pteranodon* poised ready to launch itself into ancient skies.[65] In a familiar hunched over, energetic pose, Marsh's *Allosaurus* shreds offal from a sauropod corpse. As talented as he was, the artistic Hawkins, who died in 1889, could never have produced such imagery.

In 1909, why exactly would Pallenberg's models have symbolized a "refinement" in the practice of Paleoart (while Hawkins' half century–older Crystal Palace statues seemed only deserving of derision)? What happened between 1898, when restorations had made their mark in popular publications, and 1905, the year when Pallenberg was commissioned to begin the sculptures at Hamburg?

Change had come at the dawn of a new century. In a few short years the practice of restoring fossil vertebrates had matured. While by 1900, static, two-dimensional skeletal reconstructions were "old hat," artfully done reconstructions via assembling the bones in poses suggestive of lively behavior remained fertile territory. While by 1900, increasing numbers of prehistoric animal restorations had been done, few had been engineered with such precision and exactitude as to bolster paleontologists' confidence in the result. In the eyes of Osborn, and the American Museum of Natural History staff, however, reconstructions and restorations should be devices worthy of scientific admiration as well as public praise.

Given a steady supply of new fossils unearthed in the American West, a growing number of reputable paleontologists were now confidently willing to testify on behalf

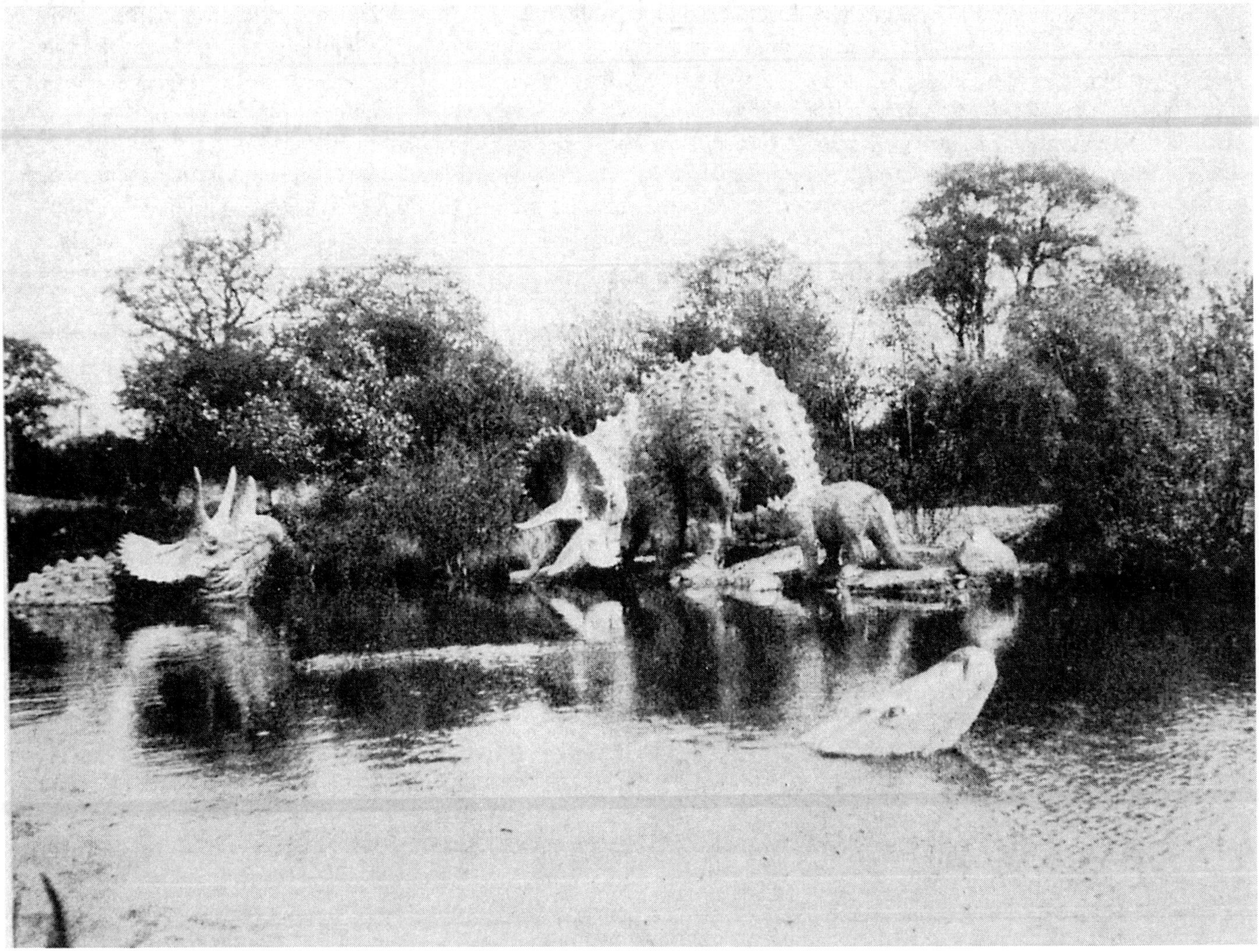

Figure 12-4: Joseph Pallenberg's sculptures of America's horned dinosaur—*Triceratops*. (Courtesy of Allen G. Debus.)

of artistic restorations. For, by then, American scientists had invented the means of confidently creating life-like restorations that were considered scientifically accurate—fitting examples of fine paleontology and fine art!

Among the early advocates was Frederic A. Lucas, then, in 1900, Acting Curator in the Section of Vertebrate Fossils at the U.S. National Museum. Lucas' 1900 testimonial, *The Restoration of Extinct Animals*, not only outlined important considerations for when it is appropriate to do restorations, but also reflected on the credibility of scientific restorations. Lucas' tone was precautionary yet instructive, favorable and supportive. Lucas wrote:

"And each and all of these wonderfully preserved specimens serve both to check and guide the restorer in his task of clothing the animal as it was in life. And all this help is needed, for it is an easy matter to make a wide-sweeping deduction, apparently resting on a good basis of fact, and erroneous.... Finally, it may be gathered from what has been said that, while the restoration of extinct animals is subject to some uncertainties, and mistakes of interpretation are liable to occur, these efforts to reproduce the living forms of past ages are not mere guesswork, but rest upon a solid foundation of scientific facts and careful deductions."[66]

The message rang out loud and clear! "Folks! America's got the fossils, as well as the artists and ingenuity to confidently restore prehistoric animals from old bones."

By 1900, American Museum scientist William K. Gregory (1876–1970), was supervising staff artist Erwin S. Christman (1885–1921), in the creation of imposing, life-size titanothere restorations. In particular, Christman's "*Brontotherium*" has been described as possessing "...formidable power and majesty...."[67] In a 1921 obituary notice, Gregory divulged how Christman had labored intently and diligently at a miniature sculptural restoration of the sauropod *Camarasaurus*, which was regarded as "...one of the most successful of all restorations of dinosaurs, yet made."[68] (See Figure 12-5.)

By 1904, work had begun on the American Museum's first mounted dinosaur assembly—an *Apatosaurus* (i.e. Marsh's "*Brontosaurus*") collected from the Bone Cabin Quarry region in 1898. The principal figure behind this activity was Henry Fairfield Osborn!

Why was Osborn able to see the possibilities of using proper techniques for restoring prehistoric animals to life, while just a few years earlier Marsh hesitated? Partly this must be related to Osborn's artistic nature. Osborn had a predilection for painting and illustration. But also, paleontology is a visual science and Osborn was influenced by the artistic "master paleontologist" Cope. Furthermore, the romantic style of Charles Knight's paintings may have also become indirectly influenced by the fact that, during the 1860s and 1870s, Osborn had befriended neighbor Frederic Edwin Church (1826–1900), an immensely talented landscape painter "...whose vast, rich panoramas of mountains and jungles were suffused with the warm, glowing light of the sublime."[69] Then, in turn, years later, Knight gained instructed from Osborn. Regardless, over the next two decades Osborn would employ staff artists at the American Museum to full advantage!

Osborn's *Brontosaurus* mount was the first of its kind anywhere in the world. (Yale University had already assembled the pelvis and hind legs of a dinosaur, and *Diplodocus* was being reconstructed at the Carnegie Museum in Pittsburgh.) There was tremendous effort expended in its reconstruction. In 1930 Gregory recalled one aspect of the undertaking:

"Dr. William D. Matthew [1871–1930] ... and Walter Granger [1872–1941] dissected an alligator, marked the areas of origin and insertion of the limb muscles on the girdles and limb bones, and then identified as far as possible the corresponding areas on the bones of the Brontosaurus; finally by means of scale drawings and paper strips representing the muscles, they endeavored to determine the course and direction of the principal muscle masses.... This was apparently the first application of the data of comparative myology to the mounting of an extinct animal."[70]

Osborn sought Knight's valued assistance in sculpting a miniature scale model from which the weight of a living *Brontosaurus* could be indirectly measured—with a result, as Gregory reported in September 1905, of 38 tons.[71] Certainly, by this time paleontologists placed great confidence in their ability to accurately reconstruct and restore creatures such as the "*Brontosaurus*," given that bold assertions could be made on the basis of a miniature scale restoration.

More museum paleoart swiftly followed. Osborn instructed his staff to begin an *Allosaurus* mount resurrected in a classic "living" pose (already expertly captured on canvas by Charles Knight and shortly thereafter to be sculpted by Pallenberg in Germany). This particular specimen, found by one of Cope's collectors in 1879, was the first carnivorous dinosaur to be assembled and exhibited. As noted in a 1907 *Scientific American* article, it was also the first ever to be mounted standing over its fossil prey, "*Brontosaurus*."[72]

Diplodocus, one of America's grandest primeval forms and the first "cosmopolitan" dinosaur, became the first prehistoric vertebrate stirring a scientific debate in which paleoart figured prominently. In August 1905, after a cast of the Carnegie specimen had been mounted at the British Museum in London, an English illustrator, F. W. Frohawk, published a note criticizing William J. Holland's reconstruction. Frohawk thought *Diplodocus*' legs should have been splayed like a lizard's, even though the Carnegie's William J. Holland had erected the dinosaur standing in an erect, elephantine pose.[73]

This triggered controversy lasting throughout the period in which Pallenberg completed his models. Along the way, skeletal reconstructions, and even Mary Mason Mitchell's restoration of a pair of incongruously creeping, lizard-like *Diplodocuses* (1910), were admitted as published scientific "evidence" in order to facilitate resolution of the controversy.[74] In 1910, alluding to some of this paleoart favored by his antagonists, Holland whimsically stated, "As a contribution to the literature of caricature, the success achieved is remarkable."[75] To this day, *Diplodocus* and "*Brontosaurus*" remain standing in elephantine poses.

As noted by Ronald Rainger, Osborn significantly influenced other renown paleontologists of the period, including Richard Swann Lull, Charles W. Gilmore (1874–1945), William Berryman Scott (author of *History of Land Mammals in the Western Hemisphere* [1st ed. 1913]) and William Barbour (1856–1947). Men such as these validated the union of paleontological art with science.[76]

Lull was responsible for designing many of the Yale Peabody Museum's dinosaur displays (after Marsh died) and is credited with a host of original dinosaur restorations. As early as 1905, however, Lull melded scientific skills with artistic visions in restoring the horned dinosaur *Diceratops*. His sculpted restoration of *Diceratops*' skull was published, together with his evolutionary views of the genus, in an *American Journal of Science* article.[77] Lull's restoration was the first to ever (accurately) portray horned dinosaurs with muscular cheeks.

Between 1903 and 1941, Gilmore supervised life-like assembly of dinosaur skeleton displays at the U.S. National

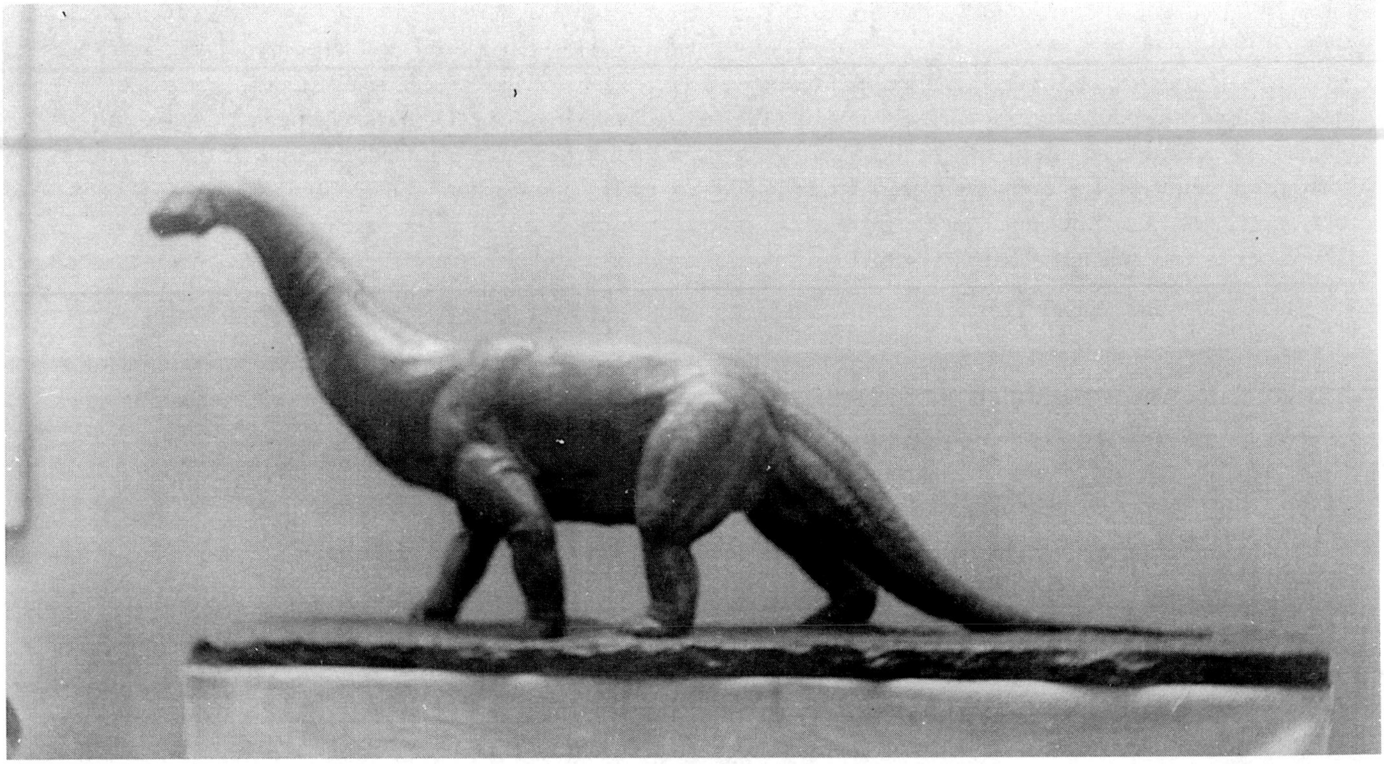

Figure 12-5: Erwin S. Christman's miniature restoration of the *Camarasaurus*. (Photograph by Allen G. Debus, circa 1940. Courtesy Allen G. Debus and the American Museum of Natural History.)

Museum. He also published important monographs on dinosaurs that emphasized pictorial representations of prehistoric vertebrates, including restorations. For example, his classic 1914 paper "Osteology of the Armoured Dinosauria" (U.S. National Museum Bulletin, no. 89), concluded with a fascinating historical discussion of stegosaur paleoart.[78] Even today his handiwork can be spotted in various natural history museums.[79]

William Berryman Scott, who years earlier had published several of Osborn's original illustrations in his scientific papers, relied on Knight and another gifted artist, Robert Bruce Horsfall (1869–1948), for illustrating his 1913 text. Furthermore, Scott briefly discussed the practice of scientific restoration, in which he remarked, "Two artists may so restore the same animal as to result in two very different pictures, and no one can positively decide between them." For the 1937 edition, this section was expanded into a subchapter entitled "The Technique of Mammalian Palaeontology."[80]

Perhaps the paleontologist most affiliated with the early development of the University of Nebraska State Museum is Dr. Erwin Barbour (1856–1947). (Incidentally, Barbour lent his good name to the ancient sabertoothed cat *Barbourofelis fricki*, the type fossil of which is contained in the State Museum collection.) Barbour was the State Museum's Director between 1891 and 1941. Barbour was a gifted artist. He illustrated his own scientific papers, as well as a book written by James Dwight Dana—*A System of Mineralogy*. In a May 1948 Memorial to Dr. Barbour, C. Bertrand Schulz stated, "Dr. Barbour was a profound and humble lover of truth and beauty."[81]

In their later years, two of Marsh's "disciples" also contributed visions of prehistoric animals. Samuel Wendell Williston, who collected fossils for Marsh and who later became a University of Chicago professor, also reveled in opportunities to combine art with science. For example, his 1914 *Journal of Geology* article "Restorations of Some American Permocarboniferous Amphibians and Reptiles" incorporated ten original restorations.[82] For his 1914 book, *Water Reptiles of the Past and Present*, Williston supplied many illustrations of Mesozoic aquatic reptiles.[83] Another of Marsh's former collectors, Arthur Lakes (1844–1917), also produced several restorations of Mesozoic saurians.[84]

Meanwhile, another skilled artist, Jules Henri Marchand (1877–1951), began restoring fossil forms in sculpture, specializing in Paleozoic invertebrates and fishes.[85] Jules, with his sons George (1903–1996) and Paul, went on to build many miniature models and dioramas (often showing marine, aquarium-view scenes of prehistory) exhibited in many museums throughout the country in the twentieth century. Perhaps their crowning achievement, however, was the life-size Gilboa Forest Devonian age restoration completed in 1924 and exhibited in the New York State Museum, a creation which, seventy years earlier, paleobotanist Unger and his artist Kuwasseg would have found remarkable.[86]

Conclusion and Afterward

Our contention is that by 1910, led by Osborn, American scientists were convinced that paleoart was a suitable means for pictorially describing prehistoric animals, not only for the public but also for the paleontological community. Other writers have referred to the paleoart "period" from 1900 to 1960s as the "Knight-dominated" era.[87] However, while Knight's visibility cannot be denied during this period, I suggest that belief in a Knight-dominated era is an oversimplification. Instead, the period from 1895 to 1910 represents the "Reformation" period in paleoart, as outlined herein. Thereafter, the popularity of paleoart and other paleoimagery escalates through the present day.

At first, scientific illustrations attempted by American artists and scientists, such as that highly artistic and quintessential observer of nature Joseph Leidy, were unencumbered by theory. Artists labored to illustrate fossils accurately and in detail while avoiding the pitfalls posed by overly speculative or erroneous reconstructions and restorations.[88] Thanks to Leidy and Cope, however, Hawkins' hadrosaur recreations represented an inviting means for introducing a bit of well intended speculation.

To civilized men of learning, the American West seemed both "untamed" and primitive, not unlike the creatures formerly inhabited this terrain. Little wonder popular artists would portray prehistoric animals of our "Old West" under attack from spear-wielding humans!

In the late nineteenth century, Time's harrowing (geological) frontier could have been subconsciously assimilated with the notion of unknown space. The American West may have seemed foreboding and mysterious, enchanting and even frightening. Under scientific direction, artists of the time could depict frightening scenes showing saurians fighting viciously or preparing for titanic struggle. Through expertly rendered contemporary paleoart it was possible to escape the civilized world and "safely" observe all that such scenes embodied and symbolized. Thus, the first "accurate," confidently done prehistoric animal portrayals were typically "dramatic" romantic scenes.

Only later, by the 1910s, do more pastoral scenes of America's prehistory become more prevalent. Such (scientific) restorations focused less on the illustrations' "emotion" and increasingly on factual (scientific) content. By then, paleontologists appreciated the power of paleoart to "accurately" model data and illustrate theoretical ideas. (Furthermore, by the late nineteenth century, techniques for reproducing artwork in books and magazines had considerably improved, relative to the technology available for printing illustrations from woodcuts half a century earlier, facilitating the reproduction of paleoimagery in technical and popular publications.)[89]

The Reformation period in paleoart was not without its embarrassing pitfalls (notably the restorations of "Piltdown Man" and "Nebraska Man").[90] The practice of paleoart was not infallible, and to some degree it was culturally influenced.[91] Yet it is refinement in the methods of reconstruction and restoration, indoctrinated by Osborn and others, that established confidence in artists' abilities to restore prehistoric animals "accurately."

Soon Reformation "fever" spread to Europe, as two artistically gifted individuals, Gerhard Heilmann (1859–1946) and Othenio Abel (1875–1946), published their (illustrated) views concerning the paleobiology of prehistoric vertebrates. Heilmann's "Our Present Knowledge About the Origin of Birds" was originally published as a series of articles in the Journal of the Danish Ornithological Association (1913–1916). This eventually was published in the U.S. in book form as the classic *The Origin of Birds* (1926). Heilmann's illustrations introduced evolutionary ideas figuring prominently during the beginning of the "dinosaur renaissance" of the 1970s.[92]

Othenio Abel's *Geschichte und Methode der Rekonstruktion Vorzeitlichter Wirbeltiere* (1925) addressed the methods of paleontological reconstruction and restoration.[93] Intended for interested laypeople as well as other paleontologists, entire chapters were devoted to the restoration of prehistoric vertebrates, such as the dinosaurs *Stegosaurus* and *Polacanthus*. In this and numerous other publications Abel incorporated many of his original restorations. Certainly by 1925, with the publication of Abel's authoritative account entirely dedicated to the subject of prehistoric animal restoration, there is indication that Osborn's Reformation period had reached a pinnacle!

Another European, Czech paleoartist Zdenek Burian (1905–1981), would become highly influenced by Charles Knight's work. He would complete his first prehistoric animal restorations in 1935, and from there go on to become the most prolific paleoartist of all.[94]

Throughout the twentieth century the paleoimagery "phenomenon" increasingly manifested itself in myriad forms—movies, toys, games, advertisements, industrial prowess, socio-political ideologies and even religious themes.[95] By the late 1900s, scientifically founded images perceived as "accurate" became popular, in turn inspiring ever newer and more refined forms of popular paleoimagery, leading to the dinosaur "fine art" of today.[96]

While paleontologist Martin J. S. Rudwick has identified the "aquarium" perspective as having innovatively inspired many nineteenth century "scenes from deep time," most certainly the "fictional reality" of the television screen, our favored window for safely eyeing the prehistoric realm, has fortified the credibility of "dinosaur" images ever since the mid–twentieth century.

Chapter 13

A Tertiary Primer

It is most satisfying when we design our own original paleoimagery. That is when personalized stories can be creatively told using images of past life. For, after all, restorations of prehistoric animals should have a purpose, theme or message. Here, we find that paleoimagery takes on many guises, shapes, and forms. The sky's really the limit.

During the latter 20th century, paleoimagery of prehistoric mammals was often neglected relative to dinosaurian imagery, possibly because fossil mammals—however impressive they may have been—still seem less exotic-looking than those ever-popular dinosaurs. And yet, as alluded to in other chapters herein, prehistoric mammals were at the forefront of scientific advancements made in Paleontology during the 19th century. They were popularized to unprecedented degree during the Victorian era, and also in America, at a time before the dinosaurs captivated our imagination. The next chapter is a firsthand account illustrating how restorations are made, what takes place behind-the-scenes before the restoration can be confidently made and, especially, what goes into a restoration, inspirationally. And you are left with an important message—one about extinction that is all too often neglected in the current "dinosaur renaissance" wave.

> The monster's thick nostrils pulsed harder this time. No scent of offal, although the sign of carrion was in the air. Wary condors spiraled ever lower in the shimmering sunshine toward unseen spoils. Rising from its haunches into a steady trot, *Dinohyus* barreled forth.
>
> Then came the awful squeals and mu· ed growling, exciting predatory instincts. Food sounds!
>
> There! *Dinohyus* squinted at a wounded, twin-horned rhino (Menoceras), denied its freedom by a slavering, snapping pair of primeval beardogs (*Daphaenodon*)! Dimwittedly driven by the prospect of another swollen belly, *Dinohyus* strutted toward the rising turmoil. One swift revolution of *Dinohyus*' yard-long face and rippling neck sent an ankle-nipping "dog" topsy-turvy into the brush. Its unwelcome companion, caught dragging *Menoceras* by a hind limb toward an underground burrow, wisely chose this moment to flee.
>
> Arighting itself, *Menoceras* shook off the pain and broke into a crippled run. *Dinohyus* snorted, splintering through a thicket toward a stately group of browsing *Moropus*. Two startled females loped gracefully to a safer vantage point before wheeling about. *Dinohyus* halted momentarily, sizing things up.
>
> The larger male stood his ground, staring down the polemical porker. In the distance, *Menoceras* was fording the stream. No time for negotiation! "Lunch" was escaping. With a hellish snarl, *Dinohyus* scraped the dusty earth with cloven heels, and charged.

Please don't insult this six-foot-tall, bison-sized prehistoric "pig" (i.e. entelodont) by crying "Suey, Suey." We'd become "chopped suey" for this Tertiary bacon-on-wheels. And I'm beggin' ya', don't "ham it up" by taunting *Dinohyus* with silly names like "Porky," "Pumba" or "Arnold Ziffel," at least whenever we're around. *Dinohyus* would slurp us down like slops! Instead, you may refer to this ancient swine rightfully as "the Terminator Pig," although at a subconscious level one can make connections to that nightmarish devil-pig you may recall from *The Amityville Horror*. Yep, this was decidedly *not* one of those little piggies who ran all the way home!

Although it most likely behaved like an enormous pig, *Dinohyus*, "Terminator Pig," "Mr. Pig," "Sir" or however you choose to respectfully refer to him, was not a "true pig" of modern variety. *Dinohyus* is the last reigning member of a group of prehistoric artiodactyls (even toed hoofed mammals) known as "enteledonts." It first appears in the mid–Oligocene of North America, 29 million years ago, vanishing by the mid–Miocene, about 16 million years ago. How surprising, considering its remarkable "resume," that this horrid hog hasn't yet made a Hollywood appearance!

Entelodonts were most likely omnivorous, and dangerous as hell. We can't imagine any self-respecting sabertooth cat of the time summoning the courage to take down an adult *Dinohyus.* So, in May 1999 we decided to take him on ourselves, first as photographic "prey," then in sculpture. Diane and I travelled to the Agate Fossil Beds National Monument of northwestern Nebraska to see the magnificent skeletons exhibited there.

With that proportionally large head affixed to an inflated "pig" body, *Dinohyus'* hideous countenance is more than just imposing. The most evident skull feature are those expanded jugals (cheek bones) jutting outward and down from behind the orbits. (*Dinohyus'* cheek bones are nearly as exaggerated as those in the Russian, Middle Permian, herbivorous mammal-like reptile *Estemmenosuchus.*) *Dinohyus'* flat and broad ventrolaterally directed jugals most likely did not serve as muscle attachment sites. They may have had a sexual or dominance display function.[1]

The bony knobs situated along *Dinohyus'* lower jaw margins, or "mandibular tubercles," may also have been a sexual feature. However, to date, paleontologists have no evidence for sexual dimorphism among enteledont species.

Dinohyus also had prominent upper and lower tusks. With this dentary apparatus, *Dinohyus* and other enteledonts could have presumably pried roots from the ground, ripped rotting flesh from bones or dispatched "fresh kill." However, it is uncertain just how frequently *Dinohyus* fed carnivorously, as opposed to herbivorously.

Often amateur paleontologists only associate horns and fossil teeth with dinosaurs when they should also be conjuring images of prehistoric mammals. Intellectually, even the sharpest of dinosaur teeth would be a dull commodity because they are monotonous ... row after row of the same old thing. Conversely, the intricately sculptured surfaces of mammalian teeth (especially on the contoured molars and premolars), like the delicate swirls or suture lines of ammonoid shells, form an important basis of mammalian classification and phylogeny.

Until my 1/12 scale miniature sculpture was completed in September 1999, only a handful of notable *Dinohyus* restorations had been attempted. Perhaps the most remarkable is the life-sized sculpture recently completed by sculptor-taxidermist Tom Shankster for the Denver Museum of Nature and Science as part of their "Prehistoric Journey" exhibition which opened in 1995. (See Figure 13-1.)

Of his masterpiece, Shankster has stated, "I like the positioning of Dinohyus ... especially its turning and wheeling stance, and how it is a nasty, gnarly, cantankerous old pig. People seem to like it, and it makes me feel good that it seems to go over pretty well."[2] It was a Denver Museum, Earth Sciences educator, Rebecca Smith, who, inspired by Shankster's handiwork, coined the name "Terminator Pig." (But, thank you, we'll take full credit for "polemical porker.")

A miniature model, done for the Carnegie Museum, was used to illustrate a 1909 scientific publication by paleontologist O. A. Peterson.[3]

Several canvas restorations have fueled our imaginations through the years, of which Charles Knight's mural, showing an assemblage of Miocene fauna is perhaps the best known. Fortunately, this impressive mural is still exhibited at Chicago's Field Museum of Natural History. Of *Dinohyus*, Knight wrote, "...they were monstrous, repulsive brutes, enormously wide at the cheeks, high-shouldered and with two heavy-clawed toes on each foot."[4]

Artists Jay Matternes, Rudolph Zallinger and William E. Scheele have also marvelously succeeded with *Dinohyus*, each in their own individual styles. Three decades ago Matternes included a striking visage of *Dinohyus* in his Smithsonian mural showing the fauna of the Agate Fossil Beds (19 million years ago). Some of you may have noticed a pair of *Dinohyus* lurking menacingly about in Zallinger's mural completed for Yale's Peabody Museum in 1967, *The Age of Mammals*. Scheele's rabid, werewolf-like critter comes right at you off the page in his 1955 children's book *The First Mammals*.[5]

Interestingly, Knight also restored a number of other entelodonts. In fact, his very first scientific restoration of a prehistoric animal happened to be an Oligocene enteledont, *Elotherium*, expertly rendered in 1894 under advisement of an American Museum paleontologist, Dr. Jacob Wortman. Yes, *the* Artist who is most famous for his visions of dinosaurs, mammoths and cavemen started his career quite humbly—with a prehistoric pig! I suppose there's a moral in there for us somewhere.

Incidentally, *Elotherium* was the first fossil entelodont ever described, in 1847. It was recovered from French deposits. Today European forms are given the name *Entelodon*.

Besides *Dinohyus*, two varieties will be most familiar to readers. One is the Oligocene *Archaeotherium*, which was roughly about half the length of *Dinohyus*. Another is *Megachoerus*, the most derived entelodont form dating from the early Oligocene, which was roughly "in between" the sizes of *Archaeotherium* and *Dinohyus*. Long ago, *Archaeotherium* was restored by artist R. Bruce Horsfall.[6] Very recently, *Megachoerus* was restored as an animatronic figure by the Dinamation company. While you'll have to delve into those dusty old books to find Horsfall's illustration, *Megachoerus* will soon be savagely stalking through your local Mall. Watch out!

There are two *Dinohyus* skeletons displayed at the

Agate Fossil Beds exhibit area. Ultimately, the fact there is any reconstructed skeleton to be seen anywhere resulted from a coin toss for the ownership of the original fossils. Richard Stucky of the Denver Museum relates that the tail end of the beast was found at Agate by a Yale University collecting team. The head was excavated by Nebraska State University collectors. Fortunately, the two parties decided to flip a coin for rights to the entire skeleton "prize." The "heads" won, and so the skeleton went to the University of Nebraska State Museum.[7]

Eventually, the task of mounting the *Dinohyus* fell to young Edwin H. Colbert, who in the early days of this century served an apprenticeship at the University of Nebraska. (Colbert also mounted the skeleton of *Moropus.*) In 1980 Colbert stated, "This rare and interesting beast [Dinohyus] had a perfectly enormous skull, armed with huge canine teeth. Behind the skull was a strong skeleton, with long legs and feet—obviously an animal adapted for running across the land at a pretty good clip. Dinohyus was and is one of the treasures of the Nebraska University Museum, and Dr. Barbour was inordinately proud of it."[8]

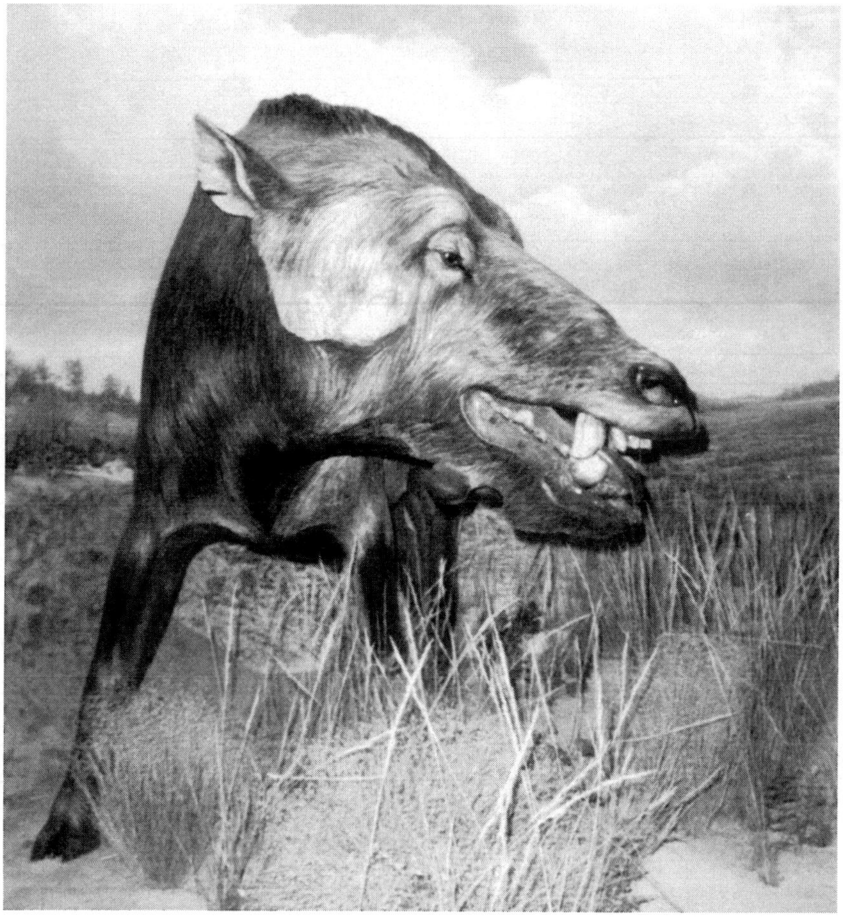

Figure 13-1: Photograph of Tom Shankster's life-sized *Dinohyus* sculpture (1995) on display at the Denver Museum of Nature and Science. (Photograph by Lynne Clos. Reproduced with permission.)

So imagine our disappointment when, during our trip to Lincoln, the exhibit case where Colbert's *Dinohyus* should have rightfully been had only the iron armature on display, minus the bones. Shoot! No quarry for my loaded camera! As it turned out, *Dinohyus'* bones had been loaned out for the task of molding and casting so the Agate Fossil Beds National Monument could place casts of the original specimen on display. Several days later, after logging what seemed like many millions of miles westward through plains and wilderness at Agate we were delighted to finally see the two dramatically posed *Dinohyus* skeletons exhibited there.

My *Moropus* is posed with left arm raised defensively, like one of three skeletons mounted at the Agate Fossil Beds exhibit center.

Okay, by now you're probably suspecting that three-toed *Moropus* is merely a "prop," providing a creature for *Dinohyus* to "interact" with in my sculpture. Or is it? Please read on. (See Figure 13-2.)

The race represented by *Moropus*, the clawed herbivores known as "chalicotheres" (or "terrible wild beasts"), are fascinating in their own right. *Moropus*, sort of an "okapi of the past" with a horse-like head, is one of those animals traditionally included in those "life through time" portrayals (such as those by Zallinger, Knight and Matternes), although without much fanfare. Perhaps the most impressive recreation is a life-sized scientific reconstruction, a "*Moropus on the half shell,*" built by Robert G. Caffrey in 1949 for the Carnegie Museum in Pittsburgh. On one side of the display, museum visitors can view the bones; on the other side, its life appearance has been restored.

The *Moropus* story goes back to the earliest days of paleontology. Baron Georges Cuvier himself was puzzled by the bones of chalicotheres found in a locality near Eppelsheim, described by Johann Kaup. Fossil footbones found there with three-toed claws were thought to be those of a pangolin (an edentate) named "*Macrotherium,*" while skulls and teeth found disarticulated from the foot bones were described under the cool-sounding name "Beast of the Gravel," or *Chalicotherium. Chalicotherium* was regarded as an ungulate, possibly related to the rhinoceros.

By 1887, a French paleontologist named Filhol suspected the clawed foot bones and skulls were associated,

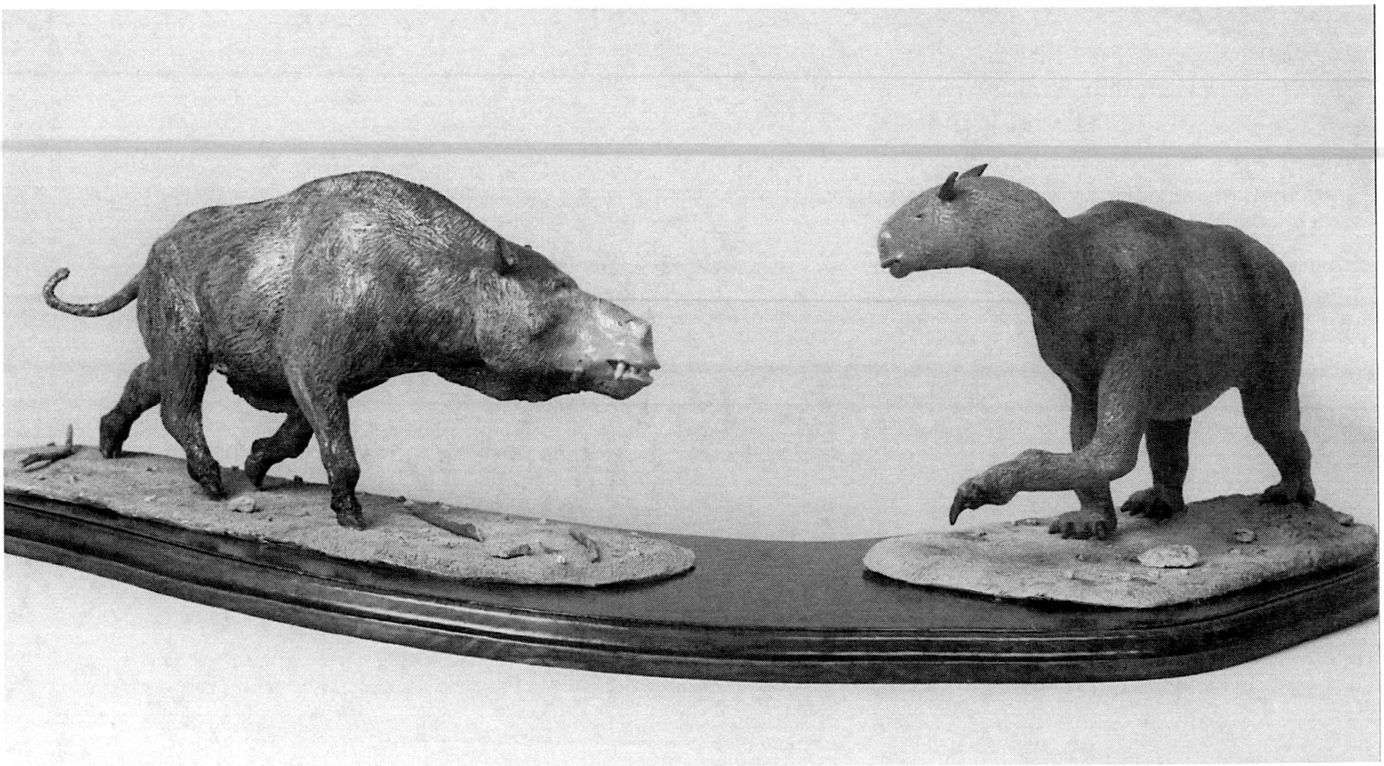

Figure 13-2: Miniature 1⁄12 scale sculpture by Allen A. Debus (1999) showing *Dinohyus* at left challenging *Moropus*.

a weird combination apparently in violation of Cuvier's "Law" of the correlation of anatomical parts. This was eventually proven when complete skeletons of *Chalicotherium* turned up, and further supported by discovery of nearly complete *Moropus* specimens from Agate Fossil Beds, as described in O. A. Peterson's 1909 paper. The great Cuvier, who had applied his Law to chalicotheres, was wrong after all![9]

Chalicotheres are often overlooked in popular accounts, an unfortunate circumstance since they are so unusual. With proportionally short legs, a broad low-slung pelvis, and long arms, *Chalicotherium*, known from Europe and Africa, had a gorilla-like gait. Paleontologists have imagined it walking with upturned knuckles like modern gorillas. Sitting on its haunches while browsing, it could rear up to 13 feet. *Chalicotherium* is thought to have gone extinct about five million years ago. However, reports of Kenya's "Nandi Bear" have stirred unfounded speculations as to whether *Chalicotherium* survived to the present day.[10]

If *Chalicotherium* was the "brachiosaur" among chalicotheres, then a North American genus, *Tylocephalonyx*, was the "pachycephalosaur," since this latter contemporary of *Moropus* had a dome-shaped bony growth over its brain case.[11] Did it bash heads with rivals to gain prominence or a mate? Miocene "Coneheads" or "punk rockers"?

Chalicothere extinction in North America has been related to the arrival of giant ground sloths from South America about six to seven million years ago. Sloths and chalicotheres resembled each other somewhat and may have subsisted on similar food sources. Some paleontologists have claimed chalicotheres used their claws to dig up roots. However, their teeth and relatively long necks are adapted for browsing leafy vegetation, like the sloths. At any rate, the extinctions of *Moropus* and *Tylocephalonyx* seem to have happened before the sloths arrived. In Asia, the last chalicotheres lived into the Pleistocene, disappearing less than a million years ago. Did prehistoric Man ever encounter them? Hmmmm.

Extinctions of the mid–Tertiary, Plio-Pleistocene and Ice Age have reduced the natural diversity of Perissodactyls (represented by *Moropus* in Allen's sculpture) relative to that of Artiodactyls (represented by *Dinohyus*). Many Perissodactyls have become endangered. Think of the rhinoceroses, for example. Perissodactyls versus artiodactyls. Odd vs. Even. What are the odds this would happen? Didn't they start out "even"?

One of the most fascinating books of the "Dinosaur Renaissance" is *After Man: A Zoology of the Future* (1981), by Dougal Dixon, concerning fauna living in the period 50 million years from now. Thumbing through this imaginative volume, it is clear the odd-toed animals will not fare well in the next geological age. Pig descendants will though.

Dinohyus and *Moropus* are relatively rare at Agate, compared to the thousands of jumbled *Menoceras* rhino bones. All are thought to have perished during a drought period. At the time of their meeting, as envisioned in my sculpture, the *Moropus* lineage had a nearly endless-seeming additional 10 million years of existence to revel in. *Dinohyus* would be gone within three million years.

As a group, artiodactyls (e.g. *Dinohyus*) never "battled" against perissodactyls (e.g. *Moropus*). Their diversities both waxed and waned with natural vicissitudes. And yet, presently, the number of modern even-toed genera overwhelmingly outnumbers the odd-toed fauna by 79 to 6.

So who really wins the game?

> Dinohyus' deadly charge had been repelled by a mighty Moropus arm swat. On the other side of the stream, a reclining Menoceros patiently licked his wounds. In the distance, Dinohyus heard her brood squealing hungrily. Time for BIG MAMA pig to head home, before some bad ol' beardog comes along to huff and puff and BLOW down her makeshift "den" in the brush. So, after digging up some tubers, Dinohyus nursed three furry piglets.

Roll a dice. You want evens? Today we'll take odds, for this day belongs to the odd-toed clan.

CHAPTER 14

Othenio Abel—Artistic Founder of Paleobiology

References in recent popular paleontology literature regarding Austrian paleobiologist Othenio Abel (1875–1946), often referred to as the "father of paleobiology," are scanty. In our rather extensive library pertaining to paleontology and dinosaurs, his name is rarely mentioned, often only in footnotes. In books in our library that do discuss some of his ideas, there is usually less than a full page devoted to Abel. In *The Dinosaur Data Book*, published in 1990 by the Diagram Group, which has a section listing many important "dinosaurologists," Abel isn't even listed. Nor is he mentioned in Edwin H. Colbert's popular book, *Men and Dinosaurs* (1968). Nevertheless, Abel wrote extensively about dinosaurs and for a time worked with the renown French-born Belgian "dinosaurologist," Louis Dollo. Therefore, let us take this opportunity to reintroduce Abel and his remarkable contributions to the art and science of dinosaurs.[1]

Perhaps his family heritage, and no doubt the influence of his father, a specialist in "gardening architecture" who taught him how to observe animals, instilled a love for nature and especially the study of botany in young Othenio.

Frustrated in his aspiration to study archaeology at the University of Vienna by strict requirements to master *seven* languages, Abel instead turned to geology under the advice of Edward Suess (1831–1914). Suess was a Viennese geologist quite famous for his influential books *The Face of the Earth* (four volumes—English publication dates—1904–09) and *The Origin of the Alps* (1875). Abel studied under Suess at the Geological Institute of the University of Vienna, specializing in geology and paleontology. In 1899 he passed his major examination leading to the Ph.D., and completed a required professor's thesis (known as a "Habilitation") in 1901. Abel had also earned a law degree but, realizing that he could not work in that field for another four to five years, joined the Austrian Federal Institute of Geology in 1900.

His first vertebrate paleontology publication was on the subject of fossil whales, a study he had initiated as a student. Abel's observations of how the mammalian skeletal frame had become evolutionarily adapted to an aquatic existence caught the eye of influential scientist Louis Dollo (1857–1931). Dinophiles revere Dollo today for his meticulous investigations of the *Iguanodon* skeletons found in a coal mine shaft in Bernissart, and also for his formulation of "Dollo's Law" (1893), a statement concerning the "irreversibility" of evolutionary change (e.g. "advanced" species do not evolve into unspecialized species). Abel, in fact, later supplemented Dollo's law.

Dollo had been seeking an assistant to study the remains of early Tertiary cetaceans. So Abel arrived in Brussels in 1900 to continue his studies under Dollo at the Royal Museum of Natural History.

Abel stated that Dollo had greater influence on his work than any other scientist. This was particularly so in the reconstruction of extinct forms, because Dollo was inspired to recreate the organism as a living animal form and not simply regard fossils as stratigraphical indicators. Although Dollo was not himself adept at restoring fossil vertebrates through the use of illustration, Abel recognized that graphical representation could greatly assist in the resolution of Paleontological problems.[2]

It is thought that Abel's artistic inclinations stemmed from his mother's side. Artistic tendencies are often favored among paleontologists. Denied an ability to directly observe their subjects, paleontologists are forced to *imagine* how prehistoric animals and the world in which they lived appeared. An ability to illustrate, sculpt or paint images of prehistoric animals provides paleontologists with an ability to overcome this obstacle.

Abel was a gifted artist, and he included many renderings of prehistoric scenes in his publications. One book, *Lebensbilder aus der Tierwelt der Vorzeit: Zweite Auflage* (1927 ed.), portrayed the evolutionary history of vertebrates through geologic time, and demonstrated how the fossil fauna of various regions and continents compared with one another.[3] The Miocene shark *Carcharadon* megalodon rises from the depths in pursuit of fleeing prey in the color frontispiece by Abel (1916), using the fish-eye aquarium perspective often employed by paleoartists for aquatic scenes. Abel published many paleoartists' restorations, including several of his own for these comprehensive presentations.

In another of Abel's publications devoted to restorations of prehistoric vertebrates, *Geschichte und Methode der Rekonstruktion Vorseitlicher Wirbeltiere* (1925), comparative studies of scientific interpretations are undertaken.[4] Entire chapters are devoted to discussion of different genera of extinct animals, such as *Stegosaurus*, and the manner in which restoration has been attempted by various scientists and associate artists.

Abel charmed students with his sketches on the blackboard during lectures, and often collaborated with academic painter Professor Franz Roubal in the pictorial and sculptural restoration of prehistoric animals. Through this novel style of presentation, Abel successfully introduced images of many North American fossil vertebrate fauna to a German-speaking audience. In the process, Abel also popularized vertebrate paleontology through use of artistic expression.

Abel was not a professional artist in the sense that Benjamin Waterhouse Hawkins and Charles R. Knight were. Nevertheless, he may be counted among the ranks of paleontologists possessing the gift of artistic expression.[5]

Abel thoughtfully considered the dinosaurs and their Mesozoic brethren in his publications. In 1912, for instance, Abel claimed that the reversed first toe on the foot of the dinosaur *Hypsilophodon* was an arboreal adaptation. This conclusion interested Danish ornithologist Gerhard Heilmann, who disagreed with Abel. Whereas Abel had predicted through his interpretations that dinosaurs' ancestors would be recognized as bipedal perchers, Heilmann claimed that even the primitive *Hypsilophodon*'s toes were insufficiently long to grasp the branches of trees efficiently. The status of *Hypsilophodon*, tree versus ground dweller, was debated through the years, leading to several mid-twentieth century restorations expressing the perching/arboreal interpretation.[6]

Finally, by 1974 the matter was settled in favor of the bipedal running interpretation.[7] Abel had been incorrect that *Hypsilophodon* hopped about monkey-like, or in a manner mimicking the modern tree-kangaroo of Australia in the branches of trees, feasting on leaves and fruits. But indeed, a lively discourse had been inspired!

Abel's speculations on the function of the cranial ornamentation evident in duck-billed dinosaurs led to Franz Nopsca's novel suggestion that the decoration had a sexual display purpose. Abel also proposed that footprints fossilized in the Upper Jurassic lithographic limestone, which previously had been attributed to the pterosaur *Rhamphorhynchus* or, alternatively, the fossil bird *Archaeopteryx*, had instead been made by the small dinosaur *Compsognathus*. However, it now seems that they were made by an extinct King Crab. During the years of the controversy concerning proper restoration of the limb posture in *Diplodocus*, circa 1910, Abel was one of the few European scientists supporting the (essentially correct) view that *Diplodocus* had an elephantine stature. (See Figure 14-1.) Abel attributed the extinction of dinosaurs to "evolutionary degeneration," a hypothetical process no longer accepted by paleobiologists, which he felt was also responsible for the demise of woolly mammoths and cave bears many millions of years later.

Abel also thoughtfully considered the life habits of fossil birds and pterosaurs. For instance, he noted that the Cretaceous flightless bird *Hesperornis* was an adept swimmer, but on land it must have progressed through a clumsy sort of abdominal flopping and hopping. And it was Abel who suggested that *Archaeopteryx* may have climbed the trunks of trees in the manner of the modern South American birds known as hoatzins. Abel offered illuminating suggestions concerning the reproductive habits of the *Pteranodon*, and even suggested that after young hatched from eggs they were cared for and fed by parents in nesting colonies. Such visions were memorialized by the spectacular artistry of Czechoslovakian artist Zdenek Burian (1905–1981), although in collaboration with paleontologist Josef Augusta (1903–1968).[8]

In his popular writings, Abel also speculated on possible historical associations linking fossil vertebrates with mythology. Was the Cyclops of Greek legend inspired by the ancient discovery of fossil elephant skulls, or were myths surrounding giants and dragons based on woolly rhino skulls? At least in one case, the connection between the creation of a dragon sculpture situated in the town square of Klagenfurt since 1636 and the discovery of a

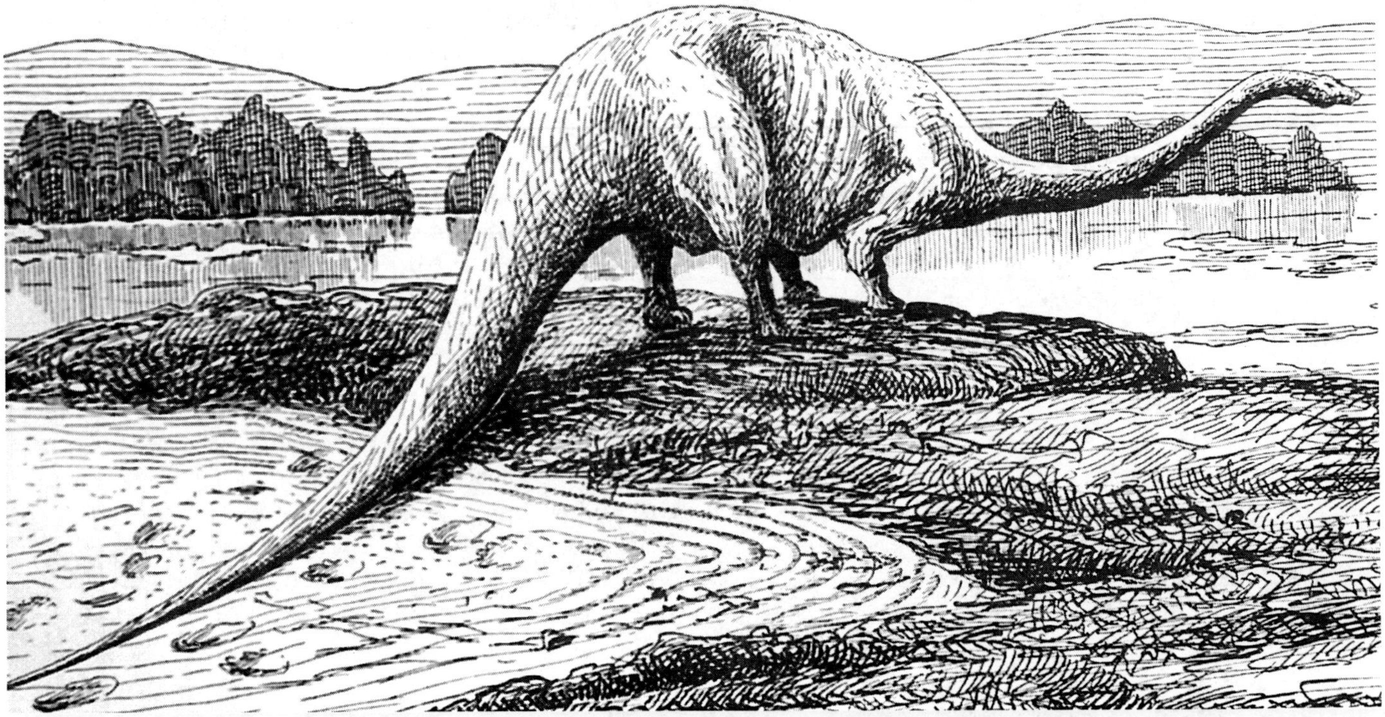

Figure 14-1: Othenio Abel's restoration of the American dinosaur *Diplodocus* (1909).

rhino skull at the nearby "dragon's cave" in 1590 was clearly established by Abel.[9]

Abel was not merely an "arm-chair" naturalist, as he did supervise excavations and collecting of fossil vertebrates, notably at Pikermi, Greece, in 1912, and at the (Drachenhohle) "dragon's cave" in Mixnitz, Styria, during 1920–23. In addition to Dollo, Abel was influenced by the writings of Russian paleomammalogist Vladimir Kovalevsky (1842–1883) and American biologist Osborn, who at the time was doing much to popularize vertebrate paleontology while contributing to contemporary evolutionary theory in the U.S.

In his *Bully for Brontosaurus* (1991), Stephen J. Gould refers to Abel as a "...dark figure who vitiated the memory of his fine paleontological work by later activity in the Austrian Nazi party," certainly a serious charge, especially today.[10] Actually, Abel was a politically conservative "Deutsch National," but in order to survive in his prominent position during the second World War, he may have made compromises with the pro–Nazis. Has the once prominent Abel become an obscure figure because of his unpopular collaboration with the Nazis? Possibly, but more directly it may be because few (if any) of his 287 publications have been translated into English.[11]

Abel earned several academic honors and honorary degrees during a distinguished scientific career. He served as the Dean and Director of the University of Vienna, conducted research at the University of Gottingen, and was the founder of a short-lived Institute of Paleobiology in Salzburg. He may perhaps be regarded as the European (scientific) equivalent of America's Osborn.[12] Certainly a scholarly treatment of Othenio Abel's life and career would be a worthy addition to the popular assortment of biographies and autobiographies of famous paleontologists and other celebrities in the field of paleontology. The fact that Abel maintained a diary and penned an autobiographical sketch in 1945 would aid anyone in this endeavor.

CHAPTER 15

Zdenek Burian's Global Visions of Prehistory

By far the most prolific paleoartist was Zdenek Burian (1905–1981). Relatively little is known of Burian outside his native land. He was born in the town of Koprivnice in northern Moravia on February 11, 1905. In 1919 he undertook an Art curriculum at the Academy of Prague. Thereafter, he met Czech paleontologist and stratigrapher Josef Augusta (1903–1968). Inspired by the renderings of Charles R. Knight, Burian produced many scenes of prehistoric fauna and flora that became known to an English-speaking audience, principally through the writings of Augusta. (See Figure 15-1.)

In 1935 Burian completed his first fossil restoration, that of two species of *Temnospondyli* (labyrinthodonts). Over the next 46 years, Burian would complete nearly 500 prehistoric animal restorations and landscapes.[1]

The Burian/Augusta association grew over the years, and today their work is best represented by six magnificent volumes—*Prehistoric Animals* (1956), *Prehistoric Man* (1960), *Prehistoric Reptiles and Birds* (1961), *The Book of Mammoths* (1962), *Prehistoric Sea Monsters* (1964) and *The Age of Monsters* (1966).[2] These were written by Augusta and featured Burian's superb paintings and illustrations. There were other books too, published in Czech, including *Life Long Gone* (1941), *Prehistoric Wonders* (1942), *Lost World* (1948) and *From the Prehistoric Deep* (1949).[3]

Other adventure and science fantasy books delving into prehistoric themes were written for Czech children during the German occupation. Burian's artistry was also featured in books written by Zdenek Spinar (1916–1995)—*Life Before Man* (1981) and Josef Wolf—*The Dawn of Man* (1978).[4]

Burian became such a highly regarded artist that in the Czech Republic his paintings are now considered national treasures. In spite of this, much of his fine art was stolen from the Burian residence shortly after his death in 1981.[5]

Although Burian's paintings have been reproduced in popular "dinosaur" books over the years, some have been (rather undeservedly) subjected to criticism. For instance, Gregory Paul states in his *Predatory Dinosaurs of the World: A Complete Illustrated Guide* (1988) that "Burian's often reproduced T. rex has a badly dwarfed head and lipless teeth—it is not good at all." Also, it is not currently believed that sauropod dinosaur genera could submerge themselves as depicted in Burian's 1941 *Brachiosaurus* painting, appearing as Plate 26 in Augusta's *Prehistoric Animals*.[6] In spite of such inaccuracies, however, it would be sheer understatement to claim that Burian's artistry was first rate.

It certainly is not possible to do justice to the numerous "paleoscenes" created by Mr. Burian in this short chapter. Instead let's focus briefly on the most successful, if not distinctive, Augusta/Burian collaboration, *Prehistoric Animals* (translated by Dr. Greta Holt for Spring Books, London). The book showcases a series of paintings that are chronologically-arranged "scenes from deep time."

This lavishly presented, innovatively designed book, with pages measuring 13¾" by 9½", contains 60 plates, 31 of which are in color. (Some of these are of such photographic quality that at a young age Allen recalls how difficult it was to be totally convinced that they were not simply photographs of animals alive today, even though such a conclusion contradicted sensibility.)

Although Augusta discusses some distinctly European landscapes that Burian has created, slightly less than half of the plates are of fauna indigenous to the North

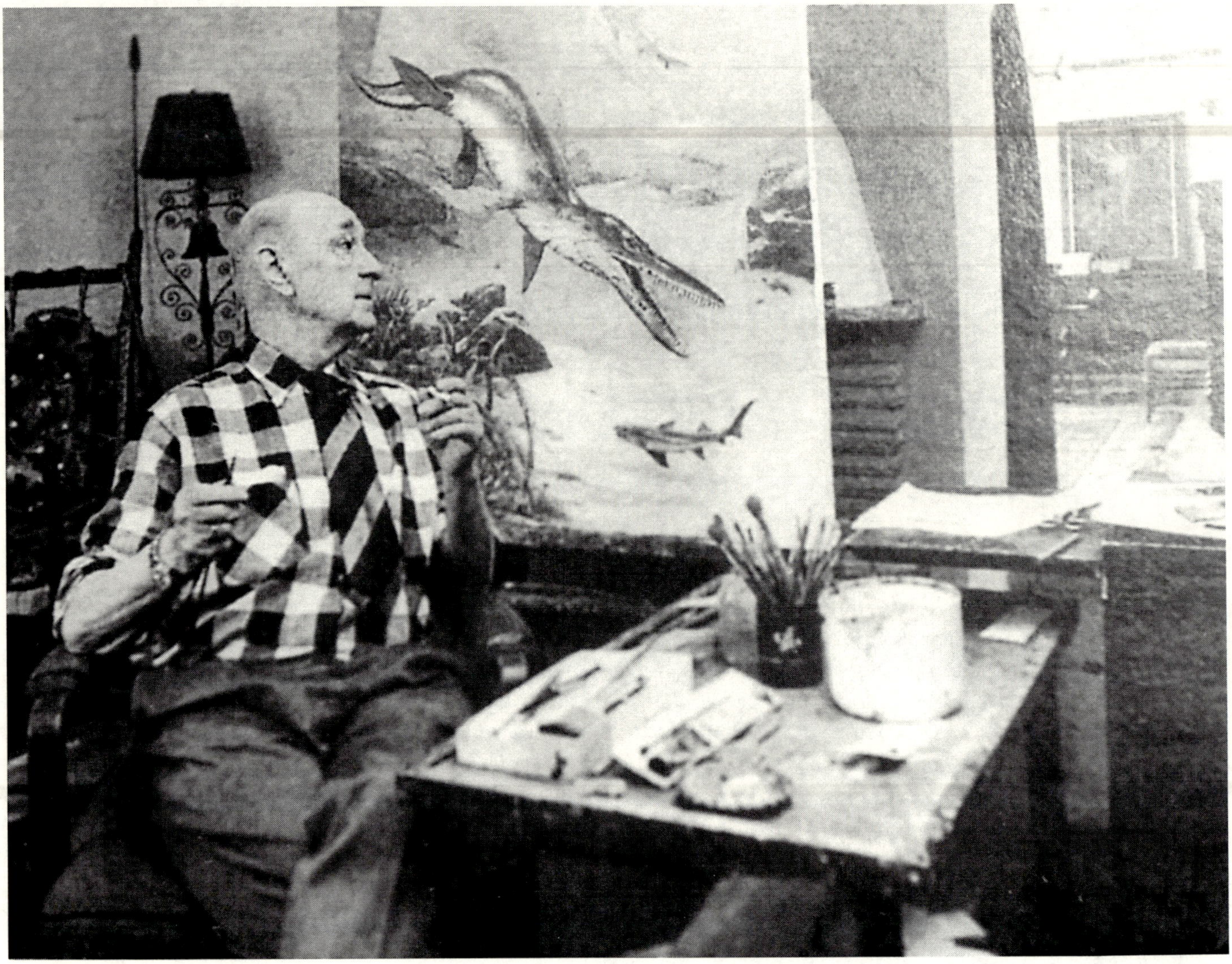

Figure 15-1: Zdenek Burian in his later years working on a pliosaur restoration in his studio. (Courtesy of Eva Hochmanova.)

American continent. Whereas Augusta's other similarly styled books are dedicated to more specific themes, in *Prehistoric Animals* the origin of life, the evolution of all organisms, paleoclimatology and the interplay between evolutionary forces and Life's history are discussed in an inspiring, majestic style. The pictorial account begins with life in *The Cambrian Sea* and concludes with a restoration of the *Cave Bear* being stalked by Stone Age hunters. The achievements of Charles Darwin are lauded, and the authors acknowledge contributions of two American paleontologists, William K. Gregory (1876–1970) and Alfred S. Romer (1894–1973), to our understanding of evolutionary principles through study of fossils.

In 1981 correspondence received from Dr. Vratislav Mazak (who established an 18 year friendship with Burian), Mazak indicated that Burian never compiled a listing of his own work. However, Mazak had "verified existence" of 1,023 oil paintings and 245 "quaches." Burian illustrated 451 books and more than 550 short stories that appeared in various magazines and journals. The total number of illustrations, as known to Mazak, was 14,000—of these, 352 being illustrations, mostly china-ink drawings.[7] In a more recent publication, *The Encyclopedia of Evolution* (1990), Richard Milner indicates that Burian's artworks exceeded 15,000 in number. But according to Burian admirer Thomas Kovacs, who met Mr. Petr Sadecky, Burian's friend and biographer, Burian completed about 20,000 pictures, or nearly one per day![8]

Burian didn't limit his talents to paleontology. The full range of Burian's portrayals includes "cowboys and Indians" themes, family domestic scenes, anthropological and sociological studies, science fictional scenes, and nature studies. He also illustrated book covers for works of fiction (translated into Czech).

Burian's daughter, Eva Hochmanova, wrote two short books documenting Burian's life. Allen had the good fortune to correspond with Ms. Hochmanova several years ago before her recent death. Her books were published (in Czechoslovakian) to commemorate the 80th and 10th anniversaries, respectively, of his birth and death. They are entitled *Zdenek Burian* (1985) and *Zdenek Burian: Pravek a Dobrodruzstvi* (1991).

Ms. Hochmanova stated that she would be appreciative of any deserving fame that can be gained for her father.[9] One significant reason for why we have so little general information at hand concerning Burian's most illustrious career can be related to the Nazi occupation during World War II.[10] Afterward, Communist repression of the geological sciences in the Czech Republic certainly stifled what would have otherwise have been disseminated about Burian.

There are worlds of paleoimagery, past and present, still awaiting discovery!

Chapter 16

The Ages of Zallinger

After the period when Charles R. Knight's paleoart loomed prominently, yet before the hot-blooded Bakkerian era, dinosaurs of human imagination thrived during the "age" of Zallinger. Rudolph F. Zallinger (1919–1995) was born in Irkutsk, Siberia, in 1919. His father, Franz, who fought in World War I, also was an artist. The young Rudolph learned art techniques from Franz.

Rudolph entered the School of Fine Arts at Yale University in 1937. For many years he held the position of Artist in Residence at Yale University, teaching at the Hartford Art School of the University of Hartford. Zallinger taught art at Yale between 1942 and 1950. During his near half-century as an active "paleoartist," Zallinger's detailed artistry captivated many paleontology enthusiasts.

In 1941, having received instruction from Richard Swann Lull, Alfred Romer and other noted vertebrate paleontologists and paleobotanists, Zallinger was persuaded to paint a mural depicting the history of terrestrial life from the Devonian through the end of the Mesozoic era on a vacant wall that needed to be "spruced up" in the Peabody Museum at Yale The resulting mural, completed from 1943 to 1947, spanned 300 million years of geological time in its 110 foot length. It won a Pulitzer Prize in 1949. A Jurassic scene from this mural was reproduced on a United States postage stamp in 1970 in a block of four stamps commemorating, rather ironically, the institution of the American Museum of Natural History.

The story of how this famous mural was completed is recounted in a short, informative book entitled *The Age of Reptiles: The Great Dinosaur Mural at Yale* (1990), by Vincent Scully, Rudolph F. Zallinger, Leo J. Hickey and John H. Ostrom.[1] Three early sketches and "cartoons" used in designing the mural are reproduced in this splendid publication. Incidentally, until the early 1980s, all published restorations of *The Age of Reptiles* were of a preliminary cartoon, due to the fact that skeletons exhibited in the museum hallway blocked direct photography of the mural.

Zallinger's magnificent effort may be the finest of its kind, but it was certainly not the first of its kind. Whereas Zallinger's *Age of Reptiles* painting traces evolutionary history in a horizontal sense, with time "flowing" toward the present in a right to left direction, over a century earlier, artists John Wichelo and John Emslie had completed similar (though smaller scale) compositions showing successive geological scenes as if taken from a single, vertically oriented stratigraphical section. In a popular publication, *The World We Live In* (*Life*, 1955), so as to "flow" with the sense of time suggested in this volume (Part V, *The Pageant of Life*), a representation of Zallinger's restoration is shown in inverted sequence.[2]

The Age of Reptiles yields an impression of time flowing continuously from the Devonian to the upper Cretaceous. However, due to the incompleteness of the stratigraphical record, geologists do not have an *unbroken* representation of this great time interval. Closer examination of Zallinger's masterpiece reveals subtle junctures situated in the painting. Trees distinguish the beginning of the three Mesozoic periods. Erupting volcanoes ominously signal the end of the Mesozoic era.

Zallinger's painting effectively illustrates that "all dinosaurs" did not live together (in time). However, the painting also produces false impressions of spatial continuity. The individual "vignettes" shown along the mural's expanse can easily be misinterpreted to mean that the restored fauna all inhabited the same continental region, but during successive stages of time. In fact, this is not the case. Fossils used as the basis of restorations have come from a broad range of localities (although the most popular scenes, representing 85 million years of time, are of

Jurassic and Cretaceous fauna, known primarily from prehistoric North America). (See Figure 16-1.)

Harry Lange's cover illustration of the March–May 1959 issue of an obscure publication named *Space Journal* obviously borrowed from Zallinger's Jurassic scene. This scene, showing sauropods in strikingly similar (but reversed) poses to those in *The Age of Reptiles* mural quietly browsing in the swamp of an alien world with a peculiar looking spaceship in the background, was included to highlight the feature topic in the issue *Other Beings on Other Planets?* This example shows the broad influences Zallinger's artistry has had on the public.

Although his choice of colors for dinosaurs, such as the blue stripes on the *Allosaurus*, in his 1947 mural was imaginatively inspired by the colors of modern reptiles, Zallinger stated that "...even some paleontologists catch themselves regarding them as true colors."[3] Zallinger's restoration of Cretaceous aquatic reptiles, published in *The World We Live In*, relied on the "fish-eye" aquarium perspective first utilized 120 years before by one of the earliest practitioners of "paleo art," William Conybeare.[4]

Another major undertaking reached fruition in the form of paintings portraying North America's age of mammals. *Life* magazine had published Zallinger's impressions of Cenozoic life history in *The World We Live In*. (Zallinger's *The Age of Reptiles* scenes had previously been issued as a *Life* magazine series between December 1952 and December 1954.) Eventually, between 1957 and 1967, under the careful advisement of noted mammalian vertebrate paleontologists, *The Age of Mammals* was substantially revised into a 60-foot-long mural in the Peabody Museum. According to Zallinger, the mural is so durable that it is likely to outlast the building! Zallinger, with wife Jean Day, also depicted the Pleistocene "invasion" of South America by North American mammalian fauna, which appeared in a *Life* magazine, January 26, 1959, installment

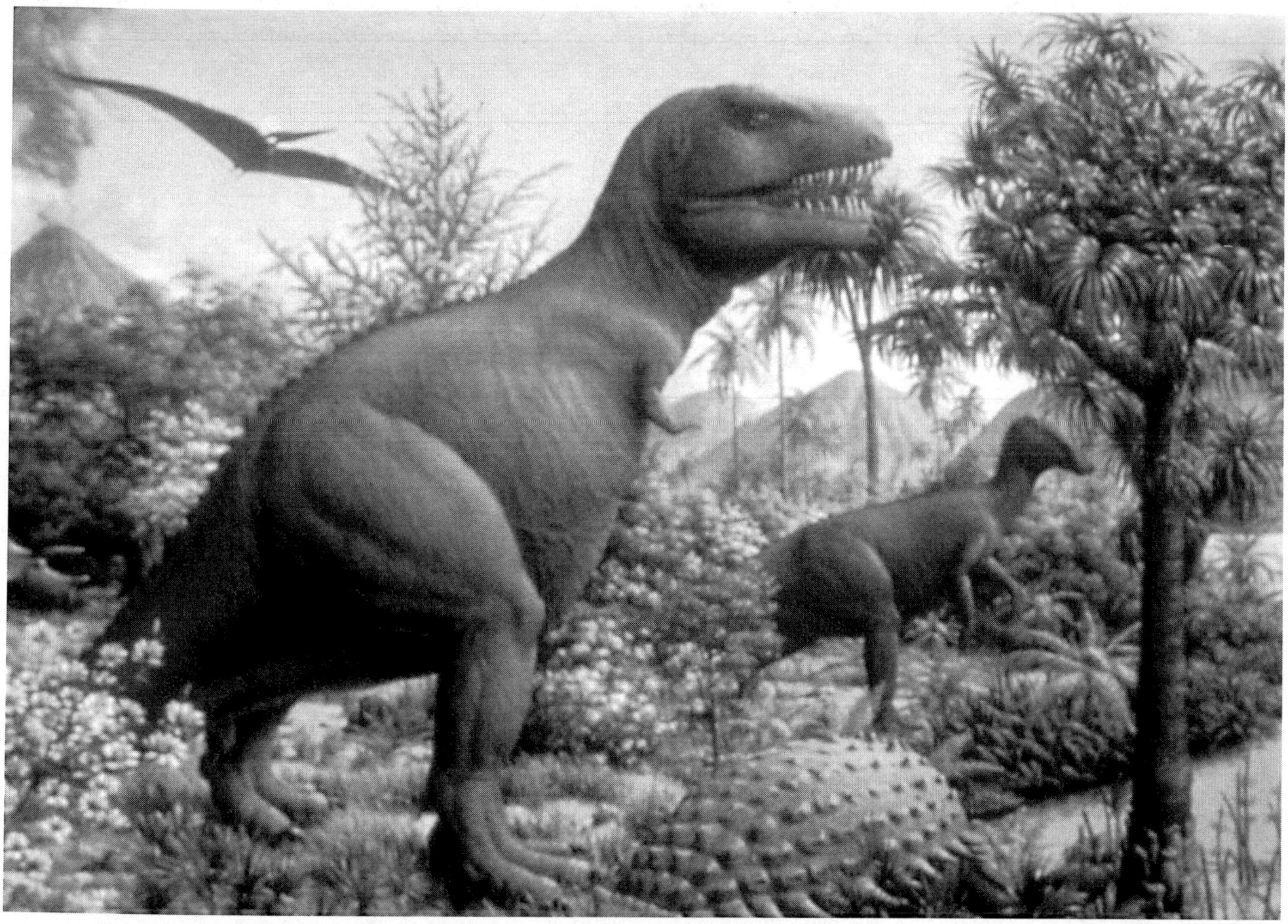

Figure 16-1: Scene from Rudolph F. Zallinger's *The Age of Reptiles*, perhaps the signature landscape painting for Man's concept of the Cretaceous. Depicted are (left to right) *Triceratops, Pteranodon, Tyrannosaurus, Ankylosaurus* and *Edmontosaurus*. Erupting volcanoes signal the end of an era. (The Age of Reptiles reproduced with permission. Copyright Yale Peabody Museum of Natural History, Yale University, New Haven, CT.)

by Lincoln Barnett featuring "Darwin's World of Nature."[5]

Zallinger also provided wonderful drawings for The Giant Golden Book, *Dinosaurs and Other Prehistoric Reptiles* (1965), by Jane Werner Watson.[6] We viewed several of Zallinger's original "Golden Book" paintings on January 13, 1990, several of which were included in the Golden Book collection, in the "Dinosaurs Past and Present" travelling exhibit (so thoughtfully and meticulously organized by Sylvia Czerkas) when it reached the Field Museum of Natural History in Chicago. Although Watson's book was intended for a young audience, Zallinger's first-rate, colorful restorations of the *Protoceratops, Diplodocus* and *Anatosaurus* still captivated us.

Although renown for his paleoart, Rudolph Zallinger depicted a diverse array of natural history and historical subjects, including scenes from Tropical rain forests and the Russian Revolution. Zallinger and his wife, Jean Day Zallinger, also an artist, together depicted scenes showing the *Dawn of Religion* on the island of Crete 3,500 years ago, when it was inhabited by Cro-Magnon Man. Several of Jean's dinosaur restorations also toured the country in the "Dinosaurs Past and Present" exhibit. One of their sons, Peter Zallinger, has carried on the family tradition of restoring prehistoric animals, as testified by his 1978 children's' book, *Prehistoric Animals*.[7]

Although Rudolph Zallinger died in the summer of 1995, his legacy lives on. Poster-sized prints of Zallinger's Peabody Museum murals adorn the walls of children's' and recreation rooms across the country. Museum postcards and several toys and model kits (e.g. Chialu, Marx and Lunar Models)[8] have commemorated the image of Zallinger's *Age of Reptiles*.

The Mesozoic and Cenozoic periods may not have actually existed quite as Zallinger envisioned them in his timeless masterpieces. However, many of us would agree that few visions of these younger worlds match the integrity and grandeur of Zallinger's. Assuredly, someday we shall be treated to a memorial volume dedicated exclusively to art in the "ages" of Zallinger!

Chapter 17

Neave Parker's Prehistoric World

Doggone it! Had Neave Parker truly been psychic he would have surely contacted us from the afterlife by now. Instead, this chapter came about through traditional means, without séances or "guiding" mediums. Many of you may have noted Parker's paleoart in popular paleontology publications but likely know little about the artist. So, submitted for your approval, here is our dedication to Neave Parker, paleontology's masterful necromancer.

Neave Parker's impressive prehistoric animal restorations are favored among an older generation of paleophiles. Thanks to Dr. William Swinton (1900–1994), recently deceased and formerly of the British Museum of Natural History and the University of Toronto, and also Dr. Maurice Burton, a mammalogist formerly associated with the British Museum, some details concerning Parker's paleoart career have been rescued from obscurity.[1]

Parker was born in 1910.[2] He was dissuaded from attending art school by his father. Instead, he was encouraged to seek the relative stability of a job as bank accountant. However, after one regrettable week in this position, Parker was kindly but firmly advised by his supervisor to try another profession. "Each day there was an error in the books and the whole staff had to stay behind until the error was found."[3] Parker became a surveyor for a short period, then enlisted during World War II.

Parker served in the Royal Air Force, in the photographic unit. It was here that he once on impulse sketched a man's face and handed it to his sergeant, asking, "Does that mean anything to you?"[4] The sergeant answered faintly that it resembled his father who had died a year before. Parker admitted that at one time he had contacted the Psychical Research Society for testing, but without significant result. Burton claimed that although in later years he sometimes questioned Parker on matters of the occult, "...he was obviously unwilling to talk about these matters ... he would dry up and refuse to discuss it, as if afraid of the subject."[5]

Burton sought Parker's assistance in the preparation of numerous color illustrations for a text entitled *The Story of Mammals*. This book would have been among the first popular books to be published on this subject. Unfortunately, despite the fact that both collaborators received payment for their efforts, at the last moment the project was abandoned by the publishers. Nonetheless, from their first association a lasting companionship resulted.

Burton, who was serving as honorary science editor at *The Illustrated London News*, also introduced Parker to the journal's editor, who was then looking for an artist capable of drawing animals. Between 1951 and 1960, Parker was commissioned by *The Illustrated London News* to produce beautiful double spreads of animals, both living and extinct.

Burton introduced him to Swinton, who facilitated Parker's restoration of many Mesozoic vertebrate forms. Swinton recalled that Parker would bring his trial efforts, done on yellow paper, for evaluation. Later, as the subject neared completion, a "finger and toe" session would be conducted to avoid problems of polydactyly (too many restored digits).

To the delight of the world's dinosaur enthusiasts of the time, Parker collaborated with Swinton in the completion of numerous restorations of Mesozoic vertebrates. A particularly striking series of articles appeared between 1956 and 1960 in *The Illustrated London News*, in which several well characterized Jurassic and Cretaceous fauna were described by Dr. Swinton and depicted by Parker. These illustrations were at a later time incorporated into the book *The Dinosaurs* (1970).[6]

Several of Parker's other magnificent dinosaur illustrations were also featured in Edwin H. Colbert's *Dinosaurs: Their Discovery and Their World* (1961).[7] A series of now out of print postcards, offered for sale by the British Museum during the late 1950s through early 1970s, bearing striking images of Parker's dinosaurs and other prehistoric animals are prized among collectors.[8]

Although artists had been striving to model particularly "small groupings" of dinosaurs since the 1850s, before Parker no artist had attempted to restore the world's "communities" of dinosaurs, as known to mid–twentieth century contemporary science. However, under Swinton's watchful eye, Parker pioneered the art of restoring entire paleo-environments of dinosaurs, restored in their probable life habitats, in regional settings, as represented in certain well known fossil collecting localities.

Interestingly, Parker's studies were completed before the effects of continental drift on paleo-geography and evolution of life forms indigenous to the rifting continents became generally accepted. Now, with the basics of plate tectonics even presented in grade school science classes, this has become a theme of scientific restoration since employed by other scientific illustrators, such as Peter Snowball and Mark Hallett.[9]

Vivid animation and near photographic quality, achieved in monochrome gouache and wash drawings, became trademarks of Parker's distinctive style in the restoration of prehistoric vertebrates. In tribute to Parker's skill, it is evident that other artists have relied on his restorations as a basis for their own prehistoric illustrations (particularly those of old favorite dinosaurs known from scanty remains, such as *Polacanthus*, *Megalosaurus* and *Scelidosaurus*) for children's books and popular accounts of paleontology.

Swinton wrote that Parker was "meticulous and wholly honest ... it was a pity that some of his drawings ... were available in the British Museum without much attribution. His original drawings for *The Illustrated London News* were on show and sale in the early 1970s ... advertised for 80 to 90 pounds."[10] Swinton would have purchased them himself, but even then the price seemed excessive. Parker was typically paid between 10 and 15 guineas for each illustration. Such a bargain! Where are Parker's famous illustrations now, dinosaur fans?

In 1989 the Ulster Museum in Belfast acquired 17 of Parker's 18 *Illustrated London News* illustrations during an auction of *The Illustrated London News* archive. In 1997 the missing 18th illustration was found and also purchased by the Ulster Museum. For each, Parker used high quality artist board, measuring approximately 28 by 17 inches, which was affixed to cheaper millboard to keep the drawings flat for reproduction at the *Illustrated London News*. Consequently, the illustrations required considerable restoration. According to Peter Crowther, Keeper of Geology of the National Museums and Galleries of Northern Ireland, the millboard had a long-term damaging effect on the paper. "Treatment involved the removal of the acidic backing millboard, and washing and deacidifying the artist's board to remove acids and discoloration."[11]

Sixteen of Parker's original *Illustrated London News* illustrations were displayed at the Ulster Museum in 1993 during a special exhibition entitled, "The Prehistoric World of Neave Parker."[12] There, patrons could admire cult-masterpieces such as *Dinosaurs of England* (*The Illustrated London News* publication date—November 30, 1954), *Dinosaurs of the Gobi Desert* (*The Illustrated London News* publication date—January 9, 1960), *Chinese Dinosaurs* (*The Illustrated London News* publication date—February 11, 1956), and *The Last Days of the Dinosaurs* (*The Illustrated London News* publication date March 5, 1960). (See Figures 17-1 and 17-2.) And, curiously, there was one canvas entitled, *Neave Parker and Dog Ghosts* (*The Illustrated London News* publication date June 23, 1956). No, we're not kidding!

Swinton stated that Parker was a great artist, although he knew little of the dinosaurs that he illustrated. Parker was also highly praised by Kenneth Oakley, a specialist on early Man. Oakley claimed, "It is quite remarkable how he [Parker] seems to grasp the details of his subject with only a minimum of help from me as if he could see into the past."[13]

Both Burton and Swinton noted that Parker had an eerie ability to "see" the ghosts of dogs, animals of which he was very fond (presumably when they were alive). Apparently, Parker who had great technical competency in restoring pictures of animals merely from their bones and skin, also on occasion would describe with great accuracy the appearance of dead dogs to their bewildered former owners. It must be emphasized that this latter "skill" did not deter Parker's illustrative capability regarding fossil animals.

In his later years, Parker was rotund, of medium height, and possessed both bushy mustache and horn-rimmed spectacles. Facially, he resembled the American President Teddy Roosevelt. (See Figure 17-3.) He was fond

***Opposite top:* Figure 17-1: *Dinosaurs of England*—135 million years ago, by Neave Parker (1954). Dinosaurs featured in this idealized grouping are (left to right) *Hypsilophodon, Polacanthus, Megalosaurus, Cetiosaurus* and *Iguanodon*. (With permission from the *Illustrated London News*.) *Opposite bottom:* Figure 17-2: *The Last Days of the Dinosaurs* by Neave Parker (1960). Dinosaurs featured in this restoration are North American genera (left to right) *Euoplocephalus, Triceratops, Anatotitan, Tyrannosaurus, Torosaurus* and *Styracosaurus*. (With permission from the *Illustrated London News*.)**

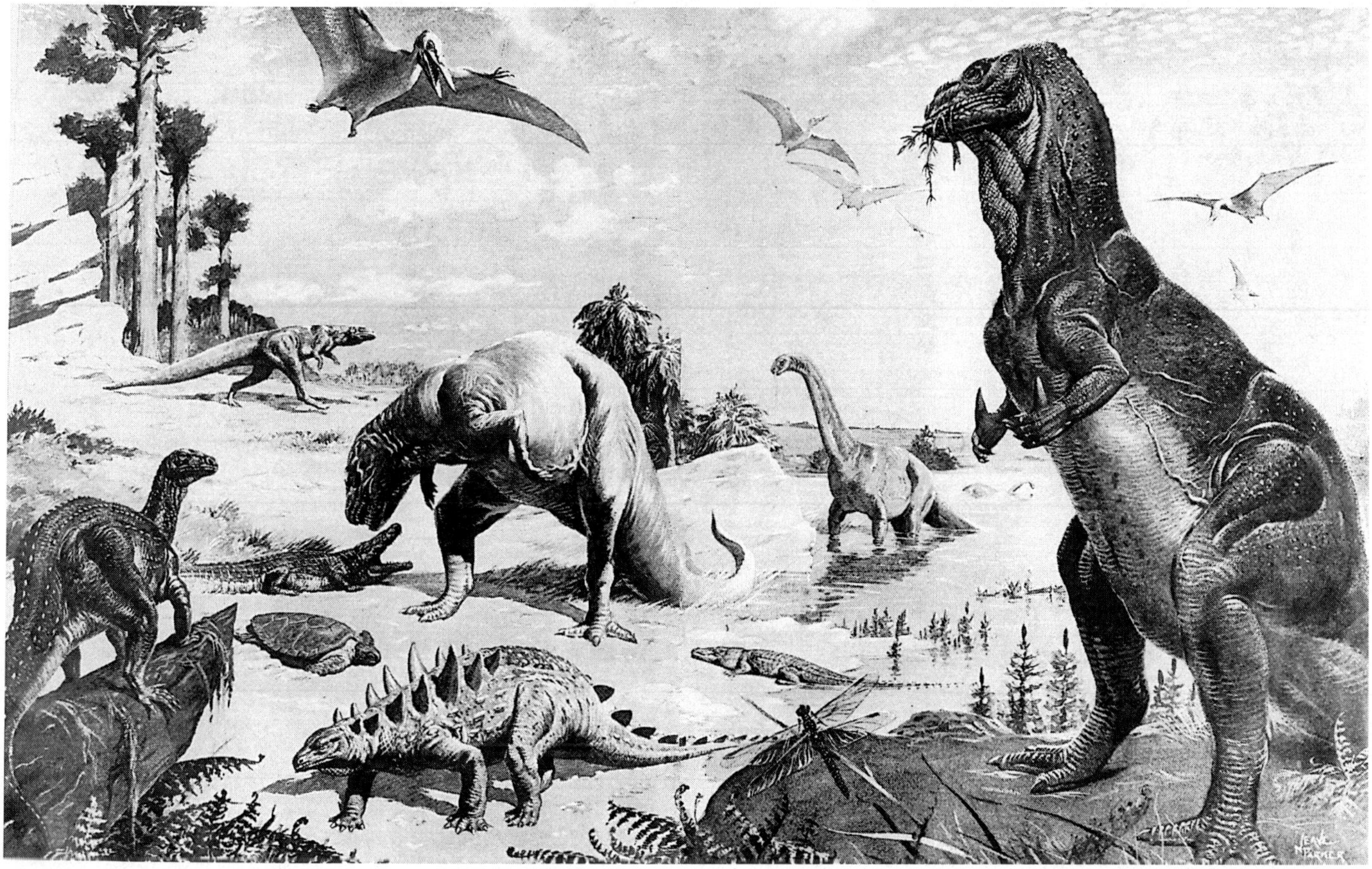

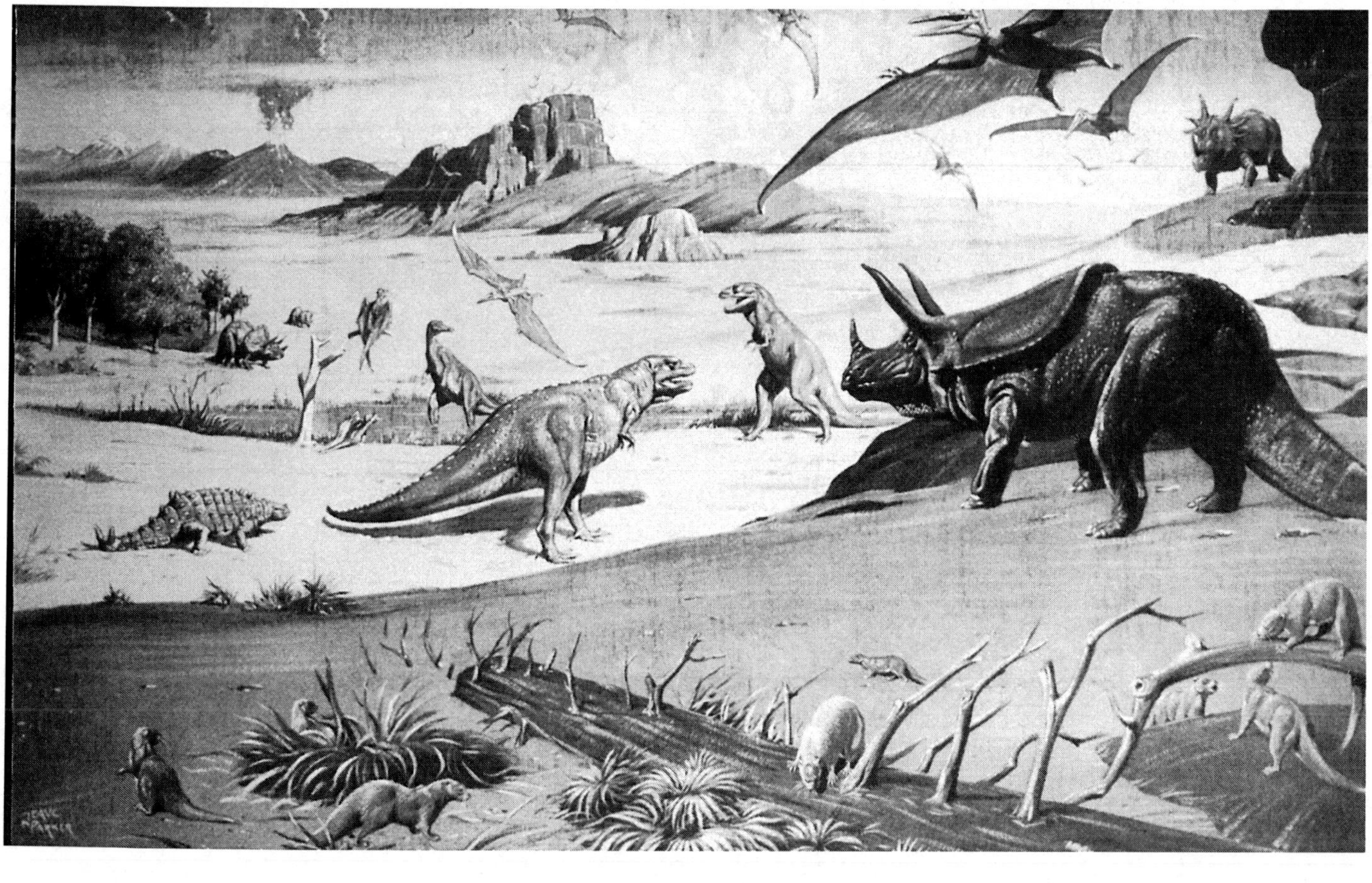

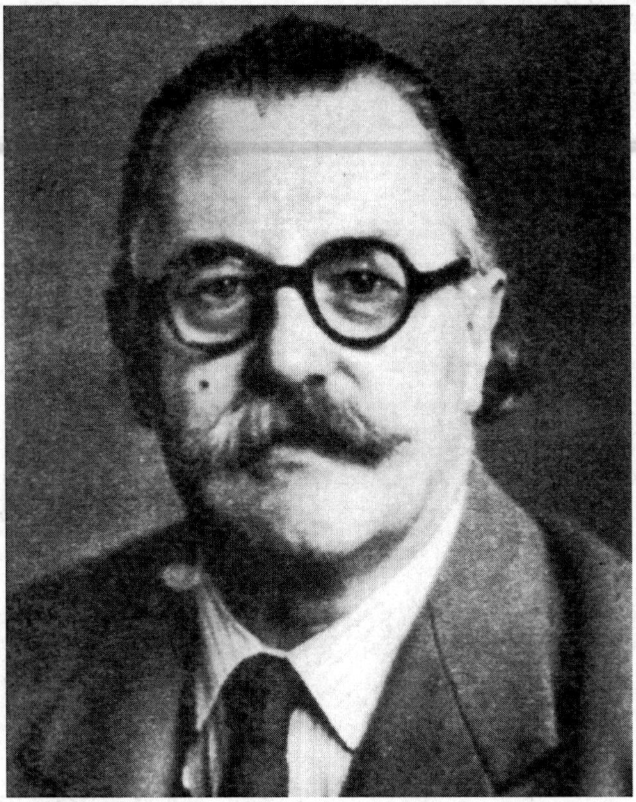

Figure 17-3: Photographic portrait of Neave Parker. (With permission from the *Illustrated London News*.)

of food and beer, though he rarely, "*lost his equilibrium.*" Parker's principal hobby was pistol-shooting, a sport at which he excelled, having won several awards, including a British Open Championship title. He also enjoyed photography and movies. In fact, he suffered the stroke that killed him at a local cinema, on May 16, 1961. He was fifty-one. According to Swinton, Parker was buried in a "West London cemetery on Friday, May 19th at 2:00 P.M."[14]

Although Parker remained a bachelor, he was fond of children. This is perhaps exemplified through the delightful animal drawings he prepared for a children's book, *Uncle Remus* (Bruce and Gawthorn, Ltd., 1956), and also a book that he wrote—as well as illustrated—for children, *The Tall Green House*, (P.R. Gawthorn, Ltd., 1948). Burton described Parker as "...a most agreeable companion who allowed intimates to refer to him as 'Toots.' It seemed absurd for a man of his capabilities, yet he carried it without loss of dignity. It did, however, seem not wholly inappropriate because of the impish streak in him that often came to the surface."[15]

Despite a few idiosyncracies, Parker was very highly regarded by his associates at the British Museum and the *Illustrated London News*. The state of scientific knowledge concerning many of the prehistoric animals Parker illustrated for scientists has changed considerably. However, his restorations do most vividly represent formerly held opinions of how such creatures appeared and lived, and must be accepted today among the other classically regarded illustrations dating from that period of understanding.

Whether it was the "ghosts" of dogs, or those of the soaring *Pteranodon* or the seafaring *Macroplata*, Neave Parker's genius for recreating the past on canvas cannot be disputed. As we gaze upon his works today, the words of vertebrate paleontologist Dale Russell take meaning, "...artists are the eyes of paleontologists ... paintings are the windows through which nonspecialists can see the world of the dinosaur."[16]

CHAPTER 18

Acanthopholis— *a "Twilight Zone" Dinosaur*

In the process of designing dinosaur dioramas, the occasional urge to merge scientific respectability and outright speculation proves simply irresistible. Even the most sensible paleoartists can be mysteriously drawn to the "Zone," that invisible boundary between scientific sensibility and the irrational. Submitted for your approval, in this article I shall confess the guilty pleasures of sculpting "fringe" dinosaurs, those dinosaurs which are known only from scanty remains.[1]

We all seem to have an uncanny devotion for that "virtual reality" of fringe dinosaurs. The first life-sized dinosaur restorations (restored by Sir Richard Owen and Benjamin Waterhouse Hawkins) were in-the-flesh, three dimensional figures founded on fossil specimens of insufficient quality. Reconstructed skeletons would have been a logical preliminary step. However, skeletons would have seemed far less dramatic. Fortunately, Victorian admirers didn't judge the Crystal Palace dinosaurs' accuracy. It only mattered that the dinosaurs of those original "Zone" masters—Hawkins and Owen—seemed "alive" and totally breathtaking.

Megalosaurus, *Iguanodon* and even *Hylaeosaurus* are better understood than was the case 150 years ago. But there is another time-honored genus, *Acanthopholis*, that continues to ba· e us as much today as it did in the Victorian era.

Thomas Henry Huxley, Darwin's "bulldog" in the often distorted tale of how those waging the war for evolutionary sciences "triumphed" over the Victorian Creationists, has the distinction of having named three of Great Britain's dinosaur genera. One of these, the *Acanthopholis*, remains one of our more reclusive dinosaurs. And yet, this peculiar dinosaur has become rather familiar to us dinosaur aficionados through the art (and imaginations) of talented "dinosaurologists." (See Figure 18-1.) For instance, many dinophiles could readily recognize Neave Parker's famous 1950s restoration, which appeared on British Museum postcards as well as in popular paleontology books.[2] However, few may be familiar with the tale of how the scanty remains of *Acanthopohlis* fell into Huxley's hands, or how the scientific status of the *Acanthopholis* has suffered in the days since its early description.

Huxley's detailed description of the fragmentary remains was originally published in an 1867 issue of *The Geological Magazine*.[3] The fossils were discovered by a resident of Folkestone in southern England named Mr. Griffiths, near Copt Point, in the fossiliferous Lower Chalk beds (Cenomanian stage—now known to be about 95 million years old). Here, in a rather complex geological association, softer Upper Greensand beds were eroding to expose the underlying Lower Chalk. On preliminary examination of the fossils, Huxley realized that they represented a new sort of reptilian creature, closely allied to the already described armored dinosaurian forms *Scelidosaurus*, *Hylaeosaurus* and *Polacanthus*. Huxley persuaded Griffiths to continue the search in the hope of identifying more material belonging to the creature. Griffiths was successful in extricating several more pieces for Huxley from the chalk cliffs, which were situated in an "inconvenient" (inaccessible) intertidal zone. A geologist, Robert Etheridge, supplied Huxley with precise details concerning the biostratigraphical horizon in which the *Acanthopholis* had been found, and noted that an *Ichthyosaurus* had been found in the same bed, although six feet lower than the position of the former.

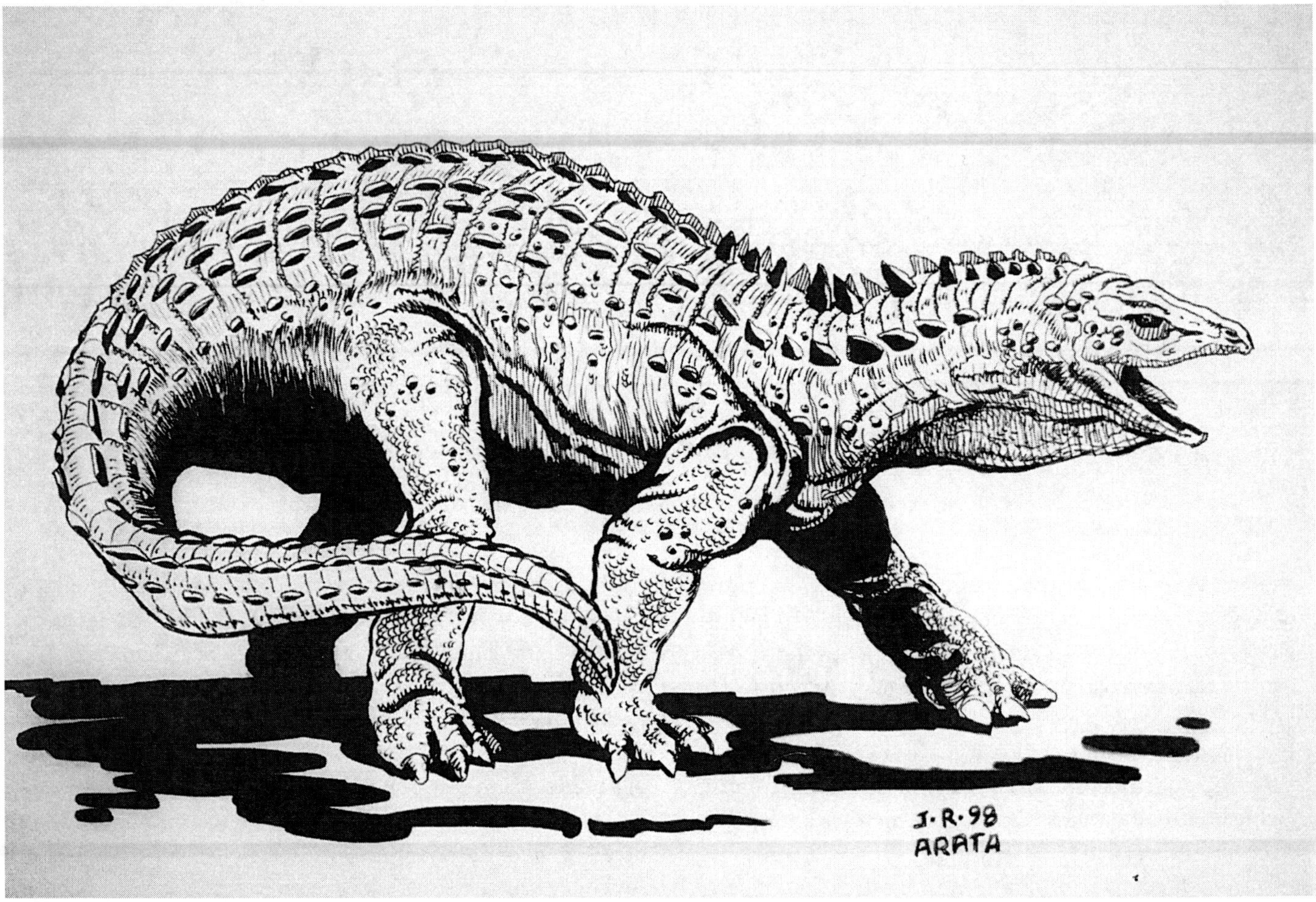

Figure 18-1: Jack Arata's illustration of *Acanthopholis* (1998), based on Neave Parker's late 1950s British Museum restoration. (Reproduced with permission.)

Only three teeth with strange color bandings; several nine-inch-long, unsymmetrically shaped spine-like dermal plates, (hundreds of) oval shaped bony scutes; dorsal vertebrae; fragmentary pieces of skull; and the distal end of a humerus were collected. However, on the basis of these remains, Huxley was able to differentiate a new variety of armored reptilian from those previously described. The teeth differed significantly from those of *Scelidosaurus*, and features of dermal armor created distinctions from the armor sported by *Hylaeosaurus* and *Polacanthus*. No pelvis (sacrum) was found. Clearly there wasn't much material on which to found a reliable description. However, the fact that this dinosaur was found in a geologically younger stratum than those in which the other known British armored genera had been discovered in (e.g. Wealden beds—Lower Cretaceous) fortified the notion that England's new dinosaur, named *Acanthopholis horridus* ("horrid thorn bearer"), by Huxley, represented a "higher" form of reptile than the others.

Between 1869 and 1871, Harry G. Seeley (who, in 1901, authored a precocious study of pterosaurs, *Dragons of the Air*) ascribed other remains to the genus *Acanthopholis*. In the process, four new species were created which are no longer considered valid. Over the next intervening century, only cursory references were made to the *Acanthopholis* in popular dinosaur books.[4] Intriguingly, *Acanthopholis* is not mentioned by name (or restored by his artist, Joseph Smit) in editions of the Reverend Henry N. Hutchinson's popular *Extinct Monsters* (1893) and *Extinct Monsters and Creatures of Other Days* (1917 ed.).

Although by 1979 *Acanthopholis* had been relegated into general obscurity, George Olshevsky restored the *Acanthopholis* as a "nodosaur" in his privately published chart titled *The Ankylosaur Family Tree*, "...only because it is figured frequently in modern dinosaur books."[5] Olshevsky wisely cautioned that there was no reason for believing that his restoration was correct, and suggested that because nodosaurs were anatomically "conservative" it is reasonable to anticipate that the appearance of the *Acanthopholis* "did not differ significantly" from the (also rather poorly known) *Hylaeosaurus*. (See Figure 18-2.) Several years earlier, in 1971, Walter Coombs claimed that the genus

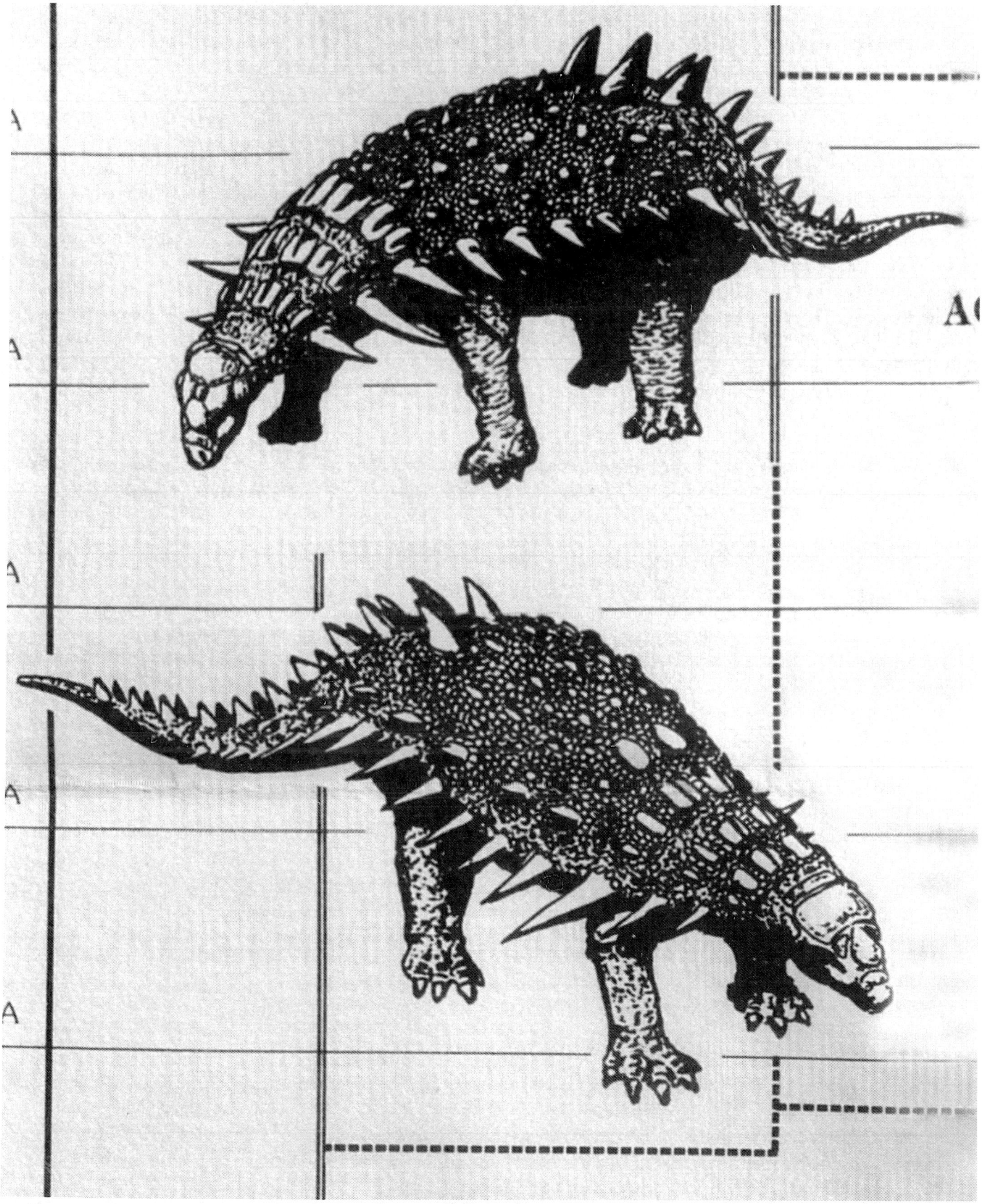

Figure 18-2.: George Olshevsky's speculative 1979 restorations of two British armored Cretaceous dinosaurs, *Acanthopholis* (top) and *Hylaeosaurus* (bottom). (Reproduced with permission.)

Acanthopholis may simply be a "nomen dubium" (dubious name). By the early 1990s, Coombs maintained that the type specimen of *Acanthopholis horridus* is valid.

Out of the late nineteenth century's (often overly enthusiastic) "mad dash" to name new fossil species and assign generic names to previously undescribed fossil remains, dinosaurs such as the *Acanthopholis* resulted. It is true that other dinosaurs have been named on the basis of even lesser material. However dubious its standing, *Acanthopholis* remains one of our "official" dinosaurs, although one must not regard any of the so far published "life" restorations too seriously. (This may be partially a reason for why there are few, if any, dinosaur "toys," life-sized sculptures or model kits of the *Acanthopholis* available today.[6])

Please permit further speculation on the matter of British armored dinosaurs, certainly not a crime in this instance, given that so much (artistic) speculation has already been proffered.

Only three summers ago, paleontologists were startled by the discovery of the 20-foot-long *Utahraptor*, from the Eastern U.S. Whether other large but as yet undiscovered carnivorous members of the raptor clan hunted in what is now southern England about the same time as the *Utahraptor* stalked its prey, is entirely speculative. And yet, perhaps some day a new European discovery will extend the (paleo)geographic range of these vicious carnivores. A low-centered body, enclosed in bony armor, such as may have been the case with the British nodosaurs, would have afforded adequate protection against "raptors," whose sickle-shaped talons were curved for downward (not upward) slashing efficiency. Why else would these armored "tanks" have evolved such peculiar and apparently efficient forms of defense? (Just work with me on this for awhile, okay?)

How interesting then, that British nodosaurs appear to be on the vanguard of the nodosaur assault (beginning with the Jurassic genus *Sarcolestes*)! Did the raptors evolve in Europe and subsequently spread to North America and Asia on the heels of evolving, armored ornithischians? Did selection pressures favor body armor over plates, driven by evolution of bloodthirsty raptors. Plates offered less abdominal protection from cruel downward slashing claws than did armored hide.

Remember that is only sheer speculation, fraught with uncertainty. Actually, an as yet undescribed Asian genus, tentatively named "*Jurassosaurus nedegoapeferkimorum*" in honor of the *Jurassic Park* movie, may predate England's *Sarcolestes* by several million years.[7] Yes, *Jurassosaurus*' age falsifies my seat-of-the-pants "theory." Yet who knows what else paleontologists will encounter in the fossiliferous British Mesozoic deposits. An even older nodosaur, perhaps? A decade ago there was the announcement of the peculiar *Baryonyx*, which is a Wealden dinosaur. And in the early 1990s came news of the discovery of new *Polacanthus* material from the Isle of Wight. Just for the fun of it, let your imaginations soar.

As a young adult I collected many Mesozoic fossils in England. Although my ammonite, sea urchin, clam and coral specimens proved immensely satisfying, there was always hope of finding dinosaur bones, certainly a possibility in the exposures I became familiar with! At that time I also corresponded with a commercial collector named "B.C." Andrews, from whom vertebrate fossils collected near Folkestone were purchased, near where the mysterious *Acanthopholis* had been found a century before. Yes, I understand the strange, sixth sense experienced by fossil collectors when a new "find" is about to be made, and the exhilaration which follows thereafter!

In my adult years I couldn't stop dreaming of stegosaurs. I sculpted over a dozen original miniature stegosaurs before penetrating into the quasi-theoretical realm, moving boldly forward into the "Zone." Doo doo doo doo, Doo doo doo doo (...piercing fife notes, followed by weird bongo drum beat)!!!!!!

By the summer of 1997 I had selected poses for a plated dinosaur scene illustrating my previously mentioned pet theory (merely a "Just So" story) "explaining" the cause of the plated dinosaurs' extinction.

So, imagine my elation when, in the spring of 1998, vertebrate paleontologist Steve Hutt reported to the media that an as yet undescribed 120-million-year-old, 12-foot-long "...creature, equipped with murderous claws, razor sharp teeth and unusually long hind legs on which it ran at high speed" had been discovered on the Isle of Wight![8] As of the fall of 2000, the Isle of Wight "velociraptor" hasn't been formally described, but during the fall of 1998 I had already seized the opportunity to self-servingly immortalize Hutt's tentative description.

The resulting sculpture, titled *Tails—You Lose!* (1998), symbolizes the plight of stegosaurs in the Middle Jurassic of Europe, with nasty raptors driving them toward the brink of extinction although in this scene I decided to portray *Lexovisaurus* in winning form). (See Figure 18-3.) In sculpting the plates and spines of *Lexovisaurus*, admittedly a fanciful arrangement, I relied principally on two (contrasting) skeletal reconstructions by paleoartist Tracy Ford. Tracy kindly offered additional insights via e-mail correspondence.[9]

My "raptor" restoration is of an entirely hypothetical, feathered evolutionary ancestor of the Isle of Wight species. Nevertheless, insulating feathers are far less conjectural with recent description of Chinese feathered dinosaurs *Caudipteryx*, *Protarchaeopteryx* and *Sinosauropteryx*.

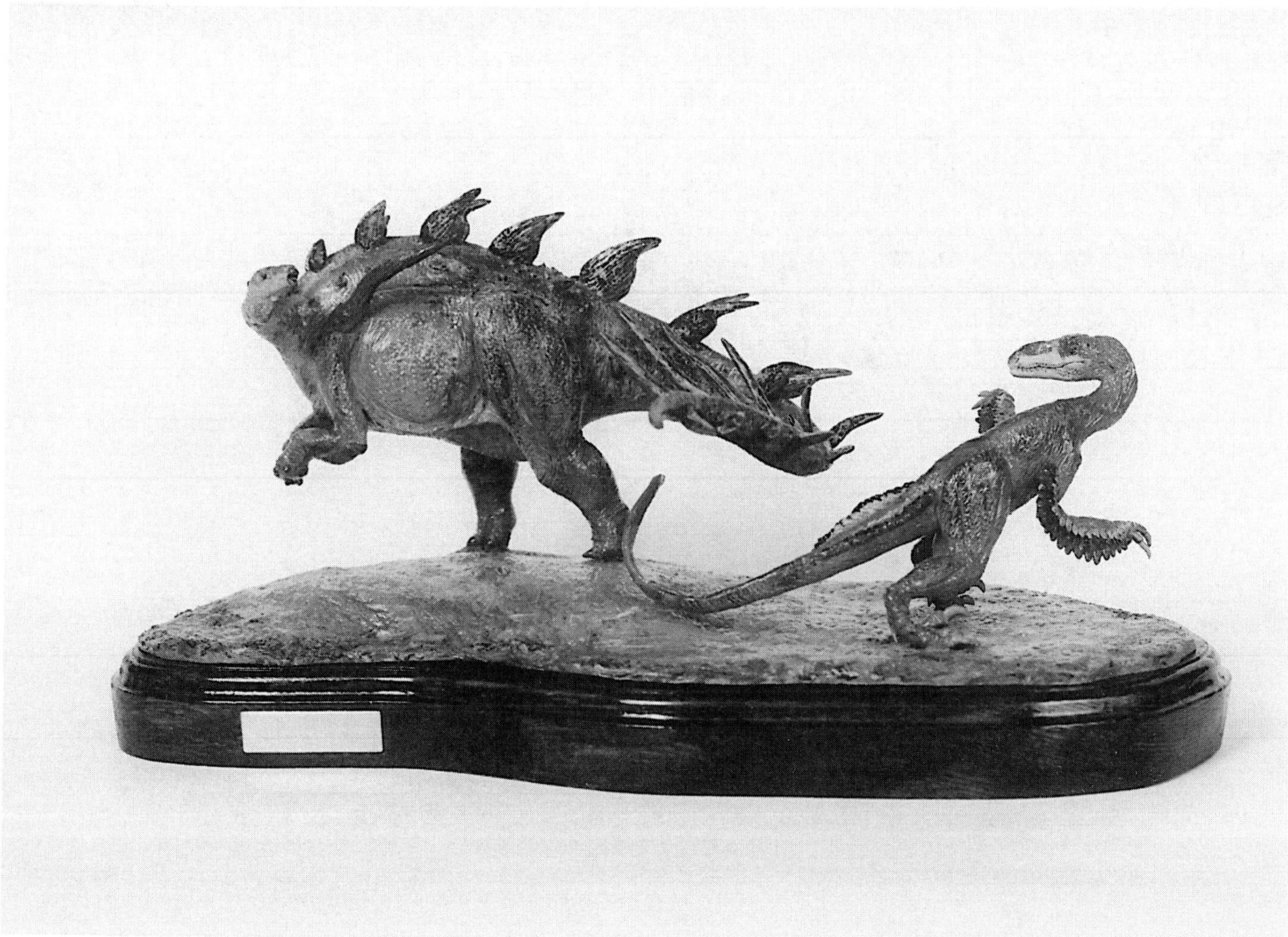

Figure 18-3: Allen A. Debus' sculpture, *Tails—You Lose!* (1998), showing speculative restoration of stegosaurid *Lexovisaurus* (left) and fictional feathered raptor. (See text for details.) Photograph by Frank Cozza.

Intriguingly, evidence for feathered dinosaurs extends back to the Upper Triassic, 220 million years ago![10]

Is this all distortion, mere poppycock, or is there a figment of reality to my restorations? Conservatively minded paleontologists—beware! Don't judge too harshly or quickly! The "Zone" works in mysterious ways! Just remember, the pleasure of working in the "Zone" is that we can be as loose and free as we wannabe (with the science, that is).

In 1999, *Acanthopholis'* status was relegated to a *nomen dubium*.[11] Is this the final verdict? Maybe, though assuredly some day (a bold prediction) the perplexities surrounding Huxley's *Acanthopholis* will be clarified in a dramatic way! *Acanthopholis*, ancient dinosaur dating from the Twilight Zone, shall rise once more from Time's tomb to stir the living!

CHAPTER 19

Bully for Laelaps

The celebrated American paleontologist Edward D. Cope wrote in a letter to his wife in March of 1897, "Mr. Knight ... has been here a couple of days and has been getting figures of ... Laelaps ... and other saurians. He is very original in his attitu*des*,"[1] about a young artist named Charles R. Knight. Their brief association would forge a legacy of modern paleorestoration! Cope's 75-million-year-old *Laelaps*, North America's first recognized carnivorous dinosaur, figuratively symbolizes how the science of "paleorestoration" matured during the late nineteenth century.

Cope described the theropod *Laelaps aquilunguis* (or "eagle-clawed terrible leaper") in 1866 on the basis of scanty remains, altogether not much more material than Buckland, Owen and Mantell had available decades earlier when they had studied *Iguanodon, Megalosaurus* and *Hylaeosaurus*.[2] Based on recent measurements of the tibia's cross sectional area, *Laelaps* may have weighed between 700 to 900 kilograms (approximately 1,500 to 1,600 pounds) in life.

On the basis of its relatively small arm bones, Cope concluded that *Laelaps* was bipedal, assuming a kangaroo-like posture while resting. However, Cope envisioned his 18-foot-long dinosaur attacking herbivores with its more powerful hind limbs. In 1867 Cope speculated that *Laelaps* may have been equipped with webbed toes to progress more efficiently while swimming, and that a warm-blooded physiology might have been required for leaping 30-foot distances in pursuit of prey.

Cope's interpretation inspired several life restorations of the *Laelaps*. *Laelaps* is portrayed as a kangaroo-like saurian in seven early illustrations, including one sketch signed by Cope and another published in the April 1869 issue of *American Naturalist*.[3] (See Figure 19-1.)

Cope and Philadelphia Academy of Sciences colleague Joseph Leidy fostered Hawkins' attempt to create a "Palaeozoic Museum" to be situated in New York City's Central Park. Under their direction, Hawkins began a full scale *Laelaps* reconstruction which was destroyed, along with other replicas, in 1871.[4] A few years later, Hawkins completed a painting for Princeton University, titled *Cretaceous Period in New Jersey* (1877), portraying a herd of *Hadrosaurus* under attack from a fiendish pack of predatory *Laelaps*. (See Figure 10-3.)

During the late 1870s, numerous specimens of the *Iguanodon* were unearthed from a coal mine in Bernissart, Belgium, which were described by the French-born Belgian paleontologist Louis Dollo. Through the work of Leidy, Cope and Dollo, Owen's erroneous interpretations were condemned.

Cope's chief rival was American paleontologist Othniel Marsh. Unlike Cope, Leidy and Dollo, Marsh conservatively avoided the practice of mounting original skeletons due to the likelihood of introducing error, especially when specimens were poorly preserved. Furthermore, Marsh published numerous *drawings* of dinosaur skeletal reconstructions, but cautioned against in-the-flesh restorations.

Unlike Marsh, Cope revelled in the opportunity to create visual images of how living dinosaurs may have appeared, often speculating on the basis of incomplete specimens. Passages in the unpublished autobiography of

Opposite top: **Figure 19-1: Restoration of the *Laelaps* (center and at top left), based on Cope's visions, published in an 1869 issue of *American Naturalist*. The two long-necked animals with flippers are elasmosaurs.** ***Opposite bottom:*** **Figure 19-2: Charles Knight's 1897 restoration of two fighting *Laelaps*. (Negative #335199. Courtesy Department of Library Services, American Museum of Natural History. Reproduced with permission.)**

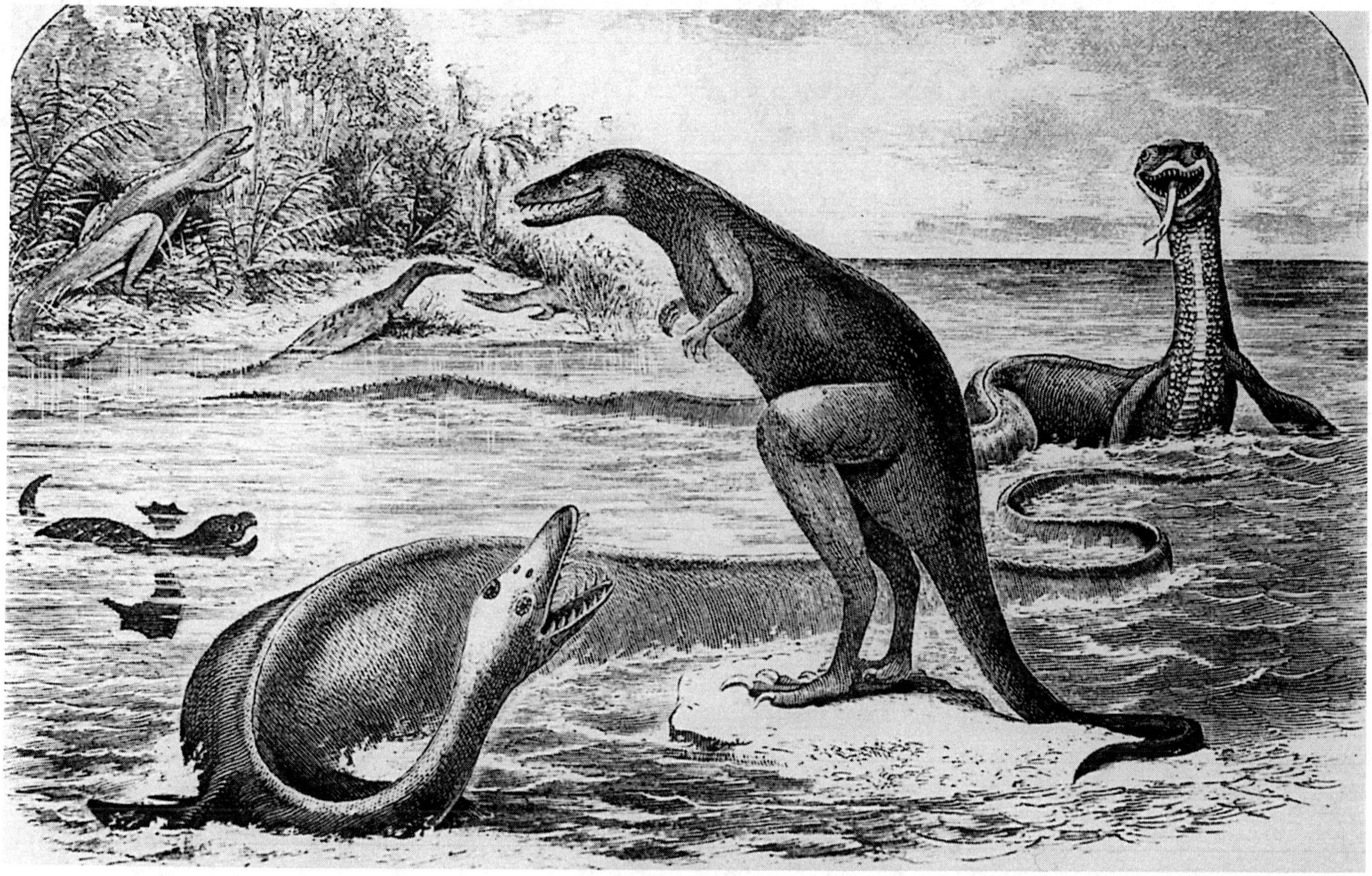

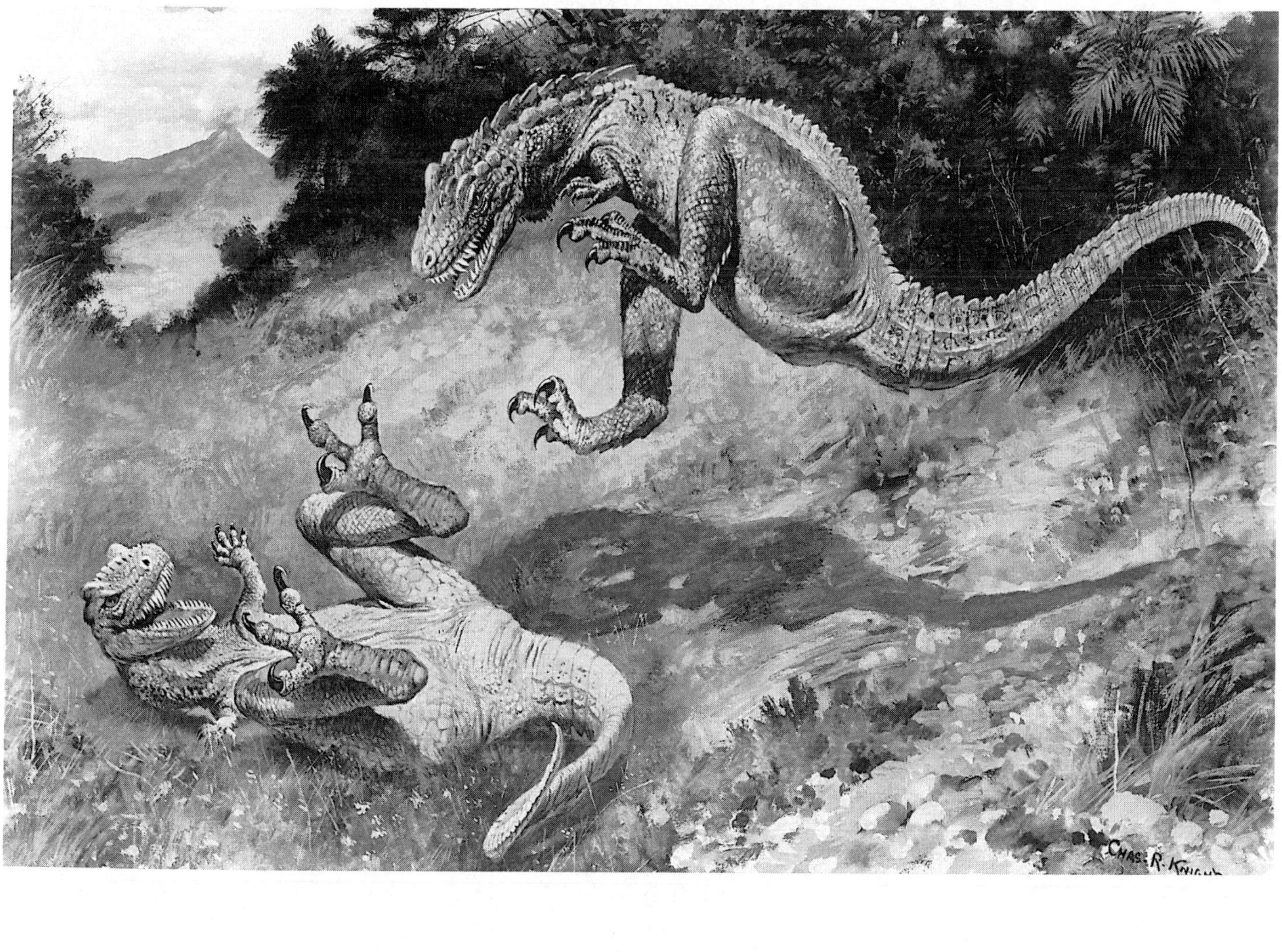
CHAS. R. KNIGHT

American artist Charles R. Knight reveal that Cope introduced the young Knight to the vast realm of prehistory. Knight had been assigned by the Century Company to illustrate an article written by William Ballou, titled *Strange Creatures of the Past*, featuring many of the dinosaurs Cope had described during his long career. Years later, Knight recalled how, "Under the spell of his facile tongue, new vistas of the life of the past opened before me in a way I had never dreamed of, and Cope drew pictures for me, pictures of the creatures I had to do for the magazine, and many others...."[5] Cope died only weeks later, on April 12, 1897. But eight of Knight's first "saurian" restorations resulting from their brief collaboration appeared in an 1897 issue of *The Century Magazine*.[6] Many of these were painted from small sculptures prepared under Cope's direction.[7]

According to Ben Creisler, the name *Laelaps* (literally "hurricane") was inspired after a mythical hunting dog which was turned to stone in the very act of leaping. However, Marsh was aware that the name *Laelaps* was preoccupied, and so renamed it *Dryptosaurus* ("slashing reptile") in 1877. Defiantly, Cope never accepted Marsh's revision. Twenty years later, Knight's famous restoration, referred to as *Laelaps* in *The Century Magazine*, clearly captured the spirit of Cope's idea. Knight's *Laelaps* made lasting impressions. (See figure 19-2.)

David Rains Wallace, author of *The Bonehunters' Revenge* (1999), suggests that Knight's *Laelaps* painting and sculpture were intended as a "parting gesture toward Marsh." Wallace claims Cope "...seems to have had [their] lifelong feud in mind when he talked to the illustrator about the 'devourer of Hadrosaurus and all else it could lay claws on,' because Knight's rendering is odd. Carnivorous dinosaurs usually are shown attacking prey, but the two in his vivid painting are fighting.... The scene recalls a titanic cockfight, and the grinning saurians give an impression of rabid joy.... Cope may have meant the prehistoric squabble as a

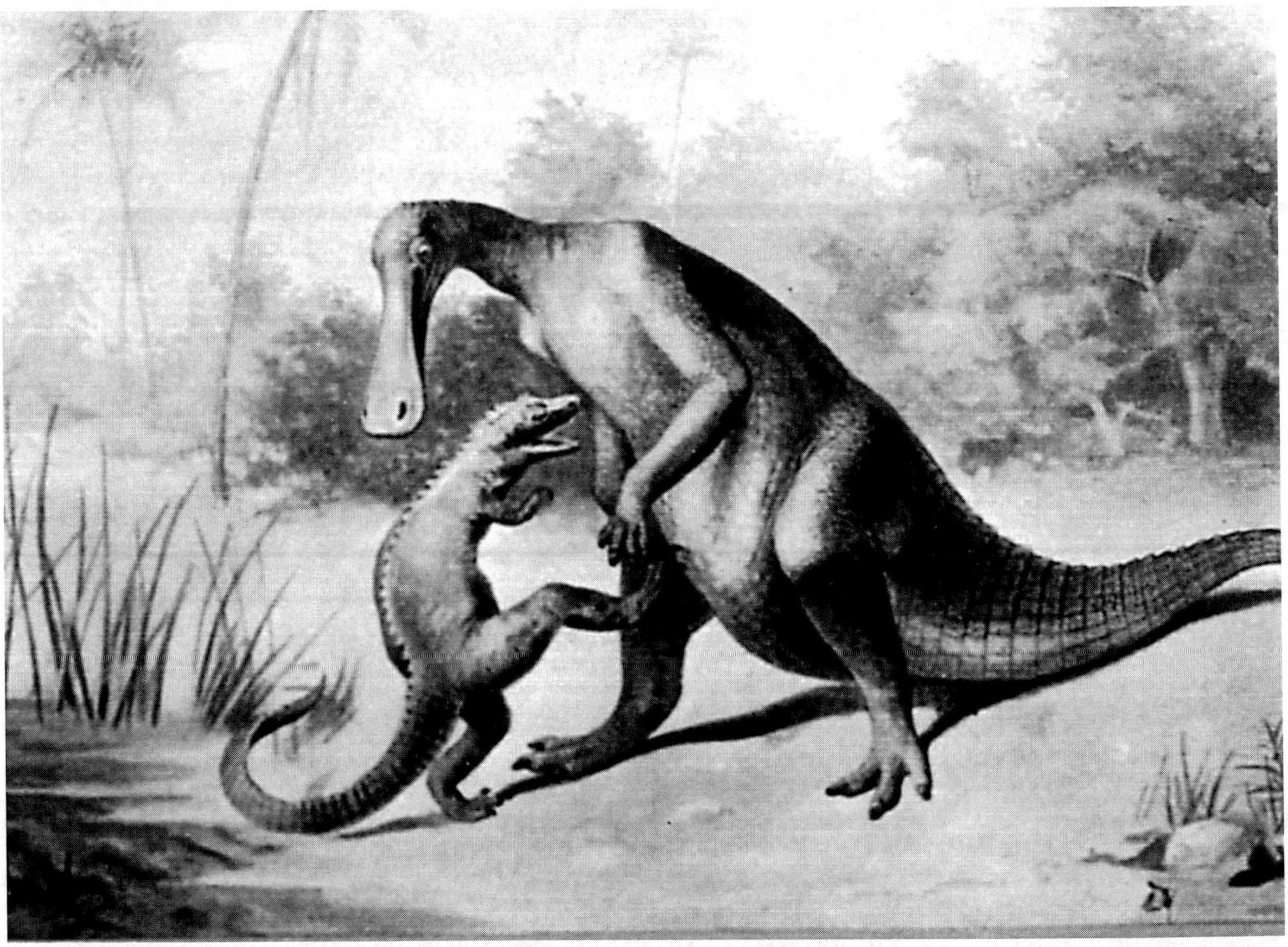

Figure 19-3: Joseph Smit's 1905 restoration of Laelaps attacking *Hadrosaurus*, evidently inspired by Hawkins' older restorations (see Figure 10-3), yet presaging the savagery of *Deinonychus* six decades later. From *Nebula to Man* (New York: J. M. Dent and Company).

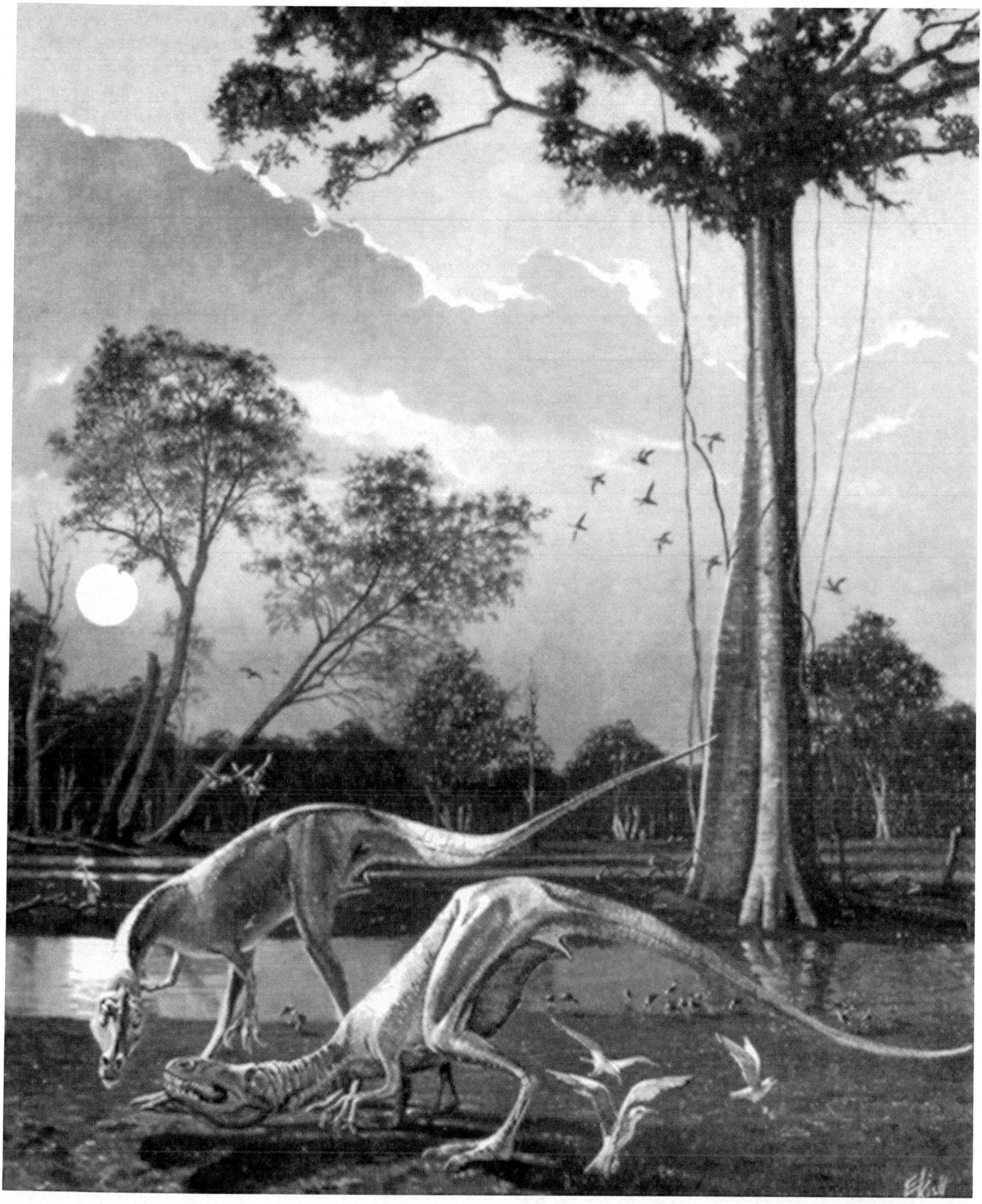

Figure 19-4: Ely Kish's 1985 speculative restoration of *Dryptosaurus*. Reproduced with permission. (See text for details.)

last "happy warrior" joke to accompany the Ballou article's posthumous nose-thumbing, although stolid Othniel Marsh surely wasn't amused."[8]

The appealing nature of Knight's restoration, coupled with difficulties stemming from the paucity of fossil evidence, may have discouraged artists from restoring *Dryptosaurus* during the twentieth century. Only four other restorations featuring *Laelaps* were completed around the turn of the century, notably one by Joseph Smit published in Knipe's *Nebula to Man* (1905).[9] (See Figure 19-3.) Then in 1988 Gregory S. Paul claimed *Dryptosaurus* may be a *unique form*.[10] Ely Kish's slender coelurosaurian *Dryptosaurus* restoration (1985) sneered at our uncertainty from within Dale Russell's *An Odyssey in Time: The Dinosaurs of North America* (1989).[11] (See Figure 19-4.)

In 1997 Ken Carpenter, Dale Russell, Donald Baird and Robert Denton reconsidered the *Dryptosaurus aquilunguis* holotype (i.e. the bones originally described as *Laelaps* in 1866). *Dryptosaurus* has been placed among a variety of theropod families over the years since its discovery. This includes Marsh's classification, the *Dryptosauridae*, which is a reflection of its uniqueness. In the new study, no consensus was reached, although three reviewers favored Marsh's opinion. Dale Russell, the lone dissenter, favored a coelurosaurian affinity. The authors claimed *Dryptosaurus* displayed maniraptoran tendencies, although diagnostic

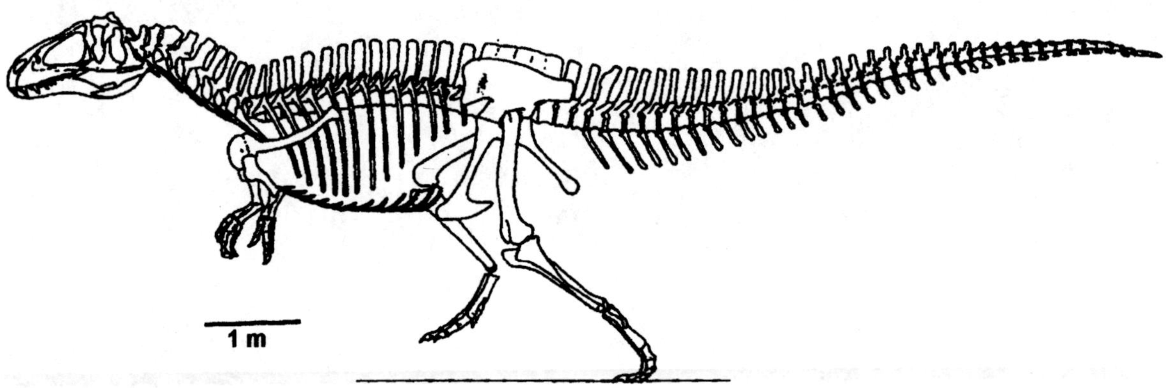

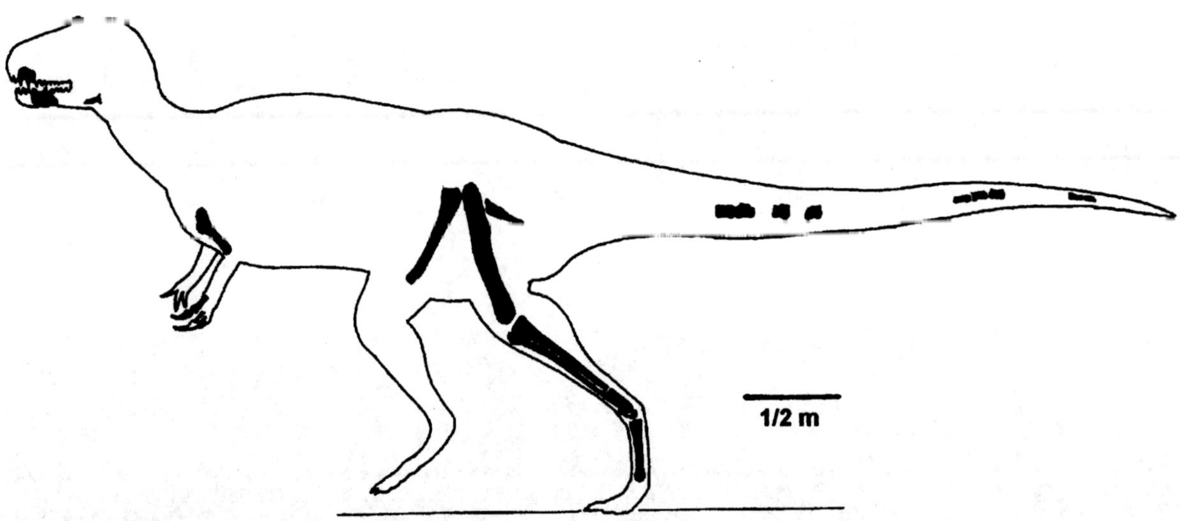

Figure 19-5: Kenneth Carpenter's speculative 1997 (top) skeletal reconstruction of *Dryptosaurus*, and (bottom) skeletal restoration showing how scant is the holotype material used for description. (See text for details.) Used with permission.

elements are missing. Cladistic analysis was not attempted, "...due to the fragmentary nature of the holotype material."[12]

A figure showing a "schematic representation of the skeletal elements" revealed the authors' collective, yet tentative, image of how this mysterious dinosaur may have appeared in life.[13] Until a more complete specimen is found, all of the Cope-inspired classical, late nineteenth century restorations are to be regarded as conjectural. (See Figure 19-5.)

Clearly, balance between (Marsh's) conservatism and (Cope's) progressive tendencies must be carefully maintained in the matter of dinosaur restorations. However, sometimes outrageous speculation (i.e. artistic license) may be scientifically beneficial, provided that the result is properly evaluated and understood. More than any other Nineteenth Century "team," Cope and Knight convincingly forged a level of success and quality in their early collaboration. Thereafter, it became easier for others, such as Henry Fairfield Osborn (who revered Cope and adorned the halls of the American Museum with stimulating paleontology exhibits) and Frederick A. Lucas (who in 1900 published a definitive essay on "*The Restoration of Extinct Animals*"), to cultivate the practice of restoring dinosaurs. And so (inspired by the wit of Harvard paleontologist Stephen J. Gould), let us proclaim, "Bully for '*Laelaps*,'" and the dinosaur restorations stemming from the foresight of those illustrious pioneers, Cope, Hawkins and Knight![14]

Chapter 20

Tickled by Feathers

In *The Last Dinosaur Book* (1998), William J. T. Mitchell claims that dinosaur images (and their associated "imagetext") may, like the dragons of folklore, fade from the mainstream. Mitchell foresees a time when the dinosaur "clade" won't seem so unique or mysterious. Then dinosaur images will no longer be resurrected as our cultural "totem of modernity." "How much mystique will remain when the creatures formerly known as dinosaurs take on their proper name of 'early birds,' and dinosaurology disappears to be replaced by paleo-ornithology?"[1] And yet, three years following publication of *The Last Dinosaur Book*, dinosaurian ideas and associated images still thrive in the media, museum displays, book stores, movies, internet exchanges and websites, and even on toy department shelves.

The prevalence and popularity of dinosaur images is analogous to that of the classic movie monsters. There is no foreseeable end in sight to their popularity! In fact, dinosaurs have already weathered their worst storm imaginable—a genetic coupling of dinosaurs with modern avians into a nightmarish apparition, the erection of an incongruous symbol—"*Daffyduck horribilis*"! For what self-respecting dinophile would want to think of his favorite Mesozoic dinosaur stock—the carnivorous lineage known as theropods—in undignified fashion, all covered with feathers? Wouldn't cartoonish pictures of feathered dinosaurs threaten to make a mockery of the bird-dinosaur evolutionary debates? Instead, diehards have been tickled by the outrageous proposition. In fact, feathers seem cool! If tarred and feathered theropods can popularly prevail, then our love for dinosaurs must be eternal.

But are the imagineers and paleoartists brooding over the "feather revolution" simply "winging it" these days? Let's survey how paleontologists have been feathering their nests recently with all manner of feathered and furry bird-like theropod dinosaurs and strange dinosaur-like birds.

In October 1998 a small but highly significant exhibition of fossils from the Lower Cretaceous, Yixian formation from Sihetun in China's Liaoning Province, opened at Chicago's Field Museum of Natural History.[2] The National Geographic display featured six stone slabs bearing exquisitely preserved specimens of extinct birds, bird-like dinosaurs and two life-sized sculptural interpretations. On view were two fossils of *Sinosauropteryx prima*, two *Caudipteryx zoui* specimens, a fossil of *Protarchaeopteryx robusta* and a slab bearing two specimens of a "genuine" bird, *Confuciusornis*. All these specimens are thought to have perished during eruptive volcanic episodes.[3] The animals died and were quickly covered in tissue-preserving ash in a lake bed. This is why so many body details, such as ingested mammal bones, developing eggs and feathers can now be discerned by paleontologists.

Two remarkable sculptures of flightless, feathered dinosaurs, *Sinosauropteryx* and *Caudipteryx*, made by Canadian sculptor Brian Cooley greeted approaching visitors along the aisle. (At first, Brian's sculptures looked for all the world like birds to me, but they are indeed dinosaurs!) Cooley's masterful *Caudipteryx* restoration continually attracts children and weary parents like moths to a street lamp. (See Figure 20-1.) One could easily get singed in the ineluctable light of knowledge! What would Sir Richard Owen, who "invented" the dinosaurs in 1842 and who throughout his brilliant career opposed Darwinian evolutionary principles, murmur to himself if he were standing beside me gazing at the *Caudipteryx* specimens?[4]

Perhaps *the* most historically significant specimen on display was the first specimen announced in late 1996 by paleontologist Phil Currie as the first "feathered dinosaur" specimen. It scored sensationally at the 1996 Society of Vertebrate Paleontology October meeting. A restoration of the dinosaur by artist Michael Skrepnick was rushed to

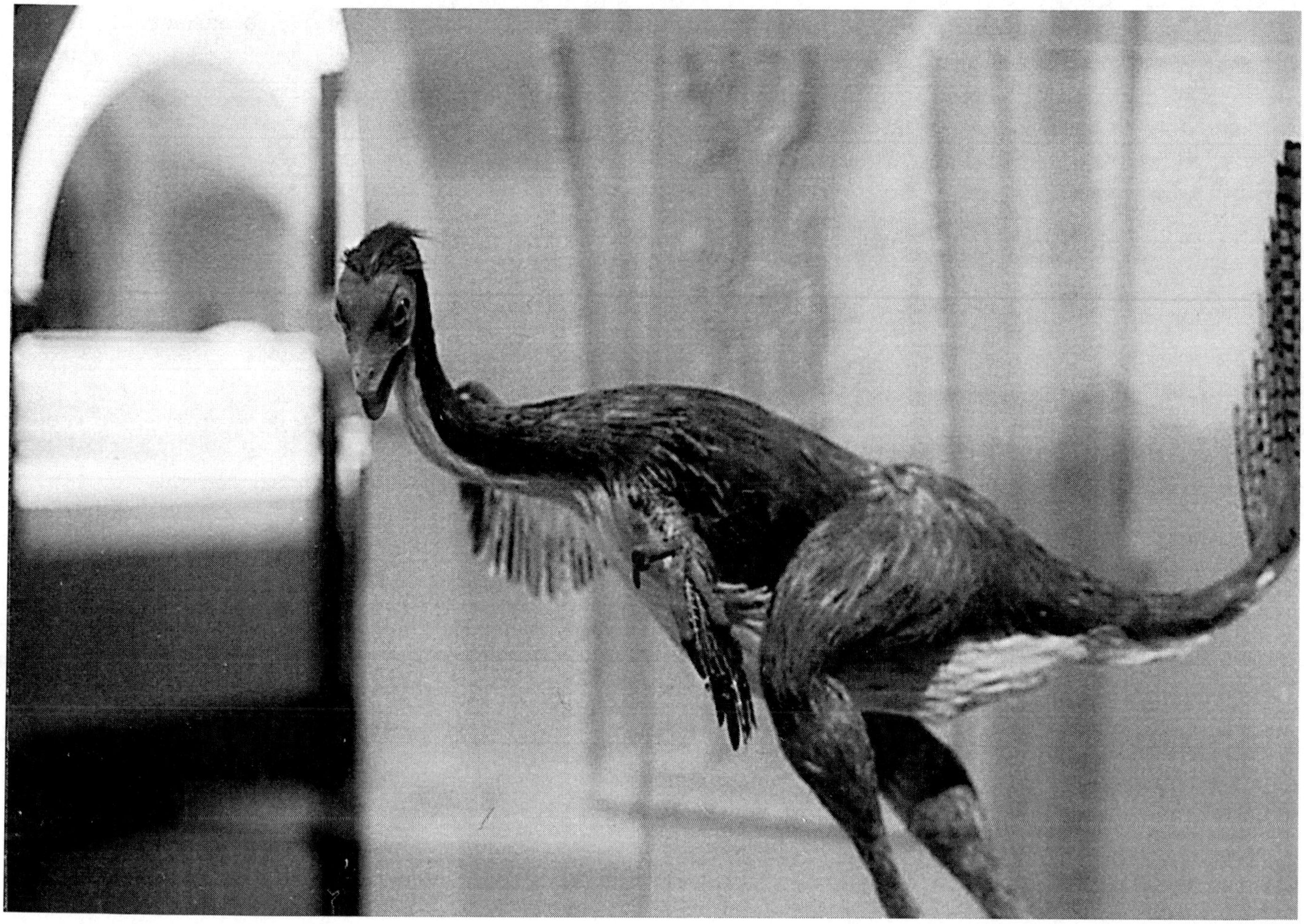

Figure 20-1: Brian Cooley's life-sized restoration of *Caudipteryx* on display at the Field Museum of Natural History in October 1998. (Photograph by Allen A. Debus, courtesy of the Field Museum of Natural History.)

print in the *New York Times*, and things skyrocketed from there![5] This bird, excuse us, DINOSAUR, is closely related to a more familiar European dinosaur genus, *Compsognathus*.

Another exhibited *Sinosauropteryx* specimen was identified as the only dinosaur ever discovered along with the remains of a small mammal "...inside—apparently one of its last meals." Here we have evolutionary contingency at its finest! That mammal could have been *my* ancestor! Thankfully, these crow-sized, hungry dinosaurs living in the real live "Jurassic Park" age of "nature red in tooth and claw" left some mammals alive to survive the K-T mass extinctions event of 65 million years ago.

"How on earth can they tell these are dinosaurs?" remarked one Chinese visitor. "They all look like big turkeys!" They did look like turkeys. How amazing that so much can be revealed from a jumble of bones. Non-flight feathers are there to behold, of course, but it would be difficult for most of us to identify them as feathers, unless, perhaps, you spied them from the corner of your eye! Thanksgiving was just around the corner, and the birds—we mean DINOSAURS—were making us hungry.

Discoveries such as these are threatening to render all those fancy books on dinosaur and avian evolution, "dead as a dodo" because the book publishers can't keep pace with the new discoveries. Old timers may wistfully remember when there "just" seemed to be the obdurate triad—*Archaeopteryx, Hesperornis* and *Ichthyornis*. In fact, scientists have described a flock of nearly 30 new Mesozoic bird genera since the early 1980s. Not since the announcement of *Archaeopteryx*, first extracted from Bavaria's Lithographic limestone deposits in the previous century, have there been fossil discoveries bearing such profound implications for our understanding of the evolution of dinosaurs and birds. Thomas Henry Huxley, Charles Darwin and Edward Cope must all be enjoying victory cigars up there in paleo-heaven! Enjoy!

Even restorations of our dear old friend *Archaeopteryx* have noticeably "evolved" during the past 130 years. (See Figure 36-4.) During the 1860s, Edouard Riou was the first to introduce an outline of the flying bird, then known from a headless specimen, into an idealized landscape scene of the "Upper Oolitic Period" for Louis Figuier's *Earth Before the Deluge* (1865). By the early Twentieth Century, *Archaeopteryx* had assumed a characteristically avian form in restorations completed by Charles R. Knight, Joseph Smit, Othenio Abel and Gerhard Heilmann. For many years, Maurice Wilson's *Archaeopteryx* restoration at the British Museum was a standard for sculptural excellence. For decades, most agreed that *Archaeopteryx* was a primitive reptilian bird.[6]

During the early 1970s, John Ostrom suggested that birds learned to fly from the ground up, and that the skeletal anatomy of *Archaeopteryx* closely resembles that of theropod dinosaurs because birds are direct evolutionary descendants of theropod dinosaurs.[7] In other words, Ostrom viewed the earliest proto-birds as winged "Evel Knievels," dinosaurs learning how to make a living in the air. Gregory S. Paul's 1977 restoration featuring scaly-looking, land-roving *Archaeopteryxes* blends avian with dinosaurian qualities.[8] However, Paul believes that birds learned to fly after adopting an arboreal existence.[9]

The theoretical framework for avian evolution was superbly captured in a painting by Mark Hallett published in the June 1985 issue of *ZooBooks*.[10] In this diagram, Hallett ingeniously symbolized the origin of flight and the evolution of birds from small running dinosaurs through *Archaeopteryx* and its hypothetical ancestors, directly to a modern form. All the figures are depicted chasing dragonflies, implying that flight arose in terrestrial, not arboreal, creatures. That same year, a paper by Richard Thulborn and Tim Hamley restored *Archaeopteryx* as a wading predator "...that held its wings as a canopy over the water while hunting."[11] More recently, however (1993), John O'Neill has introduced new images of the ancient bird. O'Neill's restoration, depicting an airborne *Archaeopteryx* sporting a jay-like crest of feathers, was published on the cover of *Science*.[12]

Without consensus, feathers have been theorized to have evolved, now both in birds *and* possibly in theropod dinosaurs, for purposes of 1) flying either from the ground up or from the trees down; 2) as protective insulation for a warm blooded physiology[13]; 3) to facilitate egg brooding[14]; 4) to serve as a feathery net in the capture of insect prey; 5) and to function as visual display plumage in mating seasons. If birds evolved from the trees down, as many ornithologists believe, from a tree climbing arboreal form, then feathers were first favored for gliding, later as an adaptation that facilitated more vigorous powered flight.[15] Conversely, if birds evolved from the ground up, feathers first evolved for insulation and were derived from bipedal, warm-blooded dinosaur ancestors. Eventually such winged dinosaurs (or protobirds) would become skilled at ground takeoffs. Paleontologist Pat Shipman claims, "Evolving Archaeopteryx was just a matter of taking the predatory behavior of a ground-based animal and finding a way to carry that behavior up into the air."[16]

Recent discovery of the Asian feathered dinosaur clan supports the ground up flight hypothesis. But news of a flying dinosaur could even outstrip the sensationalism stirred by *Archaeopteryx* over a century ago. And aren't we all a bit curious about what feathered flying dinosaurs would look like?

Paleoartists have convincingly restored dinosaur feathers so cool and sexy they could turn on any self-respecting American Bald Eagle. The new feathered interpretations have tickled our fancy too. During the summer of 2000, a remarkable diorama was unveiled at the conservatively-minded American Museum of Natural History's exhibit, *Fighting Dinosaurs: New Discoveries from Mongolia*, featuring two feathered life-sized dinosaur sculptural restorations—*Velociraptor* and *Oviraptor*, the latter guarding its nest.

Feathered dinosaurs are all the rage now! Paleoartists have nurtured the scientific upheaval in our understanding of the phylogeny of birds and dinosaurs, adding a flock of new feathered theropod images to yesteryear's plucked chicken dinosaur portrayals.[17]

A feature article in the November 1999 *National Geographic* offered glimpses of feathered theropod fauna resurrected from Asia. "Dinosaurs will never look the same. The reason: four new dinosaur fossils with stunningly birdlike bones and indications of feathers," wrote Christopher Sloan.[18] Brian Cooley's *Sinornithosaurus* sculpture, talons raised threateningly, leaped from a page right into our laps. According to Sloan, this 120-million-year-old dromaeoaur could leap but not fly. "The fossil also supports the concept that early feathers evolved for insulation or display rather than flight and adds new weight to the idea that these dinosaurs were warm-blooded."[19] There also were illustrations of a feathered therizinosaur, *Beipiaosaurus*, and a feathered oviraptorosaur. Michael Skrepnick's speculative restoration of a feathered *T. rex* hatchling standing under an unfeathered parent revealed how infectious the "feather revolution" had become![20]

Also in this same *National Geographic* issue readers would have seen a photograph of Stephen Czerkas' sculpture of a creature named "*Archaeoraptor*." The sculpture, posed as if launching into flight, was described by its maker as a "...missing link between terrestrial dinosaurs and birds that could actually fly."[21] An implication of the find was that, besides birds, the most avian theropods may also have been feathered.[22] Photographs in the article showed Stephen gluing feathers onto the outstretched arms of his acclaimed *Deinonychus* sculpture.

Blue jay-sized *Archaeoraptor liaoningensis* was thought to be more adept at flying than geologically older *Archaeopteryx*. *Archaeoraptor*'s mosaic form allegedly represented the evolutionary pathway taken as theropods descended into birds. This is because, while the specimen's fossilized arms cried out "avian," the tail appeared to be "dromaeosaur." Eventually it was realized that the slab bearing the tail end had been joined by a Chinese workman to the slab encasing the bird elements. Paleontologists questioned whether the bird and dinosaur portions had been appropriately composited.

The answer was a resounding "no." For, by early 2000, paleontologists proved convincingly that the *two* slabs comprising the *Archaeoraptor* specimen did not represent a single species. Remains from two separate genera, one Mesozoic bird and the other a dinosaur, had been composited into the ultimate "fossil fooler." Stephen's sculpture, so beautifully depicted as if spreading its wings for takeoff, seemed little more than a product of regrettable circumstances.[23] Conceptually, however, *Archaeoraptor*'s sculptural "restoration" may still be regarded as a useful hypothetical "prediction" of what paleontologists hope to find someday in the fossil record.

The latter half of the Mesozoic era is beginning to seem like an "age of birds" instead of the proverbial "age of reptiles."

Beyond *Sinosauropteryx*, one of the previous century's most significant fossil discoveries, here's a sampling of some of Mesozoic birds which have created quite a "flap" in the scientific literature, encroaching on the dinosaurs' morphological terrain.

Perhaps the strangest of all the new avian fauna is the flightless, toothy-jawed genus *Mononkyus*, which is undeniably one *strange* bird![24] *Mononkyus* created a sensation after its announcement in early 1993, with C. F. Payne's restoration even making the cover of *Time* (April 26, 1993). Bones of this highly unusual 75-million-year-old turkey-sized variety were first collected in the Gobi Desert in 1922, only to be catalogued as a "bird-like" dinosaur. Although no fossilized feather impressions were found with *Mononkyus*' remains, it likely had feathery insulation in life.

Mononkyus looks like a dinosaur from its trunk through the end of its tail. However, its most unusual skeletal features, an avian keeled sternum and stubby arms ending in knobby claws, may have evolved from theropod ancestors. The supposedly avian nature of *Mononkyus*' hand and breastbone has been debated. Were the arms used for mole-like digging, an adaptation which is difficult to visualize in birds? If this is what *Mononkyus* used its arms for, then can analogies also be made to those puny *T. rex* arms?[25] Or were *Mononkyus*' forelimbs and hands derived from an already specialized avian structure? On the basis of its specialized forelimb structure, *Mononkyus* cannot be a direct ancestor of modern birds, even though it shares more traits with modern birds than the geologically older *Archaeopteryx*.

Ninety-million-year-old northwest Patagonian fossil beds have produced another flightless bird named *Unenlagia comahuensis* (the "halfbird from Comahue").[26] According to paleontologists Fernando Novas and Pablo Puerta, the birdlike specimen suggests evolutionary links between theropods and early birds (*Archaeopteryx*). *Unenlagia* is known primarily from pelvic and leg bones, a scapula and upper arm bones, and several vertebrae. No skull, feather impressions or foot bones were found. However, most paleoartists would agree that *Unenlagia* sported ostrich-like plumage in life.

The revealing name "half bird" refers to the admixture of anatomical features found in coelurosaurian dinosaurs and modern birds. Pelvic bones display features both noted in coelurosaurian dinosaurs and also in *Archaeopteryx* and other avians that are unseen in the dromaeosaurs. Yet the pubis is retroverted as in dromaeosaurs, while the ilium is unfused. *Unenlagia*'s forelimbs bear "flapping" potential, as the upper arm bone may have extended outward (laterally) from the shoulder cavity (glenoid) and could be raised or lowered in an arc comparable to the movement in flying birds. Such a condition is absent in dromaeosaurs. However, at ⅔ the size of twelve-foot-long *Deinonychus*, the proavian dinosaur was probably not a flyer due to its relatively large size.

Novas and Puerta believe the transitional nature of the specimen suggests that the avian body plan was designed in terrestrial dinosaurs, and therefore that a flapping forelimb stage preceded gliding in airborne birds. The authors claimed, "...the new evidence ... reinforces the hypothesis that a bipedal, cursorial theropod was ancestral to birds and used its arms not only in predation, but probably also in maintenance of balance and to control its body attitude while running and leaping."[27] They also speculated on the possibility that dinosaur species evolving diminutive sizes were most able to capitalize on the osteological features evolved in terrestrial ancestors morphologically represented by *Unenlagia*. Flapping forelimb movement, coupled with flight feathers, may have opened up new niches formerly restricted to insects and pterosaurs, from the ground up.

A third remarkable bird-dinosaur "link" offering tempting glimpses into the origin of birds from dinosaurian ancestors was the 65 to 70-million-year-old fossil discovered in 1995 in northwestern Madagascar. Named *Rahonavis ostromi* ("Ostrom's menacing bird cloud") by Catherine Forster and D. W. Krause of the State University of New York at Stony Brook, Luis Chiappe of the American Museum of Natural History, and Scott Sampson of the New York College Osteopathic Medicine, certain skeletal features of the new winged bird highly resemble the 80-million-year-older *Archaeopteryx*, with *Velociraptor* sickle-shaped slashing feet "snapped on" for good measure. Imagine a raven-sized (or *Archaeopteryx*-sized) *Velociraptor*, every pterosaur's worst nightmare, and you've conjured an image of this wonderful chimera. A restoration of *Rahonavis*, shown ominously descending upon an unsuspecting mammal, recently appeared on page 32 of the June 1998 issue of *Discover*.[28]

The partial skeleton lacks a skull and several vertebrae. No feather impressions were found. However, quill knobs were noted on the caudal surface of the single right ulna that was preserved. *Rahonavis* had a bony tail and an avian synsachrum, which resembles that of *Archaeopteryx* and *Unenlagia*. The unfused condition of the pelvis is like that of nonavian theropods and other ancestral Mesozoic birds.

The single specimen is thought to be an adult, representing a mosaic of avian and theropod qualities. The peculiar feet, unanticipated in birds, are decidedly "theropod," as are other skeletal features pointing to theropod derivation.

It was debated whether the foot bones containing the sickle-shaped claw even belonged to the other avian skeletal elements. The right foot (pes) missing its claws (unguals) was found in close proximity to the rest of the skeleton. However, the lower left limb, including a nearly complete pes sporting one three centimeter-long raptorial claw on the second digit, was found approximately three feet away. Is that too far away? Yes, say Larry Martin of the University of Kansas and John Rubens of Oregon State University, who wonder whether separate dinosaur and bird specimens were sampled instead and then inadvertently combined. Denying the integrity of the "specimen" attributed to *Rahonavis* may place its "living fossil" status in question. Nevertheless, in both cases, *Rahonavis* was demonstrated to be most closely related to ancient "primitive" birds (*Archaeopteryx* and *Unenlagia*), occupying an intermediate position between troodonts and the derived, yet odd, *Alvarezsairidae*.

In a letter subsequently printed in *Science*, Berkeley vertebrate paleontologist Kevin Padian emphasized that the bird-dinosaur relationship was strengthened by phylogenetic testing performed by Forster. In either case, fossil *Rahonavis* was grouped with *Unenlagia* and other birds. Padian concluded, "As evolutionary theory would predict, the most basal birds show a mosaic of retained theropod features and new avian characters."[29]

Conceptually, *Rahonavis* isn't too dissimilar from the *idea* of *Archaeoraptor*. Yet, when found, a true missing link must not only match the expected *form* believed to be transitional between early birds (i.e. *Archaeopteryx*), it must also agree with theoretical expectations in age. It must be geologically *older* than *Archaeopteryx*. *Rahonavis* is 80 million years too young to be our "holy grail," the true link between *Archaeopteryx* and dinosaurs. So is *Sinornithosaurus*.

Dating from the Early Cretaceous, the 120-million-year-old Chinese *Sinornithosaurus* is the earliest dromaeosaurid known to science. It would be difficult to imagine a more bird-like, non-avian theropod. Paleontologists Xing Xu, Xiao-Lin Wang and Xiao-Chun Wu noted that *Sinornithosaurus* had "...attained the prerequisites for powered, flapping flight."[30] Proportionally, *Sinornithosaurus* has the longest forelimbs of any non-avian theropod, estimated to be 80 percent as long as the hindlimbs. And the eagle-sized dromaeosaurid was feathered too.

Finally! A flying dinosaur? Not quite, although sternal and shoulder elements closely resemble analogous features in *Archaeopteryx*.

Sinornithosaurus was described as a cursorial terrestrial dinosaur, yet another feathered variety fortifying the "cursorial origin of avian flight." The authors' cladistic analysis revealed that *Sinornithosaurus* was more closely related to *Archaeopteryx* than *Caudipteryx* or *Protarchaeopteryx*, yet less so than are *Unenlagia* and *Rahonavis*. In other words, dinosaurs *Rahonavis* and *Unenlagia* have more avian qualities than does the very bird-like *Sinornithosaurus*.

Nevertheless, *Sinornithosaurus*' fossilized feathers and forelimbs, so structurally similar to its geologically older cousin *Archaeopteryx*, can send minds reeling over aerial possibilities. Paleoartist Bob Nicholls' illustrated sinornithosaurs are poised suggestively leaping over tree trunks, gliding safely from unseen predators on outstretched proto-wings. (See Figure 20-2.)

In contrast to recent images illustrating the dinosaur ancestry of birds "from the ground up" are those depicting "missing" bird links having evolved flight "from the trees down."[31] Resolution of the "Archaeoraptor" forgery led to further controversy stirred after the dinosaur (tail) end of Czerkas' chimera had been scrutinized separately from the "bird" anterior and correctly described. Paleontologists were charged of being "too far out on the limb" with their interpretation.

"Archaeoraptor" was intended to champion the "ground up" hypothesis for evolution of avian flight, yet, ironically, its tail end, manifested in *Microraptor*, fortifies

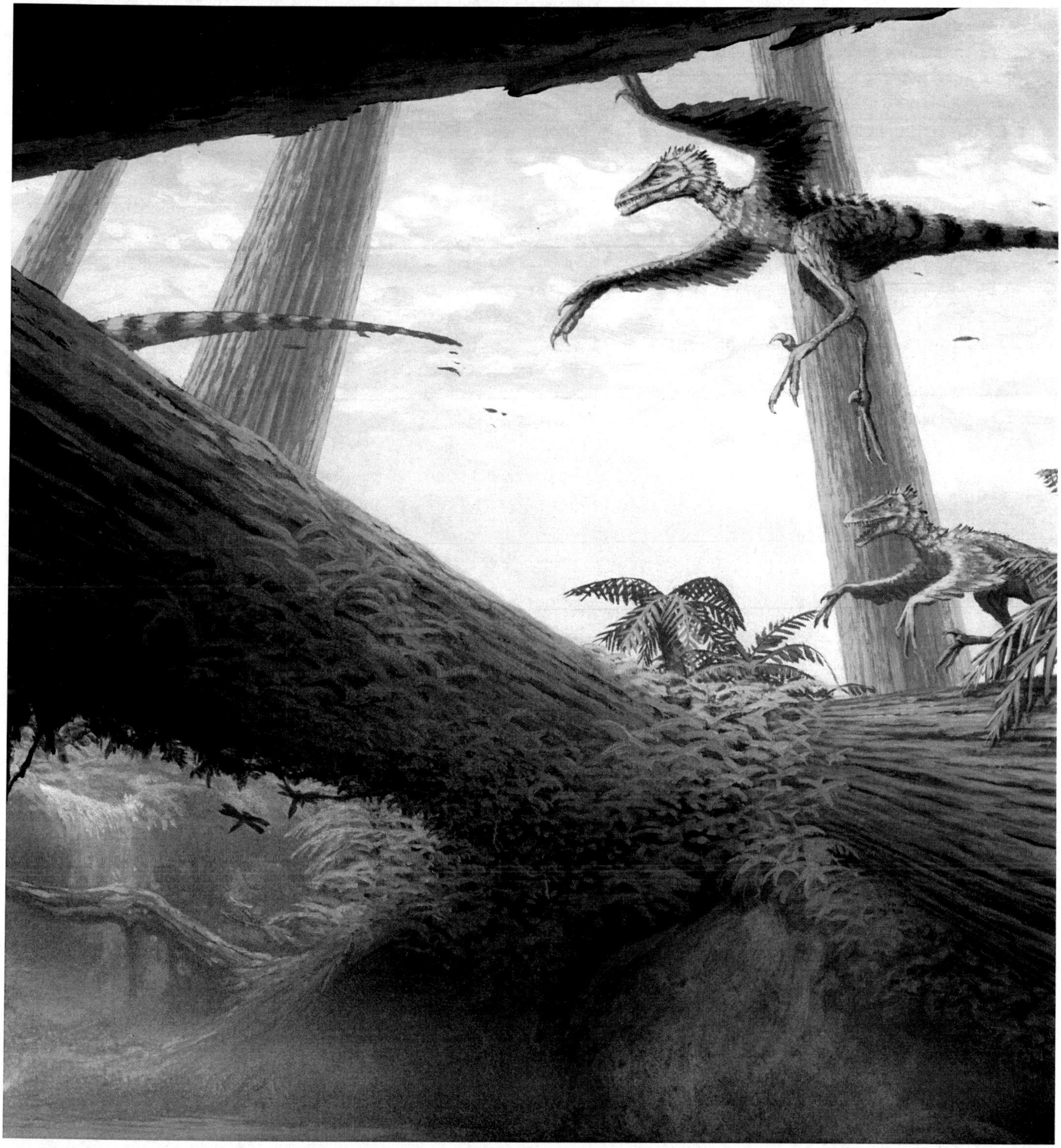

Figure 20-2: Bob Nicholl's restoration of *Sinornithosaurus*, "Above the Stream" (2000). (Reproduced with permission.)

the opposite "trees down" view. Equipped with toes suited for grasping tree branches, and "dino-fuzz" feathery integument, the 17-inch-long, 124-million-year-old dromaeosaurid *Microraptor* was hailed as the most bird-like adult specimen of a dinosaur known to science (yet!).[32] Although evidently not a flyer itself, *Microraptor*'s diminutive size matched expectations of the tree-climbing hypothesis because "proto-flyers" could not have remained airborne if they were bulky.

Paleontologists suspected *Microraptor* was an arboreal

creature capable of climbing trees. If *Microraptor*'s form is representative of the evolutionary pathway leading to flying birds, then modern birds' flight adaptations may have originated "from the trees down" instead of "from the ground up." Imagetext already associated with "ground up" newcomers *Rahonavis, Unenlagia* and *Sinornithosaurus* contrasts considerably with *Microraptor*. A restoration circulated with the *Microraptor*'s press announcement in late 2000 seemed reminiscent of Gerhard Heilmann's "Proavis," reconstructed on hypothetical grounds in 1926.[33] Except that, while Heilmann's "Proavis" was imaginary, *Microraptor* is real.

As you can tell, it's becoming increasingly difficult to distinguish between "dinosaurs" and "birds" these days, largely because the fossil record has tickled evolutionary theory with the revelation of so many apparent links and varieties in between.[34] Now paleontologists must question, *during the Mesozoic* what was "bird" and what was "dinosaur?" The old rules have simply broken down. Birds have been patiently waiting in the wings long enough. Now, seemingly reborn as dinosaurs, it's their heyday.

Having feathers or plumage used to mean "bird," but not necessarily any more. So paleoartists are becoming bold and seizing the reins The latest fad is to restore the most "bird-like" of dinosaurs with feathers, even in the absence of feathering in these fossils. Admittedly, feathers won't fossilize except under uncommon geochemical conditions favorable for soft tissue preservation.

Before the discovery of *Sinosauropteryx*, paleoartists who restored feathered dinosaurs suffered numerous indignities. In 1996 paleoartist Luis Rey, known for his colorful portrayals of lively dinosaurs, recalled, "...I have suffered somewhat with my restorations of feathered dinosaurs, becoming blacklisted by the very paleontological community here in England. However, every time they want a 'fantasy' dinosaur, they remember me (as the expert in portraying the way dinosaurs 'shouldn't' be portrayed). Sad, really because I try to be as rigorous as possible in my anatomical reconstructions and I tend to honor requests to draw featherless theropods. One must remember that ... most everything in dinosaur paleontology is hypothetical and new discoveries can shift the focus of scientific opinion."[35] (See Figure 20-3.)

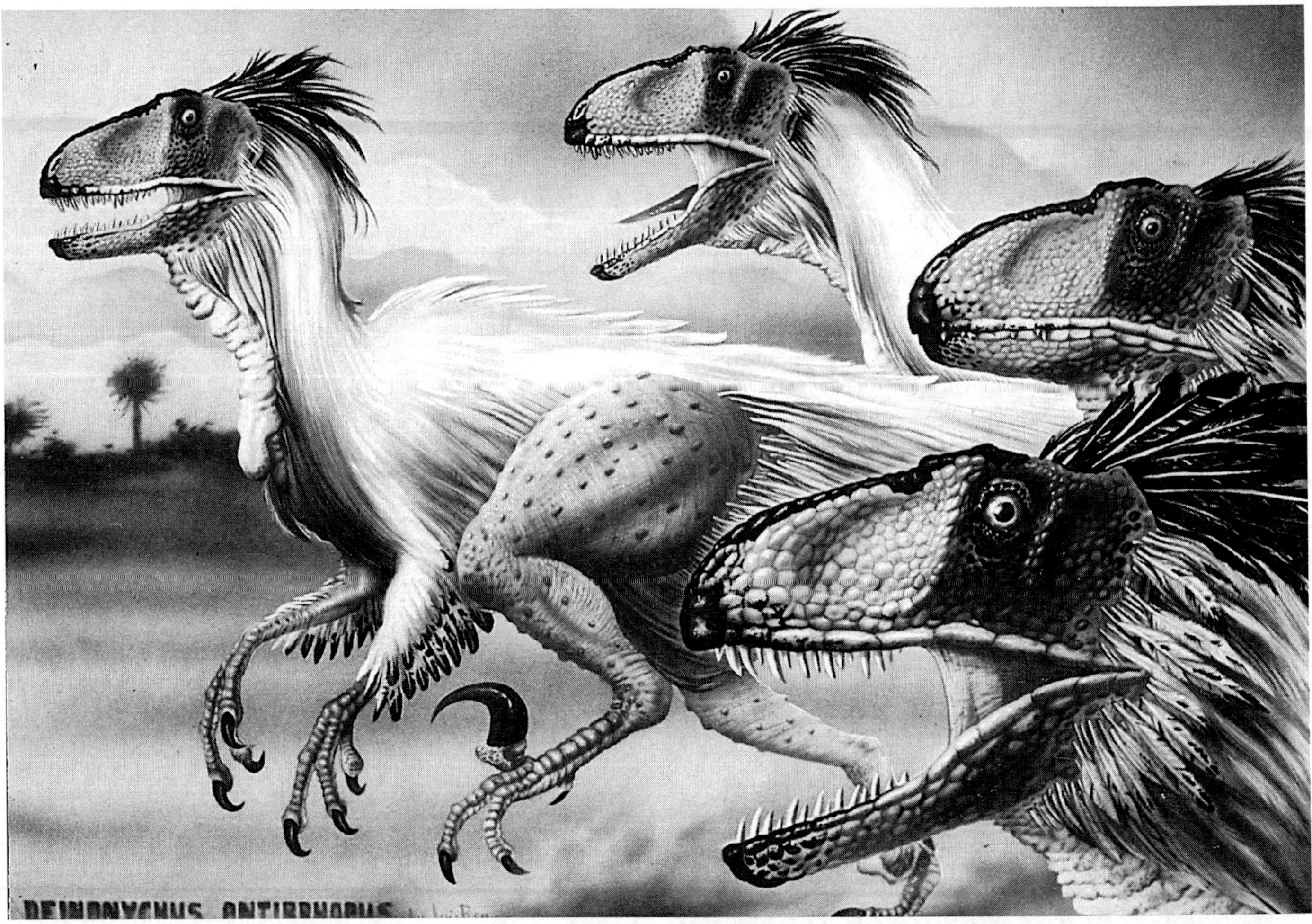

Figure 20-3: Luis Rey's restoration of a feathered trio of *Deinonychus*. (Reproduced with permission.)

How peculiar then, that the most endearing fantasy dinosaurs of the 1990s, computer animated "raptor" theropods starring in movies such as Stephen Speilberg's *Jurassic Park* (1993), the BBC documentary *Walking with Dinosaurs* (1999) and the Disney film *Dinosaur* (1999), didn't have plumage. Perhaps producers felt feathered dinosaurs wouldn't tickle our funny bones. They thought too conservatively! Or perhaps feathers would have added substantially to the costs of animation.[36]

How peculiar that some paleontologists are more out on the limb than even the producers of fantasy films! Dougal Dixon wrote a book about fantasy dinosaurs of a parallel world, *The New Dinosaurs: An Alternative Evolution* (1988), featuring images of what dinosaurs would have evolved into had they not succumbed to extinction 65 million years ago. Many of Dixon's "new" dinosaurs are depicted as feathered. Regarding their "real" Mesozoic ancestors, Dixon claimed, "There is a strong possibility that many dinosaurs were endothermic ... the swift, active, meat-eating saurischian dinosaurs were very probably warm-blooded.... Most warm-blooded animals are covered in fur or feathers to ensure adequate heat regulation.... It is possible that warm-blooded dinosaurs may have had fur or down...."[37] (See Figure 23-1.)

While paleontologists debated the evolutionary status of *Archaeopteryx*, paleoartists experimented by applying feathers to the most bird-like theropods, even though there is no direct evidence for feathering in these dinosaurs. Sarah Landry's sprinting, feather-crested *Syntarsus*, appearing in Bakker's 1975 *Scientific American* article "Dinosaur Renaissance," was in the first wave.[38] Then, in 1984, sculptor Otter Zell introduced a miniature *Deinonychus* resin toy now valued by collectors. Modeled after Robert Bakker's 1969 restoration, Zell's stylish figure had a feathered abdomen and plumed tail, evidently patterned after *Archaeopteryx*'s. (See Figure 20-4.) Since the late 1970s the accomplished paleoartist Gregory Paul has advocated restoring birdlike dinosaurs, particularly the genus *Velociraptor*, with feathers. In an homage to Knight's leaping *Laelaps* portrait of 1897, both Paul and Bakker restored a pair of fighting feathered dinosaurs, respectively, *Velociraptor* and *Deinonychus*.[39] Bakker further ruffled our feathers in stating, "So avian were these dinosaurs, it is quite probable that they had already evolved feathers for insulating their bodies."[40]

During the Early Cretaceous, southeast Australia was within proximity to the south polar region. Paleontologists speculate that in order to survive extended periods of darkness and cold, the indigenous dinosaur fauna must have evolved special adaptations, such as endothermy. Dr. John Long, however, boldly asserted that a dwarf species of *Allosaurus* found in these deposits may have been cloaked in "...fur or feather-down covering as an adaptation for cold weather...." Accordingly, the artistic Long completed the first restoration of a furry looking polar allosaur, published in the June 1990 issue of *Dinonews*.[41]

While the question of how and why feathers originated remains controversial, a recent reinterpretation of the Triassic genus *Longisquama* attempted to "tip the scales" in favor of a "thecodont" feathered ancestor.[42] Richard Ellis' restoration of the six inch long *Longisquama*, depicted with plumelike structures along its spine, adorned the cover of the April 1975 issue of *Scientific American*, the very issue heralding the age of Robert Bakker's "Dinosaur Renaissance." Now, symbolically, *Longisquama* reappears during the height of the Feather Renaissance![43] (See Figure 36-1.)

Bakker had suggested that evidence of feathers in *Longisquama* divulged its warm-blooded metabolism. In their *Science* paper of June 2000, Terry Jones, John Ruben, Larry Martin and others went one step further, claiming that *Longisquama*'s long scales were, in fact, nonavian feathers.[44] If their interpretation holds, then *Longisquama* would predate the earliest known fossil feathers (i.e. *Archaeopteryx*'s) by 75 million years! Such structures would have been of little use in active flight, but could have served as visual signals and had a thermoregulatory function.

If *Longisquama* had true feathers, there is the implication that theropod dinosaurs aren't direct ancestors of modern birds. Alternatively, it could mean that feathers evolved more than once in separate lineages, or, even less likely, that *Longisquama* is an ancestor of birds. There are many dissenters, however, who claim the new study didn't offer any falsifiable statements, and that the authors neglected to fortify their hypothesis by showing evolutionary relationships in a cladogram.[45]

Archaeopteryx, recognized for its dinosaurian skeletal features and feathers, bolstered Darwin's theory of evolution through natural selection at a pivotal time in the history of science. Now paleontologists and ornithologists understand the complexities of assigning proper relationships between Mesozoic birds, dinosaurs and other closely related fossil vertebrates. Although it will be impossible to identify an absolute, genuine Mesozoic ancestor for all modern birds, many paleontologists insist that modern birds *are* theropod dinosaurs![46] But there will always be dissenters—for instance, those professing that theropods may have descended in the form of secondarily flightless dinosaurs retaining feathers from ... bird ancestors![47] Given the incomplete nature of the fossil record, presently it would be foolish to denounce either claim. Assuredly, feathers will be a focal point of discussion in newly

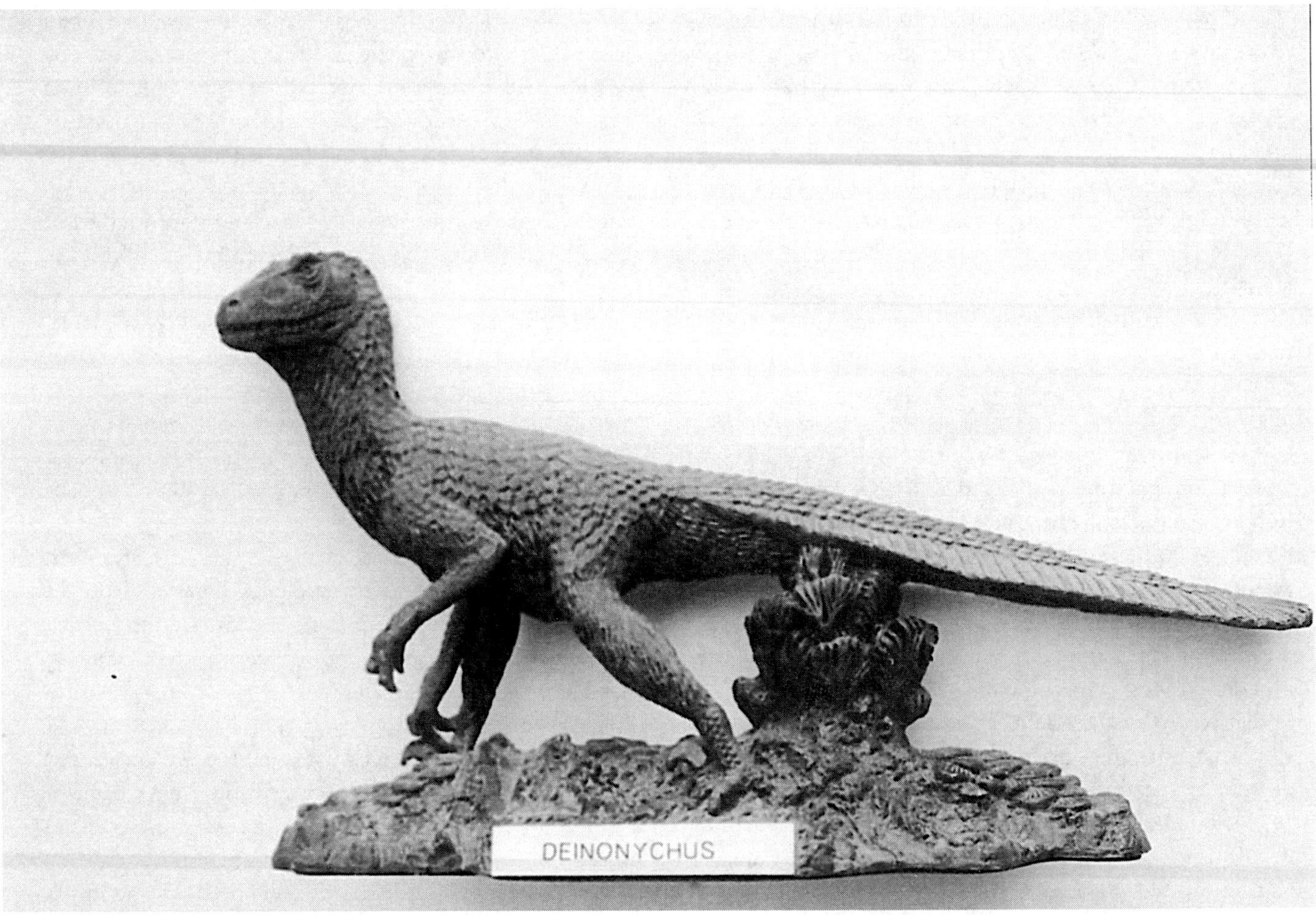

Figure 20-4: Otter Zell's miniature feathered *Deinonychus* model. (Authors' collection. Photograph by Allen A. Debus.)

constructed phylogenies. In consequence to the *Archaeoraptor* "flap," the quality of evidence will be scrutinized more carefully than ever before.

Theropods are traditionally portrayed in film and on canvas as the scariest dinosaurs, a factor contributing to their popularity because people enjoy the chills and thrills of being frightened in a safe, controlled fashion. Bats, even vampire bats and hypothetical Mesozoic proto-bats thought to have shared Mesozoic skies with furry pterosaurs, or even pterosaurs eerily restored as bat-winged pterosaurs, just don't seem as thrilling as the prospect of flying feathered dinosaurs.[48] Meanwhile, feathered *T. rexes*, "raptors" and *Rahonavis* remain the stuff of nightmares. Dinosaurs "ruled" the Mesozoic, as they do today.

Our interest in dinosaurs, plucked, feathered or furry will not readily fizzle. Scores of talented paleoartists keep the images of theropods alive and well, adding new haunting visages to yesterday's classic portrayals.[49] Like the evil vampire monsters of legend and the silver screen, dinosaurs seem destined to rise eternally from Time's crypt, often in new enchanting guises to fuel our paleoworld fantasies. Until the sun dies, there can be no "last" dinosaur book.

Chapter 21

Inside-Out Mastodon

Seeing is believing, and this is something to see! Inside a display case at the Edwin Deicke Exhibit Hall of Wheaton College, Wheaton, Illinois, stands a praiseworthy recreation of midwestern prehistory, an elephantine forerunner of the modern breed, a former inhabitant of what would eventually become Chicago's suburbia long after the glaciers receded. This exhibit will give you the "spins." For the 16-foot-long Perry Mastodon, as it is affectionately known, resides on a rotating platform offering an excellent view of the prehistoric creature, quite literally from the inside out. (See Figure 21-1.)

On October 16, 1963, perhaps conjuring a scene appropriate for the coming Halloween festivities, Marshal Erb was excavating a pond on the property of a prominent U.S. Federal District Judge, Joseph Sam Perry. It seems Judge Perry wanted to renovate his lower 40 by creating a man-made lake along the East Branch of the DuPage River near Glen Ellyn, Illinois. To Erbs' surprise, his bucket came back with what was at first thought to be a human skull, an eerie find that immediately halted operations.

Judge Perry was notified, who in turn called his wife. She left her church meeting in an astonished state, only to be greeted by a local TV news crew upon her arrival home. By the day's end, the Perrys had called Dr. Douglas Block, then Associate Professor of Geology at Wheaton College. It turns out that what was thought to be a human skull was found in the pit adjacent to a four-foot-long femur. Dr. Block astutely made the identification—a Mastodon. So, Judge Perry only needed one alibi. Why was an elephant, symbol of the Republican party, buried in a staunch Democrat's backyard?

Mastodonts first made their way into North America about 18 million years ago, via the Bering Strait Bridge, over what is now Alaska. Here the early four-tusker forms evolved into what today is known as our American Mastodon (*Mammut americanum*). Some radiocarbon studies have reported ages for Mastodon remains found in Michigan as recent as 6,000 years old, although these younger dates have been disputed. More recent information supports an extinction date of about 10,000 years ago for the American *Mastodon* in North America.

The American Mastodon resembled the Woolly Mammoth only superficially. In fact, they were as different as cats and dogs. Mammoths and the American Mastodon had differently shaped heads. Mammoth heads were higher browed. Mastodont heads were not domed, as was the case in mammoths. Mastodonts had flat, "pig-like" body contours, whereas the Woolly Mammoths had domed heads and a sloping lateral outline.

Mammoth tusks projected almost vertically downward from sheaths in the skull. Mastodon tusks protruded horizontally. Their tusks and molar tooth structures were remarkably different from Mammoths'. Males of the American Mastodont species even had two short tusks on the lower jaw (remnants from the days of the ancestral "shovel-tuskers"), while Mammoths didn't have these additional structures. Mammoths' highly cemented teeth were adapted for a grazing existence, while Mastodonts, who had knobby teeth, were browsers.

Some paleontologists believe the populations of American Mastodonts in Ice Age North America outnumbered the bison herds of the nineteenth century American plains. Yet there seem to have been more Mastodonts than Mammoths present in the states of Michigan and Illinois than in prehistoric Nebraska and South Dakota, where the opposite is true. What lured the American *Mastodon* into Illinois? The answer seems to be salt![1]

Mammals need salt or they will sicken and die. It turns out that there were abundant quantities of salt for the Mastodonts to lick in Michigan and also to the

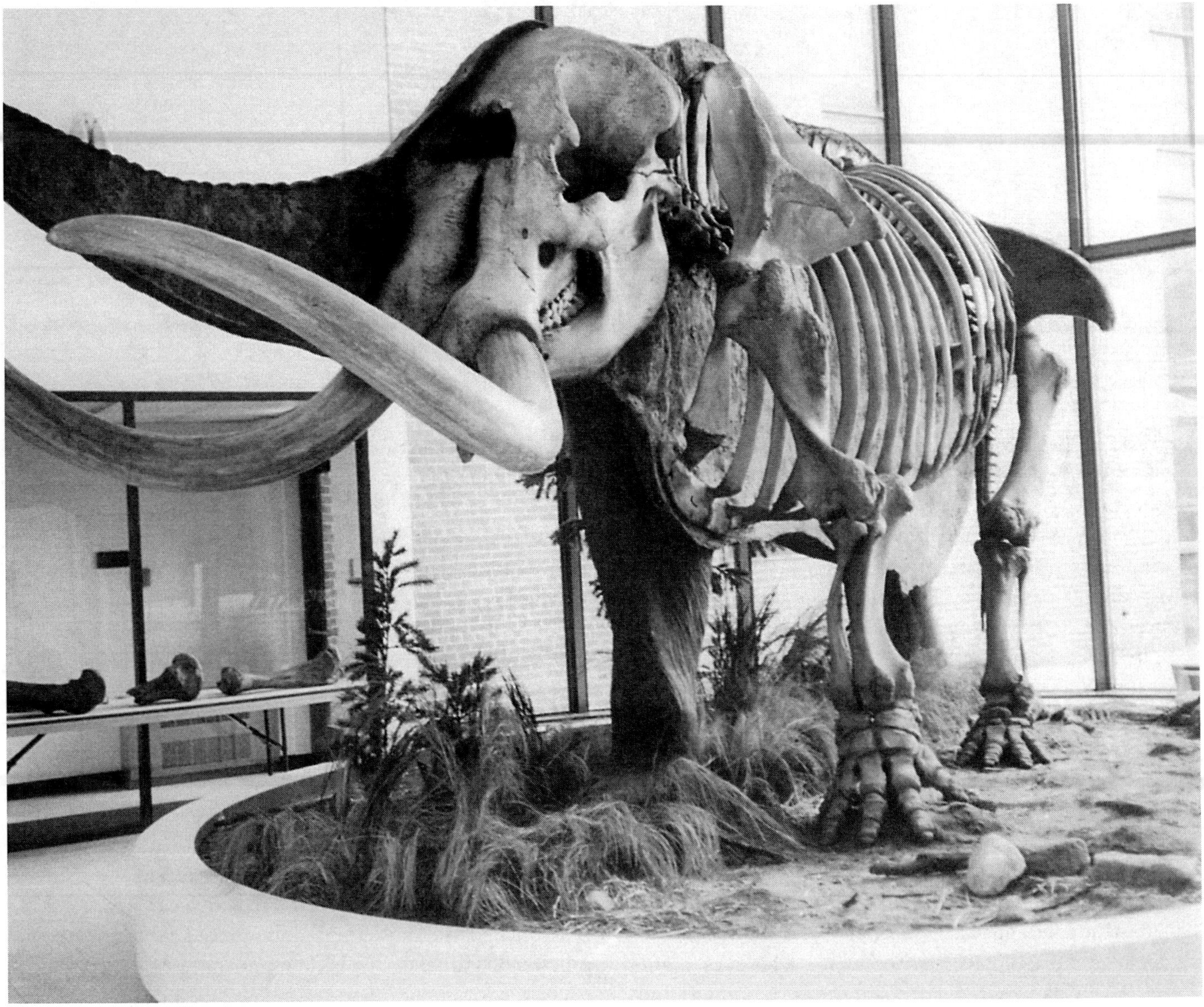

Figure 21-1: Half restoration of the Perry Mastodon on display at Wheaton College, Wheaton Illinois. (Photograph by Richard Rush 2000. Reproduced with permission.)

southeast, in what is now Kentucky (i.e. Big Bone Lick). The Perry Mastodon may have been caught passing through on his way to Michigan!

African elephants are known to visit caves where salt is plentiful to replenish their bodily reserves, so it isn't too difficult imagining the American *Mastodon* doing the same. The midwest geological formations of Michigan, with their surficial exposures of older Paleozoic salt-containing evaporative deposits, were favorable for the salt-craving elephants. They may have trekked through cold and harsh terrain, perhaps far from their favored feeding grounds and calving areas, for their precious salt. The Mastodonts can be envisioned churning up the salty soil with their tusks and, along the way, drinking saline water from encountered watering holes.

Sometimes, however, these watering holes proved to be their bane rather than their boon. Occurrences of Mastodonts are most frequent in ancient bogs. These are places which were formerly filled with glacial meltwater. As the glaciers retreated 11,000 to 12,000 years ago, some large blocks of ice became buried. After the ice melted, the water-saturated depressions, known as kettle holes, became filled with debris, salt and sand. The saturated ground supported growth of vegetation and trees, food for Mastodonts, foodstuff grazing Mammoths would leave untouched on a dinner plate.

The trees also offered protection from severe wind chill. Hungry Mastodonts, seeking refuge from the cold, were drawn to these places. During the winter, occasionally, seemingly secure in bog "heaven," a Mastodont would break through the semi-frozen surface and become hopelessly mired in the ooze. A rapid burial in the "quaking bog" minimized chances for scavengers to find the entrapped animal.

The Perrys' Mastodon was thought to be a young adult who, about 10,980 years ago, fell into a bog, a major occupational hazard for large browsing mammals back then. The Perry Mastodon apparently sank into the cool ground quickly. Because the remains were left undisturbed, nestled in surrounding oozy sediment, its skeletal remains were preserved in remarkable condition for 110 centuries. A few thousand years later, civilization "happened" and "here come da' Judge," who decided to do a little landscaping.

Realizing that police investigations were pointless, representatives from nearby Wheaton College's Geology Department offered their assistance. At the time, there were less than 30 Mastodon skeletons known from around the country, and within two weeks of digging and excavation, it was certain that the Perrys' specimen was one of the best.

Word spread. Other museums, including the Chicago Field Museum of Natural History and the University of Chicago, now vied for the precious Perry Mastodon. The Perrys, however, decided to keep their large "pet" within the community.

Along with all the warm and fuzzy good intentions came a desperate need for heavy funding to exhibit the ancient, warm and fuzzy beast. Cost estimates indicated that excavation, reconstruction and completion of the exhibit would run up a vast sum of about $100,000. This was a lot of money for a relatively small and religiously oriented private college to scrounge up, all in the name of prehistory. However, the contributions began trickling in.

The first donation came from Judge and Mrs. Perry. They generously picked up the $2,300 tab for the excavation and added another $2,700 toward its completion. The local schools pitched in too. Sixth grade classes all over DuPage County collected bottles and jars, and one Naperville youth even guaranteed his one dollar a month allowance "forever."

Having that odd $5,000, and one boy's allowance money, was a promising start, but it was obvious that much more would be needed from the community. The College began offering gifts for donations. A $10 donation would land your choice of a replica charm or 3-dimensional, freestanding metal cast of the Perry Mastodon. A $50 donation would secure a larger cast. Those offering $1,000 to $1,500 would be immortalized on a group plaque. And for donations exceeding $1,500, names would be inscribed on an individual plaque. The McCormick Foundation contributed $25,000. After this public show of adoration, the work could now begin in earnest.

The skeleton wasn't perfectly preserved, as, in all, 115 Mastodon bones were discovered. The reassembly was also hindered by the fact that not many of these creatures had been accurately assembled, and the ones that had weren't exactly in the best shape. One recreation in New York had one too many spinal vertebrae, and some said the mount looked awkward. The Perry Mastodon offered a more promising beginning, even though the left side of its skeleton was better preserved than the right.

A prominent obstacle to overcome involved reconstructing the 250 pound tusks protruding from the Mastodon's face. How, exactly, were they supported in life? It was discovered that a honeycombed section of the skull was covered by two feet of muscle attached to and running along the back. Under advisement of Dr. Donald Boardman, geology students and later pre-medical students helped to identify and sort the old bones. Next, the Richard Rush Studios of Chicago "rushed" in to do "visuals," namely drawings and sculptures.

Rush decided to reconstruct the tusks in fiberglass, reducing the weight from a rather bulky 250 pounds to a more manageable 10 pounds. With many bones still missing, Larry King, Anatomy Major at Rush Studios, sculpted the rest from scratch. He had a reputation of being able to make fiberglass look more bone-like than authentic bone. However, most of the mounted skeleton (i.e. skull, jaws, pelvis, collarbones, leg bones) is genuine bone.[2] The result was staggering.

Life-sized Mastodont sculptures are frequently encountered in North American museums and theme parks. An impressive Mastodon sculpture created by the Terry Chase Studios of Cedar Creek, Missouri "strides" across Kentucky's Big Bone Lick State Park, formerly inhabited by many large Ice Age mammals seeking a "salt lick" source. Many of North America's earliest fossil vertebrates were unearthed here in the eighteenth century.[3]

Perhaps the earliest life-sized Mastodon sculpture ever done was displayed in Albany's New York State Museum. This was an expert restoration of a Mastodon thought to be a male, known as the Cohoes Mastodon, whose skeleton was discovered in 1866 in a peat bog at Harmony Hills, New York, a few miles northeast of Albany. The Cohoes statue (partially inspired by Charles R. Knight's miniature sculpture of New York's Warren Mastodon) was expertly rendered in 1922 by Noah T. Clarke and C. P. Heidenrich. (See Figure 21-2.) It was "...a hollow model composed of layers of wire screen and Akeley's mixture (felt paper pulp, whiting, glue, and boiled linseed oil), reinforced by a wooden framework. The head and feet were modeled

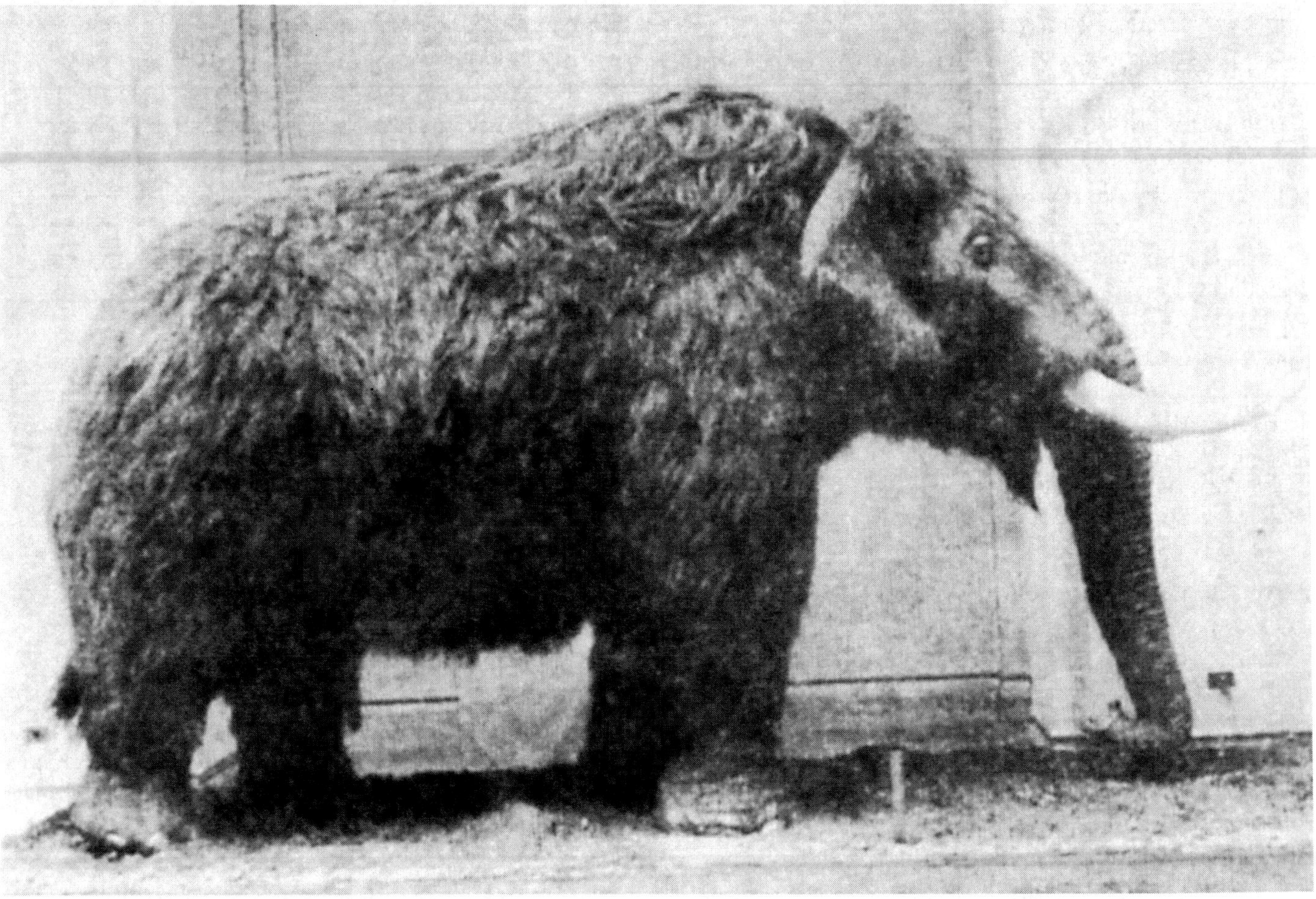

Figure 21-2: Restoration of the Cohoes Mastodon (1922). (Reproduced with permission of the New York State Museum.)

separately and attached to the main body. The hair is dyed sisal fiber."[4] This turned out to be a flammable mixture of substances.

Their creation formerly stood in between a pair of Mastodon skeletons, when the State Museum occupied the Education Building. In 1977 it was transferred to the Cohoes Public Library, where it burned four years later after kids playing with matches set fire to the fur. The now bald Mastodon was refurbished in 1986 after Hollywood special effects artist Thomas Culnan, who designed the giant Marshmallow Puff Monster for the movie *Ghostbusters* (as well as the green ghost that "slimed" actor Bill Murray in another scene), and a local taxidermist, Ted Weyhe, volunteered their services.[5]

Today, a more accurate replica of the Cohoes Mastodon may be seen at the Cohoes Public Library. This newer life-sized model of the young male Mastodon is completely "anatomically correct," if you know what we mean. (Yeah, it's a boy!) Cohoes librarian Carol Edwards comments that the lights must be dimmed in the exhibit area or else the kids might laugh at what is visible below.[6] However, since 1984, David LaPlante's sculpture of a female Mastodon and Calf has greeted patrons in the current New York State Museum location. (Incidentally, paleontologists believe that Mastodons had one of the longest mammalian gestation periods, 22 months. Mastodons may have reached maturity in 15 years.)

In the mid–1920s, Messmore and Damon created an animatronic Mastodon "robot" that later appeared during the 1933/34 Century of Progress Exposition in Chicago at the *World a Million Years Ago* display. A 1927 publication claims, "This huge replica of a mastodon, prehistoric animal, was produced at a cost of $30,000 , and among its travels was a trip, dismantled, in an express car from New York to Toronto, Canada and return. When the motors are operating, the 'animal' seems alive."[7]

In 1967 artist Paul Domke (1886–1981), fashioned a concrete Mastodon sculpture under attack from a band of concrete, spear-hurling, rock totin' Neandertals. (Of course, Neandertals, who lived in Europe during the Ice Age, would have never encountered the American Mastodon.) Domke's savage artistry may still be witnessed in northern Michigan's "Dinosaur Gardens." Another life-sized fiberglass Mastodon was created by the Louis Paul Jonas Studios of Hudson, New York. One of Jonas' Mastodon fiberglass casts may be seen in Harrisburg, Pennsylvania.[8]

The most recent entry is an 11-foot-high and 20-foot-long fiberglass Mastodon found stalking the premises of the Lake County Museum in Wauconda, Illinois. The Lake County Mastodon, also found to be an anatomically correct male (yep, we checked), was installed in April 2000. This $55,000 reproduction was created by the Terry Chase Studios. According to Kathleen Hamilton-Smith, the Mastodon is intended to "...help people visualize part of the ancient history of the area that is Lake County today.... So far, visitors love the Mastodon. Many call it an elephant until they found out more about mastodons (in the museum displays held inside the Lake County Museum). Visitors to the museum who have some background have been critical of the model being so near the prairie restoration site, as the mastodon is a forest dweller and browser. Most of these folks are satisfied when I explain that the model is purposefully not in the prairie, but adjacent to it."[9]

However, none of these Mastodon sculptures were of the "half-restoration" variety. Rush's would be the first and only fleshed out Mastodon reconstruction.

During our May 1995 interview with Richard Rush we learned that his studio has had numerous experiences in restoring prehistoric animals.[10] At six years of age, Rush owned a "museum" in the basement of his house. Later, blending interests in art and archaeology, Rush was appointed "curator" of Glenbard High School's natural history exhibit. After graduation, because his father disagreed that art could be considered a serious endeavor, Rush embarked on a chemical engineering degree. Eventually, he transferred to the Art Institute of Chicago, where he completed course work in sculpture and anatomy.

During World War II, as a member of the Army Corps of Engineers, Rush produced miniature models of the landing beaches used by American and British forces for the June 1944 D-Day assault. Rush earned a bronze star and a British Empire Medal for his efforts. The Epcot Center's Seven Seas exhibit, the Presidential Seal and even Smokey the Bear's clothing were all designed by the Rush studios.

Rush's most famous creation, the *Tyrannosaurus rex* featured in a display of Cretaceous life at the Milwaukee Public Museum, is regarded as one of the world's finest life-sized dinosaur restorations. (See Figure 26-4.) Five casts of this particular dinosaur have been made. One, the first made (in 1976), resides outside the Children's Museum in Indianapolis. An unsold cast is sometimes leased to promoters at Industrial Trade shows to attract prospective customers. (This was offered to us as we strolled by its crated parts for a "mere" $16,000. "Gee, Neighbor, nice backyard display ya' got goin' there!")

Paleontologist Donald Baird and paleoartist Greg Paul, lent their expertise to another project, a model of the Late Triassic theropod *Dilophosaurus*. Fortified by their suggestions, drawings and photographs of skeletal remains, Rush's team of talented artists, including sculptor Rodger Walshlager, recreated the 22-foot-long, life-like creature sans stylish *Jurassic Park*–like expandable frill and spitting grimace.

First, small scale casts of a miniature model were made. These were sent to Baird and Paul for critique. A miniature of the approved design was sliced into sections, which in turn were projected to the final design scale. These sections were cut from wood and assembled into a full-sized steel and wood armature. After covering the exterior of the armature with chicken wire, strips of fiberglass cloth were wrapped around the assembly and detailed skin texture was added as clay, vividly bringing it to life! A fiberglass mold was made, in sections, so that reproductions could be made and sold to other institutions. The price tag for Rush's *Dilophosaurus*, completed for Dinosaur State Park in Rocky Hill, Connecticut, in 1981, was $37,000.

With a sense of satisfaction, Rush also created life-sized restorations of two ten-foot-long Triassic vertebrates, *Rutiodon* (a phytosaur) and *Metoposaurus* (a labyrinthodont amphibian). Sculptures of these authentically detailed creatures are now exhibited at Dinosaur State Park, as well as in the halls of the William Penn Museum and the New Mexico Museum of Natural History in Albuquerque. Rush also designed a "half and half" (skeleton and skin) hadrosaur exhibit for the Academy of Natural Sciences in Philadelphia.

Life-sized "half and half" displays, or fleshed out reconstructions, of prehistoric animals are a relatively rare treat. Rush's paleoartist "hero," Benjamin Waterhouse Hawkins, is said to have invented the design in 1871. According to Donald Baird, "...he [Hawkins] invented the ... half restoration [skeleton on one side, sculptured life-appearance on the other]. But apparently his innovative museum plans were too avant-garde for the stuffy old Smithsonian, and all they would buy was a plaster Hadrosaurus and a sculptured Irish Elk.... The half-restoration was subsequently reinvented independently by J. D. Higgins in 1918 and R. S. Lull in 1919, and the two squabbled in print about the priority of invention. Benjamin Waterhouse Hawkins, who died in 1889, would have smiled."[11]

Besides the Rush Studios original productions, there is a half and half restoration of the Miocene chalicothere

Moropus on display at the Carnegie Museum of Natural History in Pittsburgh. Years ago, a half and half display of the Miocene fossil rhino *Teleoceras* greeted visitors at the Denver Museum of Natural Science, and a half and half restoration of the Miocene rhino *Menoceras* may be seen at the University of Nebraska State Museum in Lincoln. There are also several fleshed out reconstructions of prehistoric mammals exhibited at Yale's Peabody Museum in New Haven.

According to John Ostrom, a "*Monoclonius*"/*Centrosaurus* half-restoration prepared in the late 1920s was one of the earliest displays exhibited at Yale's Peabody Museum.[12] Most recently, paleoartist Gary Staab completed a half and half restoration showing, on the left side, details of the muscle, ligament and other soft tissue, and bony "innards" of the cervical and skull elements, of the African sauropod dinosaur *Jobaria*, recently discovered during one of Paul Sereno's University of Chicago expeditions to Niger. The right side is "fleshed out" in living form. (See Figure 27-4.) Another dramatically posed, fleshed out reconstruction of a life-sized *Albertosaurus*, a Hall Train production, is displayed at the Academy of Natural Sciences in Philadelphia. Our eyes rove, revealingly, from bony tail, to an abdominal region that is half flesh (right side) and half exposed "innards" (left side), to a fully restored head. Those of us who built the old "Visible Man" model kit decades ago can only dare to dream after viewing such marvelous exhibits!

However, the Perry Mastodon may be the most unique paleontological restoration ever completed by the Rush Studios. Rush had been impressed by a Jonas Mastodon displayed in Harrisburg, and had admired the Denver Museum's *Teleoceras* half restoration. So, rather than reconstructing an entire skeleton for Wheaton College as Donald Boardman had requested, Richard Rush imaginatively perceived how the Perry Mastodon could instead be reconstructed as a far more thought-provoking half body restoration display. After ordering a fresh cast of *only* the right side of the Jonas Mastodon, the bones were mounted and fitted to the reconditioned right half of the Mastodon's fiberglass body.[13]

An Open House took place on Saturday, April 2, 1966. Shortly thereafter, a letter received from one of the local school groups, commending the excellent display of the "dinosaur" (Oops!!), sent the college back to the drawing board one more time. This time, exhibitors added an instructional audio tape to explain the nature of the massive Mastodon reconstruction. Now visitors may press a button outside the display window and the Perry Mastodon slowly revolves on a platform, allowing both the reconstructed skeletal cast half and the restored half to be seen alternately and several times while the tape is playing. The narration has a quaint religious tone. Some of the original bones dug from the Perrys property are visible laying adjacent to or resting on shelves beside the Mastodon in the display rotunda; the rest remain in storage.

Judge Perry's namesake, the "Perry Mastodon," brought him much fame. He proudly displayed a fiberglass femur in his chambers. (One wonders whether he ever considered substituting a piece of Mastodon rib bone for his gavel?)

Today, visits to Wheaton College's science building seem subdued. The building is now under a massive renovation focusing more on geological than paleontological matters. Glass cases are available, waiting to be filled with new specimens, and students pass by their Pleistocene Pachyderm without much ado.

We can be thankful to Judge Perry and his love for wetland areas. We hope he enjoyed fishing there. Yet, beyond the tale of a college with determination and pride over its magnificent Mastodon, there is a bit of little known history here to dwell on as well.

The American Mastodon and *T. rex* represent America's two greatest symbols of prehistoric savagery, one long neglected, the other "ruling" the current Renaissance in dinosaurs. Perry's Mastodon has few visitors, certainly only an infinitesimal number when compared to the waves of tourists flocking to see Chicago's greatest fossil, "Sue" the tyrannosaur, displayed only an hour's drive from Wheaton in the Field Museum's Stanley Field Hall. Even the greatest restoration of the American Mastodon ever created, here in Wheaton, represents no challenge for the dinosaurs, which in peoples' minds have come to "rule" prehistory.

Few realize that for a century America's little known and poorly understood Mastodon represented the founding fathers' national conscience, pride and "virility."[14] In the late eighteenth and early nineteenth centuries, the Mastodon was a symbol flaunted by Thomas Jefferson (1743–1826) as "living" proof that the American continent's wildlife had not degenerated through time as leading French scholars had theorized. For decades, the jaws of the Mastodon, our mysterious "incognitum," perceived as the most gigantic of all prehistoric animals, symbolized America's might, primeval savagery and dominance over its forbidding wilderness.

The first relatively complete picture of the Mastodon came as a result of Charles Wilson Peale's efforts to reconstruct the bones of two individuals found in 1801 in Orange County, New York.[15] Peale's Romantic painting, *Exhumation of the Mastodon* (1806–08), focusing on the machinery used to extract bones from the earth, can be viewed as a sublime moment in the new republic's search for its missing ancient history. Peale's 1822 self-portrait, *The Artist in His Museum*, showing Mastodon fossils displayed in his Philadelphia Museum, "...was conceived as an emblematic painting and offered yet another monumental image of the mastodon bones as symbols of the nation's natural antiquity." [16]

An 1802 image of Peale's mounted Mastodon show the tusks protruding through the eye sockets. [17] Shortly afterward, by 1803, the tusks were reconfigured in an inverted orientation.[18] The upside-down tusks were thought to more correctly represent an animal capable of dispatching prey. Certainly the Mastodon's large grinding teeth, differing from those of modern elephants or Mammoths, indicated to many that the Mastodon ate flesh. The image of a *carnivorous*, rather than herbivorous, Mastodon more wholly represented America's savage prehistory and supremacy.

A more complete skeleton, now known as the Warren Mastodon, was discovered in 1845.[19] This came during the decade when dinosaurs challenged the Mastodon's supremacy. A decidedly inaccurate reconstruction of an allegedly larger variety of American Mastodon, tagged the "Missourium," was a great attraction in London, drawing crowds for 18 months until Richard Owen proved that its 30-foot-long frame had been erected with an excessive number of vertebrae and ribs. The American fraud exposed, Londoners instead understood their native *Iguanodon* to be an even larger and therefore more imposing primeval beast.[20]

But dinosaurs loomed on America's horizon as well. For, by 1868, when Benjamin Waterhouse Hawkins sketched design drawings for his New York "Palaeozoic Museum," intrigue over America's new dinosaurs, "*Laelaps*" and *Hadrosaurus*, had evidently "eclipsed" the rival Mastodon's appeal.[21] "With the rising hostility toward Darwin's ideas and the emergence of the dinosaurs, the mastodon drifted into obscurity, becoming a neglected scientific specimen rather than a national symbol."[22]

While the mastodon drifted into obscurity, during late nineteenth and early twentieth centuries, Henry F. Osborn presided over a Reformation in paleoart at the American Museum of Natural History. This Reformation program emphasized America's dinosaurs, effectively replacing the image of the Mastodon with a genuinely savage and more powerful prehistoric monster—*Tyrannosaurus rex*. For many, the impact of America's dinosaurs symbolized the violent nature of America's past, reflecting the true vision of Darwinian hostility, fueled by competition and struggle for survival, visceral themes no longer associated with the American Mastodon.

And so the Perry Mastodon stands alone during the reign of "Sue." National identity projected through our prehistory has evidently become more violent, cutthroat and litigious than the founding fathers could have ever anticipated.

CHAPTER 22

Louis Paul Jonas' Prehistoric Sculpturdermy

Full-sized replicas of prehistoric animals, especially dinosaurs, have been traditional World's Fair attractions. Paleontological displays stirred the masses, French novelist Jules Verne included, at the Paris Exhibition in 1867. Henry A. Ward's skeletal cast of *Megatherium* (one of the earliest reconstructions ever assembled), his startling life-sized Siberian Mammoth restoration and a skeletal cast of Benjamin Waterhouse Hawkins' *Hadrosaurus foulkii* graced the Anthropological Building during Chicago's Field-Columbian Exposition of 1893. A life-sized, papier mâche *Stegosaurus* sculpture, rumored to have been made of American greenbacks, greeted patrons of the 1904 St. Louis World's Fair Exposition. Three decades later, dinosaurs and other prehistoric animals invaded Chicago once again during the 1933/34 *Century of Progress Exposition*. Both Messmore and Damon's *World a Million Years Ago* exhibit and the Sinclair Refining Company's robotic dinosaur statues (equally marvelous attractions attended by my father and grandparents) fired the imaginations of both young and old during the fabled "era" of *King Kong* (RKO, 1933).

Another three decades would pass before dinosaurs created yet another World's Fair sensation. The new models, produced by the Louis Paul Jonas Studios, would be regarded as the most realistic ever built.

During the late summer of 2000, Allen visited the Louis Paul Jonas studios in Claverack, New York, and met Zella Jonas Merritt, daughter of famed sculptor Louis Paul Jonas. Although from the road, idle passersby probably wouldn't recognize the farm setting, during the early 1960s, Zella's property was a hotbed of activity, a magical place where ten Mesozoic animals, including North America's mightiest dinosaurs, were coming to life for another Sinclair World's Fair display.

On this day, however, there were only dim reminders of those days of yore. There, in the lonely warehouse once occupied by Mesozoic giants—70-foot-long *Apatosaurus*, three-horned *Triceratops*, plated *Stegosaurus* and their kind—I observed hollow fiberglass molds belonging to a few old favorites; spiky *Ankylosaurus*, *T. rex* and *Apatosaurus* head molds, and massive looking dinosaur leg molds now rest on sturdy shelving. The discovery of life-sized *Deinonychus* and *Mastodon* molds was unexpected. It may seem difficult to recognize the significance of this enchanting place.

Who was responsible for Sinclair's Dinoland, perhaps the most remarkable World's Fair dinosaur exhibition ever, and what events led to its creation? What does this union of prehistory and paleoart signify?

Louis Paul Jonas, Jr. (1894–1971) was one of five artistic brothers who excelled at taxidermy. They learned the trade in their native country, Hungary, where Louis' older brother, Coloman Jonas, satisfied demand from the Central Educational Supply House for mounted animal specimens. Eventually, Coloman had saved sufficient funds for beginning his own business, and in 1908 settled in Denver, Colorado, where the Jonas Studios first opened shop.

Louis joined Coloman in 1908, where he began an apprenticeship. Louis also honed sculptural talents and spent many hours at the Denver Zoo observing animals. He excelled both as a sculptor and in the art of taxidermy. In fact, Louis believed the two practices to be so interrelated that he referred to his trade as "sculpturdermy."

The techniques used by taxidermists for mounting an

extant animal are conceptually similar to what a paleosculptor would use in creating a life-like impression of a prehistoric animal. After taxidermists take careful measurements from a carcass, the leg, shoulder and pelvic bones are mounted, with wooden boards serving as a "backbone" and cervical column, anchoring the bones in place. Then, rather than fleshing out the body with "stuffing," an anatomically accurate clay model is sculpted around the skeleton armature. This clay model is molded in plaster. Artificial eyes and ears are added to the plaster cast. This "mannequin" serves as the body to be dressed in the preserved animal hide.

As Coloman stated in 1983, "...the real test of the artist is in the final modeling of the features of the animal, which gives the animal an aura of life or merely reflects the work of an artisan...."[1] This summation would also aptly apply to the restoration of fossil vertebrates.

Anyone who doubts the interrelation of taxidermy and paleo-restoration should study two life-size Mesozoic birds on display at Chicago's Field Museum's, "Life Through Time" display. Fully feathered *Sinornis* and *Archaeopteryx* appear like they were recently shot in a farmer's cornfield and brought to the local taxidermy shop for life-like recreation. How easy it can be to confuse these restorations' "authenticity" (in this case performed by Field Museum artists) with the handiwork of skilled taxidermists.

In years to come, Louis would study under noted American Museum taxidermist Carl Akeley and sculptor Edouard Lanteri.[2] In 1911, Akeley, who in the late nineteenth century refined the practices of taxidermy considerably, invited Louis to assist in the creation of an African Elephant display at New York's American Museum. Only in his teens, Louis was already gaining a reputation as a talented taxidermist, museum diorama builder, and sculptor. Louis was lauded for an early success, a bronze statue of a grizzly bear posed gazing into the distant Rocky Mountains, guarding her two cubs. This study, *The Grizzly's Last Stand*, symbolized mankind's progressive encroachment into the grizzly's natural habitat. It stands near an entrance to the Denver Museum of Nature and Science. When Akeley was unavailable to work on Indian elephant and giraffe displays in New York, Louis was called in his stead.

Louis appreciated that miniature dioramas took up considerably less museum space than life-sized displays, and so he concentrated on these. Eventually, Louis became a prolific sculptor of over 300 1⁄10 scale animal figures, captured in vivid poses or dramatic scenes. By 1941 he was regarded as the "world's greatest miniaturist."

In New York, Louis had befriended noted paleoartist Charles R. Knight, himself a talented sculptor. Together they conspired to build an outdoor museum featuring life-sized prehistoric animal sculptures. But their dream of collaboration ended when Knight died in 1953.

By 1955, the Jonas Studios had developed fiberglass material as a casting medium. This innovation would prove vital toward the Jonas Studios' most widely acclaimed success—the sculpting of the Sinclair Refining Company's Dinoland models for the 1964/65 New York World's Fair.

With Barnum Brown of the American Museum of Natural History and Yale's John Ostrom serving as technical advisors, in October 1962 Louis first tackled *Stegosaurus*. First he made a 1⁄10 scale miniature clay model. Sinclair approved this design. A replica of this small model was sliced and projected onto a screen so that full-sized sections of armature could be made.[3] Then, after constructing a sturdy armature out of wood and metal lathe, "flesh" was added in clay. Calipers were continually relied on for accurate proportional scaling from the miniature model. While Louis made the miniature clay models himself, he necessarily relied on a team of 18 assistants to complete the large dinosaur models. Live "pet" alligators maintained on the premises kept Louis focused on the problematic appearance of dinosaurian skin texture. Two months later the dinosaur was molded and a fiberglass cast assembled and painted.[4] (The *Apatosaurus* required 178 separate molds.) The finished fiberglass *Stegosaurus* cast was placed outdoors for a year to test its durability upon exposure to the elements.

In such manner were all the dinosaurs completed, at a cost of $400,000. Forty thousand pounds of plasticene clay was used in their making. Following a hometown parade celebrating their completion, preparations were made for rafting the models down the Hudson River by barge to the fairgrounds in Flushing Meadows. After delivery of the dinosaurs in New York, Sinclair ceremoniously staged an even grander arrival, airlifting Triceratops by helicopter.[5]

After President Johnson officially opened the Fair in April 1964, Sinclair's Dinoland attracted over 10 million visitors during its six month duration. The Fair reopened for another six months in April 1965. Visitors were treated to a spectacle of 10 lifelike dinosaurs, judged to be the most accurate ever made as full-size recreations—due to the assistance of Barnum Brown (discoverer of *T. rex* and *Corythosaurus*) and John Ostrom. Dinoland featured *Ankylosaurus, Apatosaurus, Corythosaurus, Ornitholestes, Stegosaurus, Struthiomimus, "Trachodon," Triceratops, Tyrannosaurus* and *Archaeopteryx*. A smaller Apatosaurus fiberglass replica could be seen near the Dinoland entrance. A set of "brontosaur" hatchlings and eggs was created as well. The jaw of the *T. rex*, and *Apatosaurus*' and *Triceratops*' necks were mechanized to further instill an impression of liveliness.

Little *Archaeopteryx* was stolen from the fair, and its current whereabouts are unknown. *Ornitholestes* was also stolen but later recovered from nearby Jones Beach.[6]

(A rival dinosaur exhibit, created by WED Enterprises for Walt Disney, also graced the New York World's Fair grounds. These masterful, animatronic restorations were still on display during my last visit to Disneyland's "Primeval World," in Anaheim, California, during the early 1980s.)

After the World's Fair closed, Jonas' dinosaurs, anchored to flat bed trucks, toured the United States.[7] (See Figure 22-1.) Thereafter, the nine originals were sold to various institutions. These, as well as others made after the Fair closed, were replicated to satisfy demand from a score of North American museums for their world famous dinosaurs.

Jonas' dinosaurs can still be seen today. For example, the original *Apatosaurus* and *Tyrannosaurus* guard the grounds of the Dinosaur State Park in Glen Rose, Texas. (See Figure 22-2.) Striped, duck-billed "*Trachodon*" is partially concealed by foliage near a duck pond at the Brookfield Zoo outside Chicago. The original, repainted Stegosaurus cast stands outside the Dinosaur Quarry Visitor Center in Utah. *Struthiomimus* is incorporated into a marvelous Late Cretaceous diorama at the Milwaukee Public Museum. During our travels we have seen nearly half of the 38 Sinclair World's Fair dinosaur reproductions now in existence.

The original Jonas *Triceratops* later became a "star" of the motion picture industry when it was cast in a 1978 film adaptation of Oliver Butterworth's children's novel, *The Enormous Egg* (1956). In this curious tale, a hen's mutant egg hatches a baby *Triceratops*. The dinosaur emerges and is befriended by a young boy who names his new pet "Uncle Beazley." The grown dinosaur eventually inhabits the grounds of the Smithsonian in Washington D.C. (both in the telecast and in reality). Jonas' World's Fair *Triceratops*, a familiar sight outside the Smithsonian's entrance, served in the film as Uncle Beazley's adult form. Jonas Studios also made a series of smaller, juvenile *Triceratops* models for this acclaimed NBC project.[8]

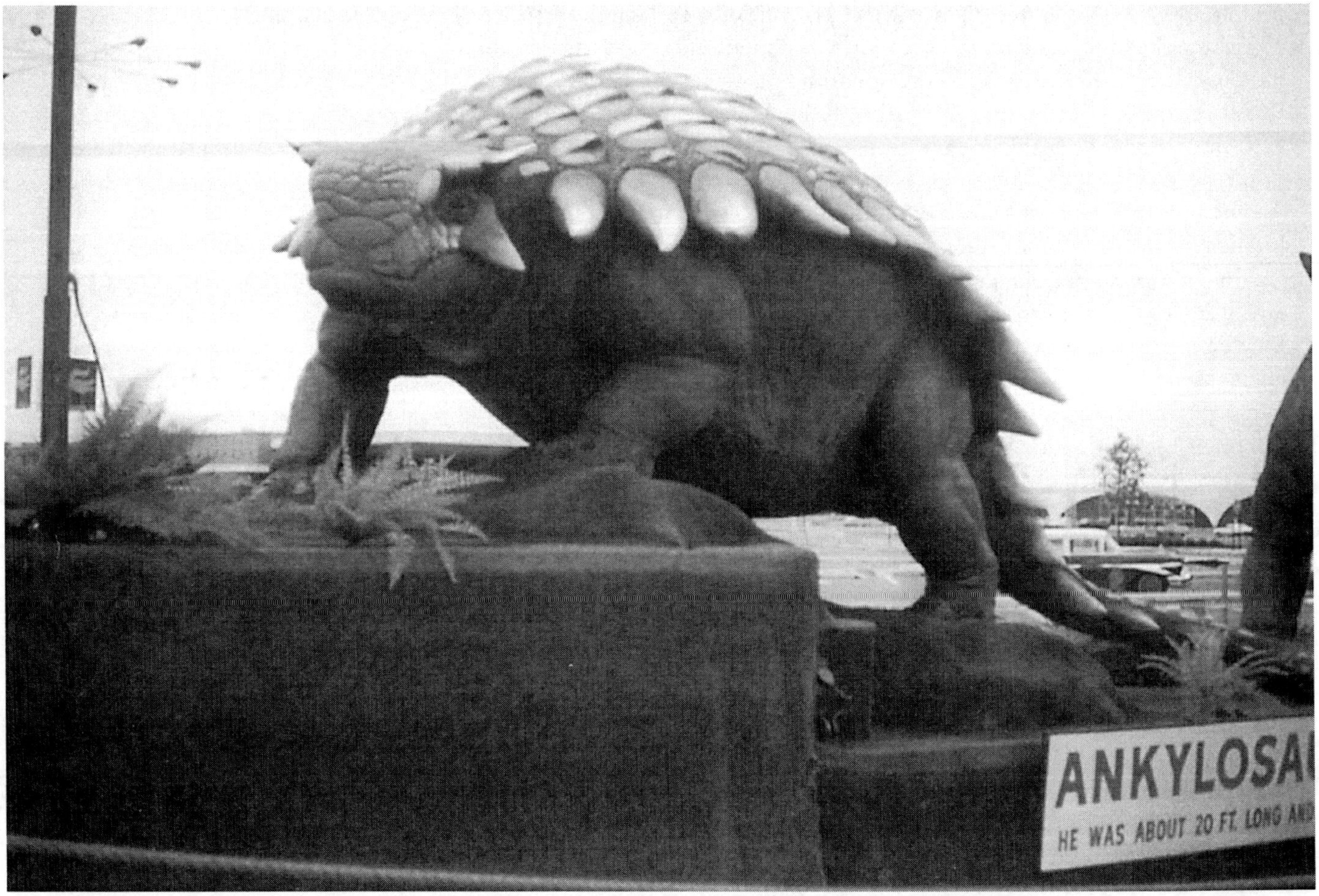

Above: Figure 22-1: Jonas' _Ankylosaurus_ on a late 1960s tour in Minnesota. (Photograph reproduced with permission of Mark Zimmerman.) _Opposite:_ Figure 22-2: Jonas' _Tyrannosaurus_ on display at Glen Rose, Texas. (Photograph by Allen A. Debus, 1984.)

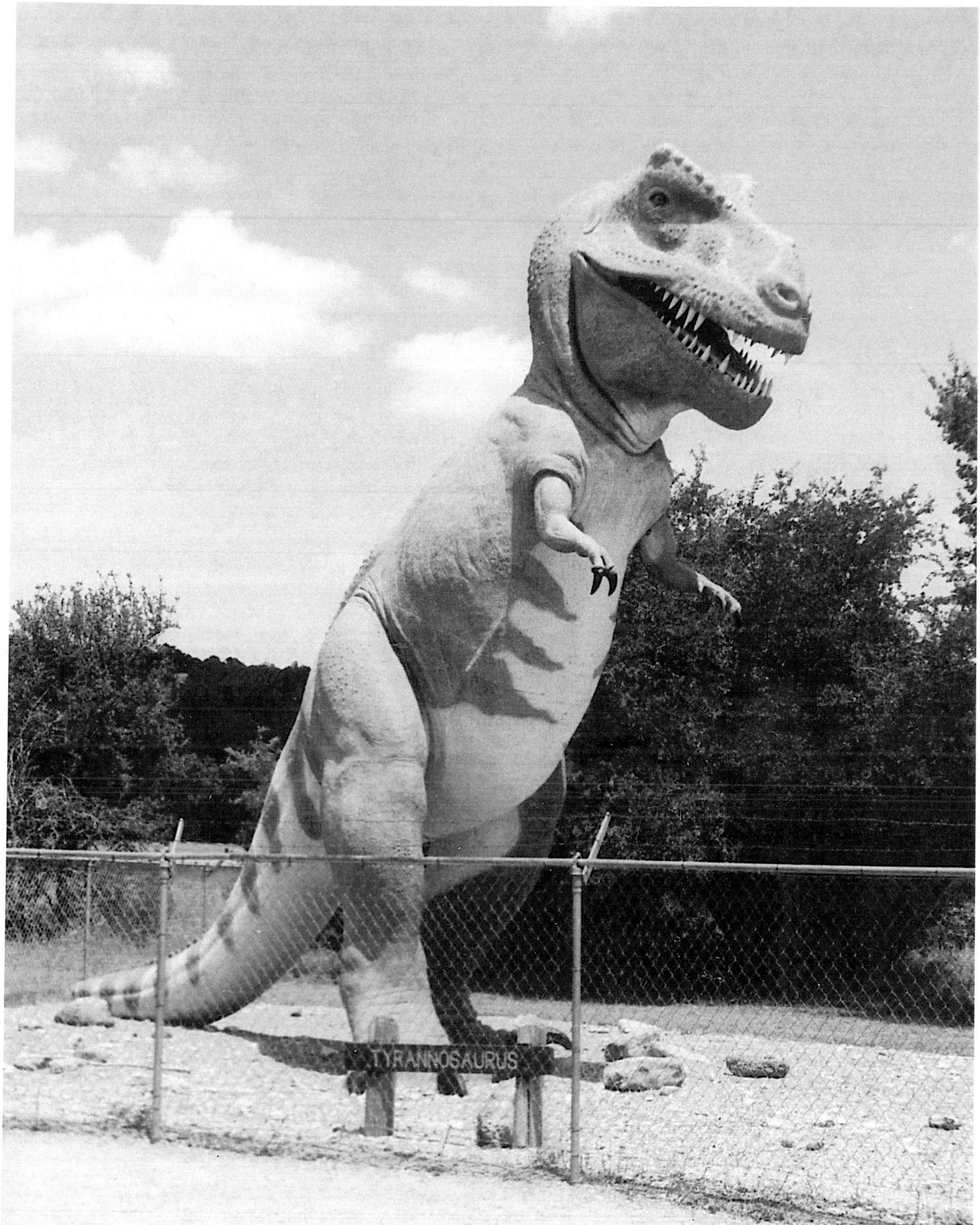
TYRANNOSAURUS

After the Fair closed, the Louis Paul Jonas Studios also created a life-sized *Coelophysis* and *Deinonychus*. Fiberglass reproductions of these handsome sculptures can be seen at a number of American museums, including, respectively, the Dinosaur State Park in Rocky Hills, Connecticut, and the Yale Peabody Museum in New Haven, Connecticut.

Miniatures are the current focus of the Louis Paul Jonas Studios. Louis' 1/10 scale miniature Sinclair dinosaur models, as well as *Mesohippus*, Mammoth and *Mastodon* replicas, have been molded and are available for purchase. Additionally, in numerous museums around the country, miniature Cenozoic dioramas prepared by the Louis Paul Jonas Studios feature Louis' expertly fashioned mastodonts, mammoths, giant ground sloths, saber toothed cats, uintatheres, giant beavers and dire wolves.

A Tribute to Louis Paul Jonas is presently displayed at the Berkshire Museum in Pittsfield, Massachusetts. On the Berkshire grounds, one notices "Wally," one of Jonas' World's Fair *Stegosaurus* casts, so named by a local youngster because Stego's brain was supposed to be the size of a walnut. Inside are "Enormous Eggs" props, Dinoland Fair souvenirs, and several of Jonas' miniature dinosaur sculptures. Jonas' dinosaurs still seem sensational.

The science and art of dinosaurs has evolved considerably since America's premier dinosaur sculptor, Louis Paul Jonas, was commissioned to complete Sinclair's Dinoland figures. For, in July 1969, Ostrom authored a Yale Peabody Museum report describing *Deinonychus*, the dinosaur that became most closely identified with a 1970s "Dinosaur Renaissance." One outgrowth of this episode in the history of science is that the poses of Jonas' *T. rex*, *Apatosaurus*, *Triceratops* and "*Trachodon*" may now seem quaint. Arguably, however, no sculptor was better prepared than Jonas to restore life-sized dinosaurs, given his expertise in "sculpturdermy." Like Benjamin Waterhouse Hawkins' marvelous concrete sculptures completed over a century earlier, Louis Paul Jonas' New York World's Fair models have left lasting, startling impressions on a generation of dinosaur enthusiasts.

Chapter 23

Speculative Paleontology—Lessons in Reverse Paleontology

As conveyed through the extraordinary visions of British paleontologist Dougal Dixon, fanciful depictions of Life's *future* history frame an irresistible virtual reality. Dixon's evolutionary extrapolations in the era "after man" cannot be considered *accurate* predictions, even though truth and beauty lie in the details. As Dixon confesses, "What I find fascinating about paleontology is the way that strict scientific study can create visions of animals and landscapes of times gone past. Imagination, but not running wild, [is] firmly restricted by scientific knowledge."[1] For paleontologist Dixon, however, time is a two-way street.

Perhaps 95 percent of the wealth of science fiction ideology published during the past 50 years can be traced to the influences of two English gentlemen, Herbert George Wells (1866–1946) and Olaf Stapledon (1886–1950). Both writer/philosophers dwelled on the future of humanity as a central theme in their stories. Wells' time travel classic, *The Time Machine* (1895), revealed his somber view of Man's future. At first the distant future seems utopian. Only later in the story is it revealed to what abysmal state Man has degenerated. We discover that twentieth century civilization was only a temporary edifice, destined to crumble. In the far future, some 800,000 years from now, evolves a second race of Man that burrows into the bowels of the earth to escape the light of day, and preys on the ancestral form.

In Stapledon's forecast, *Last and First Men* (1931), and the sequel, *Star Maker* (1937), Man persists for many billions of years, although in this enormous interval of time human evolution would produce a diverse array of descendants. Until the end, after spreading through many galaxies and parallel universes, humanity would repeatedly triumph over near extinction from a variety of causes.

Wells and Stapledon cast man as having played a vital, powerful role in shaping the future of our world, solar system and even beyond. But what if Man entirely disappeared from the universe? (Never mind the "how," focus on the "what if.") What kinds of creatures would evolve? What would they be capable of? Can the science of paleontology (which is a science probing the deep past) teach us anything about the future evolutionary "history" of terrestrial life forms, extrapolating from the present?

Fauna of Earth's far distant future are portrayed in Dougal Dixon's imaginative masterpiece, *After Man: A Zoology of the Future* (1981).[2] These are not contrived fairy tale creatures or science fantasy movie characters, such as unicorns, trolls and fire-breathing monsters. In fact, any element of ridiculousness has been eliminated through their evolutionary design. Even though such vertebrate animals do not actually exist, Dixon instructively describes how his organisms, which he refers to as his "babies," wouldn't be beyond the capability of natural circumstances.

Of course, several inherent assumptions concerning the Earth's conditions 50 million years hence are necessary to ensure the right sort of "future history" and opportunity for such animals to develop. First, Man, a victim of his inefficient use of livestock, overuse of natural resources, overpopulation and self-inflicted environmental contamination, has disappeared. Man also proved incapable of biologically adapting to the rapidly deteriorating environmental conditions, since his technology and dominance on the planet had essentially stifled his own evolutionary development. Accordingly, the Age of Man ended somewhat catastrophically.

Dixon writes eerily in past tense from a perspective

of a future time, *50 million years* beyond the present. "With the dominant life form gone the animal world entered a period of evolutionary chaos that lasted tens of thousands of years. However, man's extinction provided the impetus for the formation of many new species of animals, and his disappearance was of fundamental importance in shaping the world that has emerged 50 million years later."[3]

The Earth's geological and astronomical conditions haven't varied considerably over the intervening period since the Age of Man. Large crust shattering and mass extinction-causing meteorite strikes haven't occurred. Furthermore, in the world of 50 million years A.D., the Earth's overall climate and major topographic provinces are similar to those in the present. Continental rifting and plate tectonic activity has altered the planetary map, however. For instance, Australia has moved in a northwesterly direction, causing it to collide with Asia (forming a new lofty mountain chain), and South America has separated from Central America, becoming the world's only insular continent.

Aside from the obvious absence of Man, there are some notable similarities between the dominant vertebrate fauna of the future and those of today. The major faunal differences, however, are what makes Dixon's book a pleasure to read. "The only danger in reading this delightful volume is that some of you may reach the point where you suddenly feel saddened by the thought that the animals meticulously depicted in it ... do not exist now."[4]

After Man underscores how the ultimate expression for paleontologically founded science fiction is through visual projection. The science of *After Man* is intriguing, yet the *art* of this exceptional volume makes it a memorable treasure trove of science delightfully gone "muppet mad." Two decades following its publication, when asked about his cast's physical creation, the artistic Dixon recalled, "I sculpted a few of the creations, mostly in fairly temporary materials, to be used in publicity exercises for the books.... The artwork in *After Man* was done by some skilled wildlife artists chosen by the publishers, but at first they thought they were illustrating science fiction covers and did any old weird monsters. We had to bring them to heel, and I ended up providing detailed drawings that they colored in (almost by numbers)."[5] (See Figure 23-1.)

In 1984 an even livelier cast of life-sized, animatronic

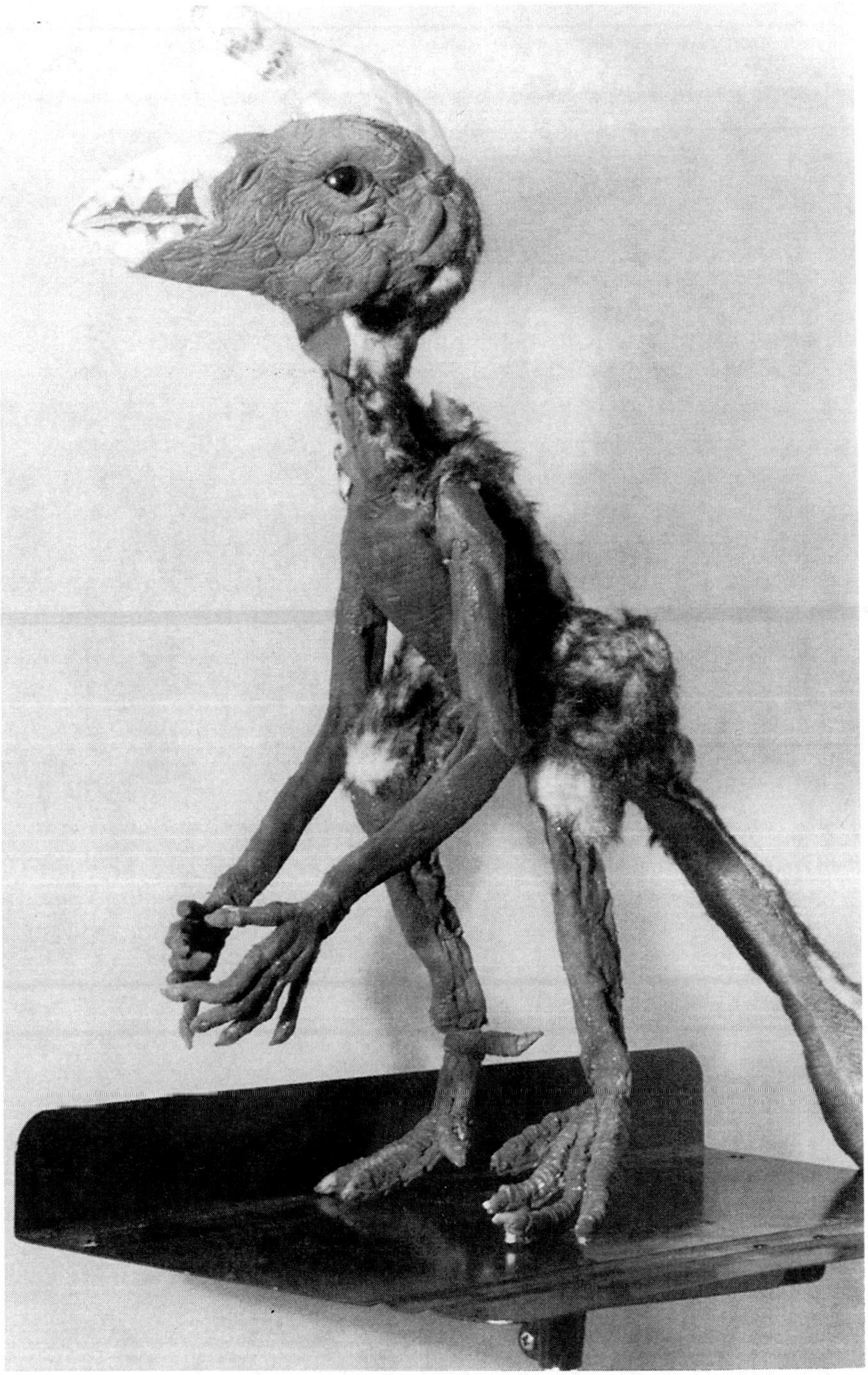

Figure 23-1: Dougal Dixon's original sculpture of the "Crack Beak," an alternate dinosaur created for his book, *The New Dinosaurs* (1988), Topsfield, Mass: Salem House Publishers. (Photograph by Dougal Dixon, reproduced with permission.)

figures were created by Tsuromoto of Japan, featuring the *After Man* fauna. These were incorporated into a traveling exhibition, "Weird Science: Creatures of the Future." The show consisted of a number of scale dioramas based on elements of the book. Dixon accompanied the show, after it crossed the Pacific Ocean from Japan, giving press appearances and lectures when it appeared in Denver, Boston and San Fransisco.[6] Peering into *After Man*'s colorful pages, one is struck by the thought that maybe these fascinating animals *do* live somewhere on the planet today, awaiting discovery by cryptozoology.

Let's take a brief mental safari through Dixon's time bridge and examine some of Nature's remarkable creations which *could* inherit the Earth 50 million years hence. We must bear in mind, of course, that due to Nature's infinite variability, and the inherent tendencies of organisms to adapt to new or changing ecological conditions, that no empty niche would be vacated for very long. Rapid ecological changes, as currently being caused both by the proliferation (and eventual disappearance) of Man, and also by other natural disturbances, are certain to result in rapid, wide-sweeping changes in species diversity.

In many cases, following the demise of Man and the large mammalian fauna hunted to extinction, new life forms arose. The immediate ancestors of many of tomorrow's bestiary were capable of survival and withstanding adversity. These groups include those hardy rats, rabbits and other rodents, and birds.

Fifty million years hence, after a long period of species radiation into vacated niches, rabbits diversified and evolved into large forest, tundra and prairie dwelling, deer-like animals. These *Rabbucks* feast off the leaves of trees, aided by their long necks, and sprint over terrain on two-toed hooves. Their chief enemies are the predatory rats, any of a variety of dog or wildcat sized, fanged carnivores descended from the small rodents of today. They stalk *Rabbucks* and other herbivorous game in packs, singling out the weaker individuals and finally stabbing them to death with their cruel canines.

The largest vertebrates still reside in the seas, but they are not whales. (*We* did them in, remember?) These stately creatures, actually a group of amphibious whale-like birds descended from the modern penguins, are known as *Porpins* and the *Vortex*. They are wholly aquatic, having entirely lost the capability of flight. They lay eggs, as do all birds, but these are incubated within their bodies. Live young are then born in the oceans. Whereas the *Porpin*, remarkably resembling the porpoise of today and the ichthyosaurs of yesterday, feeds on fish, the *Vortex* is equipped with a fibrous straining device, which evolved from an ancestral beaked form, through which plankton are sieved. (See Figure 23-2.)

Primates still exist in the tropics, though they haven't risen to technological intelligence. (This isn't a *Planet of the Apes* scenario by any means.) Instead, the apes of the future display a wide range of adaptational features. Tiger-like, some stalk fast moving game (such as the *Gigantelope*) across the African plains. Other forms, such as the *Raboon*, stride bipedally in *Tyrannosaurus rex* fashion after quarry. Such primates have effectively replaced the great predator cats hunted to extinction "long ago" during the Age of Man.

In *After Man*, technological intelligence has not reevolved, a surprise to many of us who remember Charlton Heston's role in *Planet of the Apes* (1969). Dixon believes that human-like, tool-making, "...warmongering intelligence is something of a one-off. This is the first time it has happened in 3.5 billion years, and there has not been enough time to show if it has been successful or not. Life has survived quite happily without it for that length of time. Personally, I don't think Nature will make the same mistake twice."[7]

There is a startling faunal diversity figured in *After Man*'s pages. We confront large tusked, elephant-like mammals known as *Zarander*, descended from pig ancestors that populate the jungle. A social monkey known as the *Khiffah* swings through the trees defending its thatched nest of leaves and twigs. Amphibious, hippo-like rodents occupy the swamps. By far, however, the most disgusting and nightmarish creatures are, respectively, the *Slobber* of Australia and the *Night-stalker*, native to the Batavian Island, a veritable world of bats. Perhaps it would be best not to elaborate on their revolting habits or even their appearance here.

In the new Era, certain species have evolved unusual lifestyles (that is, considering their ancestry). After all, who would have thought nature could design creatures out of rats converging on the form of wolves and dogs.

Quaint symbiotic relationships have evolved too. There is a mole-like bird, for instance, that lives entirely underground in termite nests. And from ancient antelope stock, the Cleft-like antelope has evolved. Small tick birds nest and roost between the double rows of bony ridged spines along their backs. The birds gain a warm mobile home, and the antelopes' skin is in turn protected from threat of parasitic insects. The noisy tick birds also provide a valuable "first alert" system in response to stalking predatory rats or other lurking carnivores.

If all this sounds a bit too far fetched to you, then I ask readers to keep in mind the observation that the average longevity of mammalian species seems to be "only" about 500,000 years. Looking backward from our vantage point, only two million years ago humans didn't exist, and

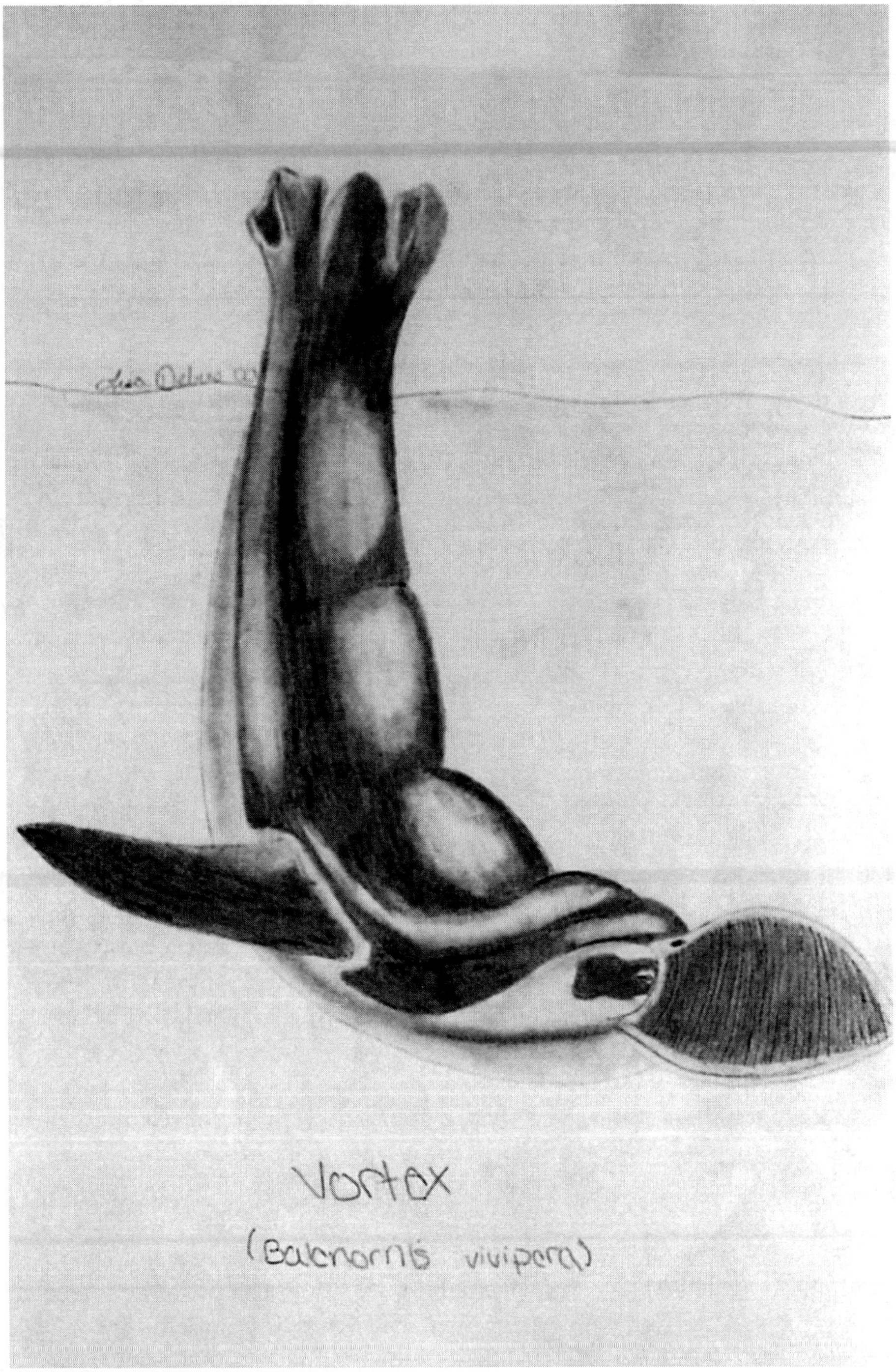

Figure 23-2: Illustration of the "Vortex" by Lisa Debus, after art appearing in Dougal Dixon's *After Man*.

Our immediate past seems incredibly brief when scaled against the vastness of Earth history. To put this in another perspective, recently it was determined on the basis of sophisticated genetic measurements that all existing human beings can trace their ancestry to a Pleistocene lady who lived about 200,000 years ago. This "mega-generation" span is only one-fourth the estimated length of time recently reported for the duration of the geologically "rapid" late Triassic mass extinction event.

Dixon's vision is certainly not how the future actually will be. It is merely a possible one. It is a vision based on the trends displayed by fossil and modern vertebrates, and extrapolated from conditions and other natural phenomena prevalent in the present Age of Man. One would agree that it is a somber as well as sobering vision.

Man will someday disappear from Planet Earth. All or most traces of our existence would be obliterated through the action of our mobile changing planet and our corrosive, dynamic ocean-atmosphere system. But on the sterile moon, Man's traces will survive to a much longer era. Hundreds of millions of years from now, an alien visitor to our solar system may well wonder what became of the race who built the relic spacecraft now stationed at Tranquility Base and other scattered lunar sites.

Or, after our demise, another species might evolve heightened curiosity, intelligence and technological capacity. Perhaps they will marvel at our fossilized remains, our monuments and relics.

strange carnivorous kangaroos inhabited Australia. A bit further back in time (to 50 million years before the present), dog-sized horse forerunners roamed the forests. Large horned vertebrate animals only distantly related to modern rhinos, and nine-foot-tall flightless predatory birds existed. Dixon's future menagerie is no more exotic than the creatures actually comprising Life's relatively recent history of the past 50 million years.

Conversely, it is also possible that we may evolve a greater sense of wisdom, understanding and appreciation for our own existence and that of our fellow creatures, such that there will be no period *After Man*.

One of the recurring motifs in contemporary science fiction concerns the technological transformation of earthbound Man into Space Age Man. As beautifully symbolized in stories such as *2001: A Space Odyssey* (1968) and

Childhood's End (1953), both by Arthur C. Clarke, just as the early paleo-humans formerly stood on the brink leading to an advanced evolutionary stage, so today do we stand on the threshold of a gateway to the stars.

We now understand that Life's "procession" was foreordained since the evolution of the earliest cells, nor was our descent from the first mammals an inevitable consequence of evolution. As is evident in the fossil record, the occurrence of repeated mass extinctions was a necessary prerequisite for the evolution of Man!

Is it possible that extraterrestrially induced or natural terrestrial catastrophes could vastly alter the current pattern of change? Will our disappearance, as anticipated by Dixon, instead pave the future path for even smarter beings?

There is another message in Dixon's book. Did prehistoric Man ever conceive of the notion that the mighty Mammoth would perish from the face of the Earth as he painted their majestic forms on the walls of his caves? Do we ever treat seriously the notion that the African elephant, the bald eagle, and species of whales could become extinct in the next century? The distant future and its depleted fauna may one day loom before us as presented in Wells' time travel story. "I looked about me to see if any traces of animal life remained ... But I saw nothing moving in earth or sky or sea. The green slime on the rocks alone testified that life was not extinct ... a terrible dread of lying helpless in that remote and awful twilight sustained me while I clambered upon the saddle [of the time]."[8] We, unlike the Time Traveler, cannot escape from or remedy our mistakes quite as readily (by fleeing in salvation toward a nurturing past).

Perhaps Dixon's *After Man* is a "ghost of Christmas future" (as in Charles Dickens' classic, *A Christmas Carol*). Hopefully, through visionary paleontologist Dougal Dixon, we may have acknowledged our (alterable) future by viewing both the past as well as one probable future.

CHAPTER 24

Portraying Paleocatastrophe

You can always tell it's a slow news week when you see articles like, "Did Comets Kill the Dinosaurs?" Here's a hot topic—who's got time for this? "Hey, what happened to the dinosaurs? Weren't they just here?" Maybe comets killed the dinosaurs, maybe they tripped and fell. What's the difference? We'll never know. We couldn't solve the Kennedy Assassination, we had films of that. Good luck with the Stegosaurus. "Round up all these reptiles for questioning, Bill. I want to talk with that salamander there. I think he knows something. Don't stick out your tongue at me, young man. I'll nail your slimy little butt to the wall."

(Jerry Seinfeld, in *Seinlanguage*, 1993, pp. 117–18)

How shall we draw the geometry of contingency? How else may we draw the history of life, so that we may come closer to meeting our ancestors face to face and may even probe pictorially into our own psyches to release the potential thoughts that lie even too deep for tears?

(Stephen J. Gould, "Ladders and Cones: Constraining Evolution by Canonical Icons," in *Hidden Histories of Science*, 1995, p. 67)

In the dramatic opening of a Walt Disney animated film we went to see last spring, a family of talking lemurs witness a meteorite collision, ending the Cretaceous period. The meteor blasts its way through the atmosphere, sending a torrent of fiery debris toward the antagonists. As even the most casual viewer realized, the few remaining dinosaurs befriended by the surviving lemur clan were doomed to extinction, while the lemurs are symbolic of our distant ancestry. The science behind the story was strangely self-evident to all. Yet, how silly would late 1990s visions of Disney's Imagineers have seemed only thirty years ago!

In the beginning, "catastrophist geologists" wrote of extraordinarily tumultuous times evident in geological history. Geological theories invoking the theme of Life's *discontinuity* were also depicted by a handful of artists. Theoretical ideas artistically conveyed included 1) Louis Agassiz's (1807–1873) Ice Age catastrophe; 2) the notion of a primordial globe *before* life at a time when the young earth's surface was incandescent; and 3) the prospect for extinction caused by atmospheric degradation (or, more precisely, an elevation of atmospheric oxygen concentration) at the Mesozoic/Tertiary boundary, as reflected in Owen's superior dinosaurian forms.[1]

Then, in 1980, when the prospect of asteroid catastrophe was connected to the ultimate fate of the dinosaurs, the art of expressing "contingency" in Life's history became an important theme in paleoart. Scientists placed on firm footing the prospect that global extinctions of the K-T (Cretaceous-Tertiary) geological horizon had been caused by the catastrophic impact of a major league curveball hurtling from space, aimed high and inside.[2] One may contend that this development represents one of the most significant innovations in the realm of dinosaur paleoart since Robert Bakker introduced his vigorously active dinosaurs in the late 1960s.

Until the mid–1980s, few artists dared to portray these pivotal episodes in the history of life. This is because most scientists claimed that life evolved gradually and species succumbed to extinction gradually, in accordance with the principle of Uniformitarianism. Species known from the fossil record were thought to have originated, evolved and disappeared gradually, through fluctuating terrestrial processes operating uniformly through time. In other words, there were no geological instantaneous "moments" that could be considered "pivotal." Time was immense, and at any given moment there was little happening that would impact the overall destiny of *many* species at once, especially instants which would be considered worthy of illustration.

Even in the early days of pre–1840s "catastrophist geology," illustrations of mass extinction events were rarely done. This was because the field of paleoart wouldn't enter its embryonic stages for half a century. Then paleoart emerged during the late nineteenth century, a period weighted by the influential uniformitarian ideal. So, instead of envisioning catastrophes punctuating Life's history, events which had most likely never happened (according to leading scientists), contemporary artists concentrated on producing the first scientifically accurate renderings of prehistoric animals.

Despite the former popularity of the "ice age" theory, in which *cold-blooded* dinosaurs were thought to have perished in an episode of global cooling at the end of the Mesozoic, there are few notable illustrations cultivating this idea.[3]

By the late 1970s, certain geological catastrophes were viewed as real events, some even regarded as extraterrestrial in nature. It is interesting to note the transformation of scientific ideas as expressed on canvas by (modern) paleoartists.[4] The fact remains that it is the 1980 asteroid theory, and no other, that inspired the art of restoring dinosaur *extinctions*.

In a mid–twentieth century example, Walt Disney's *Fantasia* (1940), dinosaurs thirst in a dusty age, an environmental condition resulting from causes left unexplained in the film. In other cases we see volcanoes and pesky little mammals doing their part to exterminate the once mighty dinosaurian race. One famous example is the Cretaceous end of Rudolph F. Zallinger's 110-foot-long mural, *Age of Reptiles* (1947) (exhibited in the Peabody Museum at Yale University), where a smoldering volcano signaling the dinosaur's impending doom is visible in the distance. A lone mammal pest, *Cimolestes*, lurks in the foreground, patiently waiting to claim its "inheritance," Earth.

Another illustration by Neave Parker, entitled *The Last Days of the Dinosaurs* (1960), blends ideas. Here we see North American dinosaur species in the center of the canvas. But smoking volcanoes foul the air and the foreground is dotted with those darn (yet "superior") mammals destined to survive! Extinction is obviously imminent, with this combination creating unholy havoc in the upper Cretaceous.[5] In 1999 Kenneth Carpenter illustrated his denunciation of the "killer rat" theory for dinosaur extinctions with a terrifying image—a cute dinosaur hatchling staring into the bared fangs of an awaiting cat-sized mammal, *Didelphodon*.[6]

More recently, and perhaps because we no longer regard the mammals-as-agents-of-dinosaur-extinction theory quite as seriously as formerly may have been the case, the theme of mammals antagonizing dinosaurs has been satirized in several illustrations. For instance, Robert Bakker's illustration published on page 96 of his *Dinosaur Heresies* (1986) relies on mammals devouring a torpid dinosaur to accentuate his theory that dinosaurs were not cold-blooded animals.[7] Then, Jim Whitcraft's 1995 *Journal of Vertebrate Paleontology* cartoon whimsically captured how the formidable *T. rex* could be so easily tripped to his death by a strange critter resembling Mighty Mouse without a cape.[8] Such devices are on par with a host of Gary Larson's similar "Far Side" sketches focusing on the dinosaurs' ultimate peril, such as the one bearing a witty caption, "The real reason the dinosaurs went extinct," while showing a trio of ("way too cool for the Marlboro Man") dinosaurs puffing on cigarettes.

During the 1960s and 1970s the supernova theory for extinction of the dinosaurs was popularized by several scientists, most notably by dinosaur expert Dale Russell.[9] Thus, the prospect for cosmic catastrophe became introduced into the spectrum of modern paleoart, as evident in Ely Kish's 1974 painting of a dying ankylosaur, singed by the searing rays of a nearby supernova, published in Edwin H. Colbert's *Dinosaurs: An Illustrated History* (1983).[10] However, Colbert was not endorsing the theory, as his caption reads, "Did the explosion of a supernova 65 mya, with consequent high energy radiation, cause the extinction of the dinosaurs? If so, why did so many other forms of life survive the crisis?"[11] Artist Don Davis also depicted a supernova irradiating unwitting dinosaurs, published in Carl Sagan's *Cosmos* (1980).[12]

The skies became more forbidding as the decade of the dinosaur came into its own. A revolution in science, in turn, sparked an entirely new art of "paleocatastrophe." The increasing level of interest in artistic renditions of the K-T boundary and other impact events can be coupled to increasing numbers of scientific publications on the subject.[13] But the decade's trendsetting image seen by millions, a *T. rex* watching comets streaking earthward under a twilight canopy of stars, appeared on *Time* magazine's May 6, 1985, cover.[14]

Since the "asteroid theory's" inception, few other images have been created of other major mass extinction events. Paleoartists steered clear, possibly because these episodes didn't feature the familiar dinosaurian cast, and because most scientific research focused on the dramatic K-T boundary event's fossil record. Also, there is more debate over whether other horizons, tentatively identified as impact events, could be less "instantaneous" and terrestrial in nature.[15]

Focusing on the canvas restorations, illustrations eerily portrayed doomed dinosaurs gazing into the heavens as a fatal streak of light shines above (or *streaks* of light, reflecting the possibility that Oort cloud comet *showers* glimmered in late Cretaceous night skies). As a *Velociraptor* thoughtfully gazes into the celestial solitude in Jon Lomberg's painting, a streaming comet shower symbolizes abrupt termination of the dinosaurs' evolutionary ascendancy toward heightened intelligence.[16] In his 1999 painting, "Impending Doom," Bob Nicholls portrayed the impending global dinosaurian disaster. A comet, dubbed the "Super Rex," streaks in the distant sky above a trio of feasting *T. rexes*, one shown with entrails dangling from blood-stained jaws. A recent entry by John Sibbick, showing duckbills eyeing a "Christmas star" appearing suspiciously as a comet of doom streaking across the heavens, was released as a 1996 Christmas card, courtesy of the UK Dinosaur Society.

Others have chosen to portray the moment of impact (or just instants before), as well as the horrorific side effects. Our favorite examples? John Dawson's painting, showing snarling dinosaurs turning away from the brilliant fireball, adorns the cover of Kenneth Hsu's book, *The Great Dying: Cosmic Catastrophe, Dinosaurs, and the Theory of Evolution* (1986).[17] Dawson's restoration shows the K-T comet blowing images of the twentieth century's two greatest dinosaur icons (*T. rex* and a sauropod) to smithereens.

Another is Ely Kish's composite painting of the K-T impact (1985), which was reproduced in Dale Russell's, *An Odyssey in Time: The Dinosaurs of North America* (1989). Russell described how Kish designed this painting, which must be regarded as the finest example of its kind, in his contribution to Volume 1 of *Dinosaurs Past and Present* (1986).[18] (See Figure 24-1.) In his 1989 book, Russell provided a description: "It seemed to hang in the sky like a second moon, or the eye of God, but no dinosaur looked at it with understanding. It suddenly swelled in the sky, and then a dark mantle spread across the firmanent. Why did it have to be this way?"[19]

Whereas paleoartists traditionally portrayed the extent of geologic time through *billions* of years, reflecting "idealized" landscapes while introducing minimal speculation, "renaissance" artists have strived to capture the brief time line of the event—moments before, through minutes or hours afterward. A sequence of depictions by several artists found in chapters 25 and 26 of *Comet* (1985) leaves little to the imagination regarding the horror and magnitude experienced through all phases of such catastrophes![20]

Geophysical events now thought to have been triggered by the impact, such as the enormous tidal wave splashing ashore in the Caribbean region in K-T time, are also envisioned. Mark Hallett's contribution to *Zoobooks*, (vol. 1, no. 9, June 1985) included a painting which instructively illustrated geophysical "domino effects" in the aftermath of impact (i.e. tidal waves and volcanic eruptions on a grand scale).[21]

Through the late 1980s and 1990s, instead of a cosmic impact, a number of geologists and paleontologists held volcanoes chiefly responsible for the K-T mass extinctions. And yet there is little contemporary paleoart reflecting the volcanic view. Douglas Henderson also depicted the Deccan Traps volcanic eruptions for *Dinosaurs: A Global View*.[22] In Joseph Wallace's *The Rise and Fall of the Dinosaur* (1987), we see Henderson's painting of dinosaurs expiring in the fallout from a volcanic eruption.[23] However, upon closer inspection of the latter, it is noted that the caption does not correspond to the illustration. Henderson restored a *local* catastrophe, the demise of a population of *Maiasaura*, which took place millions of years before the K-T event in Montana. However, the caption exaggerated the purpose of the painting, extrapolating the local to a global event. "Violent volcanic activity may have sent great clouds of ash skyward, choking plant life and small animals, and condemning the great dinosaurs to extinction." This was probably done for lack of other restorations available at that time to illustrate the volcanically founded extinction alternatives.[24]

Paleoartist Henderson has found the asteroid impact dinosaur extinction theory highly inspirational. His 2000 book, *Asteroid Impact*, introduced a number of striking new images showing the devastation caused when the asteroid fell through earth's Cretaceous skies. One powerful tableau shows a herd of *Torosaurus* being washed away by flood waters in the brilliant glare of the impact fireball. (See Figure 24-2.) An incandescent plume of fiery ejecta swells toward the drowning herd. Henderson wrote, "As the crater stopped expanding, island-sized blocks of land and walls of debris rose and fell back down.... Illuminating the early-morning sky of North America, the vapor plume was rising high above the southern horizon. Then, with a roar, the first shock waves reached the Texas coast, and the lowlands began to shake. The earthquake toppled trees and limbs crashed to the ground. The water in lakes and rivers heaved and sloshed over their banks. ...As the

***Opposite:* Figure 24-1: Ely Kish's painting, *The Extinction of the Dinosaurs* (1985). (Reproduced with permission.)**

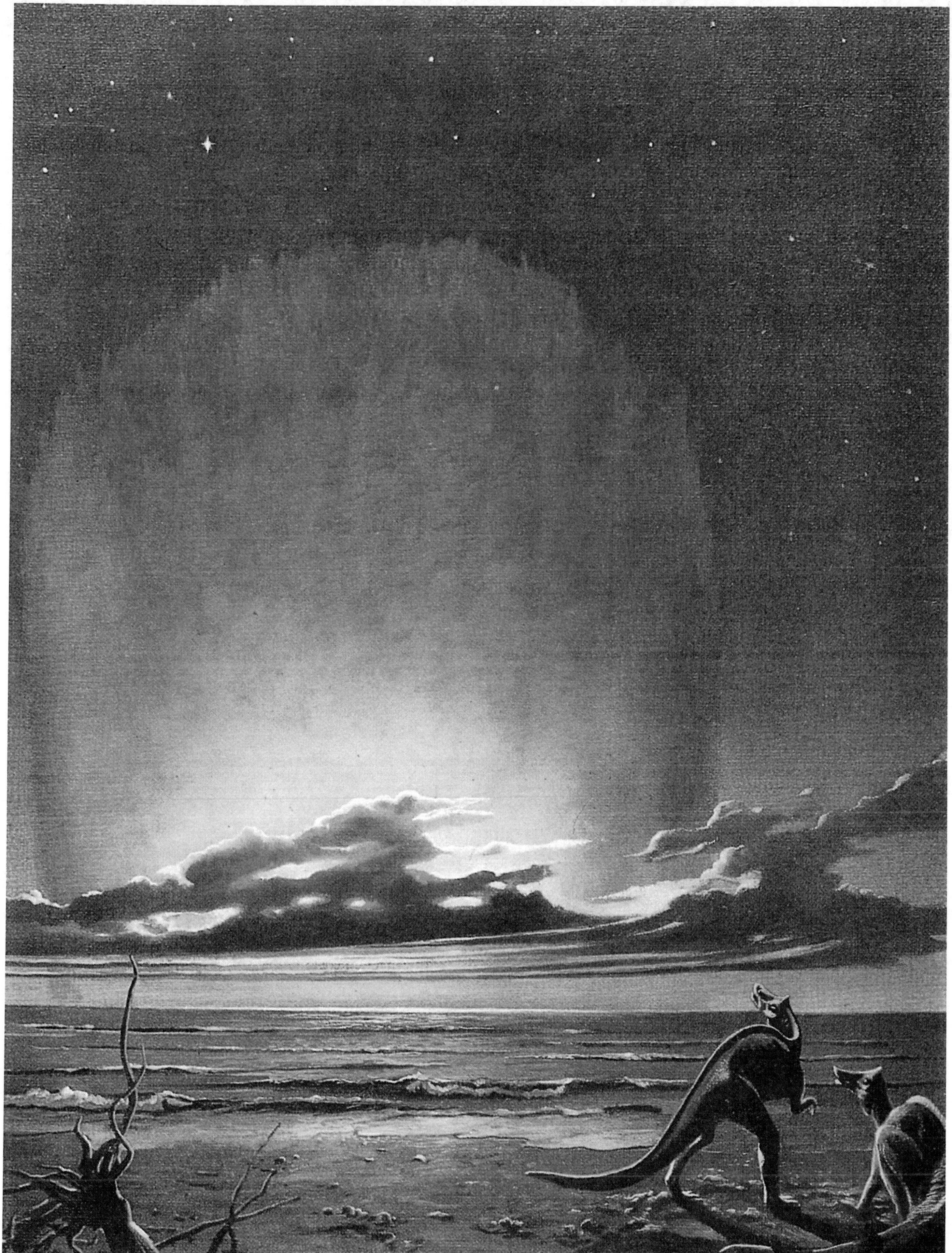

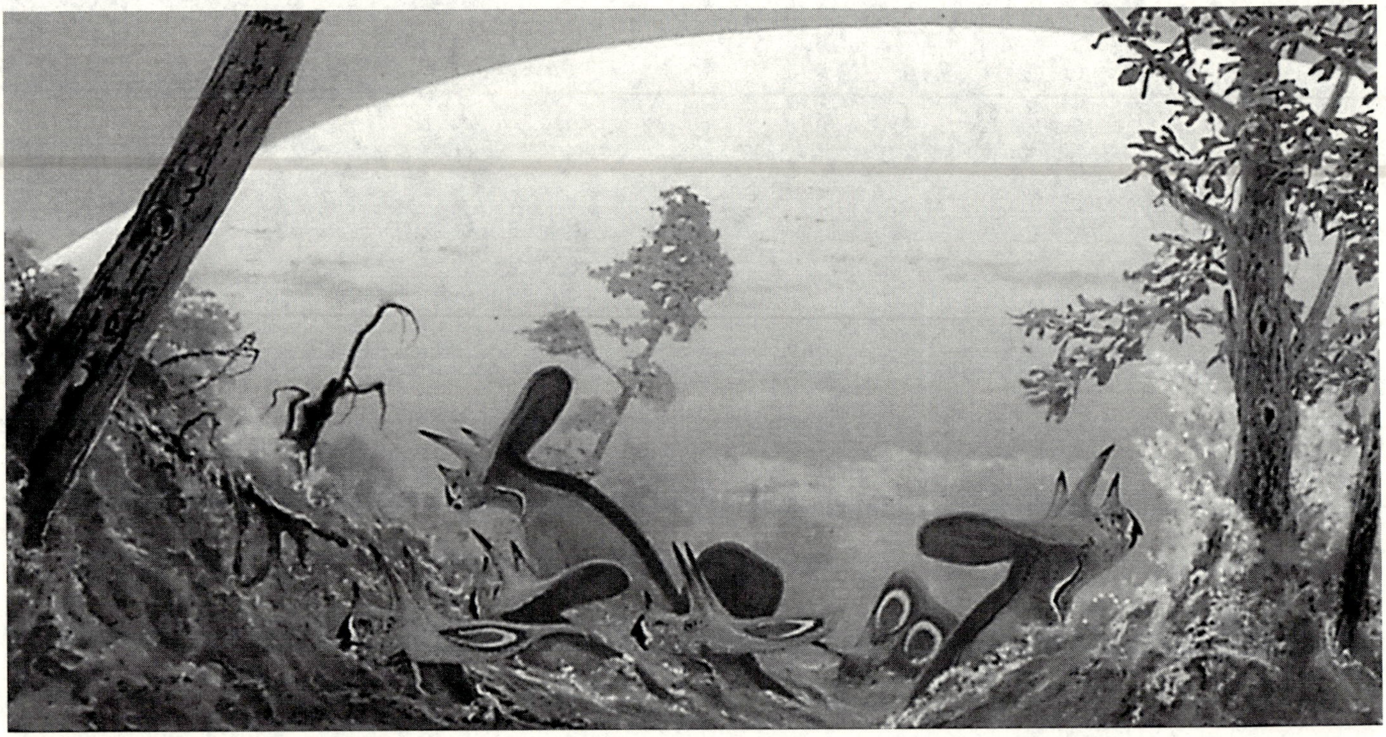

Figure 24-2: Douglas Henderson's painting showing devastation caused by the K-T fireball (1999). Reproduced from *Asteroid Impact* (Dial, 2000) with permission of the artist.

ground continued to shake, the horizon began to glow—first red, then blinding white."[25] Such soul-chilling imagery would have been scorned by scientists a century ago.

It is significant that the old paleoart masters (prior to 1950) did not sketch Life at the crossroads. The theoretical base and weight of evidence weren't there. In the eyes of prominent contemporary scientists, artists would have been wasting their time. Still, how would Charles R. Knight's rendition of the K-T boundary event have appeared? Certainly Knight's great contemporary, Chesley Knight Bonestell (1912–1981), widely recognized as the father of "Space Art," was no stranger to envisioning how *cosmic* catastrophes (both in space as well as on earth) might have appeared to human observers.[26] Yet, *terrestrial* catastrophes were conventionally unaccepted.

Why all the fuss? First there's the prospect for sudden, dramatic change in the history of our planet. The comet plunges into the seas and, it is claimed, in an instant, the entire course of future history is (unpredictably) altered. Philosophically, it is difficult to conceive that the modern biosphere may be the product of iterative, randomly catastrophic episodes!

Another possible, psychologically founded, fascination with the prospect of extraterrestrially caused mass extinctions is that for many of us, years ago, the world seemed poised on the brink of nuclear holocaust. Just as a missile bearing a hydrogen bomb would have brought sudden death from the skies, so can a large asteroid end our days at the end of its trajectory. The only *safe* way to experience the (geological) event is to witness it in artists' portrayals, or to view it from afar as astronomers did during the summer of 1994, when fragments of comet Shoemaker-Levy 9 slammed into Jupiter. From our apparently safer vantage point in space-time, one may sigh, "...better the dinosaurs than us!" Yet, let *us* remember to watch the skies!

Chapter 25

Tyrant Queen—Icon-o-saurus Rex

The gathered throng presses fearfully away as Tyrant Queen strides majestically from her pedestal onto the killing floor. "Scientists gone mad!" someone utters, and suddenly the enraged creature is in our midst. The museum corridors rumble with pounding footsteps and astonished cries. Crunching noises echo disgustingly as a victim is engulfed.

Spying Lake Michigan outside, the beast smashes through adjacent window panes. She fills cavernous lungs with breezes from our Windy City and halts traffic with her bellow! With diminutive arms she caricatures the "royal wave." The ground trembles. Horrors! Is that a volcano beginning to erupt in the middle of the Lake? Look out, Chicago! Tyrant Queen has arrived!

Well, okay, that's not exactly how things transpired on May 17, 2000, the long-anticipated day "Sue"—the majestic *T. rex*—was unveiled here at Chicago's Field Museum of Natural History. Yet, how many of us who greeted her were having flights of fancy (as if we had found ourselves cast in a dismal Hollywood "B" movie script)? Fittingly, as it turns out, the factual account is far more intriguing than any fictional encounter with Tyrant-Queen-incarnate ever could have been.

T. rex is *the* virtual symbol of the dinosaur. Adolescents identify most with *T. rex*'s electrifying persona. Science has considerably reshaped *T. rex* and the world in which it lived since the early 1900s, yet *T. rex* remains *the* dinosaur icon. Like no other dinosaur ever before, *T. rex* ("Sue") has quickly become a cultural icon in Chicago, much as a "prehistoric" fictional ape ("King Kong") became identified with New York City's Empire State Building. Only Sue is a fantasy come true, the essence of nightmares and dreamscapes.

Like it or not, nearly every dinophile feels intimately bonded to *T. rex*. However, no human being alive loves *Tyrannosaurus* more than Chicagoan John Lanzendorf.[1] John, arguably the world's foremost private collector of dinosaur "fine art," appreciates *T. rex* principally for the paleoart imagery it has inspired throughout the twentieth century. His exhibition, *Picturing T. rex—The Lanzendorf Collection*, was a magnificent accompaniment to Sue's unveiling. Lanzendorf's collection was displayed at the Field Museum for several months thereafter.

Even among the ranks of professional paleontologists, who should have the most intimate level of understanding about the toothy, bad old guy, there are many varied opinions. According to Dr. John Horner, "Speculation is healthy ... as long as it is grounded in evidence. Where you get into trouble is when you start believing the speculation simply because it appeals to you."[2] But in the case of *T. rex*'s historical paleoimagery, we find that "seeing is believing." How many of these images have already gotten us into trouble?

The fact remains that *T. rex* has been interpreted in various ways by many people over the years. For example, some regard *T. rex* as the mighty carnivorous tyrant king dinosaur scattering herds of *Triceratops* in search of a square meal. Others know *T. rex* as a polite scavenger, only feasting himself after others have dined. These and other varied impressions record the evolving history of scientific ideas concerning *T. rex*, and also reflect our cultural association with what many believe was the greatest of dinosaurs.

T. rex has claimed stature as Nature's most legendary menacing creature. Even the geological stratum where *T. rex* fossils have been found, named the "Hell Creek" formation, suggests a fitting association. What's more, as some have claimed, it took one of the most terrifying catastrophes imaginable to lay *T. rex* in its Hell Creek grave, the collision of a comet or asteroid with the earth.[3] Otherwise, if nonavian dinosaurs had avoided extinction, *T. rex* may have eventually evolved into the larger and more voracious, 60-foot-long "Gourmand." This hideous (yet hypothetical) creature, having entirely lost its front limbs in evolutionary time, was envisioned by the clever Australian paleontologist Dougal Dixon, and portrayed in his 1988 book, *The New Dinosaurs*.[4]

How is it that we have become so familiar with this awe inspiring creature? Why *T. rex* would be deserving of our adoration at all may seem perplexing. Let's consider his cultural track record.

In *King Kong* (1933) a love struck, oversized ape defeats *T. rex* in order to protect the object of his desire, Fay Wray. Sixty years after Kong tumbled from the Empire State Building, *T. rex* has virtuously turned the tables on creatures far craftier than himself, as we noticed during the conclusion of the movie *Jurassic Park* (1993), where the terribly vicious raptors meet a deserving and horrible fate wrought by a seemingly "heroic" rex.

In Robert J. Sawyer's chilling tale, "Just Like Old Times," the savage killing instincts of *T. rex* prove inferior to the nightmarish psyche of a convicted serial killer, who must serve his peculiar sentence by becoming mentally absorbed into a *T. rex* body in the late Cretaceous.[5] Mankind coupled with rex proves to be the most fiendish monster of all—a creature capable of wiping out our entire (future) existence. Yet Man is the monster instigating disaster. Rex is merely the "familiar."

Rex is thought by many to be a natural born killer, the "Wolfman" of dinosaurs. Yet unlike Man or the fiendish raptors, Rex is not hellbent on mass destruction, or at heart evil-minded. Rex can be forgiven for his ways. He hunts simply to survive.

Sixty five million years after its extinction we need not worry about being shred into splintery pieces by those cruel six inch sabers, or being swallowed whole into its belly. Alas, the only remaining threat could be that of a *T. rex* museum skull falling out of an exhibit onto ourselves.

The imaginable horror of a direct *T. rex* confrontation is riveting.[6] This is why "pseudo-dinosaurs," whose form bears obvious similarities to *T. rex*, are so popular. Make *T. rex* 20 times taller, stand him upright, have him breathe radioactive fire, and you have a "*T. rex*" as invincible as "Godzilla." Or color him purple and give him human dentures to disguise those sharp, scary teeth, and you have another version of *T. rex* known as "Barney," (which to some adults is even scarier than Godzilla). Extreme images such as these illustrate how perplexing it can be deciding exactly how *T. rex* is truly supposed to look.

Our ability to identify *T. rex* depends on *when* were you introduced to *the* twentieth century's dinosaur icon. During the century since its discovery, *T. rex*'s appearance and associated "imagetext" has changed considerably. No longer the tail-dragging, husky, cold-blood beast of the 1920s (in some cases, erroneously equipped with three-fingered hands and misplaced eye sockets), *T. rex* has transformed into a lithe, muscular, horizontally balanced, destructive horror icon.[7] Relying on cladistics principles, Dr. Chris Brochu, "Sue's" principal investigator, suspects juvenile *T. rexes* were feathered, like young birds.[8]

T. rex and related look-alike tyrannosaurids (i.e. *Gorgosaurus*, *Daspletosaurus*, *Albertosaurus* and *Tarbosaurus*) have reigned in museum displays, fantasy literature, film and those ever popular restorations through three *T. rex* paleoimagery "eras." We shall refer to these as the "Savage Rex" era (1902 to 1942), the "Lordly Rex" era (1947 to 1975) and the "Renaissance Rex" era (1979 to 2000).

The Savage Rex era is defined by Charles Knight's restorations and other salient examples largely based on Knight's artistry. Perhaps the defining restoration was Knight's early 1900s American Museum painting which standardized one of our most popularly imagined dinosaur themes—the *Triceratops* vs. *T. rex* struggle for survival. Knight's great artistic skill, coupled with Osborn's, Barnum Brown's and William D. Matthews' scientific expertise, proved a winning combination as the painting seemed accurate and "photographic" in nature. As we now know, there were anatomical errors. For instance, Knight's *T. rex* had three-clawed fingers, and the eye was placed too far forward in the skull, in front of where the eyes actually were. (See Figure 25-1.)

Two decades later Knight repainted the *Triceratops* vs. *T. rex* scene as a Field Museum mural, a restoration which has captivated many, young and old. (See Figure 7-2.) During the "Savage Rex" era, Knight-inspired *T. rexes* also grappled savagely with King Kong (RKO, 1933) and still another *Triceratops* (at the 1933/34 Chicago World's Fair). Knight and Walt Disney closed this era with vicious visions of sparring *T. rexes*. Knight's pair of battling tyrannosaurs graced his February 1942 *National Geographic* article, and Disney's *T. rex* struggled furiously with *Stegosaurus* in *Fantasia*'s (1940) "Rites of Spring" segment. Following World War II, perhaps it seemed time for a change from images of warlike, bloodthirsty primeval monstrosities. After all, war can be hell, especially if the combatants are from the Hell Creek formation.

Rudolph Zallinger's lordly *T. rex*, framed in the Yale Peabody Museum's *Age of Reptiles* mural, set an atomic age, cold war precedent for the appearance of rexes and tyrannosaurids. (See Figure 16-1.) The upright, commanding presence of Zallinger's *T. rex* is mirrored in the Field Museum's *Gorgy and Prey* skeletal display first exhibited in

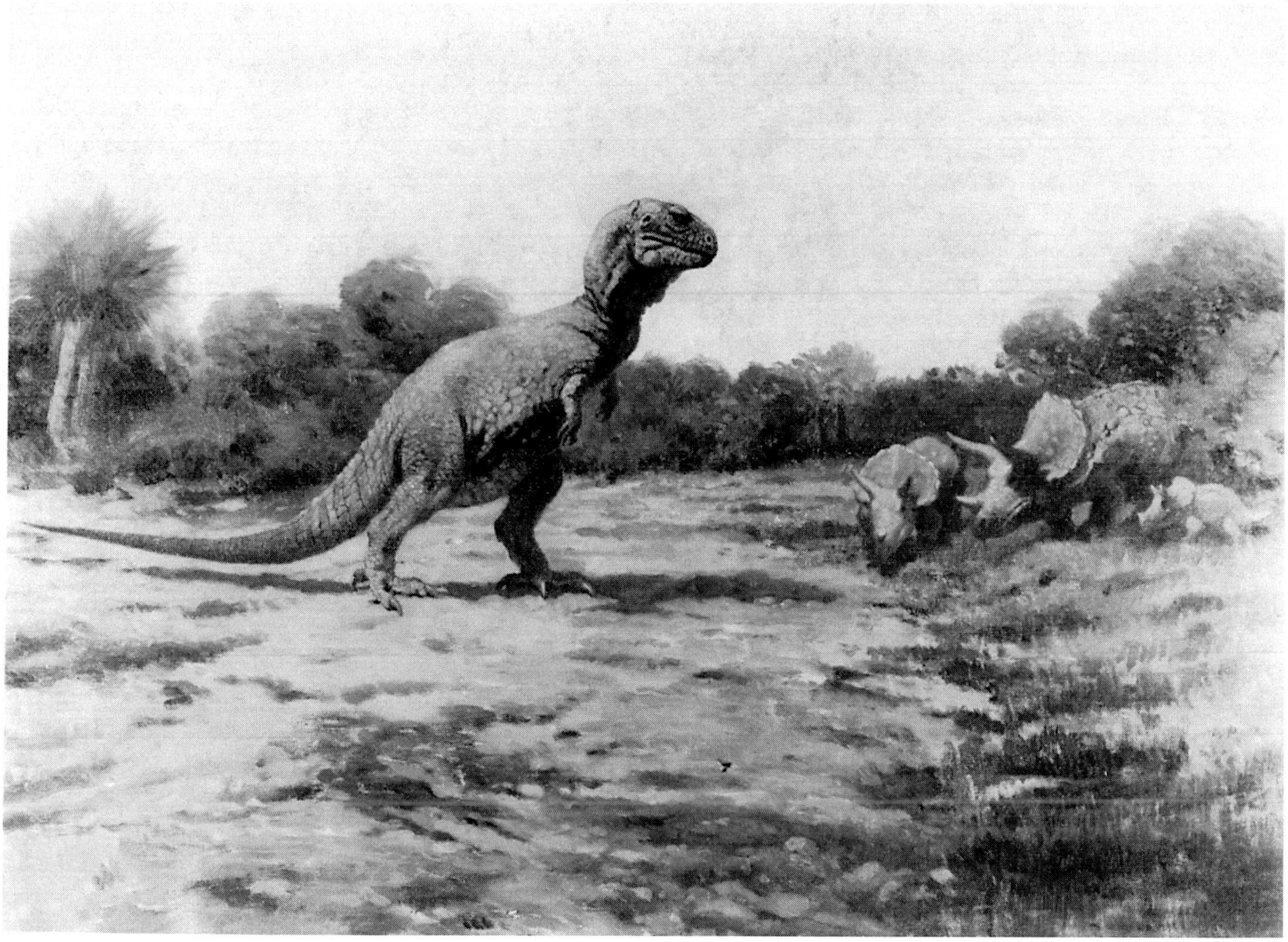

Figure 25-1: Charles Knight's American Museum painting of *Tyrannosaurus* engaging in a struggle with *Triceratops*. (Negative #35827. Courtesy Department of Library Services, American Museum of Natural History. Reproduced with permission.)

1956, and also in Louis Paul Jonas' *T. rex* sculpture commissioned for Sinclair's 1964/65 New York World's Fair Dinoland display. (See Figure 22-2.) Such rexes are the lords of all they survey, monarchs of their domain. Zallinger's *Age of Reptiles T. rex* expresses remarkable self-confidence, as its hind quarters face its old nemesis, *Triceratops*. It is not depicted charging recklessly into old three-horn, nor fearful of being stabbed in the back. Cold war era "Lordly" tyrannosaurids exude a self-confident, battle hardened, unthreatened demeanor, evidently secure in their role as Mesozoic overlords. If *T. rexes* of the atomic age, painted or sculpted by paleoartists in America, seemed majestic, it was because they reflected America's might and authority as a world power.

Cold-blooded, sluggish dinosaurs were an "establishment" to be reckoned with during America's late 1960s cultural revolution. However, new paleoimagery enlivening interest in *T. rex*'s life appearance didn't materialize until long after the Dinosaur Renaissance was in full swing. Then Gregory Paul boldly reset the standards for how tyrannosaurids should look with a 1979 illustration of a pair of cavorting warm-blooded rexes waltzing into combat against spiky *Styracosaurus*.[9] Five years later, Paul's painting of a running *T. rex* dazzled us, and in swift succession paleoartists followed Paul's mesmerizing instincts.[10] Canadian paleoartist Ely Kish conveyed her impressions of a mobilized *T. rex* snaring a duckbill in its jaws,[11] while John Gurche painted *Daspletosaurus* stepping high to avoid a styracosaur brandishing its horns in the heat of battle. Gurche's 1986 mural hangs in the Denver Museum of Nature and Science on a wall behind a skeletal cast of the American Museum's *T. rex*, mounted in an energetic, prancing pose designed by Bakker. Michael Trcic sculpted a similarly posed bronze *T. rex* a decade later.[12] (See Figure 25-2.)

Just when that popular *Triceratops* vs. *T. rex* paleoart

theme began to seem unoriginal, Paul illustrated a new twist on how the two may have combatively interacted, with *T. rex* charging in from behind.[13] Meanwhile, Paul and other paleoartists (David Peters and Bakker) founded an alternative popular paleoart theme—pitting the small airplane-sized pterosaur *Quetzalcoatlus* vs. tyrannosaurids.[14] *Jurassic Park*'s (1993) lifelike *T. rex* was a smash hit at the box office, but Matt Smith's miniature *T. rex* sculpture was praised by John Horner for being as "...closely as anything yet done on the most current fossil evidence.... His T. rex doesn't leap around like the Karate Kid, but it is agile and hostile, and real enough to scare the hell out of me.... And most importantly ... it is real ... because ... all of Matt's art and our science on *T. rex* are based on what we do with the bones we dig out of the ground."[15] (See Figure 25-3.)

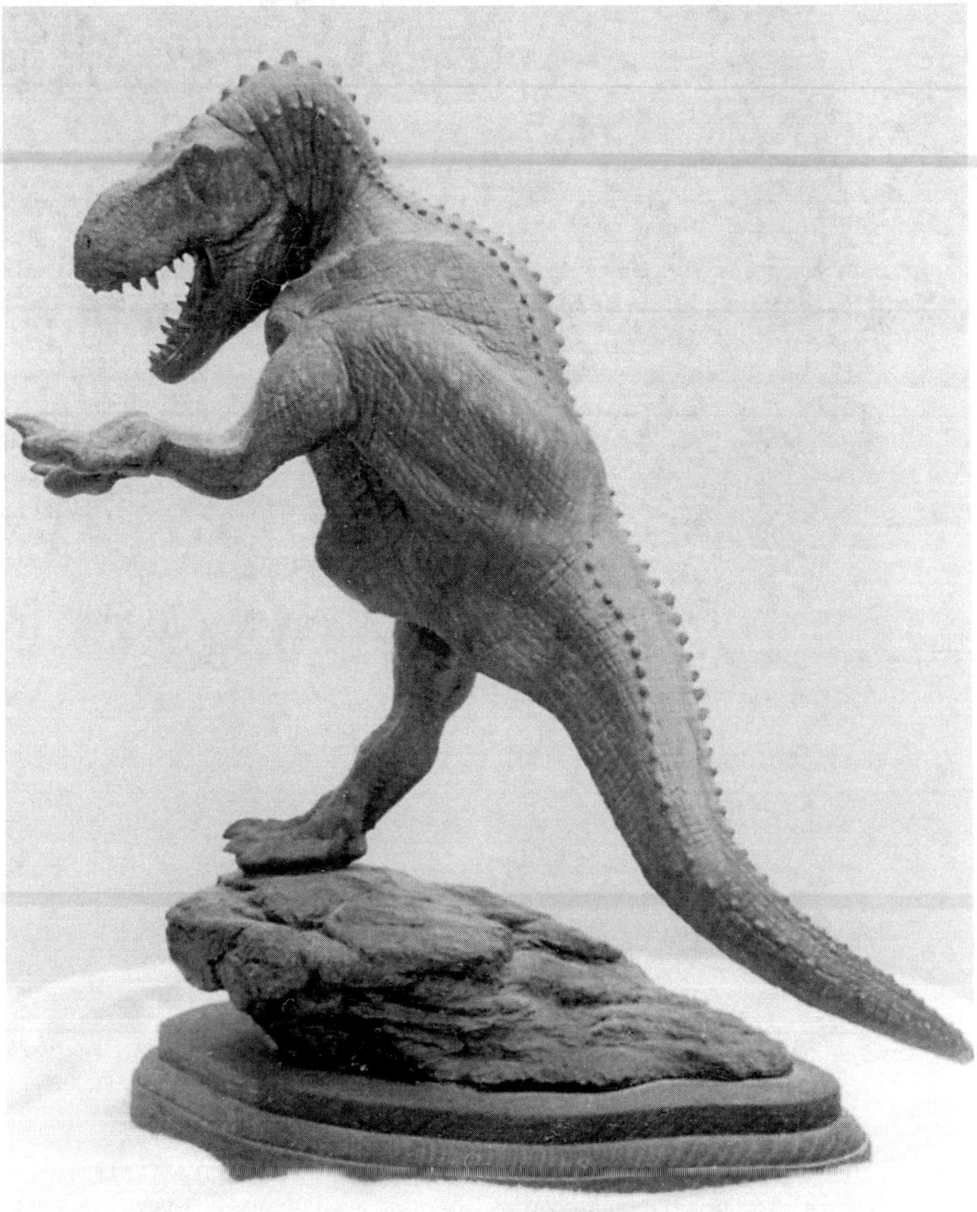

Figure 25-2: Michael Trcic's *Tyrannosaurus* sculpture (1996). (Photograph courtesy of Michael Trcic.)

Although *Jurassic Park*'s *T. rex* moved quickly across the movie screen, giving good chase to an accelerating jeep in one dramatic scene, many viewers found themselves wondering if they could outrun the monster. So, in 1995, dinosaur paleontologist James Farlow and artist Matt B. Smith offered the latest twist in the ongoing debate over how fast *T. rex* could run. Using an estimated mass of 6,000 pounds for an adult *T. rex* (derived from measurements of Matt's accurate 1/20 scale sculpture), the authors suggested a maximum running speed of 10 meters per second, or about as fast as a surging elephant. Running at top speed surely would have increased the risk of a fall, a death sentence to a relatively "unarmed" theropod. Thinking whimsically, did the early mammals learn craftily how to trip a charging *T. rex*, ensuring their survival?[16]

We all know what happened to the legal representative in the movie *Jurassic Park*, but even the most vaunted of all the tyrant lizards ever found—"Sue"—would have proved no match for what many would regard as the modern world's most fearsome "predator," the Attorney. No sooner was she chiseled from her 67-million-year-old crypt in 1990, than she was once again entombed by a legal system even her six inch sabers couldn't shear through. Is it any wonder the audience applauded when *Jurassic Park*'s legal eagle was caught, literally, with his pants down? Even the *T. rex* looked unsure of the digestibility of that morsel.

Michael Skrepnick's painting, *The Lizard Queen's Final Day* (1994) is displayed adjacent to a large photo of "Sue's" skull at the Black Hills Institute in Hill City, South Dakota. Skrepnick's imaginative interpretation tells a story of Sue's

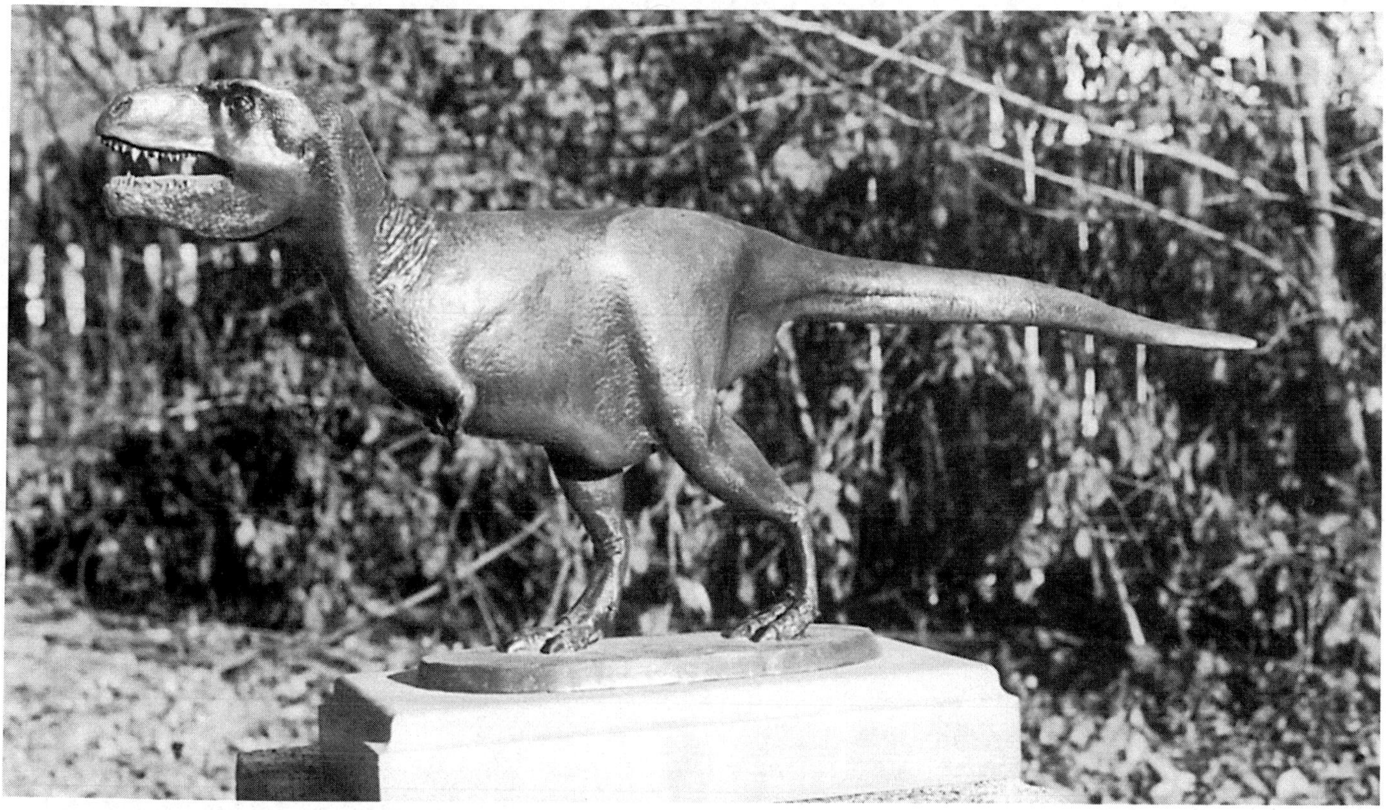

Figure 25-3: Matt B. Smith's *T. rex* sculpture (1992). In M. Skrepnick's 2001 restoration, *Science*, vol. 293, p. 852, T. rex's nostrils are closer to the lips. (Photograph by A. Debus, authors' collection.)

death, based on preliminary pathologic and taphonomic studies of her skeleton conducted by Peter Larson and Black Hills Institute colleagues. Battle-scarred Sue, having already suffered many attacks from her rexy rivals, painfully summoned the strength to fend off another attack from a marauding group of rexes. Meanwhile, her unsuspecting brood practice hunting skills using turtle bait. Suddenly it is over. Weary Sue, the juveniles and her younger mate succumb to the assault. Their bones become deposited in the sandy river bottom.[17] But that was merely a beginning, *not* the end of "Sue."

On the morning of October 5, 1997, we were astonished by the *Chicago Tribune*'s headline, "Field Museum Claims Sue as Its Own: $8 Million at Auction Wins Prized T. rex." A most excellent addition to Chicago's Field Museum of Natural History, which represented a perfect repository for the fossil. But that auction sale price of $8.4 million! Egad! What if more specimens are found? Then would *T. rex* fossils only be worth a dime a dozen, or seem as "common as muck?"[18]

No matter how you slice it, "Sue" has had a checkered history, a past only adding to her intrigue. At the time of Sue's discovery, her discoverer, Susan Hendrickson, amateur paleontologist and professional diver and treasure hunter, was employed by the Black Hills Institute of Hill City, South Dakota. Today Susan feels a mystical connection to her marvelous namesake. In August of 1990 she experienced a strange compulsion to visit the rocky South Dakota cliff in which her soon to be famous dinosaur was embedded, on tribal land, near Faith, South Dakota. After the apparent landowner, Native American Maurice Williams, accepted $5,000 from Peter Larson in exchange for the dinosaur, "Sue" was hauled away to Hill City. Over the next two years, Black Hills Institute preparators determined Sue's skeleton was 90 percent complete—in fact, the most completely known.[19]

Study of Sue's remains has been enlightening. Black Hills Institute paleontologist Peter Larson claimed, as is often the case with modern reptiles and birds of prey, the biggest and scariest of all *T. rexes* may have been "tyrant queens." Despite Peter Larson's original contention, Dr. Brochu is undecided as to whether the specimen known as "Sue" actually was a female.

Sue's brain is one foot long, proportionally equivalent to the size of bird brains. Her olfactory bulbs are "grapefruit-sized." Brochu was surprised by how exceptional its sense of smell seems to have been. In fact, smelling may have been "central to how 'Sue' perceived its role."[20] The acute olfactory sense would be of utility to both scavenging and predatory tyrannosaurs. Sue's jaws were sufficiently powerful to rip carcasses and hold struggling prey. However, even though their exact purpose remains undecided, Brochu believes those puny forearms must have been functional. Sue may have been smarter than a croc, but not as smart as a bird. In a public lecture given at the Field Museum on May 30, 1999, Brochu claimed that Sue seems to represent the Mesozoic Lord from "The Age of Olfactory Bulbs"!

AT THE FIELD MUSEUM, EVEN THE EXTINCT CAN EVOLVE!

In 1956, Field Museum workers erected the skeletal reconstructions of *Gorgosaurus* standing proudly over its prey, *Lambeosaurus*. The Field's *Gorgosaurus* was distinguished as the first dinosaur skeleton ever to be mounted without use of any external support armature.[21] "Gorgy" would be displayed downstairs in the Museum's Stanley Field Hall for nearly four decades, becoming one of Chicago's cultural symbols, until its eventual 1994 incorporation into a newer, more interactive *DNA to Dinosaurs* display, since renamed *Life Over Time*. Field Museum officials decided to replace *Gorgy and Prey* with something bigger and better. Or did they? Would *Brachiosaurus*, the "Arm Lizard," defeat puny-armed *Gorgosaurus* in an arm wrestling contest for bragging rights?

Our 75-foot-long, 40-foot-tall *Brachiosaurus* was enormous and dramatically posed. With his great head peering *Jurassic Park*–like up into the rafters, guiding second floor visitors toward the entrance of the "dinosaur hall," the giant created quite a stir but never a sensation. Though truly a dinosaur as immense as the City itself, *Brachiosaurus* was clearly not a "symbol" for Chicago, like our "Gorgy" had been. So, as in the case of the many celebrated sports figures passing through our "toddling town," on January 19, 2000, the original *Brachiosaurus* cast was "traded" to the O'Hare International Airport to make room for big name, major league "free agent" star "Sue."[22]

Now Field Museum visitors can see Sue's 67-million-year-old, 42-foot-long skeleton mounted on a regal pedestal, posed as if the great dinosaur was momentarily disturbed in the act of examining, or perhaps eating, something on the ground.[23] (See Figure 25-4.) The original skull and

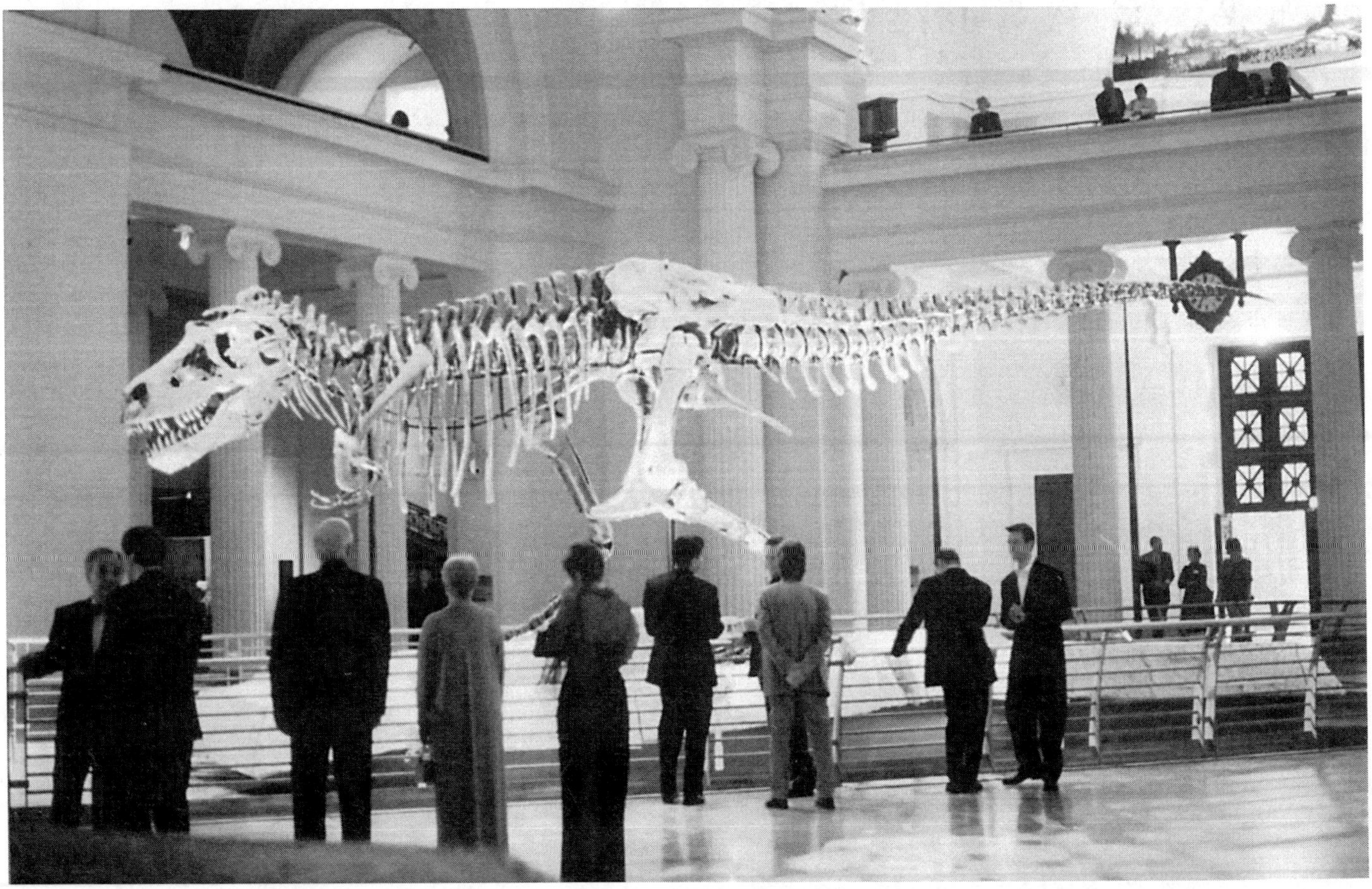

Figure 25-4: "Sue" on display at the Field Museum of Natural History. America's national dinosaur "totem" is T. rex. (Photograph by Allen A. Debus—May 2000, courtesy of the Field Museum of Natural History.)

several other exhibits are displayed upstairs beneath a new mural by John Gurche showing Sue when our icon walked the earth.

Recently, other theropods have challenged *T. rex*'s birthright to the title of world's largest theropod! Don Lessem's popular book entitled *Bigger Than Rex* (1996) made a convincing case that a Late Cretaceous genus, *Giganotosaurus*, found in 1993 in Argentina, was the world's largest meat-eating dinosaur—about one foot longer than even the largest *T. rex*![24] The news stirred media controversy, and even paleontologists found themselves swept up in the debate, as in the pages of a popular magazine—*The Prehistoric Times*.[25] Then, in 2000, another sensational discovery was reported from the eastern slopes of the Andes in Patagonia. The new, as yet undescribed, theropod, initially thought to be an evolutionary cousin of *Giganotosaurus*, had an estimated length of 45 feet. This happens to be just a tad longer than Sue.[26] But a newly found "C. rex" may be 10 percent larger than "Sue."

While there certainly must have been theropods larger than the individual specimen known as Sue, we find these "size matters" discussions rather trite. The point is that *T. rex* was big enough to scare the living daylights out of us all if we could see it alive and it was the first of its kind known to science.

Savage, lordly and renaissance-like everchanging, *T. rex* has yet to be dethroned! The "studliest" theropod of all has a girl's name—"Sue." Long live the Queen!

CHAPTER 26

The Great Dinosaur Race—Wisconsin's Dinosaurs

A wonderful book would catalog all the myriad paleoimagery of every state, isle, province and country the world over. It isn't possible to take on such a weighty task here, however, as a "test case" the authors attempted to account for all the significant forms of paleoimagery that we could track down in the State of Wisconsin. For every paleoimage we discovered, we also uncovered intriguing behind-the-scenes stories, some historical in nature and others being of a more scientific flavor. Although we sampled within but one geographic boundary, rest assured paleoimagery knows no bounds. This chapter underscores how one cannot appreciate paleoimagery to fullest degree without understanding its anecdotal background, purpose and history.

The style of this chapter reads more like a "travel log" than other chapters in this book. That is because it stemmed from our carefree, 1990s paleo-adventures of the Midwest, written at a time when, in search of paleoimagery, we boldly sought even the quaintest paleo-places imaginable.

Fall winds summon Mesozoic memories. Brisk air, transforming color, a suggestion of change, Time's flow. With passing years we indulge in reminiscing about youthful dinosaur encounters, while yearning for more. The seasonal appeal of our treasured Midwest state, Wisconsin, has beckoned to this Illinois family for decades. For young dinophiles, how enchanting seems the prospect of encountering a realistic dinosaur, even where they aren't supposed to be found, one of those lurking about in wild Wisconsin. It's sort of like tracking the Midwest's version of Bigfoot.

There would appear to be little association between dinosaurs and Wisconsin, a land known for its dairy products, "cheese heads," outdoor recreation, the mysterious "Hodag," and harsh winters. No dinosaur bones have been discovered there. The deposits are either too old or too young. Wouldn't we simply be better off visiting the nearby Field Museum to see dinosaur exhibits?

Well, there's no questioning the world class nature of the Field Museum's dinosaur displays, both then and now. But sometimes we all need a change of pace. So, when the autumn breezes blow and the fall colors lure us northward, we set our sights on that northern breed of dinosaur. For the highway is charged with excitement and possibility.

Hear us now, believe us later! Thar be dinosaurs in Wisconsin ... all the "classic" kinds in fact!

Dinophile Gary Williams has developed a classification scheme for exhibits of life-size dinosaurs (which has been modified slightly here to broaden the applicability of his classification scheme). First, there are those purely commercial enterprises where dinosaurs have been constructed to lure tourists off the road and to coax the spending of dollars in gift shops. While the artists may have promoted their handiwork as "scientifically accurate," a quick glance sometimes yields a distinct "You've got to be kidding!" impression.

Next are those samples of "folk art" one finds in offbeat places. There isn't even any attempt to make a fast buck here. Folk dinosaur art has been placed here for our sheer entertainment as we fly by on the highway. Spare metal, concrete and fiberglass have been fashioned into quaint paleo-structures that we quickly grow accustomed to.

Then there are the more realistic dinosaurs of scientific standing, which usually reside in our museums.

Here we find erected skeletons and dioramas. To be sure, there are a number of subcategories as well, such as travelling exhibits and animatronic dinosaurs.

Science, folksy and fast buck Freddie ... we love em' all. Believe it or not, Wisconsin has all of these, yet not so many of each kind to cover in a single chapter. (There is another category—dinosaurs constructed in the service of organized religion. But so far, those faithful fellows haven't popped up here yet.)

So, imagine our disappointment when a decade or so ago we drove to the Wisconsin Dells area in Baraboo for the first time in many years. We had hoped to see those concrete dinosaurs which Allen's parents had taken him to see on several September trips long ago in the 1960s—displayed at the Enchanted Forest and Prehistoric Land. Driving along the east side of the Wisconsin River in the downtown area, the life-size image of the "*Brontosaurus*" loomed large on the opposite west bank, near a bridge. The flip side of a postcard surviving from that long ago time reads, "Sometimes called the Thunder Lizard. Giant of all dinosaurs, reaching a length of 70 feet." (See Figure 26-1.)

Okay, so the "*Brontosaurus*" looked like he escaped from the Flintstones yard. Even a ten-year-old could see how unthreatening Bronto was. But he was BIG, and this is often what matters most to an impressionable child! There were others there too, best seen from inside the park area. This was the largest life-sized, realistic dinosaur Allen had ever seen. So it loomed as large as the Bronto in his youthful paleontological imagination. It was natural to associate the statue with the Field Museum's famous *Apatosaurus* skeleton.

But on this late 1980s fall day there was no "Bronto" to be seen as we drove along the river. Yet another extinction seemed to have taken place. Were we to seek solace in the Dells' Robot World, or on a "Duck" ride instead? Drat! What had happened?

The Prehistoric Land dinosaurs were the late 1950s product of artist/sculptor Dick Day. Day was a plasterer by trade and at the time lived in a Tribal village north of the park. He aspired to add the dinosaur attraction to augment fairy tale–inspired figures for the adjacent Enchanted Land (i.e. Mother Goose, Three Little Pigs, Wooden Shoe house, etc.), following a vacation to Rapid City, South

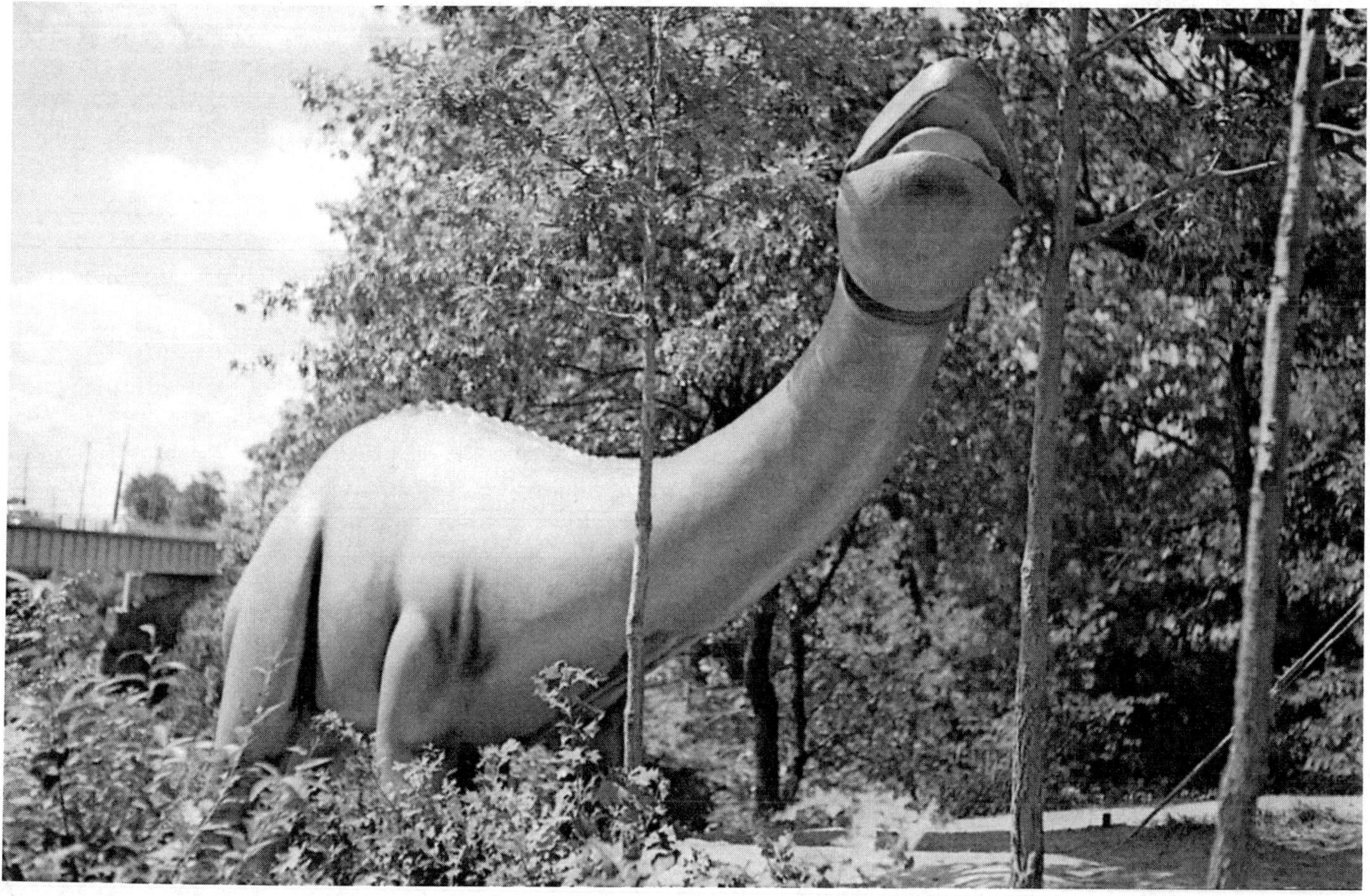

Figure 26-1: Old souvenir postcard showing Prehistoric Land's *Brontosaurus*. (Authors' collection.)

Dakota. In Rapid City, Emmet A. Sullivan's world famous concrete dinosaur statues proved influential. For, by the late 1950s (circa 1958), Day had added five dinosaur attractions to the property: *Triceratops*, *Protoceratops*, "*Trachodon*," "*Brontosaurus*," and *Stegosaurus*. The final addition was a nine-foot-tall Neanderthal Man.[1]

Contemporary, late 1950s promotional photos reveal highly attractive "cave girls" scantily clad in swim suits, sitting astride their Trachodon and Triceratops "steeds." The sight is more than enough to transform any red-blooded American male emerging from Enchanted Land into a "Big Bad Wolf" lycanthrope. But in all our old family shots there are only pictures of Allen and his brothers dressed not in scanty clothing to catch sunrays, but in warm clothing for a chilly Autumn day. (Alas, there are no pictures of "dinosaur-riding babes" in the old photo albums.) Nevertheless, the dinosaurs of yester-"Day" were a marvelous treat for once-upon-a-time grade schoolers.

What happened to them? In the fall of 1989, after nearly 31 years, the property was redeveloped into a 54-hole mini-golf haven, no doubt sufficient to impress even a Tiger Woods.[2] While tears aren't shed quite as easily for Mother Goose's fate, we feel remorseful for the dinosaurs' demise. Surely they are forever now extinct in Baraboo! Or are they? We find Wisconsin has many surprises in store for us.

Searching for a gasoline filling station, we spy a Sinclair Gas sign. Out in front, sure enough, there is a human-sized, fiberglass sauropod figure. Not quite as impressive as Day's concrete statues, but today it will have to do, especially since we're running on empty. After a photo and refueling, we are on our way once more.

A bit further on we spy another sign offering "Dining with the dinosaurs." These are the famed Dinamation animatronic variety. (Will they eat you after you've been "fattened up," hopefully before you can pay the bill?) And then a little further on we see another ¼ scale sauropod standing in a miniature golf course near the Devil's Lake State Park entrance. Golf and Dinosaur, neither threatening the other, now in perfect harmony, perhaps a strange metaphorical union. That's the way it should have been a few miles back down the road. But it's a new "Day," and there are more dinosaurs to see.

At the speed of light a message came to us from a former coworker, now situated in the land of artesian waters, the very realm of Old Style's King of Beers. Carol Witt-Smith's e-mail message read, "Yes, Al, there are dinosaurs in Wisconsin. I pass them on the expressway every time I go to Chicago."[3] Well, we had driven through LaCrosse, Wisconsin, several times before, and while the World's Largest Six Pack is hard to miss, we had never spied a dinosaur there before. Carol's message offered another clue. "They are very nice-looking and fenced in." Okay, at least no one was in immediate danger! Better hit the road. Oh, and bring the camera—just in case there are some dinosaur riding cave-babes. Beer, cave-babes and dinosaurs—what an intriguing "primeval" combination!

No cave babes after all. They've all gone extinct apparently, in lieu of the dinosaurs which are still thriving here. But having the camera turns out to be beneficial after all. There, off Highway I-90, are three dinosaur skeletons, near the intersection of Linden and La Crosse County Highway C. These are "skeletons" welded from recycled steel scrap reminiscent of Jim Gary's blowtorch handiwork—a sauropod and two toothy-grinned *T. rex* theropods. They are large but juvenile-sized.

It wasn't easy to track down the "bare bones" tale of this trio. However, we learned that the sculptures had been constructed a few years ago by Harold Hoffman. The artist died, and shortly thereafter, in 1997, they were donated by George Riniker to the Village of West Salem. The three statues were first displayed at their current location in early August of 1998. Now, Terry Hanson, the chief conspirator behind the Village's new acquisitions, is proud to have a mini version of "Jurassic Park" nearby for people to enjoy. One might say the West Salem dinos are now a sight for "saur" eyes!

Few readers may realize that when it comes down to the "wire," West Salem's dinosaurs would only win the bronze medal, with a third place finish in Wisconsin's Great Dinosaur Race. You say you never heard about this race? Want to? So, who won the race? When and where was it held? Well, depending on who one talks to, there is some dispute as to whether there was even any "race" at all.

The winner was "Mr. Ed," a full-size composite mount of the Cretaceous genus *Edmontosaurus*, now displayed at the University of Wisconsin's Geology Museum in Madison. Unveiled to the public in 1991, it was the first dinosaur *skeleton* to be mounted for display in the state of Wisconsin! (See Figure 26-2.)

The 33-foot-long, bipedal, "duck-billed" Mr. Ed, was truly a horse of a different color. "Mr. Ed" is a product of the famous Ruth Mason Quarry of northwest South Dakota. It was unearthed from the fossil rich bone bed quarry in the 1980s by University of Wisconsin volunteers, with able-bodied assistance from representatives of the Black Hills Institute. Now "Mr. Ed," undisputably *the* first dinosaur skeleton displayed in Wisconsin, remains a feature fixture at the Geology Museum.

The runner up, finishing several lengths behind (and completed about a year later in the summer of 1992), was

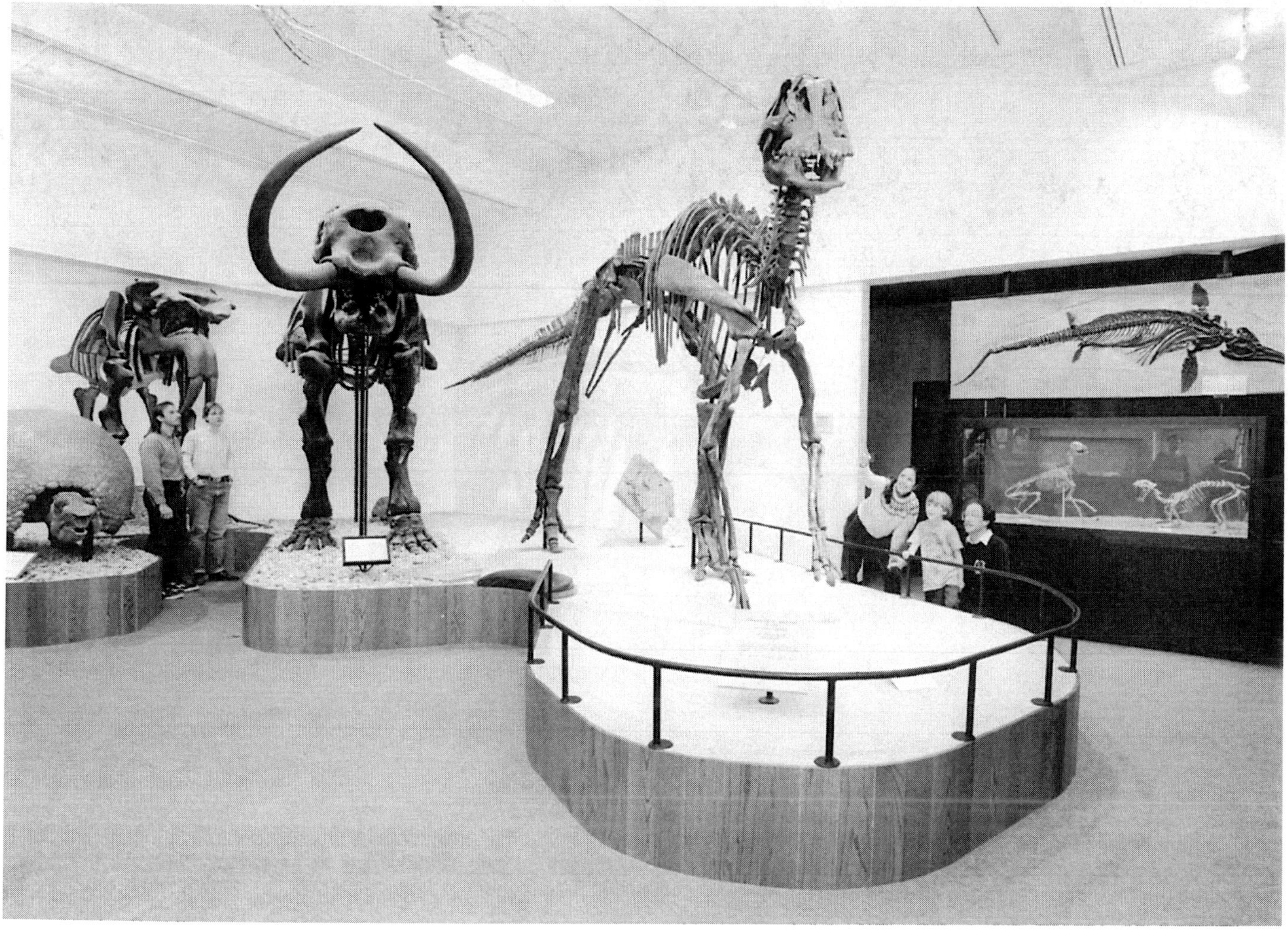

Figure 26-2: Photograph of the University of Wisconsin's Geology Museum, with "Mr. Ed," the *Edmontosaurus*, at center. Also visible (left to right) are a Ward's cast *Glyptodon*, a Ward's *Deinotherium* skull cast, a *Mastodon* skeleton, and, at right, an ichthyosaur fossil and pair of *Hoplophoneus* prepared and mounted by Joe Skulan. (Photograph copyright Geology Museum, University of Wisconsin, Madison. Reproduced with permission of Dr. Klaus Westphal.)

the Milwaukee Public Museum's *Torosaurus* skeletal mount. As a consolation prize, Milwaukee was able to claim that *Torosaurus*' nine-foot-long skull was the largest of any known dinosaur. However, while "Mr. Ed" is comprised of approximately 80 percent real bone, "El Toro" is nearly 85 percent cast.

In the fall of 1999 we asked Dr. Klaus Westphal, Director of the Geology Museum in Madison, about the Great Dinosaur Race. Klaus responded, "Well, it depends on how you define 'race.' We knew they were putting one together in Milwaukee. And so we thought it would make some nice 'hype' to see who finishes first. And yes, indeed we won. They seemed to take it more seriously than we did. As it turns out, they weren't too happy with the result."[4] It took two years to assemble "Mr. Ed" between the years 1989 and 1991.

Apparently, there was even an ad campaign sponsored by the Milwaukee Public Museum promoting their *Torosaurus* skeleton as Wisconsin's first. But the campaign had to be scrapped after "Mr. Ed" won the "trophy." When we inquired about the race during a subsequent visit to the Milwaukee Public Museum, we noted there was a conspicuous lack of recollection concerning any dinosaur "racing." Hopefully there were no "saur" losers.[5]

Milwaukee's *Torosaurus* has become a rather famous character. Discovered in the Hell Creek Badlands of northeastern Montana in 1981, this happens to be the most complete *Torosaurus* specimen ever found, especially considering its forelimbs, formerly unknown to science. It is impressive to see in person. (See Figure 26-3.) Real bone appears darker colored, while cast elements used to complete the skeleton are lighter. The majority of the cast (missing) skeleton was cast from (large) *Triceratops* material.

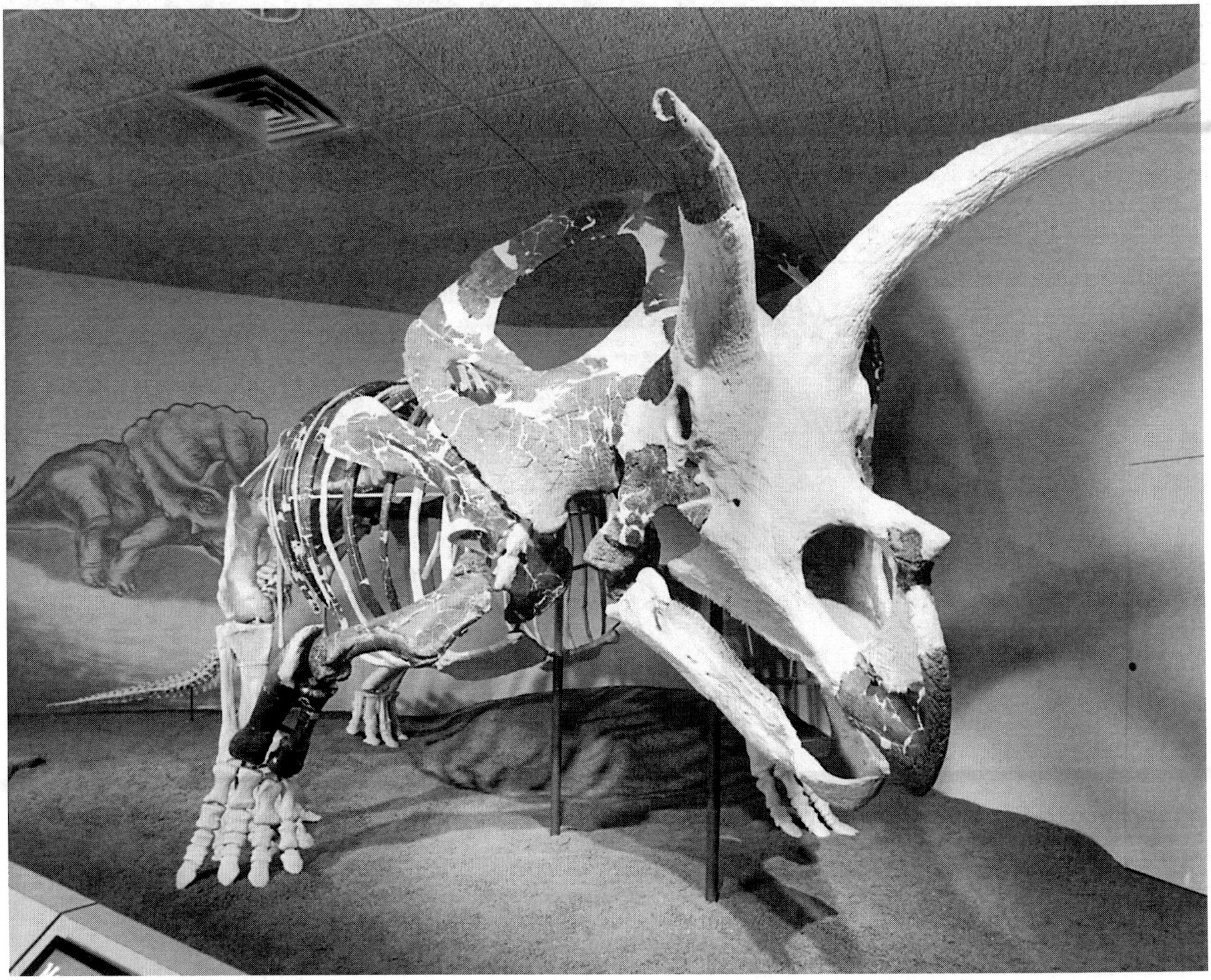

Figure 26-3: The Milwaukee Public Museum's *Torosaurus*. (Reproduced with permission of the Milwaukee Public Museum.)

The reconstructed skeleton was quickly entered into the ongoing fray concerning ceratopsian front limb stance in ceratopsians. Casts of the front limb bones were subjected to a battery of biomechanical tests in order to resolve longstanding controversy. In their 1995 paper, "The Forelimb of *Torosaurus* and an Analysis of the Posture and Gait of Ceratopsian Dinosaurs," Rolf Johnson and John Ostrom stated that *Torosaurus*' forelimbs "...moved in a somewhat sprawling manner. Furthermore, the posture and locomotion of Torosaurus and other members of the Ceratopsia must have been quite unlike that of any animal we can directly observe today."[6]

Dr. Peter Sheehan of the Milwaukee Public Museum claims, "What's right for *Triceratops* may not be for Torosaurus. Different, but related, genera could have had different styles of movement."[7] True, but in that case a matador might have an easier time dueling with El *Torosaurus* than with the presumably more agile, upright *Triceratops*. Of course, Johnson and Ostrom's 1995 conclusions offered less flexibility. Hey, guess what? Besides *Torosaurus*, there's a *Triceratops* here too.

Perhaps *the* finest dinosaur diorama displayed inside any museum institution, incorporating life-size dinosaur restorations in a hauntingly realistic Cretaceous scene diorama, lies just down the hall from *Torosaurus*. Turn a corner and there you are, smack dab back in the upper Cretaceous of South Dakota, 65 million years ago, with an array of live-looking reptilian "friends." Except the potentially friendliest of the group, *Triceratops*, is being devoured by a not-so-friendly-looking *T. rex*. Be afraid; yes, be very afraid.

This wonderful diorama, authentic right down to the most trivial-seeming details, was planned in the late 1970s and opened to the public in 1983. The central figures are a *T. rex* feasting on a dead *Triceratops*. On either side are scavenging *Velociraptors*, and in the distance stands *Struthiomimus* (a Jonas Studios replica). Young children have been known to flee from the apparition in abject terror. (See Figure 26-4.)

T. rex was cast and then slightly modified from an original sculpture made by the Richard Rush studios.[8] It may be the first life-size *T. rex* to be posed correctly—horizontally, without its tail dragging on the ground. The tableau cleverly skirts the matter of whether *T. rex* was an active predator or a scavenger. All we can decide is that he looks quite hungry, and that he must have a ravenous appetite for Mesozoic buffalo meat.

Triceratops was fashioned by Milwaukee Public Museum taxidermist Floyd Easterman. The forelimb stance is not well defined in the diorama, as the "corpse" is lying on its left side. However, one can't help noticing a series of bony, blade-like scutes embedded over the shoulders and along the backbone. Incidentally, these are features appearing in three of the earliest life restorations depicted by paleoartist Joseph Smit (1836–1929).

Smit restored *Triceratops* for two books written by the Reverend Henry Neville Hutchinson, *Extinct Monsters* (1892, 1893) and *Extinct Monsters and Creatures of Other Days* (1910, 1917), and one book by Henry R. Knipe, *Nebula to Man* (1905).[9] In all three cases, *Triceratops* is sporting a coat of dorsal armor. (See Figure 26-5.) However, Smit's *Triceratops* is far less armored than Charles Knight's *Agathaumas* restorations of 1897 (sculpture and painting), which are densely armored. By 1909, artist Josef Pallenburg had completed three life-sized *Triceratops* sculptures for Hagenbeck's Zoological Park near Hamburg (two adults and a juvenile), all of which carry dorsal armor. (See Figure 12-4.)

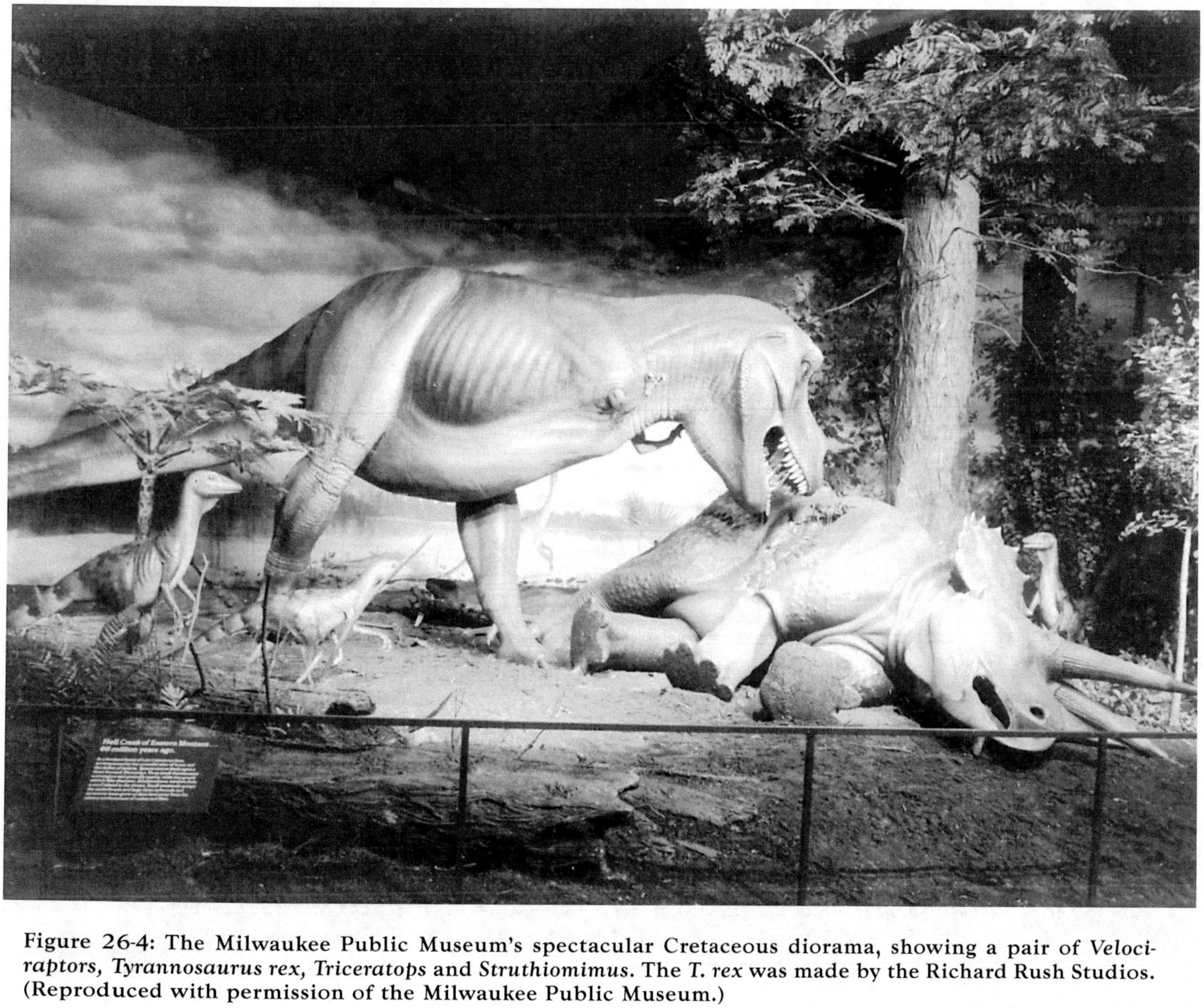

Figure 26-4: The Milwaukee Public Museum's spectacular Cretaceous diorama, showing a pair of *Velociraptors, Tyrannosaurus rex, Triceratops* and *Struthiomimus*. The *T. rex* was made by the Richard Rush Studios. (Reproduced with permission of the Milwaukee Public Museum.)

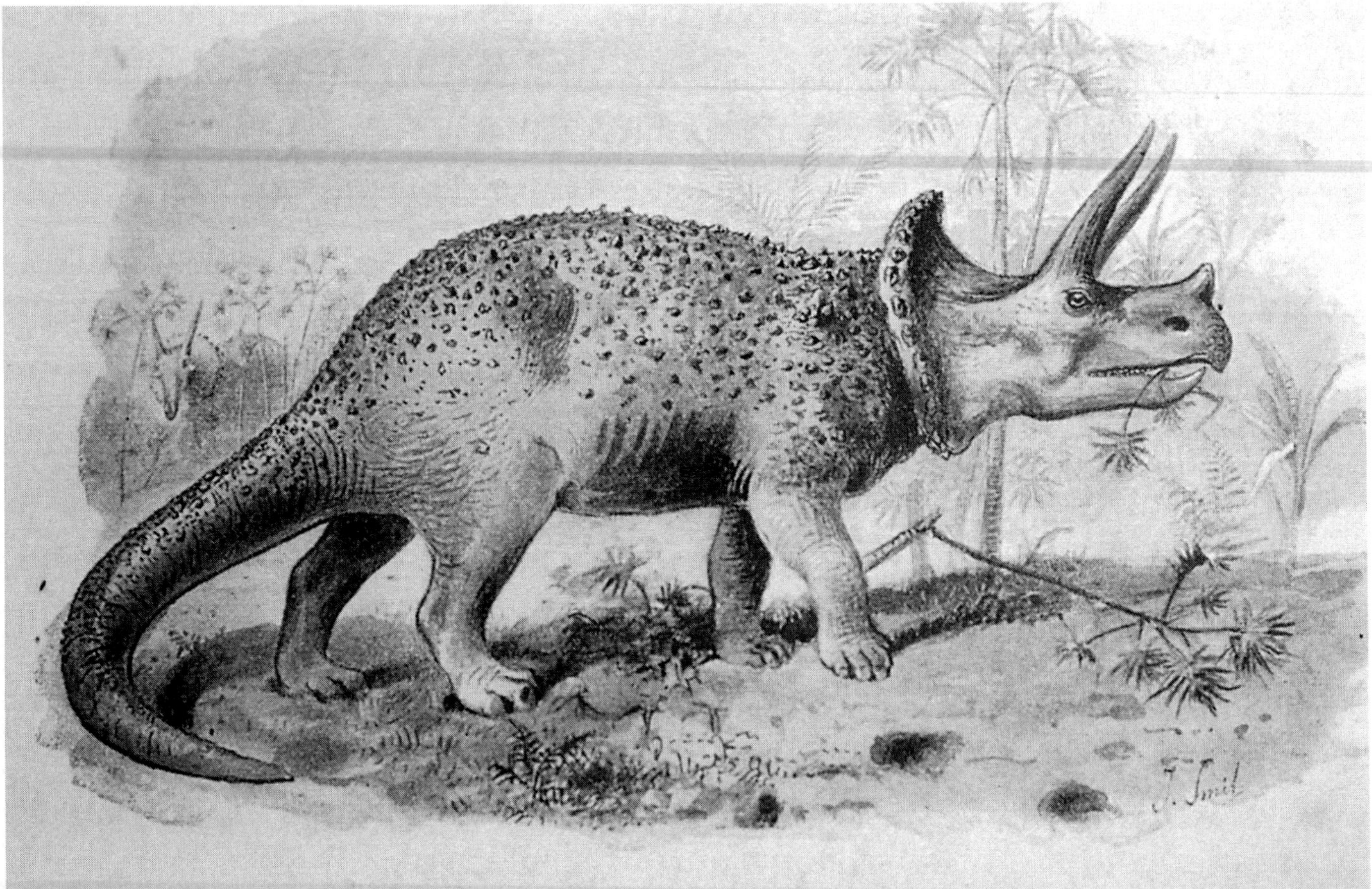

A GIGANTIC HORNED DINOSAUR, TRICERATOPS PRORSUS.

Figure 26-5: Joseph Smit's 1892 *Triceratops* restoration. From *Extinct Monsters*. (Authors' collection.)

While few modern *Triceratops* restorations show this armor, Milwaukee Public Museum staff thought there was sufficient fossil evidence to warrant their inclusion in their *Triceratops*. According to Don Glut, however, dorsal armor was originally attributed to *Triceratops* by Othniel Marsh.[10] Much later, paleontologists realized the source of the confusion. Pachycephalosaur fossil bone resembling skin nodules had become mixed in with ceratopsian material, deceiving Marsh and others into thinking horned dinosaurs bore dorsal armor.

There's another dinosaur restoration on permanent display in the Milwaukee museum, perhaps Wisconsin's second oldest "dino-acquisition" (that is, after the Dick Day/Dells dinos). For a *Stegosaurus* majestically "strides" into view through a bend in the hallway.

In contrast to the Cretaceous diorama dinosaurs which seem breathtakingly alive and savage, Paul Jonas' life-size fiberglass Stego seems accommodating and friendly, perhaps in need of a reassuring pat on his noggin. Yes, he really is an "old friend," of sorts. It's a vintage cast of the 1964 Sinclair dinosaur, *Stegosaurus*, originally displayed at the 1964/65 New York World's Fair. This particular "specimen" dates from a prior "era" in the history of the Milwaukee Public Museum, when it was displayed in the older paleontology wing prior to the museum's renovation in the early '80s.

Allen still recalls seeing the original Jonas Stego when the caravan of Sinclair figures toured the country after the Fair ended. In the summer of 1966 the Debus family traveled to Calumet City, Illinois, where the Sinclair "gang" (i.e. Jonas' *Triceratops, T. rex, Apatosaurus*, "*Trachodon*," etc.) hung out and posed for photos while still aboard flatbed transport trucks. There was even foliage and rocks strewn about the trailer floor to provide a nice "atmospheric" touch.

Another late 1960s sighting of the Sinclair entourage (Stegosaur included) was made in Brooklyn, Minnesota, by Mark Zimmerman, which Mark claims is "a scant cheesehead hat's throw away"[11] from Wisconsin. So it is likely that the Sinclair dinosaur entourage also toured Wisconsin, although city names of display points along the Wisconsin leg have been lost to posterity.

In fact, 1999 turned out to be a banner year for the dinosaurian millennial "invasion," with more dinosaurs descending upon Wisconsin than any time since the K-T impact. Between late June and September 1999, "*The Dinosaurs of Jurassic Park: The Lost World*" show opened at the Milwaukee County Zoo. Numerous fossils, skeletal casts of dinosaurs and *Lost World*, Ameritech-sponsored movie props were exhibited in the Otto Borchert Family Special Exhibits Building.

Well, not all of the dinosaurs were inside. Outside was the first life-size sauropod to ever trod on Wisconsin's soil (or, actually, pavement) since the "Day(s)" of the Dells Prehistoric Land. This was a sculpture of the Chinese genus *Mamenchisaurus*. With a 70-foot-long attraction like this standing outside the exhibit area, how could anyone not notice that the Zoo had a most unusual assortment of visitors from the past on view?

Inside it strictly educational. Dozens of skeleton casts are arranged in a life-through-time array, beginning with the late Paleozoic era through the end of the Mesozoic. Visitors are invited into the "Extinction Theater" where a short fantasy movie about the asteroid impact theory is featured.

There is barely enough room in this large space for all the cast skeletons and fossils. Among them we find familiar friends like *Dimetrodon* and *Stegosaurus* intermingled amidst more exotic varieties such as *Yangchanosaurus*, *Eoraptor*, *Monolophosaurus*, *Amargasaurus* and even a Chinese pterosaur. Perhaps the feature attraction is a full skeletal mount of the South American *Giganotosaurus*. Visitors can do some quick seat-of-the-pants comparative anatomy by gazing at the nearby displays of *Tarbosaurus* and *Albertosaurus*, while kids pose for those obligatory pix.

There's just a bit of movie hype as well. Surrounded by this wealth of curved, molded resin is a central "island" (fittingly) filled with *Jurassic Park* and *Lost World* movie props. There are several static restorations: the "breathing" *Triceratops*, the juvenile *Stegosaurus*, those pesky *Velociraptors* and, for good measure, a *Pachycephalosaurus*. Then, on a day sufficiently warm to melt those waxy "Mold-a-Rama" zoo animal figures our kids keep shelling out precious dollars for, we're back outside cooling off in *Mamenchisaurus*' shadow. Whew! Will someone please clone some mammoths? An Ice Age sounds like a good idea!

Careful what you wish for. For just a few weeks later, during an uncomfortably cooler time, Wisconsin is invaded once more, only this time by Russia, when the mighty Russian Dinosaur Exposition wheeled into the Appleton Children's' Museum. It turns out to be even grander than the *Jurassic Park/Lost World* display. This time there is no movie hype. (There are no cave babes either.)

There are 66 specimens, 33 of which are complete (i.e. 90 to 95 percent) skeletons, all from the collection of the Paleontological Institute of the Russian Academy of Sciences. As expected, this show is even more startling and exotic that the *Jurassic Park/Lost World* traveling exhibit. Russia has produced a truly impressive, hard-to-top final act. Yet we seem so much removed in time from the simpler days of idle Dells vacations, when dinosaurs were threatened by little more than a putter.

The great northern state is evidently becoming popular terrain for even the most exotic of dinosaur shapes. Picture Wisconsin, and images of fishing and dairy products typically come to mind, not dinosaurs. Wisconsin's official state fossil is a trilobite (*Calymene celebra)*, not any familiar kind of dinosaur or furry Ice Age mammal. Yet during the second half of the twentieth century, all the classic dinosaur varieties, from enjoyable kitsch to scientific display and world class diorama dinosaurs, have scored successes in the land of cheeseheads. While *T. rex* remains our favored icon, in Wisconsin "Mr. Ed" reigns as the State's "fastest" dinosaur.

In Wisconsin, the dinosaur race is thriving in the Age of Dinosaurs. "Bigfoot" lives here after all!

Chapter 27

Illuminating Fossils from the Dark Continent

Dr. Paul Sereno, revered as paleontology's "Indiana Jones," is also paleoart's "modern Osborn." Sereno is a charismatic, high-spirited phenom with a flair for the dramatic, blessed with artistic sense and an uncanny knack for finding new significant dinosaur fossils. Children flock to him during guided tours at the Chicago Children's Museum at Navy Pier. His spectacular successes on paleontology's frontier have paraded through Chicago's universities, libraries and museums, and across the country, in recent years. We've seen them emerging since the time of their Chicagoland debut. His energy instills enthusiasm on every level. "Go get 'em, Paul!"

Sereno's scientific interests emphasize intriguing evolutionary questions, using dinosaurs as a research beacon. Like Osborn, Sereno has sought answers to these evolutionary mysteries by organizing expeditions (his own) to the globe's faraway places. Sereno, the "modern Osborn," avidly promotes dinosaur science, especially our refined knowledge of African dinosaur fauna, using striking visual displays, imagery and imagetext.

Some of the most scientifically intriguing, if not downright thrilling, episodes in the annals of vertebrate paleontology have sprung from the sediments of Africa. Ranging through three eras of time, and spanning the history of our science, an array of important discoveries have come to light, including record of the mammal-like reptiles, tremendous dinosaurs, and the earliest traces of human ancestors. Now University of Chicago vertebrate paleontologist Paul Sereno has emerged from Africa's sands of time with new fossil trophies, further illuminating the nature of Africa's Cretaceous period.

Why travel into Africa's foreboding terrain when there are less risky places to dig for dinosaurs? Sereno claims there are always new opportunities for "...playing God and the thrill of creating new species." And there are other irresistible urges, the lure of freeing another lost species from Time's vault. "What are they going to look like? I don't know. That's part of the fun."[1]

Sereno is no stranger to sensationalism. Basking in the fresh glow of recent successes, culminating in his description of South American Triassic dinosaurs *Eoraptor* and new specimens of *Herrerasaurus*.[2] Sereno seemingly willed one magnificent discovery to happen after another! This time, however, the fossils beckoned from ancient Gondwana's eastern territory, in another land south of the equator, the country of Niger.

Sereno publically announced his discovery of a new ("relatively complete") African theropod, the 130-million-year-old *Afrovenator abakensis*, during a lecture at the Fermi National Accelerator Laboratory on October 7, 1994, concerning "A Saharan Odyssey: Unearthing the Secrets of a Wandering Continent." His address made national headlines, and there followed, in quick succession, televised news broadcasts, a televised documentary, *Skeletons in the Sand*, and publication of his paper, "Early Cretaceous Dinosaurs from the Sahara."[3]

Skeletal remains were cleaned and cast at the Royal Ontario Museum in Toronto. An entire skeletal reconstruction resulted as technicians speculated how the missing pieces of bone appeared. In late 1994 a skeletal cast of the *Afrovenator* and a thigh bone belonging to an African sauropod dinosaur were first exhibited at an unlikely place, the Chicago Public Library—Harold Washington Library Center. *Afrovenator*'s remarkably short torso and long curving tail yielded an impression of a skeleton flowing with life. *Afrovenator* had narrow hips and was quite barrel-chested, perhaps especially after engorging itself with a good meal! The slender nature of the skeleton, the imposing skull, the massive feet and sharp talons connected to forearms no longer than a human's captivated the imaginations of all who saw it.

Sereno's published skeletal reconstruction formed the basis of a miniature sculpture by Canadian paleoartist Garfield Minot that was displayed alongside the cast skeleton at the Harold Washington Public Library.[4] Working under Sereno's direction, paleoartist Douglas Henderson restored *Afrovenator* stalking a herd of sauropods for Sereno's January 1995 *Natural History* article, "Dinosaurs and Drifting Continents."[5] Minot's and Henderson's early restorations provided a dramatic sense of how this sleek, powerfully built and dynamically balanced creature appeared in life. Sereno envisioned an even more startling exhibit featuring mounted skeletons representing a herd of roving African dinosaurs! That day would come.

Sereno also illustrated *Afrovenator* attacking sauropod (the later formally described as *Jobaria* in November 1999). (See Figure 27-1.) Five years later, referring to his illustration and modestly to himself as a "non-professional artist," Sereno stated, "On Christmas Day 1995 ... I was inspired to sketch in pen and ink flesh renderings of two animals that I and my team had collected in the Sahara a short time before. Did these animals interact as predator and prey? Could *Jobaria* rear on its hind limbs? Did *Afrovenator* have striped body coloration? A flesh reconstruction draws upon both science and imagination. The purpose of doing a flesh reconstruction or scene painting of a dinosaur is to absorb what science has to say and then to go beyond—to depict how something might have appeared and to imagine what may once have happened.... The rendering ... attempts to capture their grace and perhaps a little piece of their world."[6]

Afrovenator was a tough act to follow, but Sereno managed to top that discovery with two more. At the University of Chicago campus on the evening of April 30, 1996, during a lecture titled "Tracking Dinosaurs on Wandering Continents," Sereno reported discoveries of two African theropods. Their remains were found in Late Cretaceous deposits formed about 93 million years ago in a river delta (now situated in southeastern Morocco).

Known as the "shark-toothed reptile," the large genus, described as *Carcharodontosaurus saharicus*, had a brain case smaller than its distantly related evolutionary cousin, *T. rex*. In his June 1996 *National Geographic* article, Sereno speculated that *Carcharodontosaurus* was "...perhaps the largest carnivore that ever walked the earth."[7] A team of artists and scientists managed to fully reconstruct the skull. A cast of this skull was dramatically unveiled to the public as the grand finale in Sereno's lecture. Paleoartist Garfield Minot restored a life appearance of the head based on the skull reconstruction. A photograph of the skull resting alongside a human skull appeared on the cover of the May 17, 1996, issue of *Science*.[8]

Sereno also spoke of another startling discovery, referred to as the "delta runner," later described as *Deltadromeus agilis*. This creature, known from a partial skeleton (including limb, hip and shoulder bones), had unusually long forelimbs and a tibia/femur ratio suggestive of fast-paced movement. In life, the slender and cursorial *Deltadromeus* grew to about eight meters in length. Its skeleton closely resembles that of a possible evolutionary ancestor, North America's diminutive Upper Jurassic *Ornitholestes*.

If you shudder at the thought of lions, cheetahs and hyaena prowling about the modern African plains, competing desperately for the flesh of hooved mammals, imagine the daily dread of a Cretaceous African herbivore, with the likes of *Carcharadontosaurus* and *Deltadromeus* roving about. Paleoartist Mark Hallett did just that, restoring *Deltadromeus* challenging *Carcharodontosaurus* for the spoils of savagery for Sereno's June 1996 *National Geographic* article, "Africa's Dinosaur Castaways."[9] (See Figure 27-2.)

While *Afrovenator* is reasonably complete (roughly 70 percent known), *Deltadromeus'* skeleton is approximately 50 percent known from fossils. Sereno claims there is no "hard line" for deciding when sufficient fossil material is available on which to base a reconstruction or restoration. In the case of *Deltadromeus*, "...we had just enough to where I was willing to try and reconstruct it. We had no part of the skull ... but there was enough of the hind limbs...."[10] The somewhat speculatively reconstructed *Deltadromeus* skeleton strides through University of Chicago's Crerar library, no doubt in search of a good dinosaur fantasy novel.

Then in late youngsters and the young at heart were entertained, instructed and fully absorbed by the "Colossal Claw" Dinosaur Discovery exhibit displayed in the Children's Museum at Chicago's Navy Pier. The central attraction was another Sereno "find," this time, *Suchomimus tenerensis* (the "crocodile mimic"), a 120-million-year-old, 40-foot-long sail-backed African dinosaur found in the Gadoufaoua desert region of Niger.[11] It is related to England's *Baryonyx* and Africa's *Spinosaurus* (although more closely to the former than to the latter).

Sereno and students sculpted missing portions of bones to complete the likeness of *Suchomimus* skeletal structure. After making a small model of how the skeleton

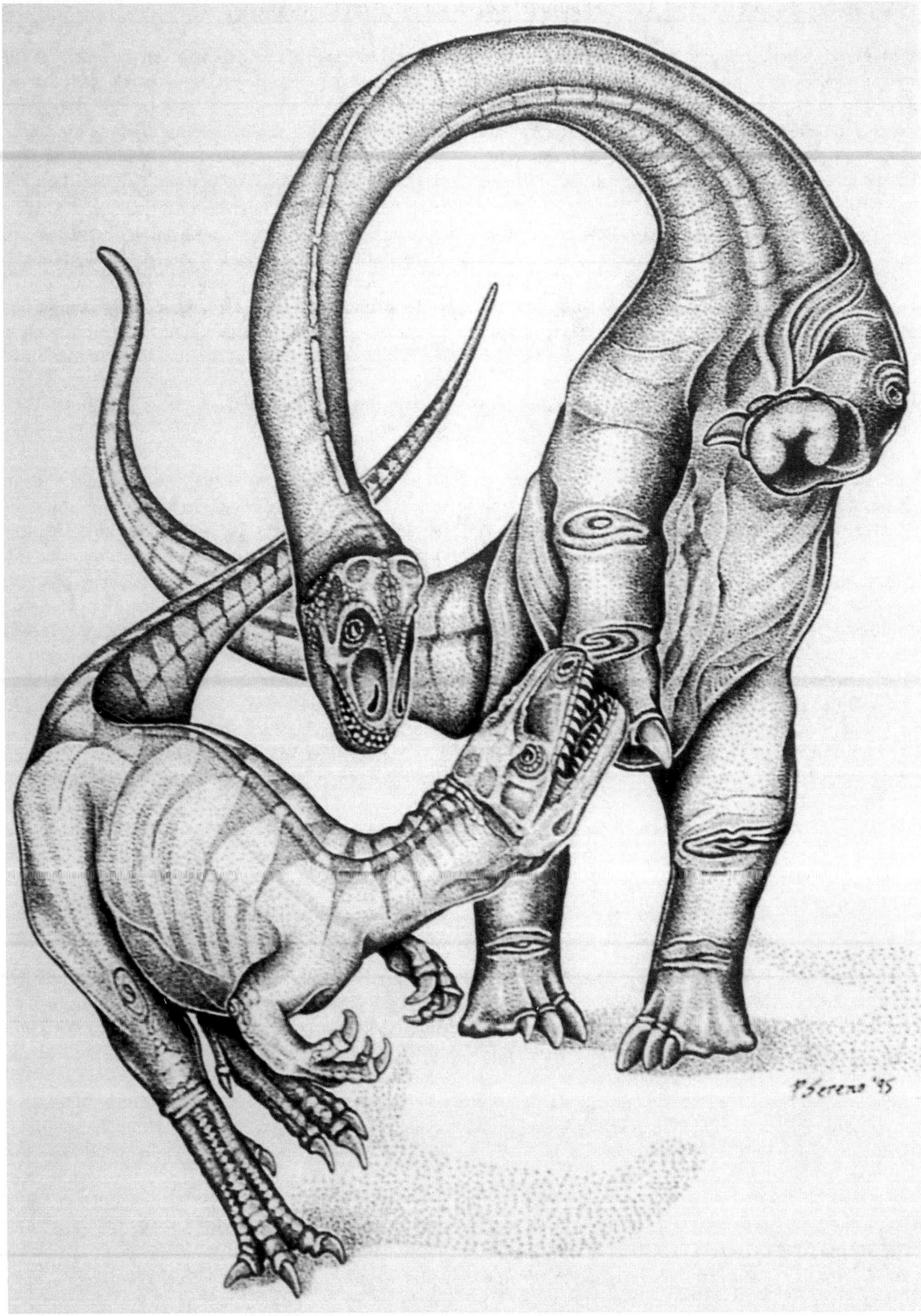
P. Sereno '95

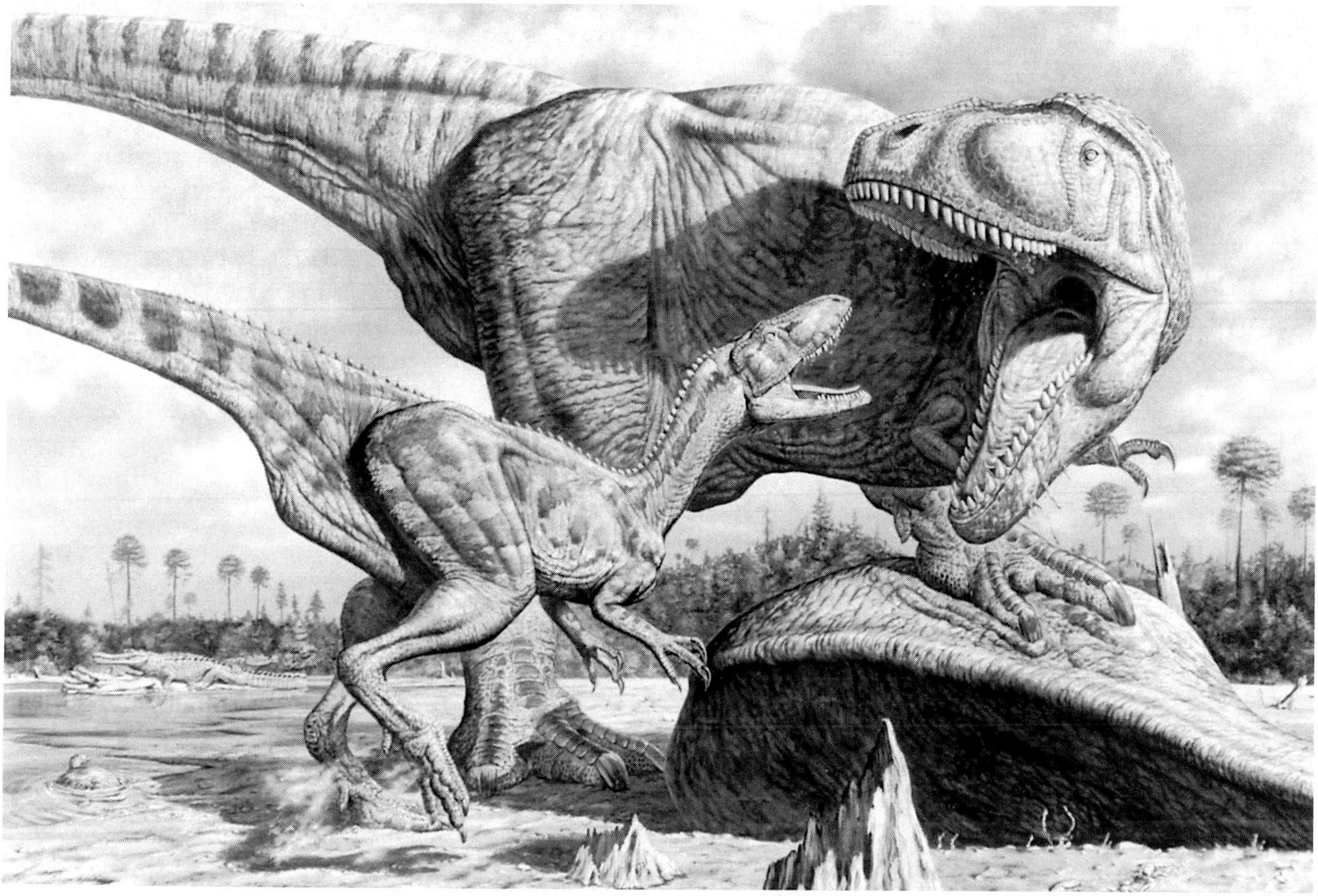

Figure 27-2: Mark Hallett's 1996 restoration of African dinosaurs, pesky *Deltadromeus* and gigantic *Carcharodontosaurus*, fighting for spoils. (Reproduced with permission of the National Geographic Society. Mark Hallett/NGS Image Collection.)

should be assembled, the miniature skeleton and reconstructed bones were delivered to a company named PAST in Drumheller for molding and casting into fiberglass.

The Navy Pier exhibition consists principally of a 36-foot-long fiberglass skeletal cast of *Suchomimus*, casts of its large thumb claw (which visitors can handle) and a videotape of the National Geographic Society documentary film *Colossal Claw*.[12] Visitors can also marvel at a mural-sized photographic enlargement of a three foot by two foot painting by Michael Skrepnick. Skrepnick restored *Suchomimus* to life in a believable, natural setting. A 1/7 scale sculptural restoration by Stephen Czerkas guarded the opposite wall. In the television documentary, Sereno, himself no stranger to the delights of sculpting dinosaurs, can be seen turning the front end of Stephen's scale armature around into a more dramatic pose suggested in Sereno's illustration and the PAST skeletal reconstruction.

As Jay Leno quipped during his November 16, 1998, monologue, "it is little wonder that a great big dinosaur would go extinct searching for fish in the middle of the Sahara Desert"! Ah, but during the mid–Cretaceous this wasn't a desert region, as indicated in Skrepnick's mural. It was a lush, tropical environment. Bipedal *Suchomimus* is thought to have dined on fish and may have even tangled with the local varieties of crocodile. The dinosaur's four-foot-long head is extremely elongated and narrow. *Suchomimus* carried an arsenal of approximately 120 teeth in its mouth. Teeth in the anterior of the jaw are sharply recurved and have a "hook-like" shape. Such a remarkable head may have been adapted for a fish-eating diet. On its thumb was a 12 inch long claw, sheathed in life and possibly of use in seizing that "catch of the day."[13] (See Figure 27-3.)

Sereno and colleagues believe that, like the elusive

Opposite: **Figure 27-1: Paul Sereno's restoration of African dinosaurs, carnivorous *Afrovenator* attacking long-necked *Jobaria*. The original restoration is owned by private collector John Lanzendorf. (Reproduced with permission.)**

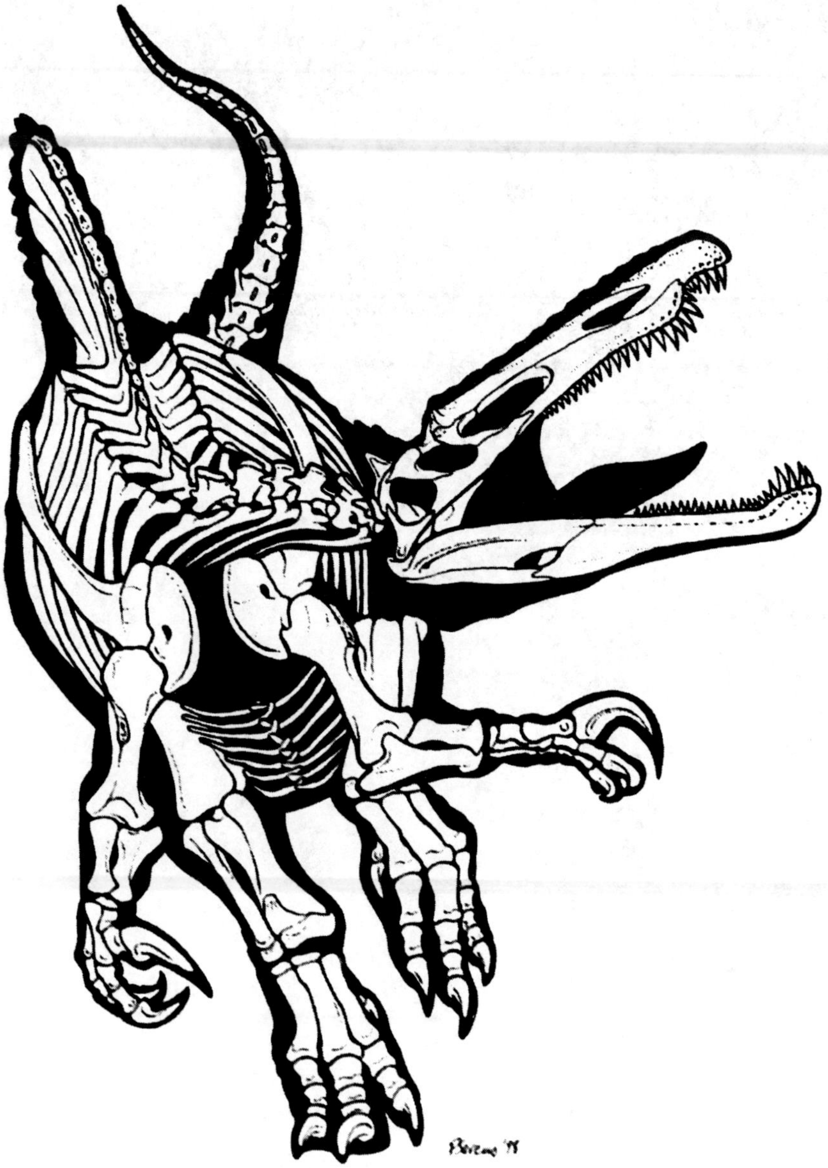

Figure 27-3: Paul Sereno's skeletal *Suchomimus* reconstruction. The original paleoart is owned by private collector John Lanzendorf. (Reproduced with permission.)

Late Cretaceous African *Spinosaurus*, *Suchomimus'* enlarged dorsal and sacral neural spines supported a sail-like crest. Only months before, paleontologist Dr. Jack Bowman Bailey of Western Illinois University published the first formal interpretation and restorations of *Spinosaurus* as a humpbacked or "buffalo-backed" reptile.[14] Bailey reasoned that fatty humps would have a thermoregulatory function in stabilizing body temperature. Camel-like humps would serve as *gigantotherm heat shields*, both preventing overheating of the animals in excessive heat and also providing storage of metabolic fuel for long migrations, or when food proved scarce. Heat absorbing sails would have only led to overheating and death.

Restorations of humped or buffalo-back dinosaurs look heftier than do restorations of these same genera as spined dinosaur, making them arguably less attractive (to humans, anyway). However, Bailey and artist Bob Johnson doubt that "...Rubenesque dinosaurs will become the norm.... [We] doubt that any single image can realistically be applied to dinosaurs; we certainly want to avoid stereotypes...." Bailey adds, "Do I think Bob's illustrations will become trend-setters? I believe its still too early to judge."[15] Discovery of *Suchomimus* has enlightened our understanding of that infamously elusive dinosaur, *Spinosaurus*. Borrowing from its geologically older evolutionary cousin, and with less speculation than before, *Spinosaurus* now can be confidently restored with its monstrous thumb claw added. (See Figure 36-3.) Instead of a tyrannosaurid head, paleoartist William Stout's new millennium *Spinosaurus* is depicted with a skull shaped for spear-fishing and gavial croc-type jaw morphology.[16]

Paleoartist Gregory Paul doubts predatory dinosaurs would have grown humps because they "...tend to keep their forms as light as possible."[17] Artistic paleontologist Kenneth Carpenter of the Denver Museum of Nature and Science claims *Suchomimus* had no hump at all, and that *Spinosaurus* used its sail for display purposes. "If the sail of *Spinosaurus* was for thermoregulation, then *Suchomimus* would have died from the heat." This was because *Suchomimus* wouldn't have been able to radiate its excess body heat because, according to Carpenter, it had no sail.[18]

However, Sereno evidently remained skeptical over the proposition that spinosaurs could have survived in a hot paleoenvironment without a heat-dissipating sail. Consequently, Skrepnick's and Czerkas' life restorations of *Suchomimus* suggest a sexier "finback" design. It's a wonder a spinosaurid has yet to appear in a Hollywood dino-monster movie![19]

Suchomimus lived about 100 million years after the extinction of a North American phytosaur, *Rutiodon*. Disregarding the skull openings, the shape of *Rutiodon*'s head is effectively the same as *Suchomimus*. In fact, they are *astonishingly* similar. It is uncanny, considering they are entirely distinct species separated in time by so many millions of years![20]

A year later, another impressive display featuring

Sereno's African dinosaurs invaded Navy Pier's Crystal Gardens solarium. While many smokers would still walk a mile for a Camel, Paul Sereno will run 26 miles for a dinosaur (something Osborn never would have done!). For Sereno participated in the 1999 Chicago Marathon to help raise funding for this (free) exhibit. Here, during a six week period, over 150,000 visitors were treated to the remarkable sight of "Dinosaur Giants," featuring his prized seventy-foot-long *Jobaria tiguidensis*.[21]

Jobaria, known from 95 percent complete composited remains, was an Early Cretaceous sauropod, "long neck," "vegesaurus" type of dinosaur that lived in what is now Niger 135 million years ago. Although extinct, *Jobaria* may be thought of as a "living fossil." This is because it had a number of distinguishing ancestral characteristics, more recognizable in sauropods that lived some 30 million years earlier during the Jurassic period than in contemporaneous Cretaceous sauropods seemingly more derived or "advanced" in their skeletal structures.[22]

Sereno and company believe the lineage represented by *Jobaria* simply evolved more slowly than other dinosaur genera. In other words, this genus' evolutionary design must have been extremely well engineered in order for "Jobarian" traits to have persisted so long, geologically. So much for the notion of ponderous and evolutionarily backward sauropods ultimately destined for extinction because of poor adaptational design. Quite conversely, these guys were among the best adapted terrestrial animals of all time. In evolution, "stasis" translates as "good."

Sereno published a flesh-outlined skeletal reconstruction of *Jobaria* in his November 1999 *Science* paper, which was used as a basis for several restorations and skeletal reconstructions in the *Dinosaur Giants* exhibit. The Crystal Gardens exhibit really brought them back to life in all their glory.

A tableau from deep time, a single moment in prehistory, has been reconstructed and restored in grand style. One doesn't even notice the dinosaurs at first. The spacious, glass-lined, biosphere-like structure is a Crystal Palace–type haven in a bustling metropolis. We experience the sound of calling birds and, a sight for "*saur*" eyes, ivy-covered palms and ferns. Natural and accented light filters playfully from above. Musically choreographed programmed fountains playfully fire darts of water across our path, passing harmlessly overhead without spraying or splashing visitors. The recorded music is of sufficient taste to calm even the savage beast. And then emerges the Beast himself, in the form of the "African Hunter"—*Afrovenator abakensis*!

Afrovenator's 27-foot-long cast skeleton "menaces" two cast *Jobaria* skeletons, displayed a few feet over across the tiled walkway. One *Jobaria* is 60 feet long, and the other, a juvenile, is a "mere" 40 footer. The high-vaulted ceiling is a necessity, for the 60-foot-long skeleton is rearing to a dizzying height, thwarting *Afrovenator*'s bloody "attack." The *Jobaria* skeletons are artistically positioned in life-like stances. There is an implied analogy to a herd of elephants.

You can almost feel the tremor of pounding, tree-trunk thick forelimbs, hear the African Hunter's strident war cry and the sauropod parent's defiant bellow. Imagine the massive forms struggling against death's inevitability. Paleoartist Mark Hallett has painted the dramatic scene. His restoration hangs on a nearby wall. The caption reads, "Driven by hunger, 135 million years ago, a pair of *Afrovenator* emerge from the underbrush to attack a young *Jobaria*. The scene is based in part on a graveyard site that preserved skeletons of juvenile and adult *Jobaria* as well as bones and teeth of *Afrovenator*. Jagged grooves on the bones of the ribs of the juveniles offer telling clues of a lethal encounter." The fossils reveal the gory outcome played out on the African continent long before the days of roving lions and wildebeest.

Michael Skrepnick's painting shows two *Jobaria* standing bipedally erect, dueling with the large thumb claws on their forelimbs. One has slashed its rival. Skrepnick's painting vividly depicts the conifer forests which formerly were situated along the bank of the broad river where the giants lived and died (in an ancient flash flood). The fact that juvenile and adult bones were found together suggests *Jobaria* was a social creature, existing in herds, as one caption reads, "...like the large plant-eating mammals of today." Sereno's original illustrated reconstructions of *Jobaria*'s skeleton were also on view.

Also on exhibit was a bronze 1/15 scale statue of the rearing *Jobaria*, sculpted by Michael Trcic. This five-foot-long bronze is posed bipedally erect, like the 60-foot-long skeleton. (See Figure 35-4.) A caption states, "...[The] rearing posture is controversial. We think Jobaria could rear because its heavy muscular tail shifts center of gravity [mass] closer to its hind legs. By bending slightly at the knees, it could balance its mass on its hind legs. By rearing, Jobaria could feed high in trees or challenge rivals with its thumb claws." Paul Sereno has illustrated this argument on signage adjacent to Trcic's bronze statue.[23] In deciding the pose for the 60-foot-long *Jobaria* skeleton, Sereno, in fact, made observations of a trained elephant rearing on its haunches. Sereno's illustrated reconstruction of *Jobaria*'s skeleton is also on display.

Another magnificent artist, Gary Staab, has produced a life-size restoration of a portion of *Jobaria*'s neck and head. (See figure 27-4.) Whereas most sauropods had more neck vertebrae, *Jobaria* only had 12 (a "primitive" or ancestral condition). Staab has sculpted the bones, air sacs

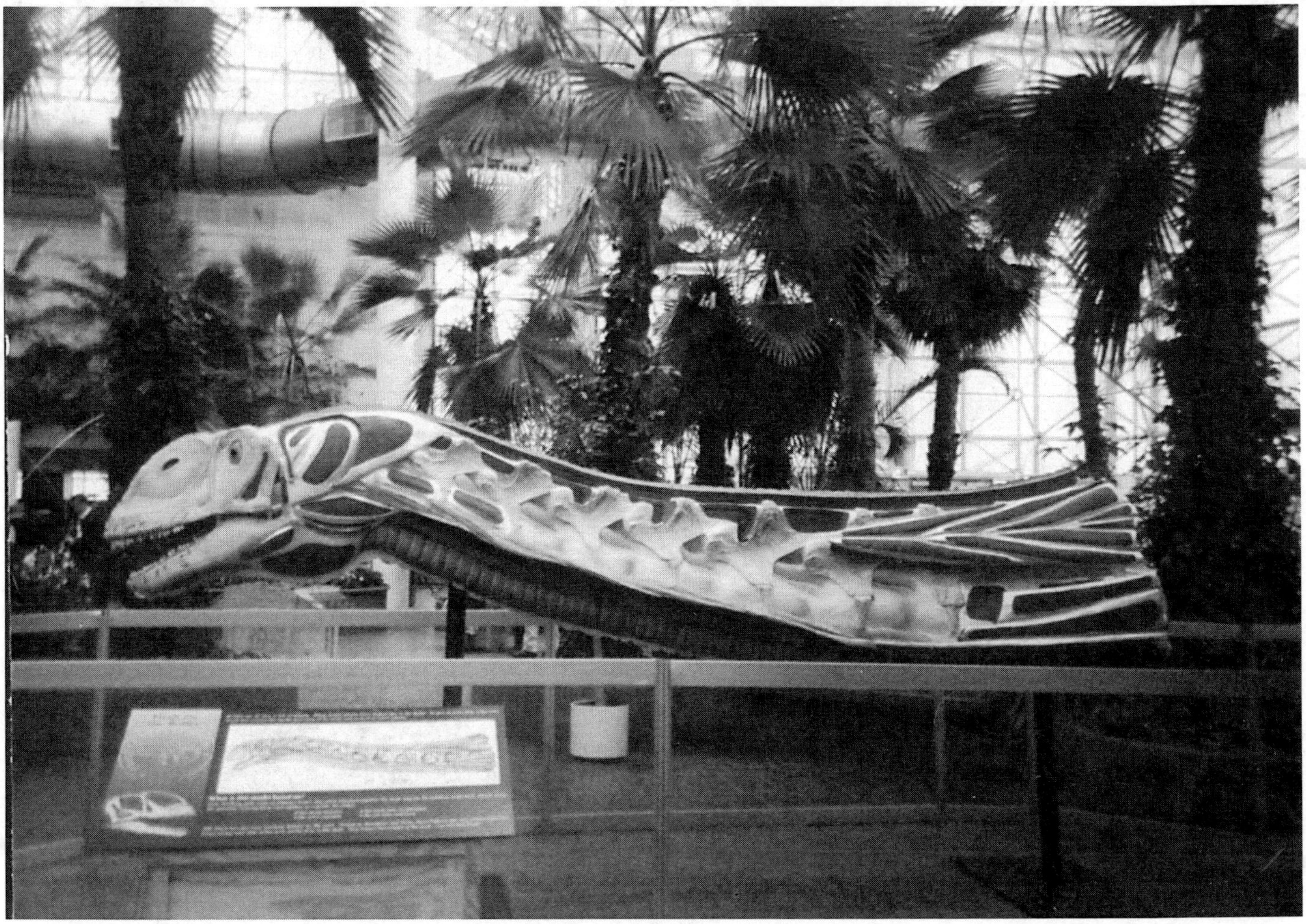

Figure 27-4: Gary Staab's half restoration showing the "innards" of *Jobaria*'s neck. (Photograph by Allen A. Debus, 2000.)

and muscles on the left side, and on the right we see the "living" animal. A nearby caption clarifies that, "Air sacs are very unusual features in a neck. Found in dinosaurs and birds, these balloon-like sacs extend from the lungs along each side of the neck all the way to the skull to reduce the weight of the neck."

What is the significance of these discoveries? Sereno is falsifying steadfast notions that Gondwanan fauna was inherently similar throughout the late Mesozoic of South America and Africa. African Cretaceous dinosaurs bear closer similarities to those of the northern continent, Laurasia, than to South America, although with differences. For instance, *Jobaria*'s skeletal design is more closely related to North American Late Jurassic camarasaurs than of Gondwanan titanosaurs. Prominent Jurassic Laurasian lineages, such as allosaurs and sauropods, suffering waning diversity during the Cretaceous persisted in Africa. In fact, they evolved to new heights during the Cretaceous, as evident in *Jobaria*, *Afrovenator* and *Carcharodontosaurus*.[24] The observed evolutionary patterns may have been influenced by the presence of northern land corridors leading into Africa even as South America drifted to the west.

Descendants of theropod fauna thought to be indigenous to North America evolved into even more remarkable predators on the African continent. Prior to geographical isolation and evolutionary radiation, ancestors of the African *Carcharodontosaurus*, the South American *Giganotosaurus*, the new South American carcharodontosaurid and North American *Acrocanthosaurus* may have ranged broadly across the globe during the Late Jurassic or Early Cretaceous through as yet unidentified land corridors. In the case of *Deltadromeus*, there is indication that Late Jurassic coelurosaurs must have existed on southern continents. By the early Cretaceous, with the land bridge severed, distinctive isolated populations emerged. Sereno concluded, "As a consequence, there is no recognizable ... unity among dinosaur groups that persisted on Gondwanan continents during the Cretaceous. Rather, these ... dinosaur groups experienced different fates on each southern continent as isolation increased during the Cretaceous."[25]

Making great discoveries is a human preoccupation. But paleontologists become most godlike in their production of cladograms, reconstructed evolutionary trees showing the diversification and origination of species through inconceivably long periods of time. While such diagrams can oftentimes seem stuffy and visually unappealing, Sereno's style of presentation incorporates paleoimagery to the fullest extent allowed by scientific convention. For instance, silhouettes and reconstructions of Mesozoic birds and dinosaurs highlighted evolutionary trees illustrating Sereno's acclaimed article, "The Evolution of Dinosaurs," published in *Science*, June 25, 1999.[26]

Sereno's displays have steadily evolved into more visually exciting, elaborate affairs, rivaling what Osborn was capable of engineering at the American Museum of Natural History a century ago at far greater expense. His use of paleoimagery enhances scientific reports to a degree further than most of his contemporaries would dare to go, tempting more prospective science *readers*. And there will be more discoveries to look forward to. On the heels of massive *Jobaria* and *Nigersaurus*, a geologically younger, 110-million-year-old, African sauropod, Sereno will soon be announcing discovery of a small armored African ornithischian dinosaur.[27]

Sereno is a trendy guy at the forefront of a trendy science. A century from now who will be regarded as the "modern Sereno?"

Chapter 28

Hypsilophodon—a "Super" Dinosaur

Faster than a speeding bullet! Able to leap tall branches in a single bound! Is it half-bird, half-dinosaur? No—it's *Hypsilophodon*, the Super Dinosaur! Judged by the qualities which humans find appealing in dinosaurs, such as large size, fierce disposition, or unusual associated paraphernalia (e.g. plates, spikes, horns), *Hypsilophodon* was not an imposing creature. Yet, for decades, this little fellow was a source of intrigue as scientists deliberated evolutionary connections between dinosaurs and birds.

Hypsilophodon was "recruited" into a nineteenth century debate waged chiefly between Richard Owen, who believed dinosaurs possessed certain mammalian qualities, and Thomas H. Huxley, who thought dinosaurs were more closely related to birds. Owen had described remains discovered in 1849 on the coast of the Isle of Wight as a juvenile *Iguanodon mantelli*. Years later, however, referring to specimens discovered in the same thin sandstone stratum by the Reverend William Fox in 1868, Huxley demonstrated that Owen had been careless in his 1855 description. Arguing that, "...the teeth of this reptile leave no doubt as to its distinctiveness from Iguanodon...," Huxley named the new dinosaur *Hypsilophodon foxii* ("Fox's high-ridged tooth") in 1869.[1]

Hypsilophodon was a seven-foot-long herbivorous creature. The anterior margin of the premaxillary bone was roughened and beak-like, and the premaxillary was outfitted with several teeth. Huxley identified striking similarities between the pelvic structures of birds and *Hypsilophodon*. He would use such skeletal features in classifying dinosaurs as "Ornithoscelida" ("bird leg"). Huxley also believed that the dinosaurs' hind limbs "wonderfully approached" those of birds, praising E. D. Cope for independently reaching similar conclusions. Huxley even proposed that dinosaurs were warm-blooded.[2]

In 1882 John Hulke noted that merely on the basis of its additional (fourth) functional toe, *Hypsilophodon* should be distinguishable from the genus *Iguanodon*. Fortified by Caleb Barlow's skeletal reconstruction of *Hypsilophodon* as a quadrupedal creature, Hulke asserted that "*Hypsilophodon* was adapted to climbing upon rocks and trees."[3] Joseph Smit's 1910 restoration blended this possibility with Marsh's 1896 description of *Hypsilophodon* as a bipedal dinosaur.[4] (See Figure 28-1.)

In 1905 Baron Nopsca noted the series of ossified tendons extending along, and increasing in number toward, the end of *Hypsilophodon*'s tail. He also remarked on the bird-like character of the base of the skull. Nopsca also detected what he interpreted as "thin and flat, feebly punctured bony plates," similar to the dermal armor of stegosaurs.[5]

During the early twentieth century, two artistically gifted scientists, Austrian paleobiologist Othenio Abel and Danish ornithologist Gerhard Heilmann, deepened our understanding of the evolutionary ties between dinosaurs and birds. Were traits of the most "primitive" (i.e. ancestral) dinosaurs still preserved in *Hypsilophodon*?

Abel thought that *Hypsilophodon*'s toes could bend backward in apposition to the hallux, an indication that the herbivore lived in trees and was capable of grasping branches like a bird. Abel concluded that the ancestral condition of all dinosaurs was preserved in *Hypsilophodon*—the earliest dinosaurs had evolved as tree dwellers. By the Lower Cretaceous, although *Hypsilophodon* still perched in trees, most other dinosaurs had lost this ability and lived exclusively on the ground.[6]

Initially Heilmann generally concurred with Abel. He concluded, however, that *Hypsilophodon*'s foot was adapted for a terrestrial existence and might have had a *secondary* arboreal function. Abel praised Heilmann's 1916 *Hypsilophodon* restoration in which the dinosaur was

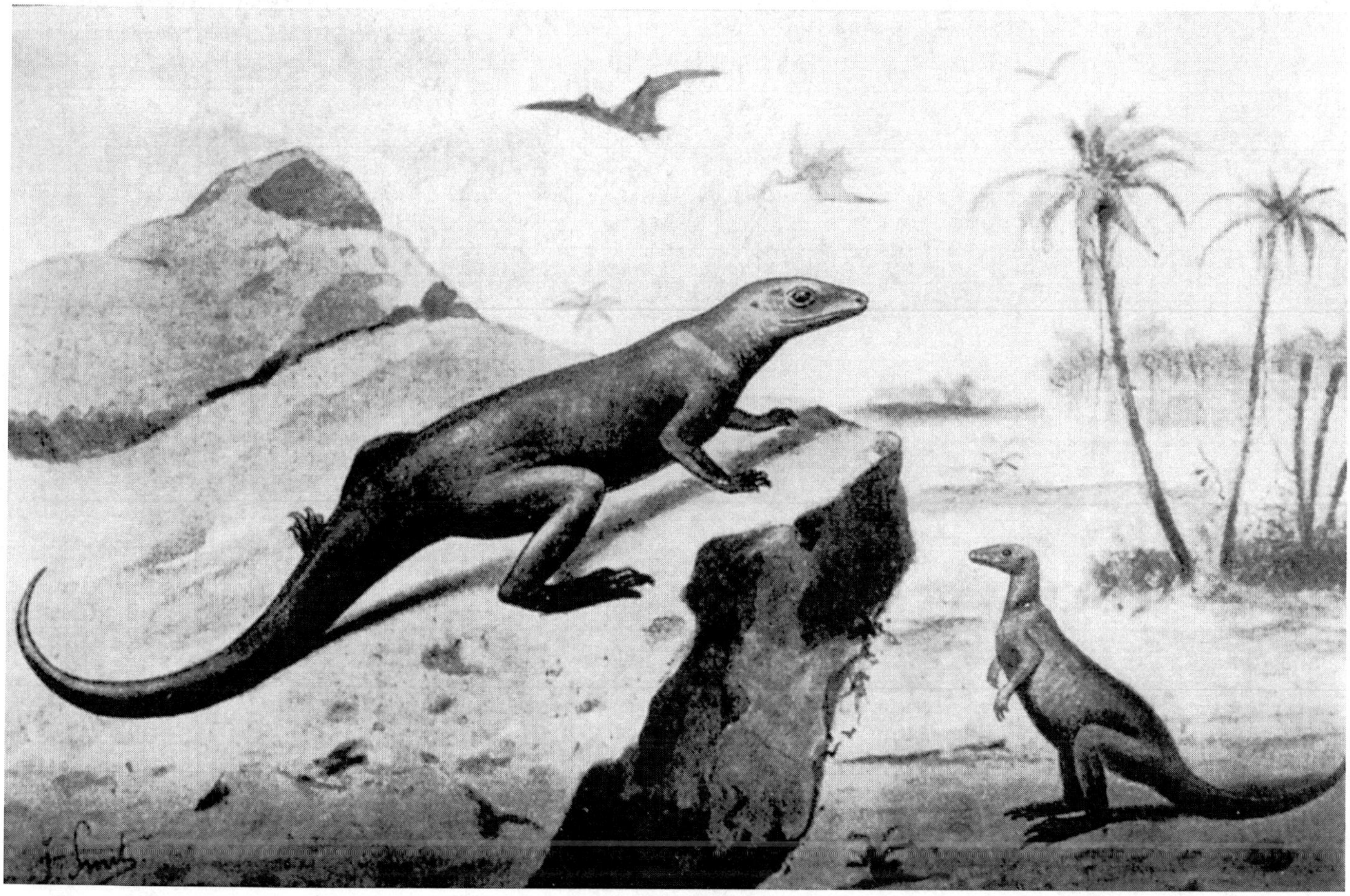

Figure 28-1: Joseph Smit's 1910 *Hypsilophodon* restoration showing both quadrupedal and bipedal views. From H. N. Hutchinson's *Extinct Monsters and Creatures of Other Days*, New York: D. Appleton and Company, 1917 ed. (plate opposite page 173).

depicted clinging apelike to a tree branch. But because Abel discerned important similarities between the skeletal morphologies of *Hypsilophodon* and modern tree kangaroos (*Dendrolagus*), his 1922 *Hypsilophodon* life restoration was patterned after a color plate of Bennett's tree kangaroo.[7] (See Figure 28-2.)

In his *The Origin of Birds* (1926), Heilmann no longer believed that *Hypsilophodon* was arboreal. Heilmann felt that Abel had, perhaps unwittingly, distorted the configuration of *Hypsilophodon*'s toes. Now referring to Hulke for the proper orientation of the foot bones, Heilmann concluded that *Hypsilophodon* was incapable of grasping branches in a bird-like manner. Citing Louis Dollo's researches, Heilmann suggested *Hypsilophodon* might have been quadrupedal (although presumably not precluded from occasionally climbing a tree, as shown in his 1916 restoration). Moreover, he decided that the dermal armor noted previously by Nopsca would have been uncharacteristic of an arboreal creature. In Heilmann's opinion, ornithischian dinosaurs weren't arboreal.[8]

Heilmann's compelling arguments were neglected for nearly half a century, as William E. Swinton, then of the British Museum of Natural History, popularized the arboreal theory through the artistry of Vernon Edwards (1936), and Neave Parker's 1950s *Hypsilophodon* restorations. Swinton proposed that when in danger, *Hypsilophodon* escaped into the trees where its rigid tail provided balance as the dinosaur moved through the branches. For a slow-footed dinosaur inadequately protected by undeveloped dermal armor, tree limbs were a safe haven.[9]

Then, in the early 1970s, Peter M. Galton disproved the arboreal theory and established that hypsilophodontids were fast-moving bipedal dinosaurs.[10] He discounted the evidence of dermal armor and suggested that *Hypsilophodon*'s rigid tail actually served as a "dynamic counterbalance" while the animal scampered over the ground. A restoration by Robert Bakker, published in Galton's 1971 *Lethaia* article "Mode of Life of Hypsilophodon," reinforced the new view.[11] (See Figure 28-3.)

Bakker's bounding hypsilophodonts presaged new discoveries, further validating Galton's analyses. In 1978 John Horner's field crew discovered nesting sites at a

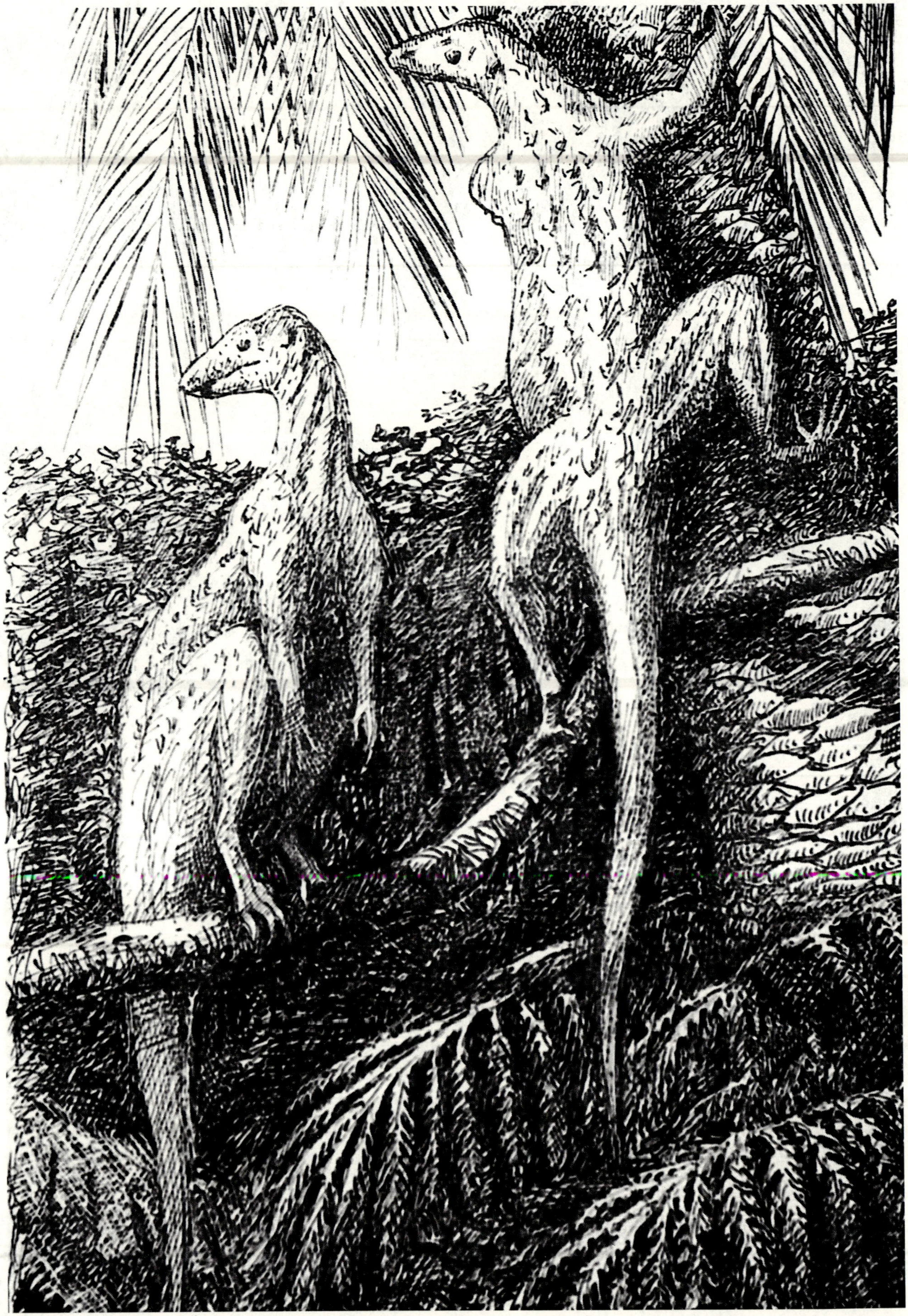

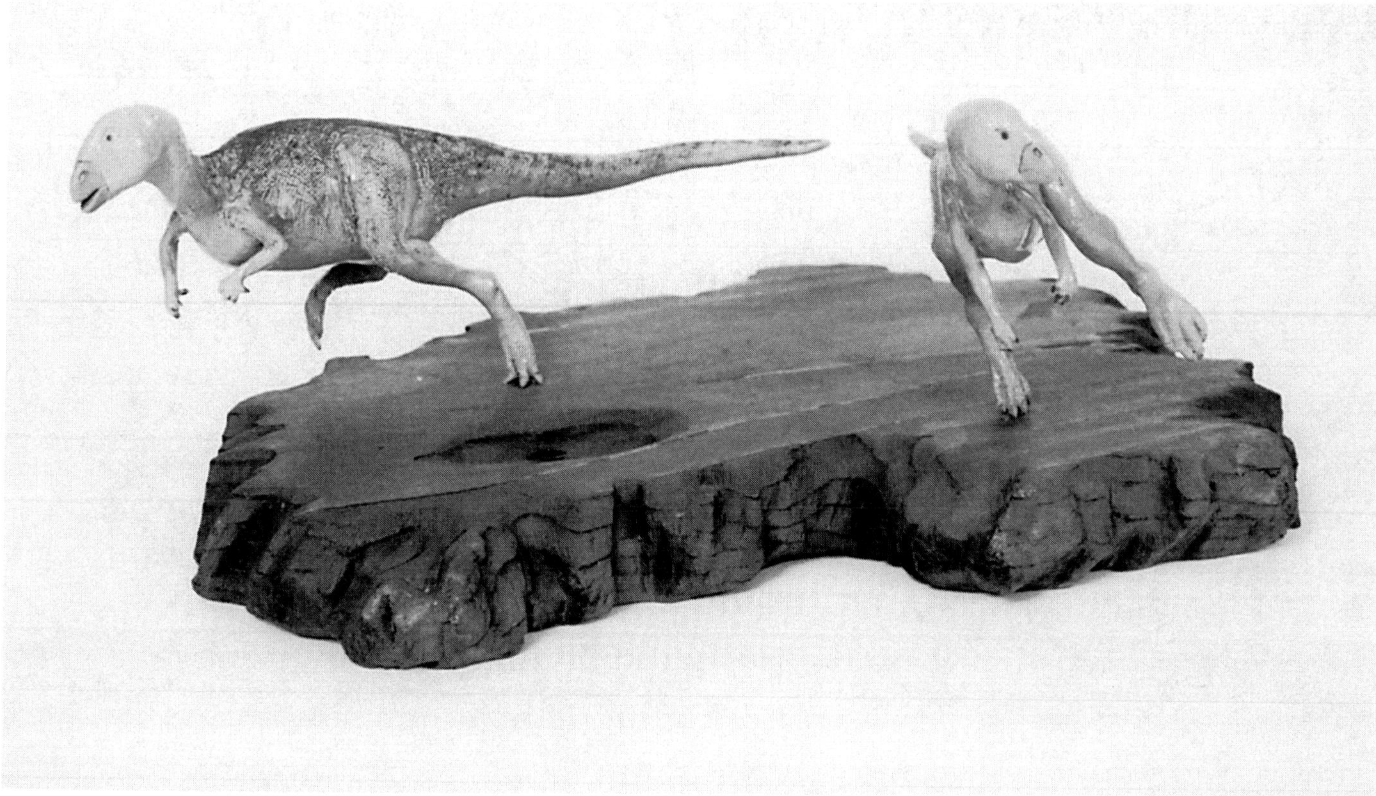

Figure 28-3: Sculpture by Allen A. Debus (1994) showing sprinting hypsilophodonts—after Robert Bakker's 1971 restoration.

fossiliferous Montana locality dubbed "Egg Mountain." These nests became associated with a hypsilophodontid named *Orodromeus* ("mountain runner"). Paintings of the gregarious nesting dinosaurs were completed by artist Douglas Henderson to illustrate Horner's descriptions of the Willow Creek Anticline discoveries.

In 1987 Horner discussed his association with Henderson in the restoration of the Egg Mountain nesting area: "In 1982, I wrote a short article about the discoveries at Willow Creek Anticline ... in which artist Doug Henderson attempted to reconstruct particular habitats, including Egg Mountain. The information collected over the previous three summers was used for a restoration. At that time there were several aspects about the nesting site that were unknown, and as nearly always the case when information is lacking, the interpretation and restoration were somewhat incorrect. The island was depicted as being rather large and covered with large trees. And the hypsilophodonts were restored as much too lean and described as carnivores."[12] (See Figure 28-4.)

Discovery of a hypsilophodont embryo, described in 1988, seemed to seal Horner's impressions of what life was like in an *Orodromeus* nesting ground. Then, in 1996, Horner and David Weishampel retracted their conclusions concerning the hypsilophodont embryo, which turned out be a theropod embryo of the genus *Troodon* instead.[13] In late 2000 Henderson reflected on his 1982 painting: "I should note that the *Orodromeus* image was done in 1982, well before accurate details of the nesting site were known. As a consequence, the only vestige is as an illustration of speculative notions as they stood in 1982. The animal in the image is a hypsilophodont before being given the name *Orodromeus*. The anatomy is quite wrong. In time, it was determined that the nesting site was actually that of *Troodon*, not *Orodromeus*."[14]

Recent episodes of our understanding of the hypsilophodontids began in 1987, with discoveries of *Leaellynasaura* and *Atlascopcosaurus*. These Australian dinosaurs, described by Thomas H. Rich and Patricia Vickers-Rich in 1989, were collected near Victoria, which, during the Early Cretaceous (106 million years ago), was only 15 degrees from the South Pole. Rich and Vickers-Rich noted that, "...neither cold nor dark of night stopped the *hypsilophodonts*." Hypsilophodonts seem to have thrived in a polar environment.[15] Super dinosaurs after all? In order to have survived the freezing temperatures to which they may

Opposite: Figure 28-2: Othenio Abel's 1916 *Hypsilophodon* restoration. (See text for details.)

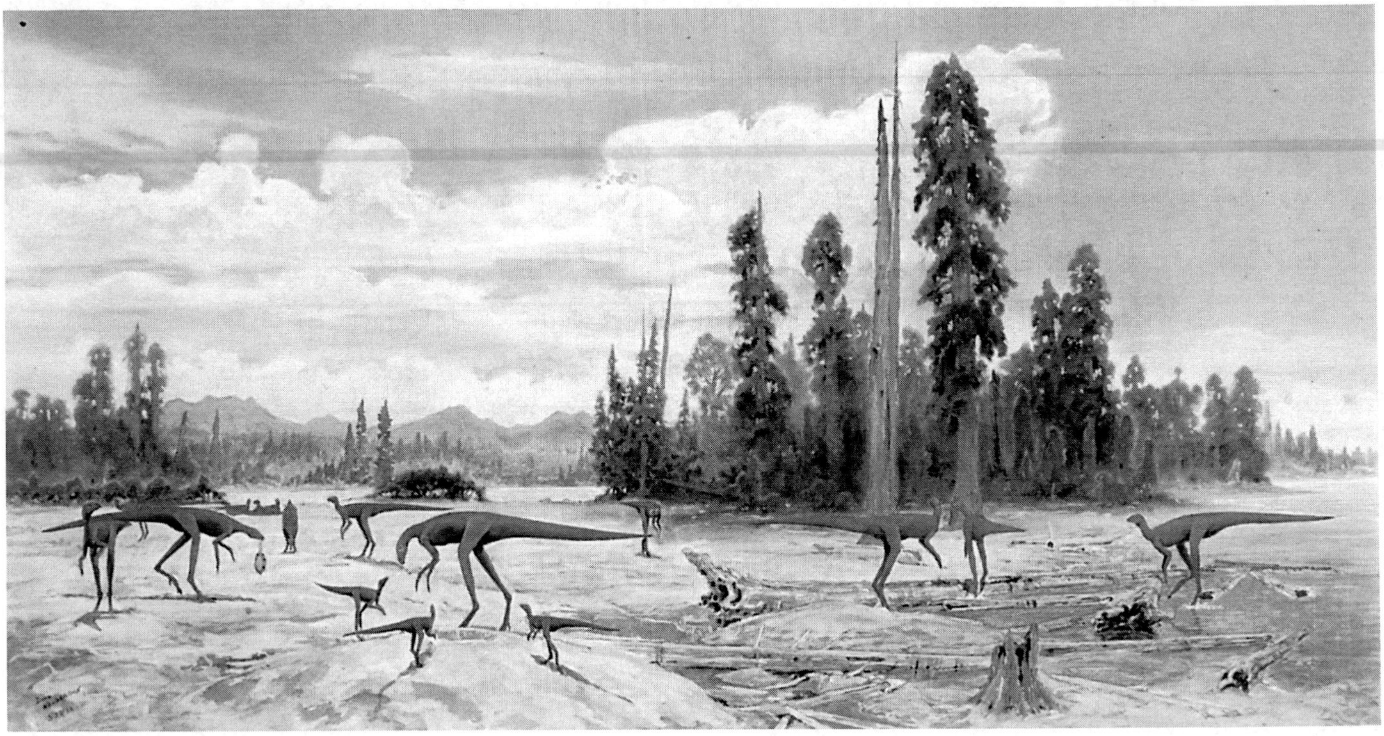

Figure 28-4: Douglas Henderson's 1982 restoration of *Orodromeus*. See text for details. (Reproduced with Mr. Henderson's permission.)

have been subjected, the Australian hypsilophodontids may have evolved a warm-blooded physiology. Hibernation during the colder periods is also a possibility.

Hypsilophodontids may have thrived in southern polar twilight, yet they certainly weren't "heartless" creatures of the night, as borne out by an early 2000 announcement—the discovery of a fossilized heart. The heart was preserved as an iron concretion found in the thoracic cavity of a hypsilophodontid, nine-foot-long *Thescelosaurus*, that lived in what is now Montana 66 million years ago. While disagreement lingers over their conclusions, fossil heart researchers believe *Thescelosaurus* had a four chambered heart, suggestive of an "intermediate-to-high," bird-like metabolism.[16]

Huxley suggested a dinosaurian evolutionary heritage for birds in referring to *Hypsilophodon*'s skeletal anatomy. Heilmann later denied that birds could have been derived from ornithischian dinosaur stock. Some lines of evidence suggest that "birds" evolved more than once, or that *Archaeopteryx* itself may not have been the direct ancestor of all modern birds. As with many speculations involving bird-hipped *Hypsilophodon*, these suggestions "fly" in the face of convention.

Paleoartists have already begun treating us to their visions of feathered hypsilophodontids.[17] Ken Carpenter has stated, however, "It is difficult to know ... how far we can go with the bird analogy in illustrations of dinosaurs."[18] But Man will always go where dinosaurs lead.

Chapter 29

High Stakes Dinosaur— Las Vegas' Robotosaurs

Life's a gamble to be sure, but even if you don't gamble, the odds are that you'll really enjoy the dinosaurs of Las Vegas. On our recent trip to the "sin city," we encountered 17 different prehistoric animal figures situated in two locations, the Las Vegas Natural History Museum, about the only place besides bathroom stalls where we didn't spy slot machines, and the "Adventuredome" at Circus Circus Casino. The best of all are the mechanical robot creations, the liveliest dinosaur recreations known to man. Hey, these "robotosaurs" look like a million bucks, and they don't come cheap. Get *real* lucky at the craps table though, and you could afford one!

Past the north end of the Strip we find the Las Vegas Natural History Museum where several statuesque dinosaur recreations guard their turf, indoors. After moments of contemplation, subtle movements can be detected in the dinosaurs. Just our imagination? Has our presence somehow resuscitated them? A *Triceratops* head shield bows soundlessly, a tail arcs gently upward, a thick neck supporting a monstrous-looking *T. rex* head lowers imperceptibly. The great maw opens...

We become absorbed in a hominid evolution wall chart, replete with head sculptural restorations of our distant ancestors. A mirror cleverly indicates the "current state" of human affairs at the end of the evolutionary lineage. "Gee! Don't we look so much more handsome than *Australopithecus boisei*?" we find ourselves musing ... only to become startled by a *Jurassic Park*–sized *roar* coming from behind! The *T. rex* bellow knocks our socks off!

The life-sized "dinosaur" restorations are arranged in a life-through-time display directed from the right side of the entrance. Visitors pass by unmechanized Coal Age and Permian fauna, including *Eustenopteron*, *Dimetrodon* and *Edaphosaurus*. Then suddenly, after striding a few paces past a *Herrerasaurus* model, you're in the Late Jurassic period. The nicely constructed *Allosaurus* actually used to be a *T. rex*. What? No, this isn't retro-evolution. We'll get to that part later.

One Cretaceous diorama incorporates three *Troodon* banding with a possum thief, together stealing eggs from a nest in a lush forest. The adjacent walls are nicely painted to suggest contemporary paleo-landscapes. Near the *Troodon* there is a *Parasaurolophus*, shown rushing into a scene of two primitive mammals (sculpted) raiding her nest. The only dinosaurs which appeared to be mechanized the day of our visit were Cretaceous varieties: *T. rex*, *Triceratops*, *Deinonychus* (which occupies center stage in the middle of the room) and the armored Mongolian dinosaur *Talarurus*. In the "Young Scientist Center," children can operate a robotic baby dinosaur.

There are fossils and fossil casts sprinkled here and there, including a skull of the husky, 50-foot-long croc, *Deinosuchus*, and the most complete Mastodont head and tusks ever found. An impressively mounted *Smilodon* stands near the entrance to the Young Scientist Center. Yet, undoubtedly, the Las Vegas Natural History Museum's unheralded robotosaurs steal the show.

Mechanized dinosaurs have been show stoppers ever since their invention 75 years ago. The very first "robotosaur," a 48-foot, 9-inch-long "brontosaur," was built in New York City during the early 1920s by George Harold Messmore and Joseph Damon.[1] Their dinosaur was nine feet three inches high and weighed 4,000 pounds. During design phases, Messmore and Damon consulted with American

Museum of Natural History paleontologists, lending an air of authenticity to its construction. (See Figure 29-1.)

Years later, Messmore's son, Francis B. Messmore, recollected that, "The dinosaur's body was built over a rattan framework with a spring construction for flexibility. Inside the dinosaur stood a table which carried 10 motors and 10 gear boxes to control the various motions. It had movement in the tail from the base and then secondary movement in the middle of the tail. The same applied to the neck. In addition, the eyes, mouth and nostrils (which dilated) had movement. The ears were depressed so there was no possible movement there. To coin a current phrase, the Messmore and Damon dinosaur was state-of-the-art. When the dinosaur opened and closed its mouth, its tongue mounted on a spring, lolled there. It moved its neck, its hips rose up and down, it had breathing action and could raise and lower its shoulders. When it raised and lowered its hips and shoulders, it caused the four feet to rise off the ground slightly."[2]

People naturally wondered how Messmore and Damon's brontosaur had been made. Until their patent was approved, however, the inventors remained cagey about its design. To beguile copycat inventors, a number of popular articles appeared showing (incorrectly) how the brontosaur's innards operated. For example, the October 1932 issue of *Popular Science Monthly* included a diagram suggesting that the mechanical movements were manually operated by technicians from inside the brontosaur body.[3] When the U.S. Patent Office issued the patent to Messmore and Damon in early 1933, their brontosaur had already forged early phases in its illustrious history.[4]

In the 1920s Messmore and Damon's brontosaur was featured in a prehistoric act unveiled at the Roxy playhouse in New York City, where it "walked" on stage and "acted" with a sacrificial dancer who was placed in the dinosaur's mouth. The dinosaur raised the actress up into the air and later lowered her to the stage floor. Thereafter, the brontosaur toured many Canadian and U.S. cities. However, its greatest billing proved to be in *The World a Million Years Ago* exhibition opening at the Chicago World's Fair in 1933, where it was complemented by a cast of other life-sized animatronic prehistoric animals created by Messmore and Damon for a spectacular life-through-time display. Although the Sinclair Oil Company also had mechanized dinosaurs displayed at their Fair exhibition, Messmore and Damon's patented mechanization was far more sophisticated.

In later years *The World a Million Years Ago* animatronic figures toured the country, eventually reappearing at the New York World's Fair in 1939/1940. Francis B. Messmore recalled that during World War II, "President Roosevelt asked Mayor LaGuardia of New York City to put on a great parade showing New York's support of the war effort. And so a *World a Million Years Ago* dinosaur went to war on the

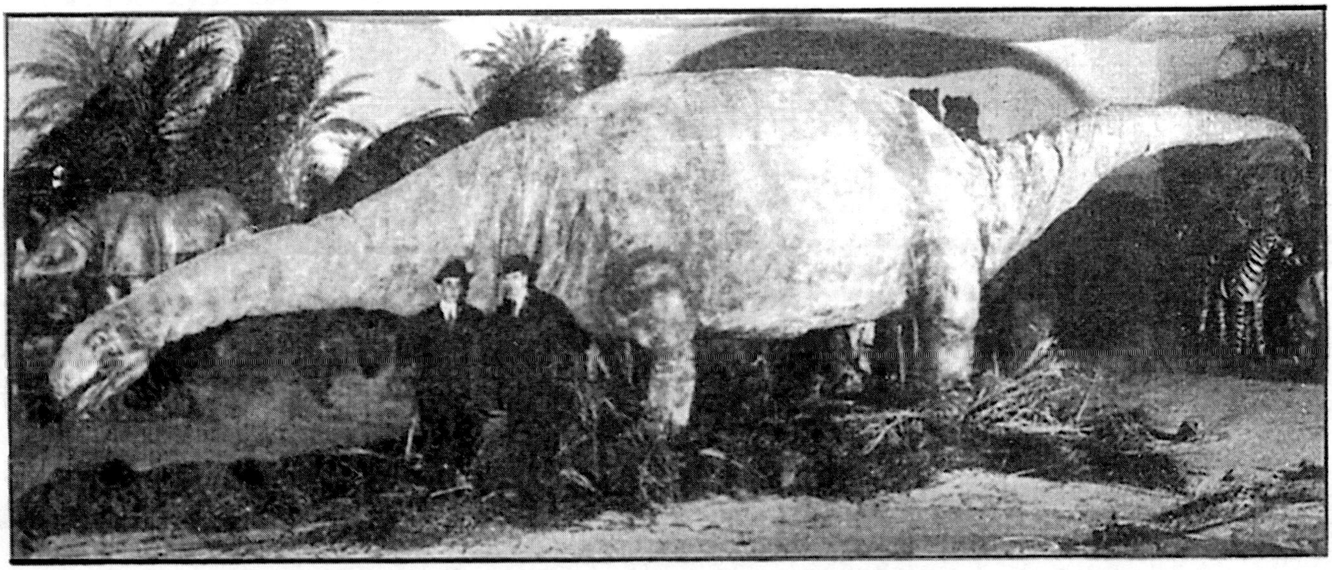

HE CAN SHAKE A HOOF AND WAG HIS TAIL

Mr. Dinosaur carries ten motors and speed-reducers that make him a lively reincarnation.

Figure 29-1: Messmore and Damon's mechanical "brontosaur," as it appeared in 1925. From *The Literary Digest* (November 1925).

lead float, called the Axis War Monster." A rare photograph shows Messmore and Damon's brontosaur leading the 1942 parade, all bedecked in a suit of spikes and armor resembling a reptilian tank with a Nazi swastika raised over its hips. A German soldier rode the formidable apparition.[5]

Messmore and Damon's brontosaur and *Dimetrodon* starred in a 1951 television episode of *Tom Corbett, Space Cadet*, when they were encountered by the ship's crew on a "prehistoric" planet in the Alpha Centauri system.[6] The robotosaurs were filmed unmoving in the broadcast, an effect of the spacemen's paralyzing ray guns. *The World a Million Years Ago* figures also made appearances in Japan and Paris before their retirement in the late 1970s.[7]

Not as familiar to dinophiles as Messmore and Damon's popular "robotosaurs," the Las Vegas Natural History Museum's nine-year-old dinosaur and robotosaur exhibit was created by the Boyd Foundation in memory of Sam A. Boyd, who owned several Las Vegas casinos. Mr. Boyd assisted the museum in obtaining its first animatronic dinosaur several years ago.

However, Marilyn Gillespie, the Las Vegas Natural History Museum's Director, didn't hold the same enthusiasm as we did for their dinosaur robots. Yep, seems there's an old vendetta brewing here. Apparently, the builders of several robots made the models too quickly, proceeding directly to the final assembly from preliminary designs without consulting Ms. Gillespie. One builder, a former Disney worker, evidently had never done an assignment like this before, and so the initial results were "off."

Eventually, one creation, their original *T. rex*, was sentenced to the guillotine (of sorts). The top half was cut off and discarded, and the lower half eventually united with a head and torso that looks more like an authentic allosaur. The resulting *Allosaurus*, formerly *T. rex*, was the museum's first dinosaur robot.

Yet the Museum staff felt it was essential to also display a *T. rex* "battling" the *Triceratops* robot. So another *T. rex* was commissioned after they realized its horizontally oriented vertebral column could be easily accommodated under their eleven foot ceiling. (The original "*T. rex*" was incorrectly posed in an upright position with its tail on the ground.)

Ms. Gillespie dislikes their version of the armored ankylosaur robot *Talarurus*. The head of the spiky brute is situated over its nestlings while its tail moves gently to and fro behind, really a friendly "wag" instead of a threatening war club swing. So what's the problem with this fellow, who looks realistic enough? For one thing, he is a bit too spiky-looking. Secondly, the motors for this and other models prohibit complex movements as we have seen in the Kokoro and Dinamation robotic creations. Furthermore, *Talarurus*' head apparently was fashioned from a cow prop, without museum approval. So Ms. Gillepsie still views *Talarurus* with a critical eye, brooding over its "bovine" genesis. Still, at a $15,000 production cost, this model is a bargain, considering the very positive influence it will have on visitors, especially children.

The museum is not publically funded, so obtaining the dollars needed to finance this dramatic exhibit was no minor task. The museum attracts sixty thousand visitors annually. The real value of the dinosaur hall, according to Ms. Gillespie, is that it is "...great for young children who need to visualize what the animals looked like."[8] Children may have difficulties comprehending real fossils or fossil casts, but, as we climbed back into our socks to pitter patter past the room's menacing predators, we understood how life-like these restorations can seem.

There was another natural history museum situated in Las Vegas, 10 miles south of the extant institution. The former institution boasted a magnificent collection of dinosaur skeletons and mechanized Dinamation dinosaurs. The museum proved immensely popular, and it *was* in a better location, but it folded due to "mismanagement" difficulties. (They rang up a hefty $2 million debt.) The exhibits had been leased, Dinamation repossessed their robots. One was immediately junked because of its 13 year age, and others were sold because they were technologically inferior compared to their more recent successes. The skeletons, really casts made by Jim Madsen, were returned to his Utah address. So the former Las Vegas natural history museum institution went bankrupt in 1989 after only two years of operation. The paleontology exhibits were liquidated. In its wake, Marilyn and a partner convinced the City of Las Vegas to create a new corporation in 1990, from which the new museum sprang in 1991.

In a city boasting recreations of just about everything past, present and future, from Great Egyptian pyramids to "futuristic" *Star Trek* settings, it's only fitting that we find ourselves plunged into a Cretaceous period somewhere. With visions of robots ("Danger, danger, Will Robinson!") prancing through our heads, we made what seemed an incredibly long trek from our hotel to Circus Circus. Everything is built on a grand scale here. In a building as humongous as the Adventuredome, even a brachiosaur could get lost. No wonder dinosaurs are mere side attractions.

Many tourists have at least seen pictures of the Circus Circus tyrannosaur posed near the roller coaster rails. But few know much about this wonderful exhibit. It took many strides before we spotted our quarry. We were very surprised to find an entire dinosaur "fauna" lurking here.

To dispel misconceptions, the "tyrannosaur" is a *Tarbosaurus* (not a *T. rex*), the best tyrannosaur robot "alive." (See Figure 29-2.) There are also two *Protoceratops*,

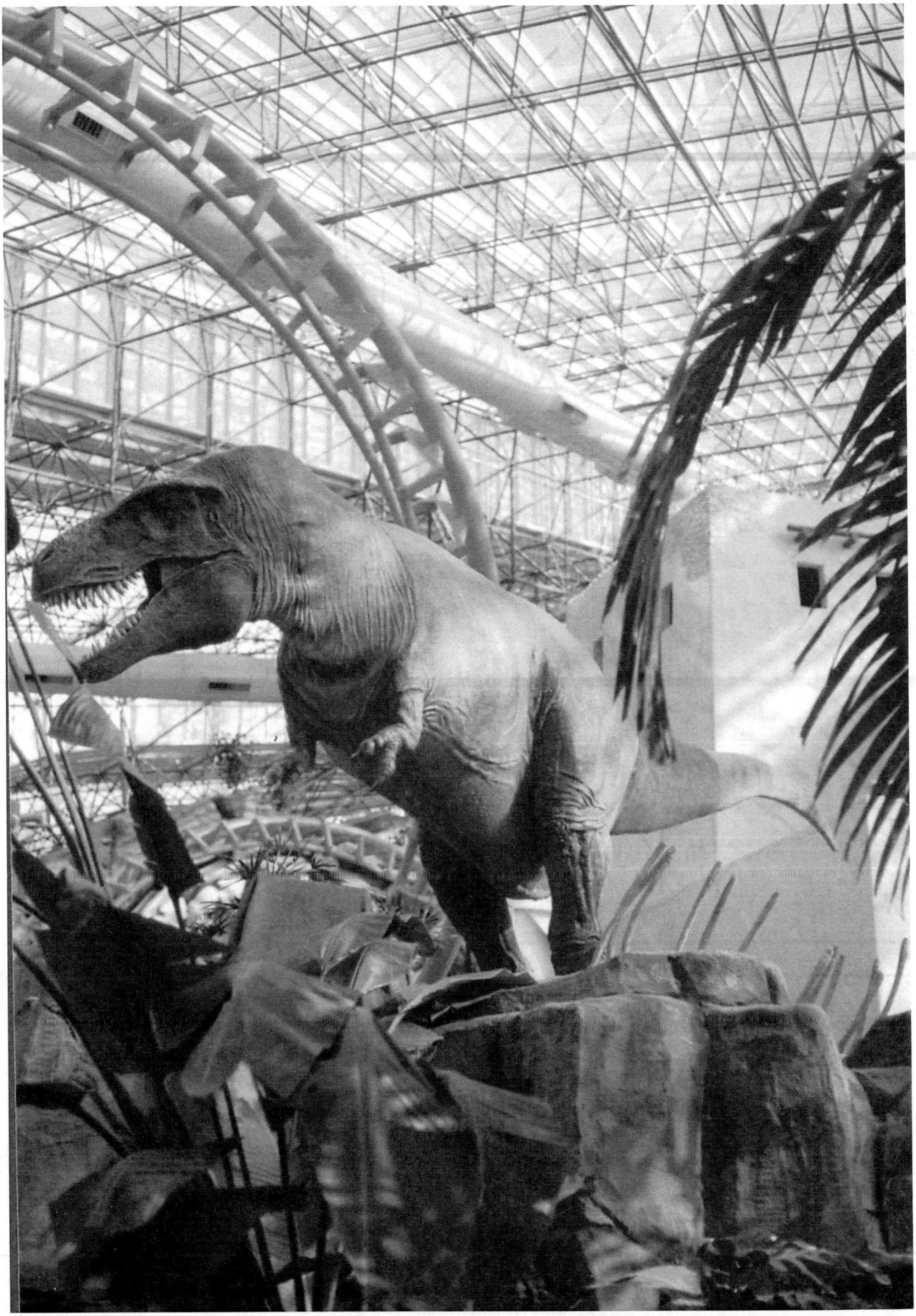

an *Oviraptor* holding an egg, two *Velociraptors* (one in repair), a *Corythosaurus* and a *Dilophosaurus* (with head movements timed to spit water at "voyagers" on the nearby water ride). We learned from Circus Circus' Quality Control Engineer Richard Larmouth, that these computerized robots were made by a company named Creative Presentations (formerly of Valencia, California, no longer in business).

Nearby, in the *Passage of Time* Center, one can view five painted murals by artist Sjana Nanni showing geological stages in the age of dinosaurs. In the final segment, a fiery asteroid hurtles earthward. The dinosaurs may have met their extinction in the *Passage of Time*, but outside, thanks to Richard, they still seem very much alive!

Richard used to be an aircraft engineer before lending his considerable talents to the mending of machines, rides and (since no one else can do it) dinosaur robots. So he runs the dinosaur repair shop and intricately performs regular maintenance on the mechanisms and foam skin. These creatures run daily (except in the evening), so every once in a while they need more than just a simple oil and grease job. Sometimes they even need "face lifts" or "surgical transplants" when certain motors become corroded or "arthritic."

The "brains" of the operation are consolidated in a room hidden from view underneath the flooring where *Tarbosaurus* twists, gyrates, snarls and bellows. Here in the delicate nerve center, dinosaur movements and earth shattering sounds are controlled by electronically digitized "cards" read by computer. Never before have we seen roboticized hand, wrist, elbow, throat and facial movements, the latter including eye blink, nostril flare and lip curl, this convincing.

All motion is regulated by digital and analog commands. If not for "Dr. Richard," the *Tarbosaurus* would have stopped functioning years ago. For even if the computer relentlessly orchestrates dinosaur motion, they won't move if the "flesh is weak." Joints and gears do deteriorate.

We saw how ugly it can get in Richard's repair shop, where he stole precious moments from his workday repairing a damaged *Velociraptor*. The foam skin had been removed, revealing the metal frame and mechanical parts, which were in places still covered in a fiberglass shell (to give the dinosaur body its shape).

We'll spare you the details of what he had done to the hydraulic and pneumatics systems, whatchamacallits and doohickeys. Of greater interest to us were Richard's artistic modifications.

The original *Velociraptor*'s head had begun to decompose, so Richard had to restore, mold and then recast a sturdier, more authentic looking head and jaws. He had made noticeable improvements in the reconstructed teeth and eyeballs.

Not only do the dinosaur moving parts wear out periodically, but visitors sometimes vandalize them. For instance, a *Protoceratops* eye was recently plucked out for an eye-popping souvenir. So Richard keeps an ample supply of his own specially made eyeballs ready to use in repair work, although now he keeps a watchful eye on his dinosaurs.

The *Tarbosaurus* cost over $3 million. Other robots were less expensive, although we could buy several large houses with the money it took to purchase them. So it is more than worthwhile having an expert engineer on call to repair the valuable attractions. According to Richard, if he hadn't insisted on the value of maintaining the dinosaurs, Circus Circus would have discarded them long ago. Richard is a dying breed of engineer; without him Circus Circus would seem to be just another hotel casino.

The dinosaurs remain "alive," yet are threatened by the tendencies of predatory Man, a species who would condemn them to the scrap heap to clear room for more rides or slot machines. We watch intently as the *Dilophosaurus* robot spits carefully timed volumes of chlorinated water into the water slide before the oncoming human passengers. (People complained about getting spat on, so Richard had to modify the aim.)

As Chevy Chase discovered, acting in his comic movie role as "Clark Griswold" in *Las Vegas Vacation*, the area around Las Vegas is risky territory indeed. Whether facing bankruptcy after a bad night of gambling, or extinction after a bad roll of the cosmic dice, one must tread cautiously here. Charles Sandberg of the U.S. Geological Survey claims that a comet struck an area about 130 miles northwest of modern Las Vegas 370 million years ago, triggering a localized mass extinction. Fortunately, bad luck "Clark" didn't show up during *our* Las Vegas Vacation—Diane came out about $60 ahead, and the dinosaurs still roam triumphantly in all their primeval glory.

In a city that boasts everything from "flying Elvises" to the Great Egyptian Pyramid, Las Vegas' unique assortment of robotosaurs would seem unlikely attractions. But, since anything goes here, perhaps it's only fitting we could also find ourselves suddenly plunged deep into the Cretaceous period. To many dinophiles, especially those younger in age, growling, snorting, mechanically moving dinosaurs seem more convincing than all the other kinds of restorations.

But are they here to stay? You can *bet* on it!

***Opposite:* Figure 29-2: Circus Circus' animatronic *Tarbosaurus*. (Photograph by Allen A. Debus, 1998.)**

Chapter 30

Ernie's Paleozoo

We only hope that 50 years from now our enthusiasm for paleontology is no less diminished than that of the eighty-eight-year-old Ernie V. Nelson, creator of southwest Oregon's Prehistoric Gardens. "I never really grew up," claimed Ernie, referring to his interest in dinosaurs. To our left, the statue of a monstrous *T. rex*, accurately proportioned after the American Museum's skeleton (as it was formerly exhibited as an erect biped), grinned menacingly out of time's antiquity.

Why would anyone ever want to model dinosaurs? In order to avoid the a· iction, Ernie advises, simply *never* start. And yet, he did—with marvelous results, as testified by over 20 full scale prehistoric animal sculptures inhabiting the scenic Prehistoric Gardens in the misty midst of southern Oregon's coastal rain forest.

It was gratifying to finally meet Ernie, an artist with over four decades of dinosaur designing experiences. Ernie has been an elusive fellow. In 1993, when we first spied his *T. rex* from our car window, he was unavailable because it was his birthday. The following year, when we arrived with our two children, we were informed that he was attending a relative's funeral. (Our kids scampered into the wonderland gleefully shouting, "Look at all the dinosaurs in here!" even before we could pay their fare. We had to chase them down before they were "eaten.")

This time, during our 1995 visit, Ernie was in nearby Brookings, shopping for a new typewriter. So, after conversing with a granddaughter, Kiki Halcumb, we arranged to meet with the elusive Mr. Nelson the following morning.

Our meeting with Ernie became a delightful affair. Many of you may have caught glimpses of Ernie in a 1994 video, *Dinosaurs: Next Exit.*[1] He is a slim, mustached gentleman, a little hard of hearing but with an agile mind. He converses enthusiastically about his favorite subject—the study and restoration of dinosaurs.

We were astonished to find that many of his views on dinosaurs were as anachronistic as using a typewriter would seem in these modern times! He clings to opinions that dinosaurs were cold-blooded, that birds are not descended from dinosaurs, and that catastrophes have no place in the theory of dinosaur extinctions. Ernie read Adrian J. Desmond's *The Hot-Blooded Dinosaurs: A Revolution in Palaeontology* (1975) several times years ago and claimed it presented insubstantial evidence supporting the hot-blooded dinosaur theory. He remained unconvinced even after attending a lecture by Robert Bakker. Apparently, at age 87, he wasn't about to "evolve" either.

Extinctions caused by asteroid impacts? Well, it seems scientists merely feel obliged to concoct some new and sometimes preposterous theories just to get their names in the paper every now and then. Ernie mentioned "those Creationists" with disdain. Yet, clearly, "evolutionist" Ernie was adhering to mid–1960s views. We chuckled at this, but it was a pleasure discussing (not debating) these topics with him.

Although in advertisements for the Prehistoric Gardens one reads that the prehistoric animal sculptures are scientifically accurate representations, such a statement would cause some critics to voice other opinions. One must remember, however, that these sculptures were constructed several decades ago, at great expense, and that the tenets of dinosaur science have since changed. (Although, in a philosophical sense perhaps, historical "change" may not always necessarily imply "advancement.") As we are aware, the scientific mirage of "accuracy" shifts episodically when it comes to dinosaur restorations. We are all entitled to heartfelt opinions when it comes to dinosaurs, which is probably one of the reasons why paleontology is so enjoyable.

Consequently, we refrain from criticizing Benjamin Waterhouse Hawkins' remarkable nineteenth century sculptures and restorations, and we can most certainly admire Ernie V. Nelson's! Both artists were pioneers of their time, and they worked hard for their results. Furthermore, the Prehistoric Gardens accomplishes just what Ernie primarily intended it do, which is to provide a safe and fun form of entertainment, as well as provide a steady source of income. According to one regular visitor, "...a trip to Prehistoric Gardens is akin to a religious experience."[2] Ernie has succeeded in this endeavor ever since his park opened on January 1, 1955.[3]

Ernie's enchantment with prehistory began one day long ago (sometime around the year 1920), while gazing through an issue of *The National Geographic Magazine*.[4] Ernie gazed longingly at Charles R. Knight's restorations, perhaps dreaming of someday recreating the world of the dinosaur even more realistically.

Thirty years would pass before this dream could be realized. Ernie's family returned to Oregon (from Minnesota), and young Ernie became employed in a variety of occupations. He toiled as an accountant for 13 years, worked at a forgery as a tool manufacturer, and also made money on the side as a cartoonist. He was sufficiently talented to be offered a position at the Walt Disney Studios, although it would have paid less than he was making as an accountant.

Eventually, in 1953, Ernie situated his family in Port Orford, Oregon, where, guided by his love for paleontology, he designed and sculpted his first restoration, the *Elasmosaurus*. Through 1954, work continued on a host of other figures—*Triceratops*, *Stegosaurus* and old *T. rex*. The Prehistoric Gardens made $18,000 its first year, after opening on January 1, 1955.

After a busload of tourists sauntered through his gift shop, Ernie wistfully spoke of how the mighty *Brachiosaurus* was constructed. Interestingly, small models were not used as a starting point. According to dinophile Riff Smith, "Only the *T. rex* was produced as a small scale model before the actual large structure was built."[5] Instead, Ernie relied on a multitude of drawings.

Completed in 1969, the immense *Brachiosaurus* statue was, at the time, perhaps the most colossal dinosaur sculpture ever made. The concrete monster, whose lofty, crested

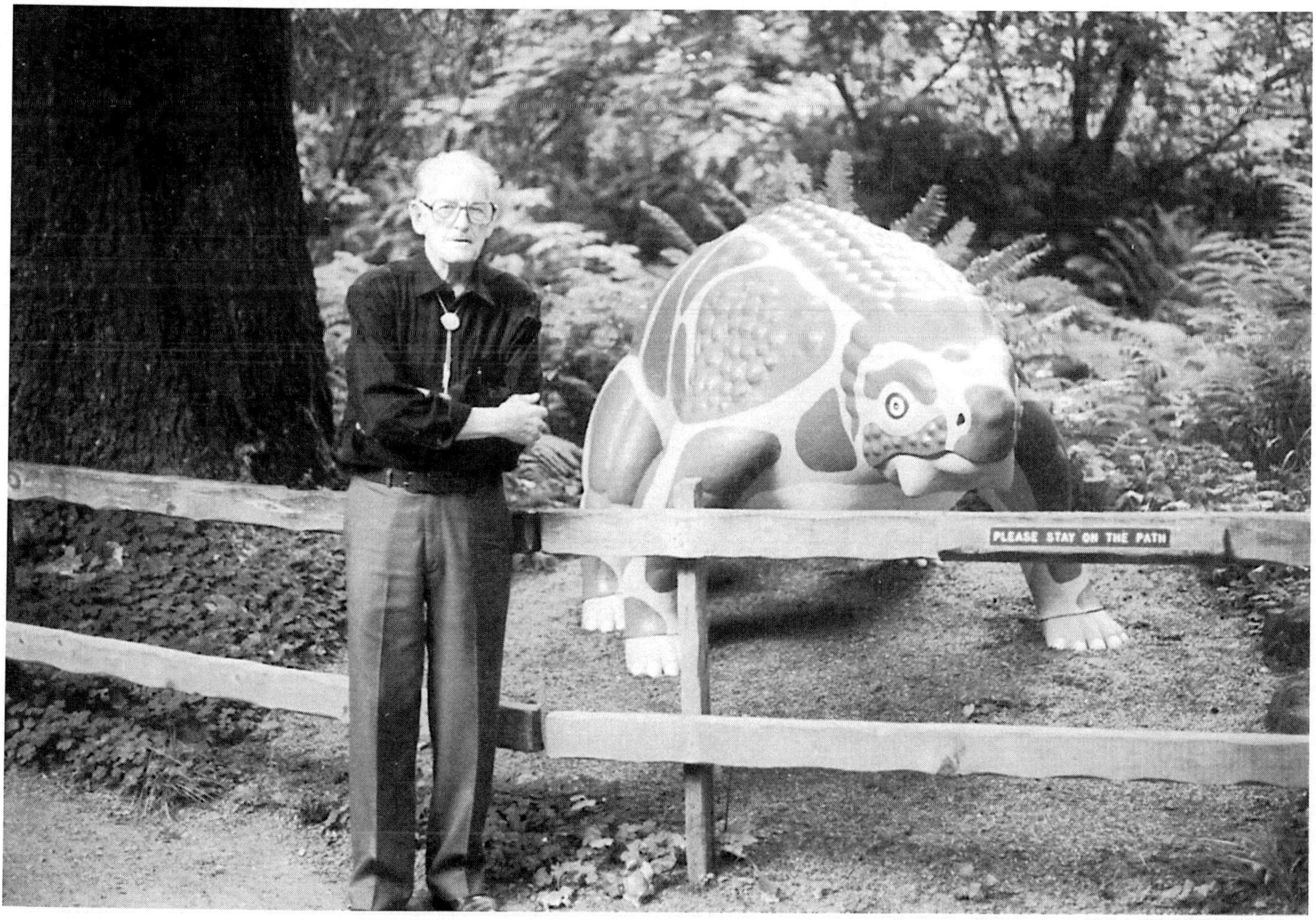

Figure 30-1: Ernie Nelson standing next to his *Bradysaurus*. (Photograph by Allen A. Debus, 1995.)

head peers through the treetops, is internally reinforced with a crisscrossed network of steel. The head was shaped using a wooden frame, covered by metal lath. Ernie hired a welder to build the framework, according to his design. Finally, a concrete finish was applied by a professor of sculpture, Walter Pritchard.[6] And naysayers said it couldn't be done!

We took another tour through the Prehistoric Gardens, which is instructively organized in the traditionally popular "life through time" format. Some of his creatures are painted rather garishly. Yet, if this were not the case, they would blend into the lush rain forest foliage. Our favorites? Definitely the *Brachiosaurus*, whose size and dimensions favorably compare with the skeletal cast of this dinosaur now exhibited at the Field Museum in Chicago. Also impressive are the *Archelon*, the *Stegosaurus*, a pair of egg-robbing struthiomimids and a feathery *Diatryma*, which Ernie mentioned attracted quite a stir when it was first exhibited. According to Ernie, it was a labor of love to sculpt the feathery texture of this large predatory Eocene bird. It's a good thing Prehistoric Gardens was completed before vertebrate paleontology's "feather revolution"!

Ernie continues to design dinosaurs, though he questions whether or not he would ever have the time to witness the final, sculptural results. According to artist Jack Arata, who visited "Ernie's Zoo" in late 1998, Ernie Nelson sculpted the prehistoric mammal figures for another prehistoric theme park known as "Thunderbeast Park" in Chiloquin, Oregon Thunderbeast Park, featuring 12 life-sized replicas of Tertiary and Pleistocene fossil vertebrates, was built in 1962. It closed in the early 1990s after its owner died. Kiki Halcumb claims her grandfather, Ernie, was embarrassed by the appearance of his prehistoric mammals, and so he rarely spoke about Thunderbeast Park.[7]

Children now question where the "raptors" are in the exhibit. Ernie answers quite astutely by pointing to the *T. rex*, "There's a raptor, the largest of them all." Even so, he has contemplated adding a *Velociraptor* to the exhibit. The most recent creation was *Bradysaurus*, an eight-foot-long pareiasaur from the middle Permian of Africa, which he posed with for a photograph. (See Figure 30-1.) Ernie was not enthused about the 1993 movie *Jurassic Park*, which he feels introduced too many dinosaurian inaccuracies (such as the fact that the starring dinosaurs must have been borrowed from a *Cretaceous* park instead).

Ernie Nelson is rightly proud of his accomplishments. We share his vision. How wonderful it would be to live in such a beautiful setting and spend one's life designing dinosaurs, which people would pay to come and see! Never idle, Ernie is busy researching his family's ancestry. And when the muse strikes, he even dashes off a bit of poetry. Ernie professes that one must enjoy work to avoid feeling like a "slave." Through his fascination with paleontology, it may be said that Ernie V. Nelson has solved the riddle of life![8]

Chapter 31

Building Life-Sized Dinosaurs

Next to CGI and stop-motion animated dinosaurs of the movies, our best bet for experiencing real, "living" dinosaurs the life-sized sculptural representations. We *need* to see dinosaurs *alive,* and, like no other restorations, full size sculptures fulfill our fascinations for the primeval.

Public enthusiasm for dinosaurs took root over a century ago with the earliest displays of life-sized prehistoric animal restorations, a tradition thriving today at many museums, zoological theme parks and outdoor dinosaur galleries. Mounted skeletons tease us with bare bones details of how dinosaurs may have appeared and lived. Conversely, life-sized sculptures dramatically reveal what it's like to stare down a live menacing dinosaur that's ready to bite your face, just like in the movies.

Life-sized "theme park" dinosaurs, some as big as buildings, are "strange attractors." More than any other characteristic, the sheer enormity of many dinosaurs and prehistoric animal genera seizes our attention. So when sculptural restorations are made at life-size scale, even if liberties were taken in their construction, or if anatomical accuracy was sacrificed for expediency of design, people admire the results.

Who can resist the timeless atmosphere of a Dinosaur theme park? Such parks can represent scientific ideas, some antiquated, now frozen in time. The Crystal Palace "dinosaurs" at Sydenham, England, come to mind. Others are meant to commemorate scientific achievement through collaborative exploration, such as the sculptures of Upper Cretaceous Mongolian dinosaur fauna exhibited at Poland's Valley of the Dinosaurs. Other parks are fabricated to be nothing more than tourist attractions, built to tease dollars from loose wallets. Oh how we love to be photographed next to those enormous constructions!

Mounted skeletal reconstructions are always impressive, but restorations built to life proportions displayed along "primeval" trails in forested dinosaur theme park tourist traps lure us in, stirring our emotions. Benjamin Waterhouse Hawkins built the first outdoor dinosaur theme park, while Michael Crichton created the most infamous fictional dinosaur theme park of all, a "Jurassic Park" brought to bloody life by Stephen Spielberg on the silver screen in 1993.[1]

During the 14 decades spanning Hawkins' and Crichton's creations, dozens of artists have resourcefully erected dinosaur parks "inhabited" by scores of wooden, fiberglass, concrete, plaster, rubber, styrofoam and bronze oddities. Such manifestations seem more recognizably "dinosaurian" than others, perpetuating the myth of "accuracy" in paleontological restoration while underscoring paleoimagery's universal appeal.

Accuracy in Life-Sized Restorations

According to Austrian paleobiologist Othenio Abel, extinct rhinos influenced medieval perceptions of mythical beasts, namely dragons and unicorns. For example, one statue designed as a dragon fountain situated in the town square of Klagenfurt in southern Austria was inspired by the skull of a Pleistocene woolly rhinoceros (*Coelodonta*). The skull was found in the year 1355. Sculptor Ulrich Vogelsang then began "restoring" the fossil animal as a winged, quadrupedal dragon in 1590. Vogelsang died in 1601, but the enormous sculpture, impressively showing the dragon battling mighty Hercules, was completed in 1636.[2]

Does Vogelsang's dragon count as the "first" life-sized prehistoric animal restoration, as Abel suggests? Maybe not, for scholar Adrienne Mayor claims that long before, ancients of the Mediterranean region discovered fossils of prehistoric vertebrates and made crude attempts at restoration, reconciling the petrified remains with their gods and legendary heroes.[3]

When it comes to "accuracy" in building life-sized dinosaurs, is there a magical key to success? Typically one associates "accuracy" with museum restorations, not necessarily with roadside tourist traps. Consider a life-size *Allosaurus* restoration exhibited at Lincoln's University of Nebraska State Museum. This is the handiwork of former museum exhibit technician Roger A. Vandiver. It was built as part of a mid–1970s "Renaissance" revamping of the dinosaur hall. Vandiver and cowriters stated, "The prime objective of any reconstruction is to represent an extinct animal accurately and to make it believable to the viewer. It should look right to both the casual museum visitor and to the critical eye of the scientist. The pose should be such that it appears to have momentarily stopped; that it will move again in a second. The total configuration of pose, texture, color, and expression should convey a definite character such as all real animals have, but is disappointingly absent in many reconstructions. The experienced museum visitor has undoubtedly been tempted to laugh at one time or another when standing before a poorly executed example."[4]

For posterity, Vandiver chronicled stages in the creation of his imposing theropod. Starting with a small scale plaster model, Vandiver erected a welded L beam armature, over which galvanized sections of metal were affixed to form the body shape. One-quarter inch mesh hardware cloth was applied, followed by a trowled-on mushy, mâché formula "skin" made of plaster-perlite and white glue, which contained asbestos for workability. Scaly texture was sculpted in before the paste set. Then the allosaur was painted in layers of acrylic so as to "...give the right amount of sheen to simulate its reptilian appearance and heritage."

Vandiver also believed that, "Today museums cannot and do not allow speculation to dominate the predominance of fact and reason; in fact, they usually bend over backward to play down sensationalism in their exhibits. However, even though museums today are very careful to exhibit only those things that have been well documented, the public still must realize that reconstructions are only approximations based on the evidence available at the time."[5] Vandiver's *Allosaurus* is convincing, certainly not overly speculative, and grandly suggests how "flesh" should be applied to bones.

But is Vandiver's the current "formula" for establishing museum accuracy in restorations? How "accurate" is Vandiver's "renaissance" dinosaur in this revitalized age of feathered dinosaurs? Surely it must be more accurate an image than was Vogelsang's "restoration" of *Coelodonta*. Certainly Henry Fairfield Osborn would have been impressed by Vandiver's objectivity and conviction.

Another popular museum restoration, a life-sized *Ceratosaurus* sculpture originally built in 1960 for the Cleveland Museum of Natural History's dinosaur display, was demolished during the 1970s when the exhibits were upgraded to reflect a newer "renaissance" science of dinosaurs. Perhaps museum officials no longer believed their *Ceratosaurus* to be sufficiently accurate for display purposes. During removal, recalls vertebrate paleontology curator Michael Williams, heat from a blowtorch ignited a fire on Ceratosaurus' tail, causing smoke to issue in dragon-like fashion from the head region, nostrils and mouth.[6]

So dinosaurs of science revert to mythical fire breathing dragons. The shapes of time are shifting. In the end, paleoimagery of past ages merges with that of the present day. Accuracy in paleontological restoration will forever be a moving target, laden with subjectivity and artistic license.[7]

Osborn's Invisible Hand

Usually, dinosaur designers and paleoartists claim the underlying goal is to recreate an accurate visage of ancient fauna. Their messages reflecting natural history ideas, and the meaning or imagetext behind the erected statues becomes obscured with the passage of time.

The accuracy of outdoor theme park dinosaurs may often be questionable, yet how many of us recognize Osborn's invisible hand in their making? How many of us realize that our favorite theme park dinosaurs owe their origins to the tradition of revivifying prehistoric animals established by Osborn over a century ago at the American Museum of Natural History? Dinosaur theme parks may be regarded as an individual artist's attempt to construct an outdoor menagerie of prehistoric life in "realistic" settings, reflecting or mimicking the displays one might find inside museums. Dinosaur theme parks of the twentieth century, in fact, evolved as an unanticipated indirect "byproduct" of Osborn's influential American Museum displays.

Dinosaur "parks," comprising anywhere from one to over a dozen prehistoric animal sculptural restorations, grace the exterior grounds of many natural history museums worldwide, notably Moscow's Palaeontological Museum, Canada's Royal Tyrrell Museum of Palaeontology in Drumheller, southwest California's George C. Page Museum of La Brea Discoveries, the New Mexico Museum of Natural History in Albuquerque, the Utah Field House of

Natural History State Park in Vernal, and Taiwan's National Museum of Natural Science in Taichung. Prehistoric animal restorations adorning museum grounds are expected to be based on the most recent or contemporary scientific standards, even though the sculptures are destined to become anachronistic.

But, assuredly, prehistoric animal sculptures have migrated far beyond the confines of museum walls and the outside grounds, well beyond the parking lot spaces to the furthest, most out of the way country roadsides—as that billboard broadcasts to oncoming motorists, "Dinosaurs—next exit!"[8] Yes, here is yet another spacious theme park where we find sculptures of dinosaurs and other prehistoric animals of every persuasion, some fairly accurate and some, well, not as accurate. Yet, in all cases, they are loved by their makers and patrons alike. And isn't that what really counts? Let's survey a few of the world's foremost and most treasured dinosaur parks, emphasizing those not described in previous chapters.

Rapid City's Dinosaur Park

South Dakota's most famous dinosaurs are the famed menagerie of Dinosaur Park occupying the summit of Skyline Drive in Rapid City. These creations were constructed by the City of Rapid City and the Works Progress Administration (Project no. 960) in 1936. Not hard to get to, you can spy the large "brontosaur" from the downtown area. Just follow the roads aiming for that large hill over there. (See Figure 31-1.)

Dinosaur Park must rank as the oldest "dinosaur theme park" in America, discounting Hawkins' "Palaeozoic Museum" of the 1870s, which was never completed. Although the Dinosaur Park was leased for many years by the Sullivan family, the City of Rapid City gained property rights in 1967. Thereafter, following extensive repairs to those many steps, the land became a city park, while the concession stand continued to be operated by Mrs. Sullivan.

According to South Dakota geologist John Paul Gries, these full-sized figures were based on miniature models once exhibited at the American Museum of Natural History.[9] The concrete dinosaurs were designed by sculptor Emmet A. Sullivan. Workers Frank Lockhart and George McGraw built the dinosaurs. On July 21, 1990, the park was placed on the Register of Historic Places.

Two of the smaller figures, a *Protoceratops* and *Dimetrodon*, greet gift shop customers. But it's a looong climb up that hill to join the real party upstairs. The dinosaurs occupy a vantage point resting on Mesozoic formations. Half

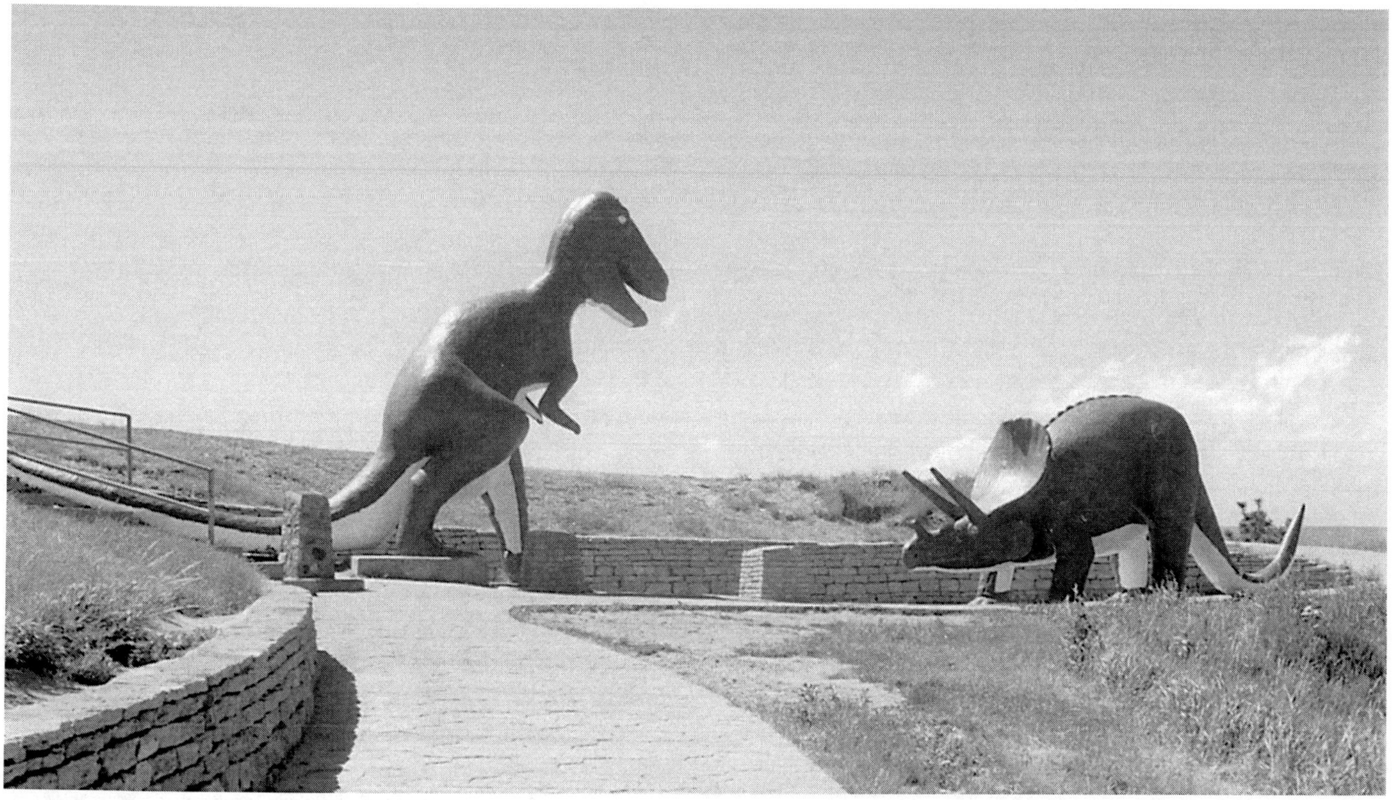

Figure 31-1: Perhaps the world's first life-sized sculptural recreation of Charles Knight's mythic *T. rex vs. Triceratops* battle, still visible at Rapid City's Dinosaur Park. (Photograph by Allen A. Debus, 1999.)

a century ago, dinosaur bone fragments were found in the vicinity of where the statues now stand. The site attracts many youngsters and their avid parents.

The *T. rex vs. Triceratops* scene was inspired by Charles R. Knight's famous Field Museum mural.[10] Nearly all of the kids we noticed couldn't resist climbing on the huge *Triceratops* head, making it difficult to snap a few clear photos. *Triceratops* represents the earliest form of reptilian "buffalo" to have inhabited the region. (In fact, paleontologist Othniel Marsh misinterpreted the first pair of *Triceratops* horn cores he ever saw, revealingly naming them *Bison alticornis*.)

Apatosaurus is statuesque, a near replica of another famous dinosaur displayed in the city of Wall, known as the Wall Drug dinosaur.[11] Over the years the length of the park's "Brontosaurus"/*Apatosaurus* has actually decreased due to damage to the figure. Originally 80 feet long, the tail was broken in past years (it is now mounted on concrete blocks) and not fully restored, reducing the figure's length to a "mere" 60 feet. Park postcards chronicle this change, with older ones showing the "Brontosaurus" tail flat on the ground and crediting its length as 80 feet, while more recent vintage postcards show the tail mounted on the blocks and give the revised 60-foot-long dimensions.[12]

"Duckbilled" *Anatotitan*, for that is what it is rightfully called now, peeks downhill through tree branches as visitors climb the many steps to greet him. Of course, this creature was known as "Trachodon" in 1936. Well, the name can be changed to protect the innocent, but nobody can easily change that backbone and tail structure incongruously bent into upright postural stance. Hey, he's quaint (even if his poor tail "bones" would be broken standing vertically like that).

Stegosaurus is evidently a juvenile of his kind. He is proportionally diminutive relative to the rest, which are adult-sized, and his tail spike region is lumpy-looking but void of any child-endangering spikes. Its plates are double-rowed, paired. With the exception of the two ceratopsians, all the dinosaurs were restored with upright leg posture. (*Triceratops*' arms are a little bowed outward at the elbows, but his rear legs appear columnar.) All things considered, the aptly named Dinosaur Park turned out to be an enduring achievement![13]

CUBA'S VALLEY OF PREHISTORY

Little wonder why Cuba's "Valle de la Prehistoria," nestled in 32 hectares of Cuba's Baconao National Park, perhaps the world's largest dinosaur theme park, is also one of the world's best kept secrets. It's in southeastern Cuba! Nuff said! According to a Fodor's 1997 guide to Cuba, there are 243 life-sized prehistoric animal sculptures here, representing 59 species altogether. That's about ten times as many "antediluvians" as Benjamin Waterhouse Hawkins produced for the Crystal Palace in the early 1850s. That's nearly as many as the combined population of prehistoric animal figures "inhabiting" the world's next three largest dinosaur theme parks, "Ogden's Eccles Dinosaur Park and Museum" in Utah, Germany's "Dinosaurierpark" in Munchehagen, and Ontario's "Prehistoric World."[14]

A few dinophiles may already be rather familiar with the images of certain statues displayed here. If you happen to own a copy of *Dinosaur Stamps of the World* (1991), by Stuart Baldwin and Beverly Halstead, you'll see that several Cuban postage stamps are reproduced on pages 42 and 43.[15] There are 13 stamps shown, each including a facsimile of a park figure. (The stamps feature the following prehistoric animals: *Pteranodon, Apatosaurus*—being attacked by *Allosaurus, Monoclonius, Iguanodon, Stegosaurus, Corythosaurus, Tyrannosaurus*—two stamps, *Triceratops, Edaphosaurus, Styracosaurus, Euoplocephalus* and *Saurolophus*.) For all you paleo-philatelists, that leaves open the possibility of over 200 additional postage stamps to complete the set! Then, one wonders, how many more figures may have been added to the park since 1997? Has this become an out of control "dinosaur breeding" experiment?

There are more than just dinosaurs here. One postcard in a collectible series reveals a Cenozoic route through "primeval terrain" past majestic mammoths, sabertoothed cats posed fighting over downed prey, fleet-footed prehistoric horses, several human-sized cavemen and an imposing, Paul Bunyan-esque, 39-foot-high, axe-wielding Cro-Magnon male. (See Figure 31-2.) The largest displayed figure in the park is a dinosaur, an impressive 89-foot-long *Diplodocus*; yet, the most menacing creatures on "display" are the live goats found foraging amidst the statues.

According to Rogelio Oran Fontaine, a park representative, the statues were built by eight Santiago de Cuba sculptors between the years 1983 and 1989.[16] Their objective was to illustrate Darwinian evolution by representing life from the Paleozoic, Mesozoic and Cenozoic eras. Great care was taken to create life-sized replicas as accurately as possible, relying on references, principally Zdenek Spinar's popular *Life Before Man* (1981), a book featuring painted restorations by the Czech grandmaster paleoartist Zdenek Burian. (Fodor's 1997 estimate of 243 figures was apparently overzealous though, because in 1998 Fontaine claimed there are "only" 214 figures here.) It is a marvel that so many prehistoric animal figures could be sculpted in such a relatively short time. Benjamin Waterhouse Hawkins would be jealous.

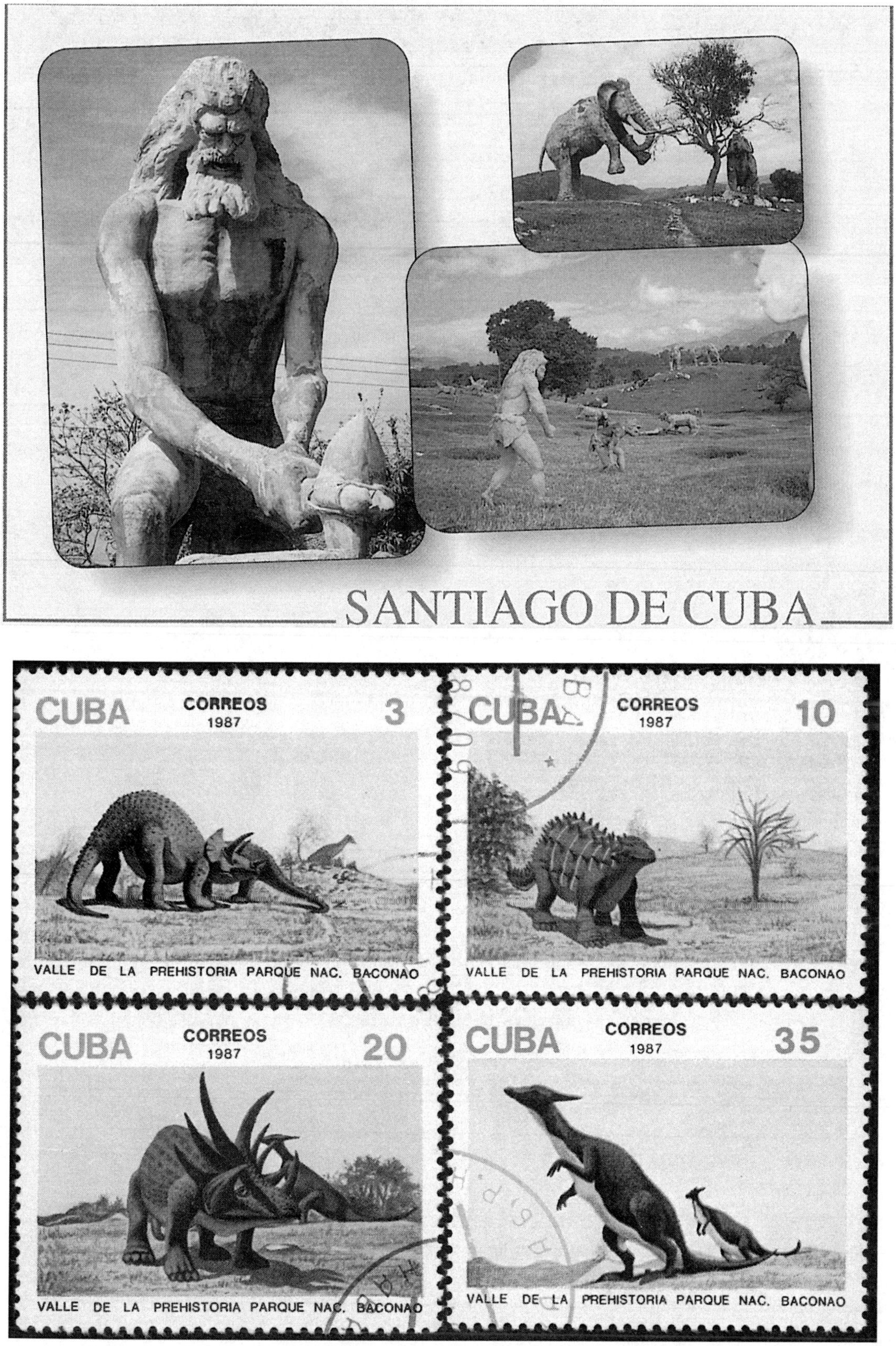

Figure 31-2: *Top*—Image from a rare souvenir postcard showing giant Cro-Magnon and displays of other Cenozoic creatures in Cuba's Prehistoric Valley. *Bottom*—Cuban postage stamps showing samples of dinosaurs (including *Triceratops*, *Saurolophus*, *Ankylosaurus* and *Styracosaurus*) displayed in Cuba's Prehistoric Valley. (Authors' collection.)

DOMKE'S DINOSAUR GARDENS

Northeast Michigan's "Dinosaur Gardens" is a true-to-form "Jurassic Park," situated among 40 acres of majestic pine trees, wild flowers and rivers. In this setting, most would expect to encounter bear, deer, moose or the occasional bobcat. Instead, visitors find themselves taking a most unusual sort of "trip down memory lane," one conceived in the 1930s by paleosculptor Paul Domke (1886–1981).

Few, however, would anticipate the curious admixture of ideas instilled in the wonderful creation known as "Dinosaur Gardens," an attraction located in Ossinike, Michigan, and a dinosaur park very infrequently mentioned in books or articles on the topic.[17]

An advertising brochure bills Dinosaur Gardens as "The World's Finest Prehistorical Zoo," perhaps a bold claim considering the competition existing today. On the back of the brochure another phrase adds, "18 Hole Mini 'Dino-Golf' Under Construction," a pairing often found in tourist areas from Florida to Arizona.[18]

Domke was born a few years before Benjamin Waterhouse Hawkins and Richard Owen died. He was a self-trained sculptor whose interests in natural history and dinosaurs compelled him to recreate the "prehistorical" world in this terrain, which, he felt, bore resemblance to that witnessed in the age of dinosaurs. Signage posted along the trail identifies birch, trembling aspen, red, white and Norway pine, and other trees growing amidst luxuriant ferns. Domke researched his designs carefully, studying skeletons of prehistoric animals at the Smithsonian Museum of Natural History in Washington D.C. and in European museum displays. He also visited the La Brea Tar Pits in southern California.

It has been said that Domke would begin his year-round sculpting work as early as 4:00 A.M. Domke would prepare his secret formula with a mixture of cement mixed with oil, lime and deer or human hair, the latter obtained from a local barber shop After drafting young apprentices to do the hauling of sand and gravel, and for the necessary digging of bases, he would build the steel frames from the detailed sketches he had made.

After draining the cedar swamp in 1930, Domke began sculpting his first scene (1935) showing the La Brea Tar Pits in the Pleistocene. In this grouping a lone *Smilodon* straddles a mammoth carcass caught in the tar, while fending off a pair of Dire Wolves. The *Smilodon* is correctly restored with a shortened tail, an unusual accuracy for most dino-park restorations.

Until his death in 1981 he applied his energies to the "Garden's" development, dreaming of additional creations he could build. During a 35 year stretch, until his last dinosaur scene was completed in 1970, Domke completed 22 prehistoric scenes, some comprising multiple figures. One exhibit, especially, displays Domke's considerable talents. His marvelous *Styracosaurus* comes complete with wrinkles and muscle tone. In fact, many of his creations have realistic looking skin texture and folds. But *Styracosaurus* is one of the very best we've seen anywhere.

According to Dinosaur Garden's current owners, Wilma McCourt and her son Frank McCourt Jr., Domke's sculptures were built for eternity. Their construction is very sturdy, designed to survive even an earthquake. Maintenance work usually only involves touching up the paint every now and then.

Domke's 60,000 pound "Brontosaurus" took two years to build from 1949 to 1951, but once you've recovered from the astonishment engendered by his 85 foot length, it actually looks quite friendly. (See Figure 31-3.) If you've ever wondered what it would feel like to be a sauropod snack, all one must do is hike up the dozen or so stairs into his innards. What you will encounters is certainly unexpected. Inside this amazing specimen one notices a portrait of Jesus situated above a representation of the great beast's heart. Domke added the words, "*The Greatest Heart that Ever Lived.*"

"Brontosaurus" is the second large statue to be seen entering the grounds. The first is the 30-foot-tall statue of Christ presiding over the highway. Christ and "Brontosaurus" convey Domke's religious message, a most peculiar blend of ideas. One discovers this park isn't inherently about golf and dinosaurs, but dinosaurs and *religion* instead!

An explanatory sign reveals that Domke combined images of dinosaurs and Christ "as a testimonial to his belief that Christ was the master planner of an Earth that included dinosaurs." Another sign makes the proclamation that, "The seven days in which God created the world lasted several million years, a period during which the dinosaurs evolved." Here evolution is reconciled with creation science. It's hard to fathom such contrasting ideals and symbols, but at Dinosaur Gardens the imagery is juxtaposed for all to see.[19]

ONTARIO'S LE MONDE PREHISTORIQUE

Seemingly out there in the middle of nowheresville, in the not so bustling town of Morrisburg, Ontario, motorists will find one of the world's foremost dinosaur theme parks. The park's English name is "Prehistoric World." This fascinating park has been designed as a striking

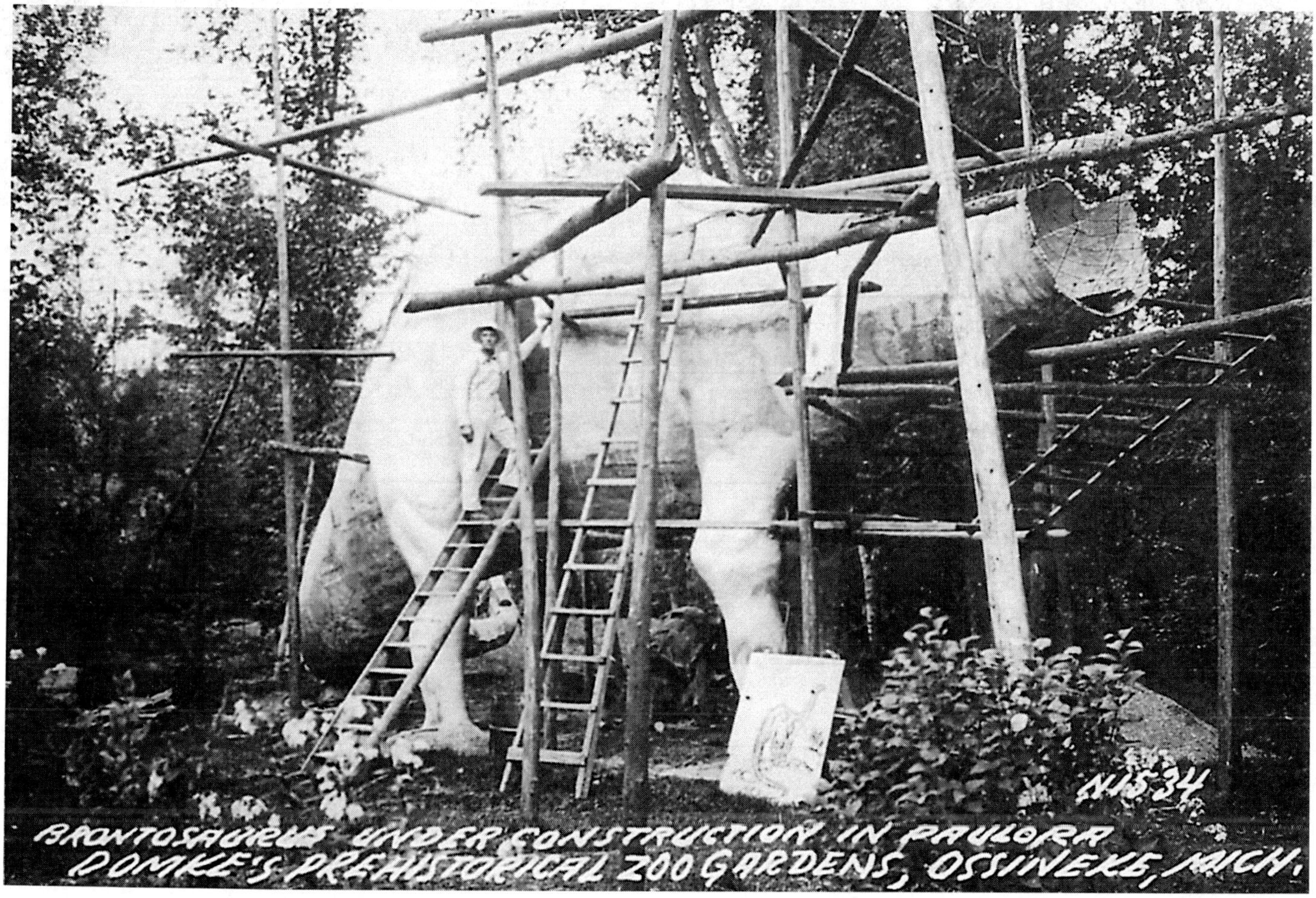

Figure 31-3: Paul Domke busily at work on his "Brontosaurus" in Ossineke, Michigan, as shown on this rare souvenir postcard.

portrayal of life, captured vividly in cement, through geological time. On the day of our visit,[20] the parking lot was jam packed with 25 cars. Inside, many children waltzed happily from one dramatically posed display to another.

Here visitors will find nearly 50 prehistoric animal sculptures, each amazingly and lovingly fashioned from iron rebar and cement. Two brothers, Serge and Paul Dupuis, are the sculptors. They have managed to finish all the statues displayed here since 1981 (including several which had to be demolished after trees fell on them during the winter ice storm of 1998). The brothers have also sculpted a volcano and *Pteranodon* cliff roost. Most of the statues were completed in their first eight years of operation, as a result of extended "marathon sculpting sessions." It may take two to three months to build a twenty-five-foot-long figure, although their gigantic, 100-foot-long *Apatosaurus* was completed in a two year interval.

Accuracy is their goal. Accordingly, Prehistoric World's dinosaurs are selected from fossils that are known from relatively complete skeletons. Yet, in conversing with the gifted duo, one can detect a sense of artists' remorse that several of their earliest dinosaur creations would no longer pass scientific scrutiny.

The Dupuis brothers proceed directly to building full-sized sculptures without relying on an intermediate step, the construction of miniature models ordinarily used by museum paleoartists for scaling purposes. Once they have agreed on the "concept," the Dupuis brothers can confidently produce the intended dinosaur or prehistoric mammal three dimensionally. This may seem rather daring, but let the eye judge. Close examination of their handiwork quickly establishes trust in their paleoartistic instincts.

These statues are simply marvelous. Are they the best in the world? Well, how do you define "best?" In terms of accuracy, many of the displayed figures would be considered "museum quality." The variety of prehistoric animals is astounding. While many dinosaur theme parks of the atomic age focus on the classic groupings (typically North American dinosaurs popularized by Charles Knight's and Rudolph Zallinger's paintings), an unusual diversity of exotic prehistoria may be seen here, courtesy of the Dupuis' talents. Yes, the typical cast of dinosaurs lurks here through the winding, forested "primeval" trails (including a standard, larger than life *T. rex* vs. *Triceratops* scene). But there is also an odd assortment of dinosaurs from other continents, figures one will be hard pressed to find elsewhere.[21]

At Prehistoric World visitors will find a cosmopolitan array of statues, including the Asian plated dinosaur *Huayangosaurus*, the Asian club-tailed sauropod *Shunosaurus*, the South African stegosaurid *Kentrosaurus*, the British armored genera *Polacanthus* and *Scelidosaurus*, the Australian armored genus *Minmi*, the European sauropod *Cetiosaurus*, and the African iguanodont *Ouranosaurus*. (See Figure 31-4.) Antarctica's magnificent Early Jurassic theropod genus, the crested *Cryolophosaurus*, may be their next intended design.

Paul Dupuis claims their chief inspiration has been Canadian paleoartist Ely Kish's paintings. Perhaps their finest accomplishment is a life-sized statue of the dinosaur *Hypacrosaurus*, posed and painted exactly as in Kish's painting. The Dupuis measured the bones of a museum skeleton before beginning this masterpiece. Besides Kish's, the Dupuis have also admired Charles Knight's paleoart.

Most "dinosaur" theme parks wouldn't be expected to have such a startling collection of creatures representing the Tertiary and Pleistocene epochs. It is possible that more prehistoric mammals inhabit Prehistoric World than in any other dinosaur theme park. So far, only about 30 acres of their 153 acre property have been occupied with prehistoric animal statues. The spirited Dupuis are resolute with their targeted completion of over 100 statues someday, and they certainly have the space for achieving that goal. Barring more damaging ice storms, "only" 50 statues to go!

Due to prior confrontations with visiting creationists, Serge and Paul no longer offer guided tours of their park to creationist groups. They feel that Prehistoric World bears testimony to the *fact* of evolution, and that there is no divine intelligence driving evolution's blind course. But within Prehistoric World's grounds there is much intelligent design inherently evident throughout the winding forested trails taking us through three eras of Life's history. For this is an active dinosaur theme park, proudly expanding into another millennium through the tireless efforts of Paul and Serge Dupuis, master paleoartists.

INTO THE VALLEY OF DINOSAURS

This subheading sounds appropriate for a science fiction or fantasy tale. In the tradition of any worthwhile fantasy story, the making of Poland's *real* Valley of the Dinosaurs (distinct from Cuba's "Prehistoric Valley") only came about after intrepid explorers braved treacherous lands and prehistoric reptiles. The Valley of the Dinosaurs is allied to a zoological park situated in Chorzow, Poland, and dedicated to the paleontological investigations of Mongolia conducted jointly by a Polish-Mongolian team between 1963 and 1971. In order to commemorate their scientific discoveries and successes, a dinosaur park was built, "inhabited" by full-sized models of genera collected in the field.

The American Museum expeditions to Mongolia of the 1920s employed the services of male scientists exclusively. However, in unprecedented fashion, several female Polish paleontologists including Zofia Kielan-Jaworowska (an expert on Mesozoic mammals), Teresa Mayranska (an expert on armored dinosaurs) and dinosaur specialist Halzka Osmolska, headed the Polish-Mongolian team. Other participants included the Mongolian paleontologists Barsbold and Dashzeveg, and a Pole, Wojtek Skarzynski (himself a paleoartist). Together they braved vipers, blinding sandstorms and shadeless 100 degree temperatures. Plaster and dust was wind-blown into their eyes, water proved scarce nearly everywhere, and when water *was* present there were fierce fly and mosquito attacks.[22]

By 1962, a line of demarcation through the Gobi Desert separated Soviet-oriented Outer Mongolia from China. Missile launching bases were situated in the Gobi sands. Yet, as has been the case for thousands of millennia (or at least since the days that ancients claimed griffins protected gold deposits in the Mongolian wilderness), fossils eroded from weathered rock surfaces, exposing the creatures of other days to the sun's scorching rays.

The Polish-Mongolian investigations uncovered nearly two dozen Cretaceous fauna new to science. Some of the most remarkable skeletons they uncovered included theropod *Tarbosaurus bataar*, sauropods *Nemegtosaurus* and *Opisthocoelicaudia*, armored genera *Saichania*, *Shamosaurus*, *Tarchia* and *Pinacosaurus*, and ceratopsids *Bagaceratops* and *Protoceratops*.

A pair of eight-foot-long fossil arms was their most puzzling discovery of 1965. These fossil arms, so far attributed to this mysterious dinosaur, were later named *Deinocheirus*. Then in 1971 they topped that remarkable find with the unearthing of one of the world's most spectacular fossils—the famous fighting pair of dinosaurs *Velociraptor and Protoceratops*. In 1983 paleontologist Edwin H. Colbert described this incredible fossil. "The two skeletons were in their death poses, with the hands of *Velociraptor*

***Opposite:* Figure 31-4: The Dupuis' life-sized *Ouranosaurus* sculpture. (Photograph by Allen A. Debus, 2000.)**

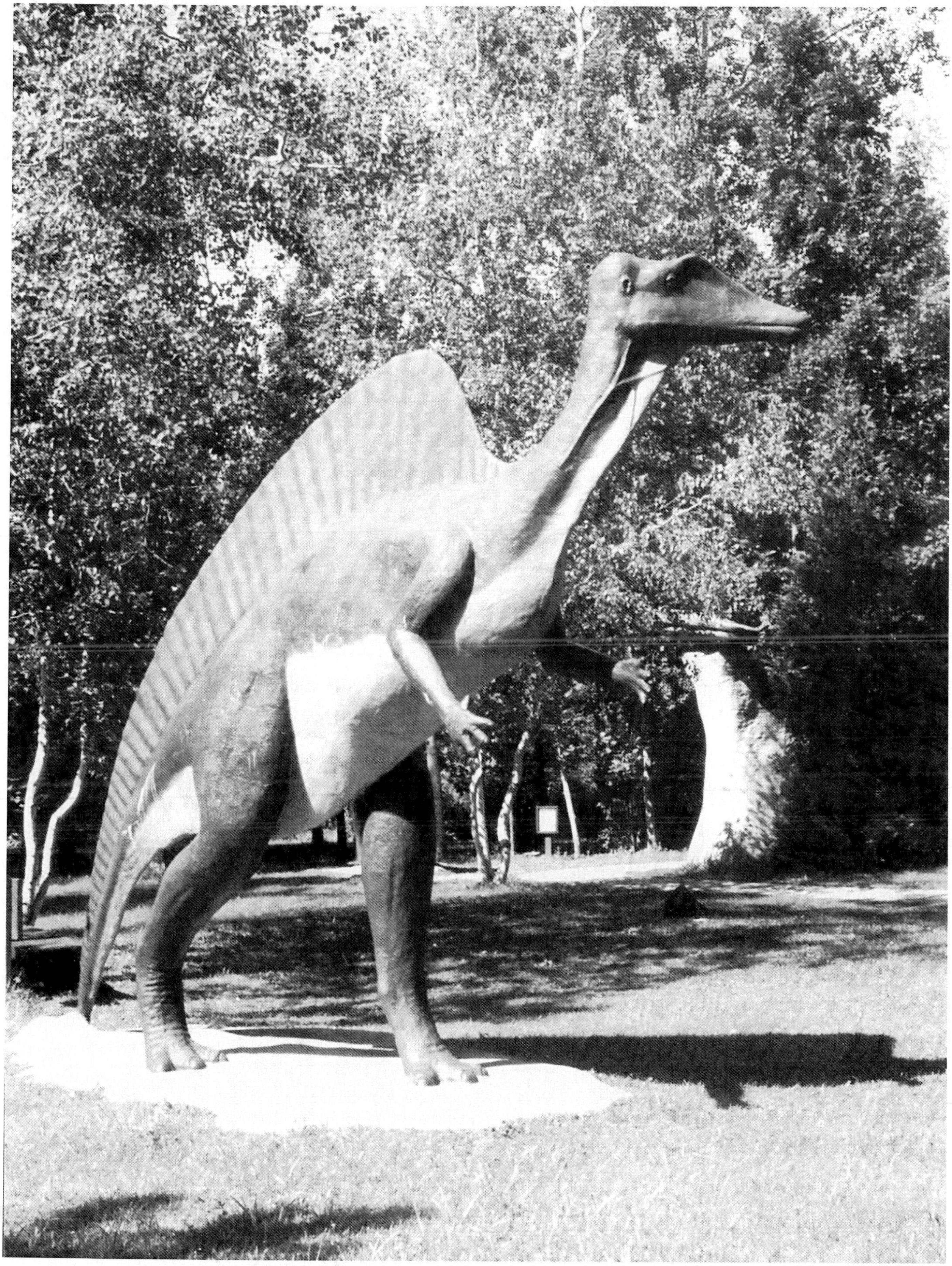

grasping the skull of *Protoceratops* and the sickle claw of the foot positioned as if to tear into the body of *Protoceratops*. It would appear that *Protoceratops*, in turn, had buried its beak within the body of *Velociraptor*. The two antagonists ... seemingly had killed each other and died, locked together."[23] Their bodies were quickly covered by windblown sand in the arid paleo-landscape.

Additionally, a total of 150 mammal skulls dating from the Cretaceous were discovered. In retrospect, the Polish-Mongolian expeditions unearthed a major haul from the Gobi sands, representing a brilliant victory for the paleontological sciences. In 1980 a Polish commemorative stamp was issued featuring an in-situ *Tarbosaurus* skeleton lying alongside a map showing the Polish-Mongolian collecting sites. But the scientific successes were to be commemorated in an even more spectacular fashion, through the construction of life-sized replicas of Mongolian dinosaurs excavated by the Polish-Mongolian paleontological team.

The "Valley's" dinosaur figures were sculpted in cement by talented Silesian artists, scaled from Mr. Skarzynski's preliminary models crafted in plaster or papier-mâché. The models represent genera native to the Mongolian area during the Cretaceous period. When the park opened on July 22, 1975, there were 16 models displayed, comprising nine dinosaur genera. One additional sculptural display was added shortly after the park's opening.

Twelve-year-old Karol Sabath, now a respected vertebrate paleontologist at the Institute of Paleobiology—Polish Academy of Sciences, ran exuberantly into the park following the conclusion of inaugural ceremonies. He was the first official visitor. In 1990 Mr. Sabath related, "...there were also plans to extend the scope of the exhibit (additional open-air models of other Mongolian Upper Cretaceous fauna, e.g. oviraptors and crocodiles, lizards, turtles, dinosaur nests, exposed dinosaur skeleton) ... and to build a pavilion at the entrance to the Valley ... to present an exhibit about the paleontological expeditions and the fossil specimens that underlie the restorations.... But the lack of funds had not permitted such a development of the project (final number of models was intended to be 33). The total cost of the project was several hundred thousand dollars."[24] Polish-Mongolian team members Professors Maryanska, Osmolska and Kielan-Jaworowska served as consultants in designing the "Valley's" dinosaur figures.

In the "Valley" visitors can see the large tyrannosaurid *Tarbosaurus* chasing two ornithomimids—*Gallimimus*. Another *Tarbosaurus* has "...stopped for a moment and raised its head to look behind...." A *Gallimimus* is digging for food. There are also statues of *Nemegtosaurus*, *Saichania* and two *Saurolophus* (a 21-foot-high adult standing bipedally and a juvenile posed quadrupedally). *Nemegtosaurus* tussles with *Tarbosaurus*. *Protoceratops* run and pull vegetation with horny beaks. *Sauronithoides* is shown grasping prey, a multituberculate mammal, in its forelimbs. And those two famous fighting dinosaurs, *Velociraptor* and *Protoceratops*, are frozen in time here, although restored in their final living moments. At the time of its making, no set of dinosaur statues had ever been built in such dramatic, interactive poses. Such animation is difficult to achieve in concrete. (See Figure 31-5.)

In 1972 a former park director, inspired by Pallenberg's dinosaur statues exhibited at the Hagenbeck's Zoo in Hamburg, signed an agreement with the Institute of Paleobiology to proceed with the dinosaur project. Accordingly, the Valley of Dinosaurs display lies on the grounds of a zoological park, the Silesian Zoological Gardens, built on reclaimed coal mine property. Ironically, however, the man-made Valley of Dinosaurs is being threatened with extinction. For the dinosaurs are often vandalized. To make matters worse, environmental pollution caused by atmospheric emissions and periodic flooding of the "Valley" weathers the concrete.

Poland's Valley of the Dinosaurs is the world's most unique dinosaur theme park, as it was constructed to commemorate scientific achievement. Karol Sabath lovingly does all he can to curate the steadily deteriorating displays. The dinosaurs are in good hands! One of the world's finest dinosaur sculptors, Krzysztof Kuchnio, may someday be enlisted to renovate the existing statues and also to enrich the park with new designs.

Karol was the "Valley's " first visitor, but he assuredly won't be the last.

SAURIERPARK

The cosmopolitan interest in life-sized dinosaurs is further exemplified by Franz Gruss' "Saurierpark," located in Germany's Kleinwelka. Franz, originally a concrete worker and janitor by trade, first tried his hand at his newfound "hobby" in 1978, when he built two dinosaurs—*Stegosaurus* and *Triceratops*, sculpted at half size. Since then he has completed over 60 full-sized restorations for his private garden "Sauriergarden," although he too aspires to reach the magical 100 milestone someday. As of the early 1990s, another 40 were displayed in the city-owned Saurierpark, which opened in 1981. (See Figure 31-6.) Each of Gruss' life-sized sculptures are fleshed out in concrete. They are carefully scaled from miniature design models. In design stages he consults with a paleontologist, Dr. Boehme, befriended at a Berlin Science Museum conference.

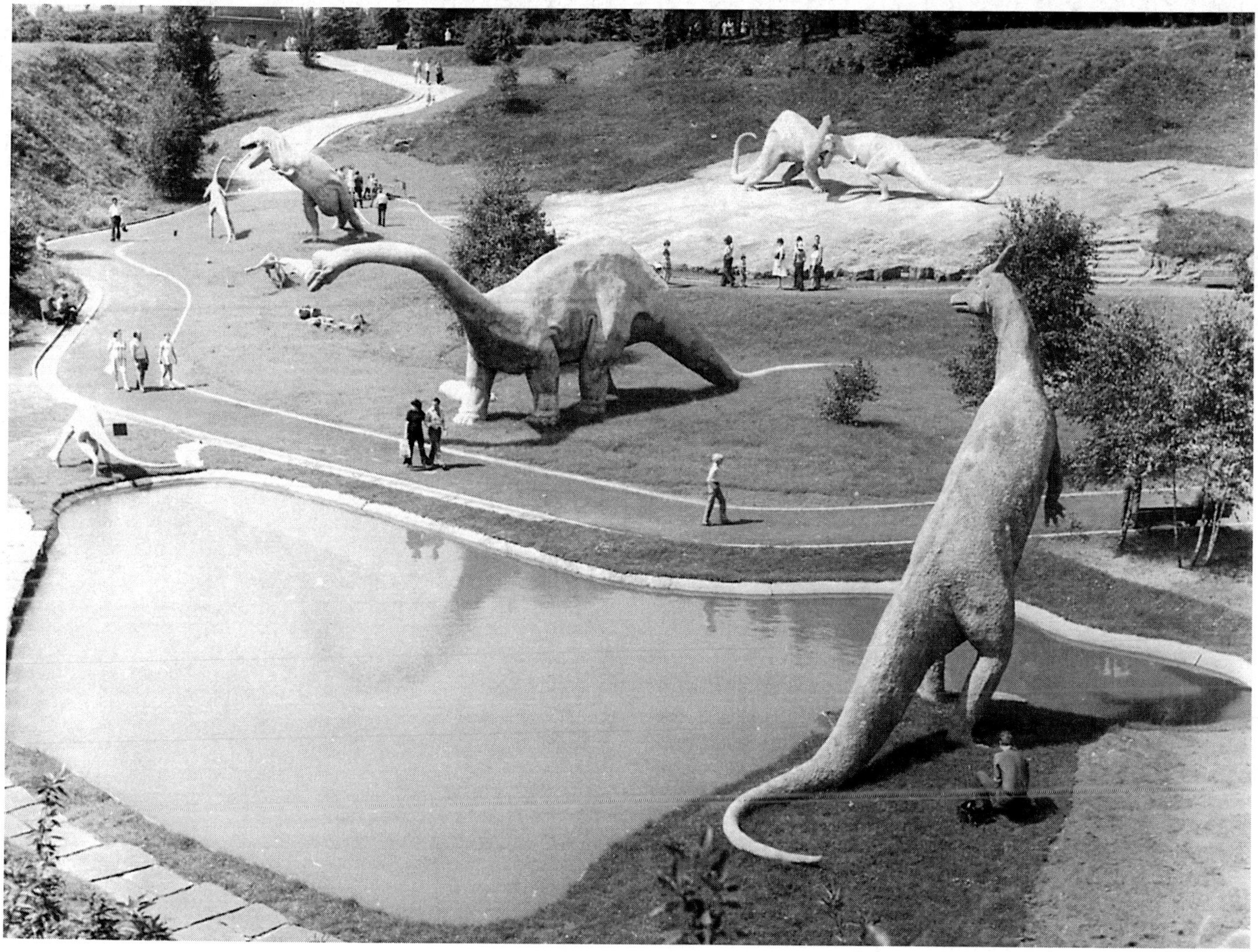

Figure 31-5: An early 1970s panoramic view of Poland's "Valley of Dinosaurs." (Photograph by Wojtek Skarzynski. Reproduced with Mr. Skarzynski's permission.)

Saurierpark was introduced to a western audience in a Society of Vertebrate Paleontology *News Bulletin* issue.[25] But, through translators, Franz has corresponded with the authors. The primary artistic inspiration for Gruss' labors happens to be Zdenek Burian. Like the organization of Burian's most influential paleoart, presented in Joseph Augusta's book *Prehistoric Animals* (1956), Saurierpark is also arranged as a life through time display. Gruss has also sculpted a dragon, which stands among the dinosaur restorations.

Through the 1980s, additional concrete prehistoric animal statues were erected, emphasizing dinosaurs of North America and England. Many displayed genera are the "classical" theme park breed (e.g. *Triceratops*), but visitors will also encounter several animals represented as fossil skeletons in German museums (e.g. *Kentrosaurus, Spinosaurus, Brachiosaurus,* and *Plateosaurus*). Subjects unusual for inclusion in a theme park, the prehistoric hominids *Ramapithecus, Homo habilis* and *Australopithecus*, reflect the sculptor's interests in primate evolution.

Franz has labored to illustrate impressions of dinosaur behavior. For instance, *Diplodocus* stands in a marshy setting beside a nest filled with eggs. A beautifully sculpted *Archaeopteryx* is seen climbing a tree, perhaps suggesting the "trees down" arboreal theory for origin of flight in birds. A pterosaur with a 36-foot-long wingspan, *Quetzalcoatlus*, skimming the surface of a lake, has just pulled a wriggling fish from its tranquil waters. And then there are the expected combat scenes, most dramatic of which are two battling tarbosaurs.

Gruss believes that dinosaurs were warm-blooded, even though the savagery depicted in some of his sculptures is of the "cold-blooded" variety. He speculates that the reason why warm-blooded dinosaur genera grew to such enormous sizes may have been because "...the animals possessing a lower force of gravity due to anatomy had to

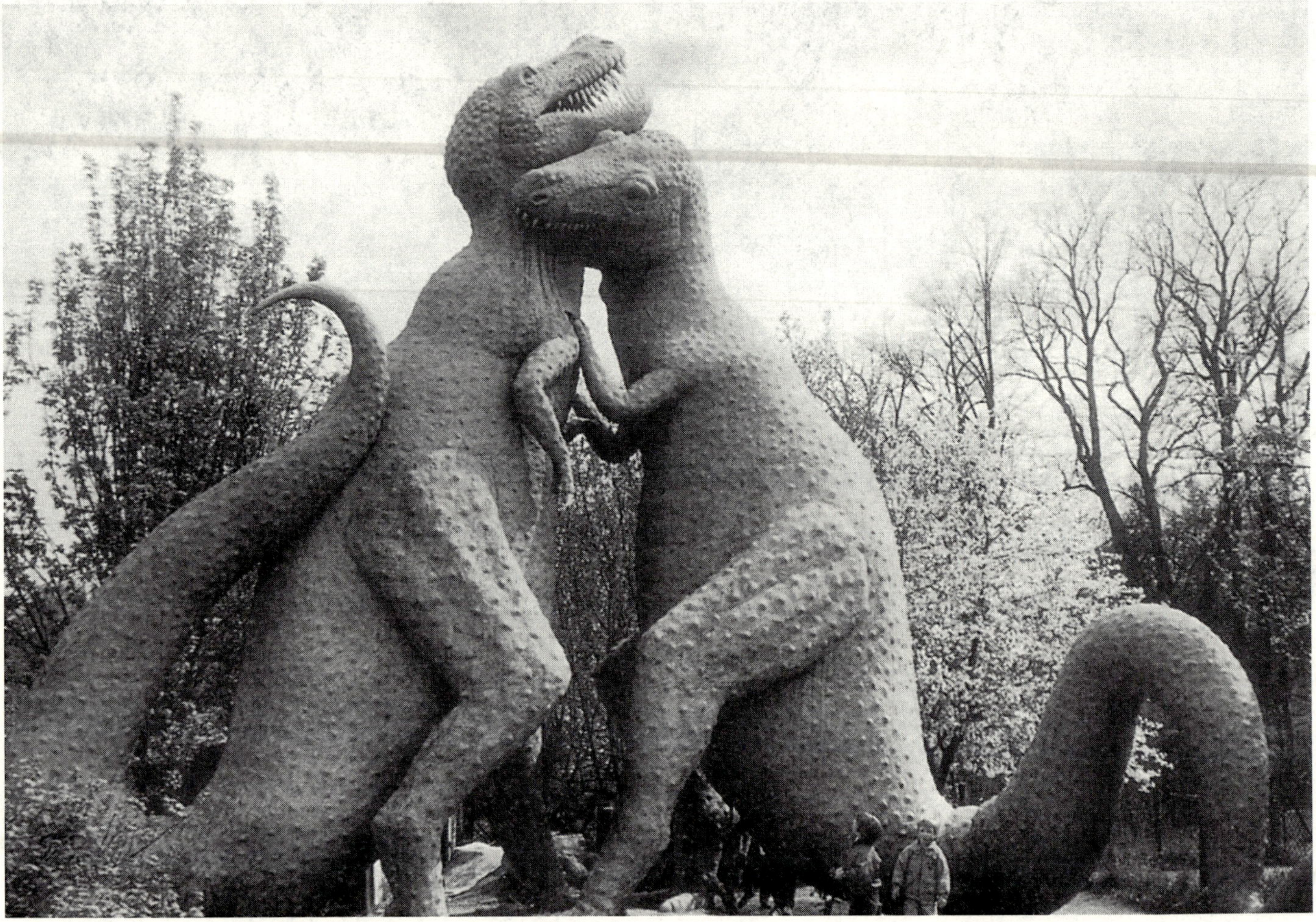

Figure 31-6: Franz Gruss' fighting tarbosaurs. (Reproduced with permission.)

counter this with more activity ... this is only theory and not yet confirmed by scientific study."[26] Well, okay, but perhaps the talented paleoartist should focus on his art and not be troubled by such theoretical details.

Apparently Kleinwelka is rather remote, yet Franz envisions someday constructing a hotel for welcoming visitors—designed as a dinosaur. The "dino-tel" would be built ten times the natural size of a dinosaur, with an observation deck on its head. If ever built, it would be the world's largest dinosaur sculpture! One wonders if the hotel would be shaped like a tyrannosaurid or sauropod.[27]

George S. Eccles Dinosaur Park

Prehistoric animal figures displayed at the George S. Eccles Dinosaur Park and Museum constitute an outdoor museum and "Mesozoic zoo." Here, in a pristine, almost hidden, heavenly setting near the base of northern Utah's Wasatch Mountains, visitors can commune with over 100 life-sized dinosaur and prehistoric animal sculptures. A promotional video boasts, "You'll not likely see more dinosaur figures in any other park," (which is true, unless you've been to Cuba's dinosaur park). The video also states that the dinosaurs have been "...captured in high quality sculpture and a living museum."[28]

The Eccles dinosaur park first opened in the early 1990s. The outdoor theme park provides educational opportunities for aspiring dinophiles. Funding for each of the exhibited sculptures has been privately donated, either by individuals or foundations. Donors' names are indicated on park signage. The cost of installing the sculptures has ranged from $2,000 to $80,000.

Eccles' sculptures are made either of fiberglass or concrete. Most are the independent handiwork of sculptors Tom Prevatt or Matthew Cosman. Recently, a David Thomas fiberglass sculpture of a juvenile *Camarasaurus*, praised by park officials for its anatomical accuracy, was supplied by James Madsen's Dinolab.

Park administrator and vertebrate paleontologist Brooks Britt is proud of the global diversity of prehistoric fauna captured by Eccles' paleoartists. Yet there are discernible patterns in this variety. North American genera are emphasized, particularly those native to the western United States and Utah. Also, the cosmopolitan nature of Mesozoic life is biased toward recent discoveries, favoring 1980s and 1990s headliners such as South America's *Giganotosaurus*, *Amargasaurus* (with a sail adorning its long neck), *Carnotaurus*, England's "big claw"—*Baryonyx*, and even North America's *Utahraptor* and *Maiasaura*. (See Figure 31-7.)

The dino-trail adjacent to the Ogden River offers breathtaking scenery and sights. Many of the statues are dramatically, rather than statically, posed. The effect is so convincing at times that visitors can easily become beguiled into thinking they've strayed onto a "real" *Jurassic Park* movie set.

The paleoartists have certainly earned their wages. Only a little imagination is needed to bring their tableaus to life. Thunder reverberates and trees are heard splintering up there just around the bend. *Carnotaurus* silently dashes around a tree from the lake. To the left, an allosaur forcefully rips into a mother diplodocid protecting her young. To our right, Ginsu knife–footed deinonychi slice into tenontosaur hide, while ostrich-like ornithomimids flee in terror. All around one hears snorting, sniveling, shrieking noises. These are sounds of "living" dinosaurs (not somebody's kids on the loose).

An aquatic combat diorama staged in a pond encircled by parkway would have thrilled Jules Verne over a century ago. Here *Elasmosaurus* engages the ichthyosaur *Ophthalmosaurus*, poised leaping over a rock. This modern scene reminds us of stylish Victorian era paleoart, when ichthyosaurs and plesiosaurs symbolized prehistory.[29]

Eccles is thriving. Donations permit continual expansion. In the fall of 2001, a museum will be assimilated into the park, augmenting the popular children's' educational center. Brooks Britt acknowledges that the earliest sculptures displayed here were not as accurate as those installed more recently. Nesting *Pteranodons* are disproportionally large, for example. The single prehistoric mammal diorama, featuring a cave bear attacking *Smilodons* while Dire Wolves close in for spoils, doesn't really fit at Eccles, North America's consummate *dinosaur* theme park.

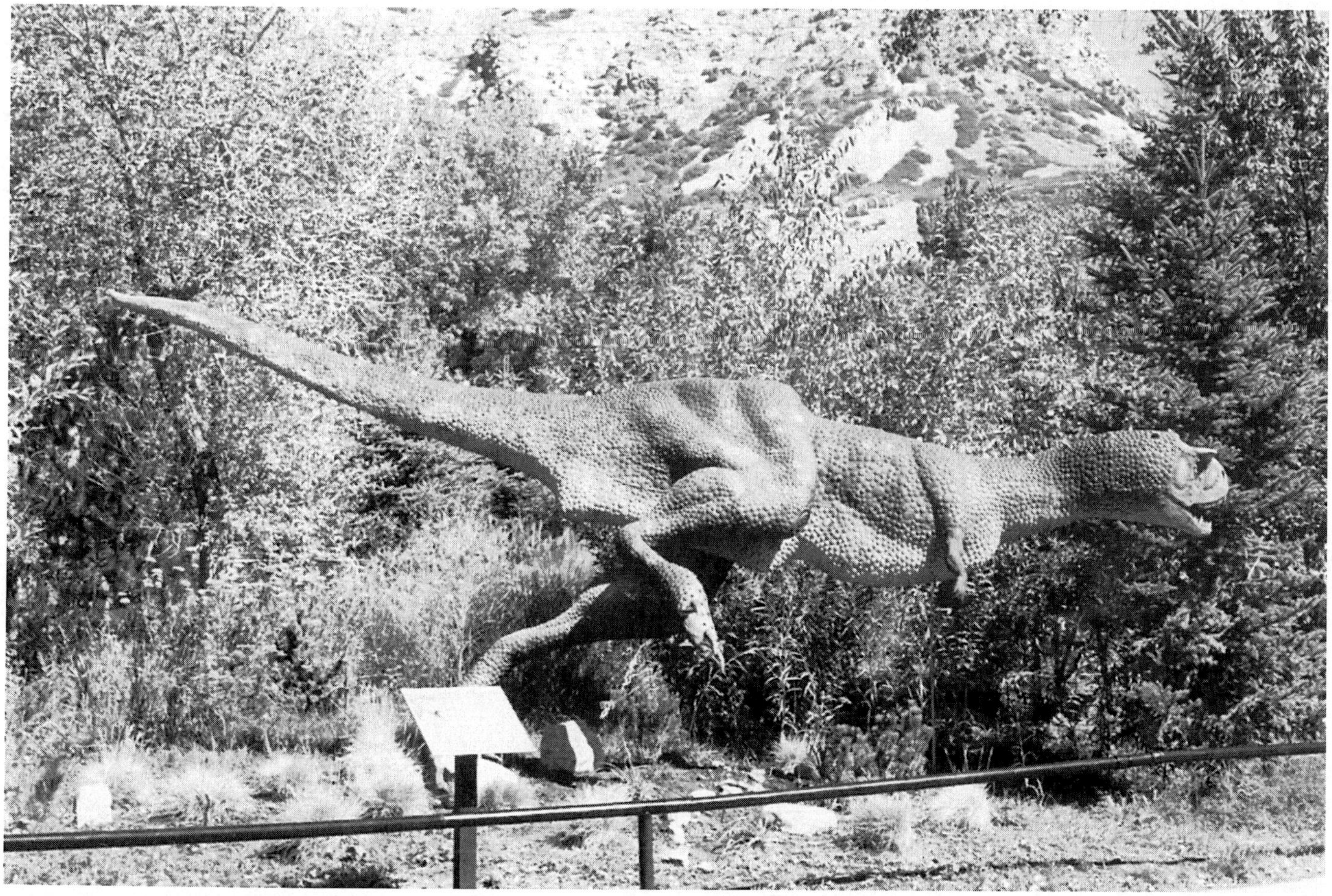

Figure 31-7: Sprinting *Carnotaurus*, one of the many life-sized, dramatically posed dinosaurs displayed at Ogden's George S. Eccles Dinosaur Park. (Photograph by Allen A. Debus, 2000.)

MUNCHEHAGEN'S DINOSAURIERPARK

In spite of Franz Gruss' herculean efforts, the most densely populated dinosaur theme park in all of Europe is Munchehagen's Dinosaur Park. Here, over 100 prehistoric animal figures have been displayed along a winding 1.5 mile trail, organized as a "life through time" interactive "open air" exhibit. Dinosaurierpark's restorations illustrate the sort of creatures of long ago that left their footprints in mud. The 130-million-year-old tracks are preserved in the park within an enclosed structure. Inside this dinosaur monument fossilized tracks of sauropods and juvenile iguanodonts may be seen.[30]

Munchehagen's forested "yellow brick road" leads tourists from the Devonian period, the "age" of fishes, to nearly half a billion years later in the recent Ice Age. However, as expected, the lion's share of prehistoric animal figures are of dinosaurs, most impressive of which is the gigantic *Seismosaurus,* with its neck directing visitors to the restrooms. There is also a three-horned *Torosaurus,* and a *Diplodocus* standing tripodally like the American Museum's skeletal display. Barbara Slater, who visited the park in 1998, found Munchehagen' sauropods absolutely awe inspiring. "Seeing the giant sauropods as they might have been in life, long necks towering above the trees, was truly a sight to make us catch our breath.... The Munchehagen park believes theirs is the first life-restoration of a sauropod in a tripodal stance."[31]

Other dramatic displays feature a recreated scene showing Mongolia's "fighting dinosaur" pair, entombed together for all time, *Velociraptor* and *Protoceratops.* There is also an area devoted to large flightless birds of the Tertiary, and a scene depicting Neandertal men defending themselves from a cave bear's attack. Along the way there are many panels and signage informing the public about the animals on display and what life was like in those former times. A *Tyrannosaurus vs. Triceratops* life-sized diorama attracts visitors to the park entrance. Prehistoric life restorations offer unique educational opportunities here at Munchehagen. The park benefits from the cooperative assistance of the Institute of Paleontology in nearby Hannover.

The figures on display here largely come from either one of two sources. When the park opened in 1992, there were several dinosaurs displayed (many of which have been reproduced for a relatively new park in Florida known as "Dinosaur World").[32] Then a young talented artist, Krzysztof Kuchnio, arrived on the scene in 1994. Mr. Kuchnio, now arguably the world's foremost sculptor of life-sized dinosaurs, has set Munchehagen's outdoor dinosaur displays at world class level.

Kuchnio was lured into the mystical world of paleoart at Munchehagen, first skillfully painting dinosaur scenes. He confesses having been particularly inspired by the paleoart of Doug Henderson, John Sibbick, Mark Hallett and Zdenek Burian. After Kuchnio was asked to repair some of the existing sculptural exhibits, he "...often noticed deficiencies in the extant statues."[33] So he decided to build several of his own design. These now rate among the world's most remarkable prehistoric animal sculptures. Beginning with six *Archaeopteryx* sculptures for one Munchehagen enclosed display, he proceeded to life-sized sculptures of *Coelophysis, Deinonychus, Diplodocus, Plateosaurus, Brachiosaurus, Euparkeria, Euplocephalus, Stegosaurus, Iguanodon* and others. (See Figure 31-8.)

The sauropods at Munchehagen are Kuchnio's. So are many of the horned dinosaurs and theropods, and a scene of *Deinonychus* ripping into an iguanodont. Kuchnio initially sculpts a miniature model, leading to the casting of a life-sized fiberglass replica. A common circumstance among dinosaur theme park artists is the lack of available fossil material to study in the critical preliminary design stages. As meticulous a performer as he strives to be, Kuchnio laments that, "...rarely I've got the possibility of profiting museum pieces, and the major part of reconstructions are made only on the basis of description and illustration in books."[34] So the accuracy of his Munchehagen models is partly dependent on his consultations with Deltew Thies, a professor of paleontology at Hannover's Institute of Paleontology.

Kuchnio is in his mid-thirties and has high aspirations. He has only been working in this field for less than a decade, yet has already sculpted 50 life-sized prehistoric animals. Before he retires, will he set the record for completing the most original life-sized prehistoric animal sculptures? The race is on!

Kuchnio dreams of building a single diorama comprising from ten to twenty prehistoric animal statues. His fine work is proudly displayed in several European paleontology institutions and certainly will receive further deserved attention in years to come. For Kuchnio is a master among the modern breed of life-sized dinosaur "renaissance" builders.

It all started here at Munchehagen!

HALF-BAKED ODES TO PREHISTORY?

Surely there are theme parks of lesser majesty than the few examples mentioned here. For every "good" park there are many of questionable caliber. Dinosaur theme parks' reputation has been disparaged by the authors of *The Encyclopedia of Bad Taste* (1990): "Dinosaur parks are

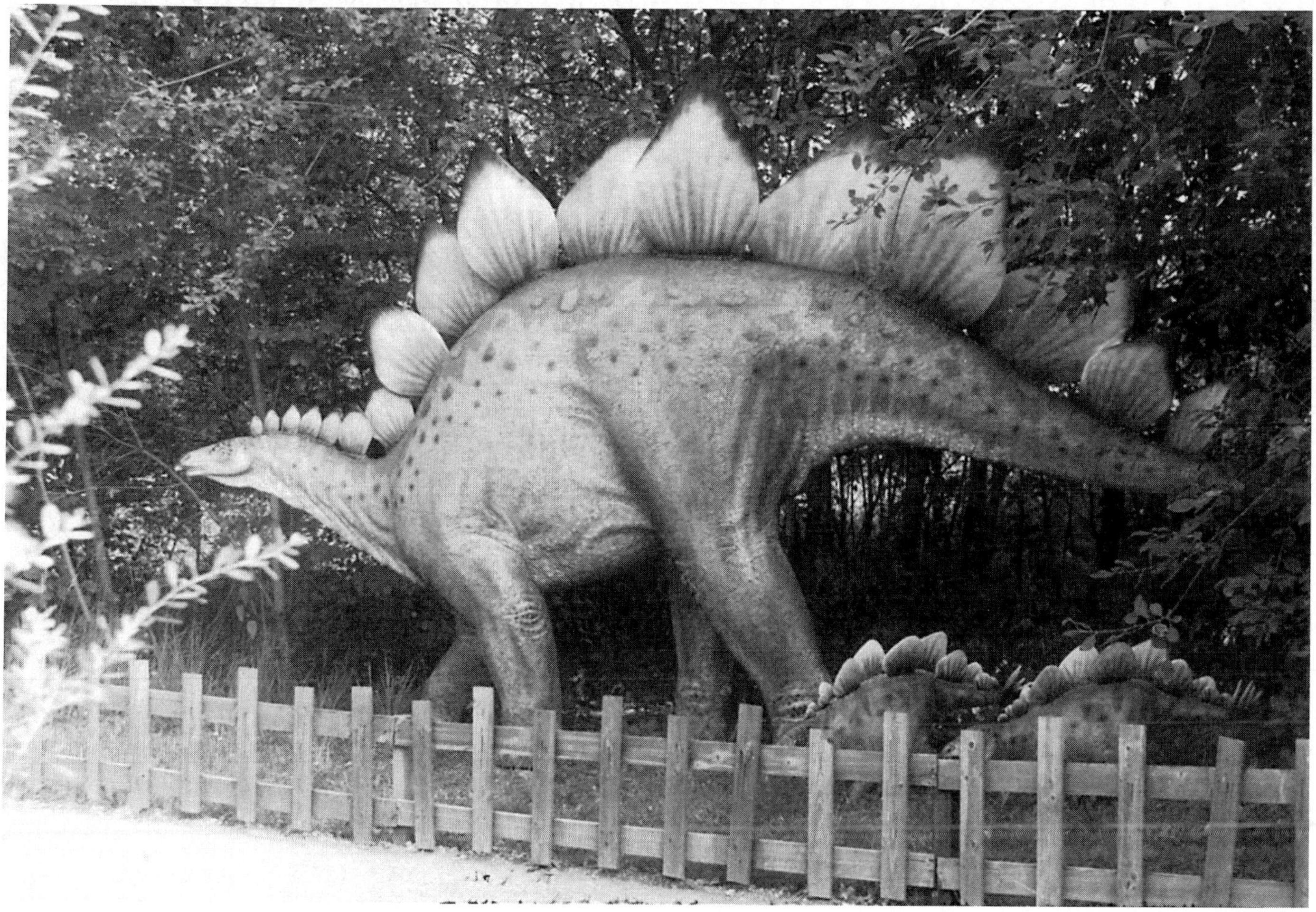

Figure 31-8: Krysztof Kuchnio's life-sized sculpture of the *Stegosaurus* and babies. (Reproduced with permission.)

half-baked odes to prehistory, most of them created by roadside entrepreneurs who combine their enthusiasm for triceratops with audacious showmanship and a promise that visiting the park is educational."[35] One bad apple spoils the whole bunch.

True, you can't expect to find the hucksterism inside museum walls that can be encountered at an outdoor dinosaur theme park, and the science is always better. Yet dinosaur theme parks, a neglected or perhaps "devolved" expression of Osborn's love for edifying museum restorations, continually reel us in.[36]

Hey! Read that signpost up ahead: "Dinosaurs—Next Exit!" Here we go again. You know the drill. Kids, put your shoes back on. Grab the camera. We're goin' in!

Such symbolic paleoimagery cannot be underestimated. Osborn's legacy!

Chapter 32

Those Incredible Shrinking Dinosaurs

The artistry of miniature scale dino-diorama building was perfected long before the invention of Super Sculpey, the modeling clay most often used by today's generation of paleosculptors. In fact, the first "modelers" were museum technicians and "paleoartists," not "garage kit" builders and hobbyists. The preferred materials and finishing techniques may have changed since the early days of dino-diorama building, yet the inspirations and objectives of the pioneers were similar to those of today's generation of garage kit builders. "Make it look alive and as accurate as possible, and let the art tell a story convincingly." Those are the time-honored guiding principles.

It is recorded that during the 1820s a sculptor named Sir Francis Legatt Chantrey (1781–1841) "modelled" the earliest significant plesiosaur fossils, found in 1823. Michael Taylor states that, "Although this, strictly speaking, implies that Chantrey was sculpting a model or replica, perhaps of clay, I suggest ... that Chantrey (or rather his assistant) was casting the specimen in plaster."[1] Well, yes, maybe, but perhaps he really was preparing the first miniature sculptural restoration of a plesiosaur instead. Does the practice of creating miniature prehistoric animal sculptures date to the 1820s? Do the murky origins of dino-dioramas predate even Owen's 1842 naming of the "Dinosauria"? Wild idea, huh?

Any of us who have sculpted or modeled a miniature dinosaur or admired the handiwork of others, understands that feeling of satisfaction scale models and replicas of dinosaurs and other prehistoric animals can bring. Small-scale dinosaur replicas are the "bite-size" prehistoric animal equivalent of miniature railroads.

Life-sized dinosaur models can take on a breathtaking life of their own, but small models delight us with their diminutive proportions. Small models and dioramas can also be a Curator's delight, because miniature exhibits also conserve precious museum display space and are less costly to build. They can be constructed in far less time than large restorations, and with more bang for your buck too.

Even if Chantrey's plesiosaur wasn't the first, artists made the earliest strides into miniature prehistoric animal diorama building well over a century ago. Benjamin Waterhouse Hawkins should perhaps be credited for sculpting the world's first miniature dinosaur sculptures. Hawkins' models served as guiding "tools" for accurately scaling his life-sized Crystal Palace concrete statues, many of which remain on display at Sydenham, England. Five of Hawkins' small-scale models were reproduced as plaster replicas and marketed for sale through Henry Ward's *Catalogue of Casts of Fossils from the Principal Museums of Europe and America* (1866).[2] In the late 1860s Hawkins also built several small test scaling models intended for the Paleozoic Museum planned for New York's Central Park. Regrettably, both life-sized and miniature models were demolished in 1871.

Dino-philes have seen photos of the nineteenth and twentieth century's greatest paleoartist, Charles R. Knight, sculpting miniature prehistoric animals in clay. We have noted 20 of his masterful creations, many of which were available for sale as plaster replicas from the American Museum of Natural History years ago.[3] So, as a creator of small-scale prehistoric animals in the (approximate) 1/10 and 1/20 scale range, Knight outshone even the nineteenth century's greatest paleo-sculptor—Hawkins.

An artistic vertebrate paleontologist, Charles W.

Gilmore, respected for his researches into the paleobiology of *Apatosaurus*, plated and theropod dinosaurs, also dabbled in miniature paleo-sculpture. A note in the 1948 edition of *The National Cyclopaedia of American Biography* states that, "Creating dinosaur models was his hobby, and golf and motor trips to historical places his chief recreations."[4] However, Gilmore's "hobby" seems to have bordered on preoccupation. For Gilmore sculpted a number of small-scale dinosaurs that were reproduced as plaster replicas and sold as museum displays. Several of his classical dinosaur casts still are exhibited in natural history institutions. They are also cherished by collectors of vintage, cold-blooded dinosaur era figures.

Knight's, Gilmore's and Hawkins' sculptures could be purchased as educational tools or as decorative gift items. Sculptural representations such as these could be placed on display in paleontological halls adjacent to skeletal reconstructions. Yet these artists can't be categorized as diorama-builders, per se.

Besides Gilmore, Richard Swann Lull was another respected vertebrate paleontologist who was also a talented sculptor. Several of Lull's dinosaur miniatures remain on display at the Yale Peabody Museum (e.g. *Podokesaurus* and *Yaleosaurus*). A Peabody Museum archive photo shows a contemplative Lull assisting in the construction of Yale's miniature Jurassic dino-diorama (featuring *Stegosaurus*, sauropods and allosaurs) during the mid–1920s, a formative period for this popular style of paleoart. Yale's Jurassic diorama was removed from exhibition in 1966.[5] (See Figure 32-1.)

From individual sculptural studies of prehistoric animals, fleshed-out dioramas were the next logical step. A presumption with any "accurately" depicted diorama is that the restored period's flora and paleoecology is also sufficiently understood, and that paleoartists can confidently portray how dinosaurs interacted with each other and other contemporaneous fauna. Dioramas may illustrate more about scientists' perceptions of dinosaurian

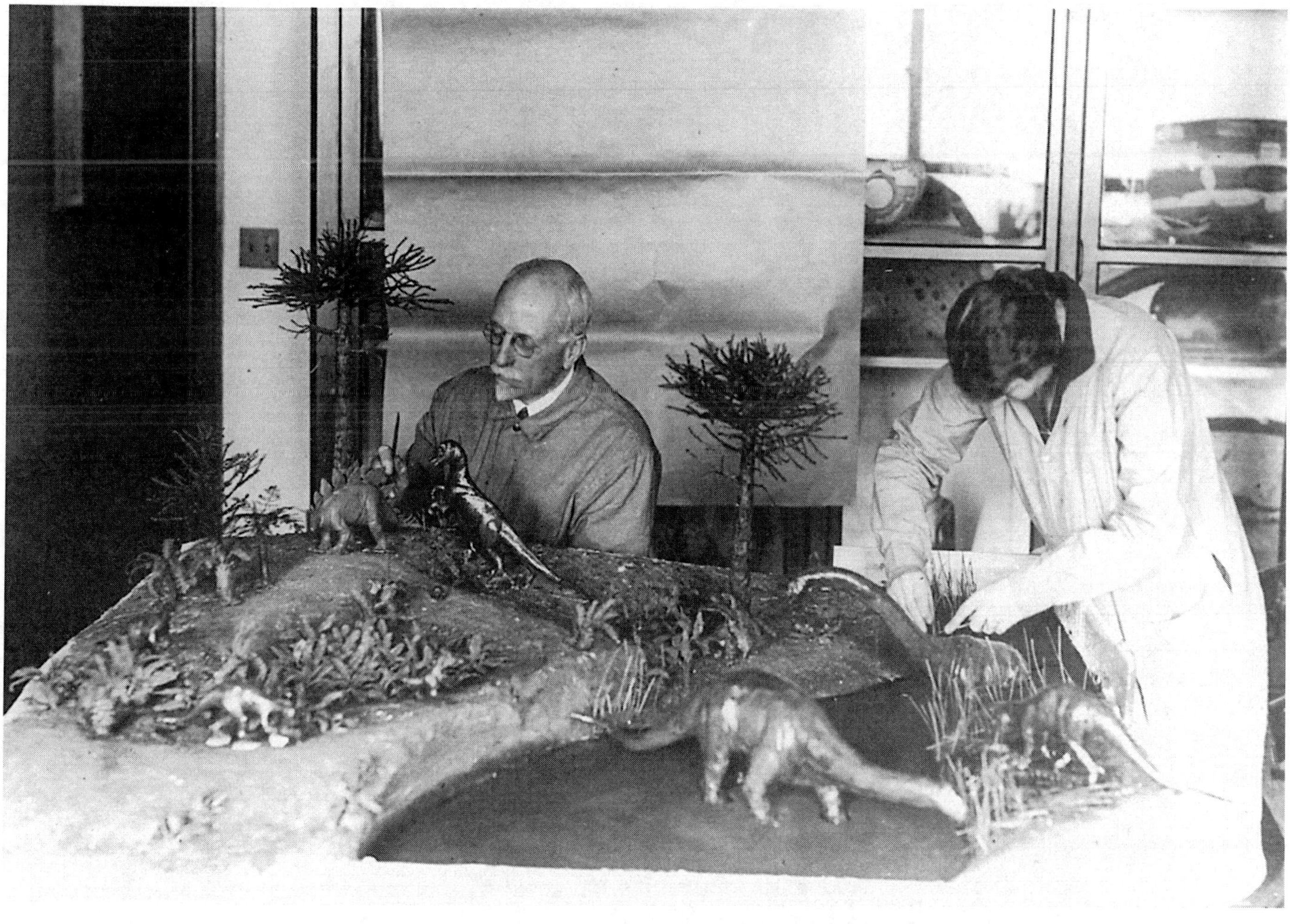

Figure 32-1: The Jurassic age diorama under construction at the Peabody Museum of Natural History. Richard Swann Lull is seated at left. (Courtesy of the Peabody Museum of Natural History, Yale University, New Haven, CT.)

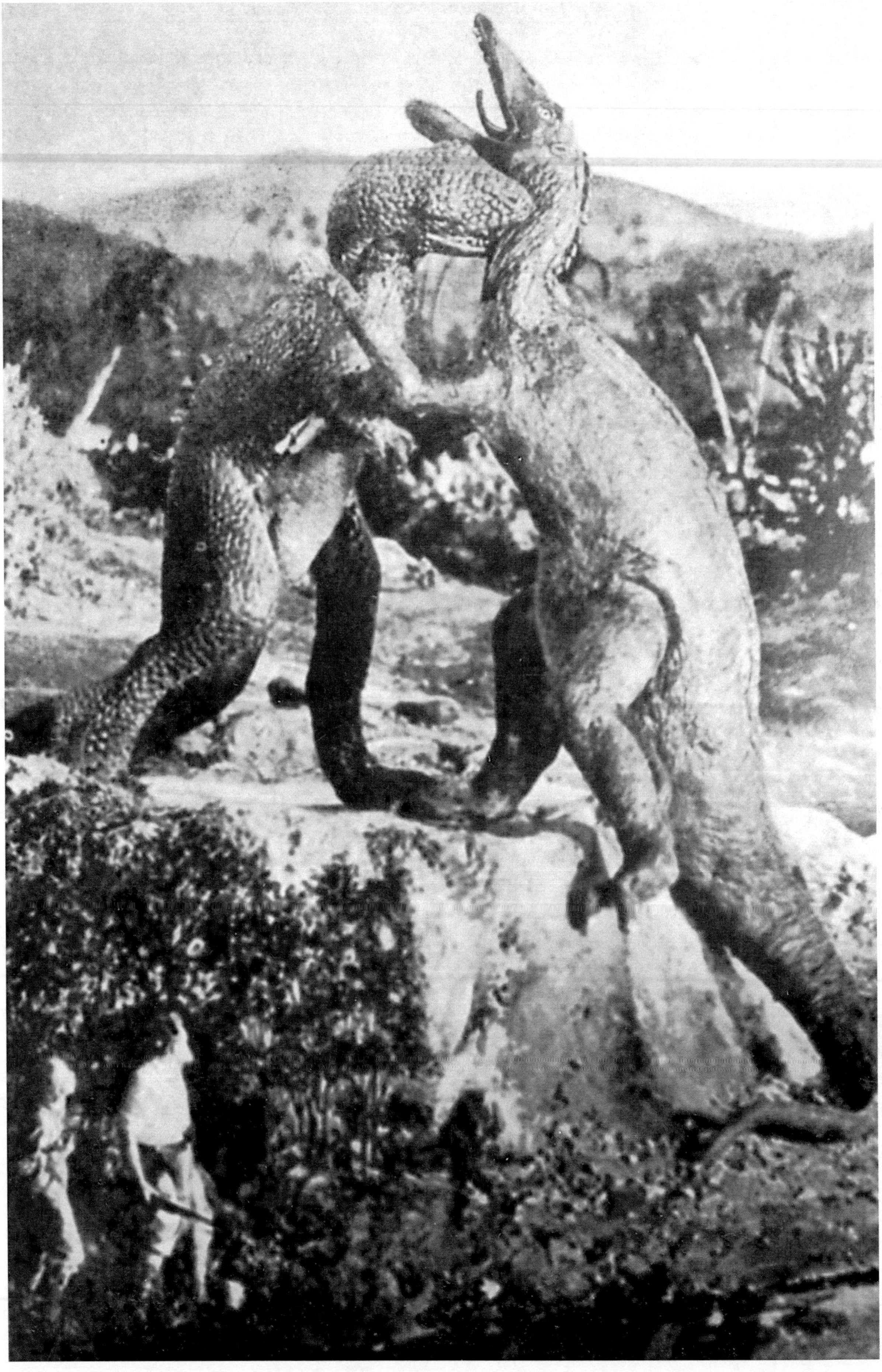

behavior and social tendencies than would a sculptural restoration of an individual animal.

Skillfully done paintings and murals can lure our spirits into—and beyond—the canvas (in a "through the looking glass" sort of fashion), but a three-dimensional miniature museum diorama or restoration must seem even more appealing to our inner childlike nature. Dioramas instinctively attract the attention of modelers, hobbyists and toy lovers—those who would like to dive in and "play" with the figures on display.

It's tempting to muse whether the early miniature museum dino-dioramas represented a static manifestation of the public's fascination with movie special effects. Masterful stop-motion animation artists Willis O'Brien and Marcel Delgado had enormous success creating small dioramas incorporating dinosaurs and other prehistoric animals that "lived" following projection of their images onto film. "Trick" photography, use of incredible sculptures and lavish miniature sets captivated the hearts and minds of any and all who subconsciously yearned to revel in those marvelous primeval fantasy "playgrounds." (See Figure 32-2.)

Sci-fi movies offered "actualization" that could never be experienced by viewing a museum restoration or diorama, no matter how cleverly engineered. Although the dinosaurs in such films as *The Lost World* (1925) and *King Kong* (1933) are scientifically inaccurate (in some respects even by contemporary standards), the "primeval" landscapes engineered by O'Brien and Delgado may be considered "living dioramas" executed with flair and perfection. And, to some degree perhaps, dinosaur sci-fi movies also fostered the practice of museum dino-diorama building, as well as production of the earliest dinosaur playland toys.[6] While the first dinosaur "toys" may not have been influenced by museum dioramas, it is possible that the concept behind playset dioramas (such as the 1950s Marx playsets and dinosaur model kits), and the stimulus for avocational dino-diorama building, were.

By the 1940s, master wildlife sculptor Louis Paul Jonas had developed a reputation as the world's foremost wildlife "miniaturist." His small-scale animal figures and dioramas became valued museum displays. Jonas' early successes with extant wildlife eventually inspired him to create a portfolio of prehistoric animal figures and dioramas. The prehistoric line included his marvelous ⅒ scale dinosaurs, originally serving as the basis for modeling a life-sized dino-diorama constructed for the 1964/65 New York World's Fair Sinclair Dinoland. Jonas' miniatures are still displayed in many North American museums

By the late 1950s, miniature dino-dioramas had invaded many museums worldwide. And less sophisticated "dinoramas," manifested in the form of playsets and toys had replicated beyond the museum confines onto children's bedroom and living room floors, creating a mini, pre-Bakkerian "renaissance" of their own, an unprecedented invasion of saurian "Mini-Me's." According to a reporter writing for a 1958 edition of *This Week Magazine*, considering all the dinosaur toy playsets then available to kids, America experienced its first "Dinosaur Renaissance" during the late 1950s. "The American assembly line has already outdone nature," stated one contemporary museum curator.[7]

During the late 1960s it was possible to admire miniature creations of master sculptors Maidi Wiebe, George and Paul Marchand, Robert Butsch, Erwin Christman, F. Konig, Franz Roubal, Jay Matternes, Vernon Edwards, Arthur Hayward and many others. Their work must have (at least, perhaps, on a subliminal level) influenced today's generation of "miniaturists," many of whom weigh in regularly in the pages of a popular dinosaur magazine especially devoted to the hobby of prehistoric animal modeling, collecting and paleoart, the *Prehistoric Times*.

Besides the few individuals mentioned here, there must be many "unknown" diorama sculptors, long-lost names and people we'll never know although their work survives under museum display plexiglass, or is preserved in museum storage, archive photos or on the faces of collectible postcards. (Nor is it our intention to list here all the wonderful dioramas either presently or formerly on display.) Thanks to their efforts and visions, miniature dino-dioramas have become a highly appreciated and, traditional museum display form.

What can we look forward to beyond the days of Super Sculpey? Can the task of dino-rama building be simplified? Well, now computerized industrial scanning technology offers a means of accurately scaling down images of real fossils to desk top scale. A technique known as "stereolithography" creates shrunken resin casts of the original figures.[8] Imagine the possibilities for dino-modeling presented by these innovations!

Welcome to the future. The days of the "Incredible Shrinking Dinosaur" have only just begun.

Opposite: Figure 32-2: Still from *The Lost World* (1925) showing undaunted Professors Challenger and Summerlee observing a death struggle between Marcel Delgado's animated *Allosaurus* and *Trachodon*. Note how this scene, in which diminutive humans interact with large dinosaurs that in reality are small "puppet" models, reciprocates with Figure 32-1, in which humans design diminutive models of creatures that in reality were giants. Small dinosaur sculptures can easily lure humans into the Mesozoic era's enchanting realm.

CHAPTER 33

Joined at the Hip?

The lower Cretaceous British dinosaur *Hylaeosaurus* (the "Wealden Lizard" or "Fossil Lizard of Tilgate Forest") was a charter member among Sir Richard Owen's "Dinosauria" of 1842, the second fossil saurian to be classified as a dinosaur. Announcement of the first specimen came from Gideon Mantell during the summer of 1832 following its discovery in a Tilgate Forest rock quarry. Mantell inferred that this specimen had been transported fluvially over a great distance, suffering extensive mutilation along the way prior to deposition in a river delta. No skull was found with the remains. Mantell claimed excitedly, "I have made a grand discovery in the grit of Tilgate Forest. A mass of stone blown into fifty pieces or more by the quarrymen was found full of bones. They reserved it for me, and with much difficulty and great labor I have succeeded in uniting and clearing a slab 4½ by 2½ feet exhibiting twelve vertebrae, eight in place, with many ribs, coracoid bones, omoplates, chevron bones, etc. and several of those curious dermal bones which support the scales."[1] Mantell's "grand discovery" would prove perplexing to paleontologists for over 150 years!

Mantell described the *Hylaeosaurus armatus* in 1833.[2] Fifteen-inch-long dermal spines were also noted in the fossil. The dermal spines were positioned in the fossil at right angles to the vertebral column, above the level of the coracoid bones.

Another significant specimen (the third documented *Hylaeosaurus* discovery) was also found in the Tilgate Forest during the Autumn of 1837. Mantell had to procure this specimen from parish laborers who were "...unacquainted with the increased value of carefully extracted specimens."[3] This specimen consisted of several distinct spinal column sections. Mantell noted how dermal spines associated with this specimen matched those discovered with the first. Other remains attributed to *Hylaeosaurus* by Mantell included a perfect 18-inch-long scapula and 16-inch-long humerus (from the Weald clay near Bolney), four fused sacral vertebrae, and (possibly) a tooth found in the Tilgate Forest.

Although Mantell tried to rename *Hylaeosaurus* in honor of Richard Owen, in 1844 the two dinosaur authorities disagreed over the nature and function of the dermal spines. While noting certain crocodilian (and chelonian) affinities in *Hylaeosaurus*' skeletal anatomy, Mantell objected to Owen's suggestion that the spines represented an internal "monotrematous modification" of abdominal ribs.[4] Perhaps Owen sought to demonstrate that *Hylaeosaurus* was a superior (advanced) reptile converging on the mammalian form, as he had claimed in the case of the two other saurians he had assigned to the "Dinosauria," the *Iguanodon* and *Megalosaurus*. Such arguments would help in refuting the hypothesis that animals "progressed" or evolved (in a Lamarckian sense) through geological time.[5]

When viewed under the microscope, a section of *Hylaeosaurus*' 15-inch-long spine was found to have an internal texture identical with that found in *Hylaeosaurus*' scutes, or those "curious dermal bones" originally announced by Mantell. Therefore, Mantell claimed, the dorsal spines were (exterior) dermal, sternal features which, in life, must have resembled the scaly integument and dermal patterns found in modern iguanas. Owen was wrong about the (monotrematous) condition of *Hylaeosaurus*. Instead, to Mantell, *Hylaeosaurus* bore reptilian, not mammalian, affinities.

Furthermore, Mantell speculated, the iguana-like spines must have been positioned in a fringe or crest descending along *Hylaeosaurus*' spinal column. Possibly *Hylaeosaurus*' crest had been pushed to one side of its body after its carcass had come to rest in a sedimentary deposit. Clearly Mantell had become so comfortable with the iguana paradigm that he could not accept the orientation of the fossilized bones as direct evidence of its living form.[6]

Although living iguanas only grow to five feet in length, with one-inch-long dermal spines situated along the spinal column, Mantell estimated *Hylaeosaurus'* length to be 20 to 30 feet. He thought the body was wider than it stood vertically on its relatively short limbs, and that its wide flat tail was long, slender and flexible, contrasting considerably with that of the *Iguanodon,* which possessed a tail "greatly developed" in the vertical dimension. The skin of Mantell's *Hylaeosaurus* was covered in bony scutes, horny tubercles, and possessed a single row of thin, angular spines extending down its back, which formed a serrated crest.

Perhaps the earliest life restoration of *Hylaeosaurus,* a watercolor titled "Reptiles restored, the remains of which are to be found in a fossil state in Tilgate Forest, Sussex," was prepared for Mantell by artist George Scharf in 1833.[7] (See Figure 3-2.) Five years later, artist John Martin incorporated *Hylaeosaurus* into his 1838 restoration, *Country of the Iguanodon,* commissioned by Mantell as a frontispiece for his popular text, *Wonders of Geology* (1838).[8] (See Figure 4-1.)

Several years later, paleoartist Benjamin Waterhouse Hawkins completed a full-sized replica of the *Hylaeosaurus* under Owen's direction for the Crystal Palace exhibition at Sydenham. Owen deferred to Mantell's vision of the *Hylaeosaurus.* Hawkins' sculpture, like Scharf's then 20-year-old restoration, bore a series of spines along its vertebral column and tail. The exhibition opened to the public on June 10, 1854.

Hawkins needed to create a head for his *Hylaeosaurus* sculpture, and judging from the appearance of his fanciful model, he seems to have relied on the iguana for anatomical inspiration. (Seventeen years later, in fact, Hawkins resorted to the same strategy for Joseph Leidy's *Hadrosaurus.* See Figure 8-5.) Hawkins also favored the iguana-like crest of vertebral spines still visible on his Crystal Palace restoration, and appears to have been further influenced by Owen in the selection of an upright, pachydermal limb posture. (See Figure 33-1.) With two obvious exceptions, Hawkins'

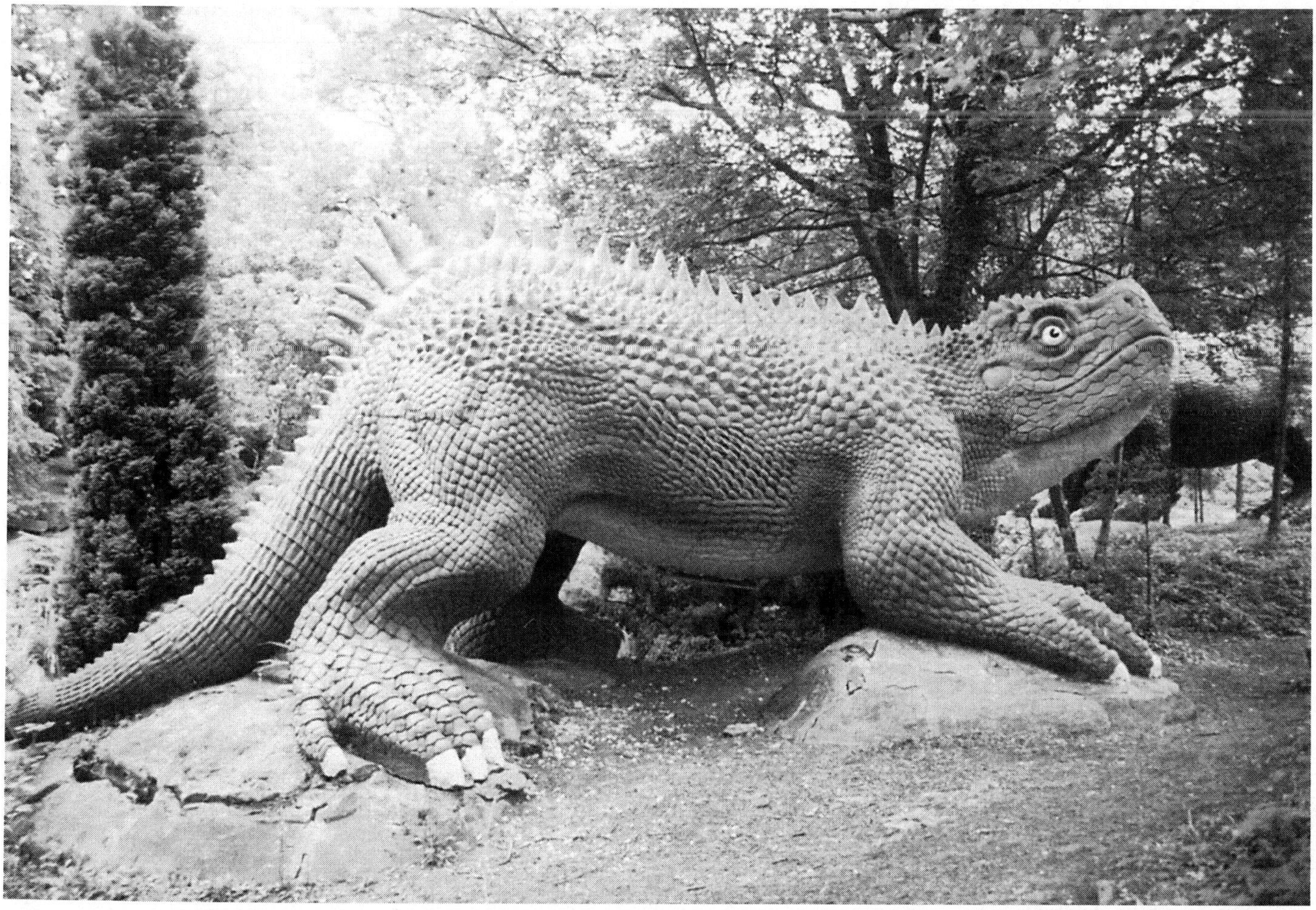

Figure 33-1: Benjamin Waterhouse Hawkins' *Hylaeosaurus* sculpture displayed at Sydenham, England. (Photograph by Steve McCarthy, reproduced with permission.)

Hylaeosaurus appears similar to his *Iguanodon*. *Hylaeosaurus* has a dermal crest lacking in *Iguanodon*, but the latter restoration had a misplaced rhinoceros-like nasal horn not seen in the former. (See Figure 6-2.) *Hylaeosaurus* also appeared in a souvenir wall poster created by Hawkins in 1862, and an entire herd is evident in his painting *Jurassic Period in England*, completed during Hawkins' residence at Princeton University in the mid–1870s.

In referring to episodes in Victorian palaeontology, popular books on dinosaurs usually include photographs of Hawkins' most famous recreations at Sydenham, *Iguanodon* and *Megalosaurus*. Of the dinosaur fauna sculpted by Hawkins for the Crystal Palace, only the *Iguanodon* is well represented in the fossil record. Images of *Megalosaurus* and *Iguanodon* are popular subjects for paleoartists. We find that *Hylaeosaurus*, however, has been largely neglected.

Hawkins' design was copied a handful of times by other artists during the next half century. Artist, Edouard Riou restored a spiny, iguana-like *Hylaeosaurus* for the 1860s editions of Louis Figuier's immensely popular book on prehistory, *The World Before the Deluge*. (See Figure 33-2.) In the Preface written for Reverend Henry Neville Hutchinson's popular paleontology book *Extinct Monsters* (1893), Henry Woodward noted the inaccuracies in Hawkins' Crystal Palace dinosaurs.[9] In his text, Hutchinson stated, "Mr. Waterhouse Hawkins' model at Sydenham, near the Iguanodon, was based on the above discoveries, which are insufficient, and is far from the truth."[10] Hutchinson's artist, Joseph Smit, did not prepare a restoration of the *Hylaeosaurus* for this book, although his restoration of a related dinosaur, *Scelidosaurus*, looks remarkably how paleontologists might restore the *Hylaeosaurus* or even the *Polacanthus*, today. (See figure 33-3.) Then, in the early twentieth century, artist F. Long relied on Hawkins' restoration for his own rendition of the *Hylaeosaurus*, included in an obscure set of German souvenir poster cards issued by the Cocoa Company of Hamburg-Wandsbek. Evidently, Mantell's "headless dinosaur" was a relatively unpopular subject for early dinosaur artists.

By the turn of the twentieth century no new *Hylaeosaurus* discoveries had been made to augment the specimens known to Mantell and Owen, making it impossible

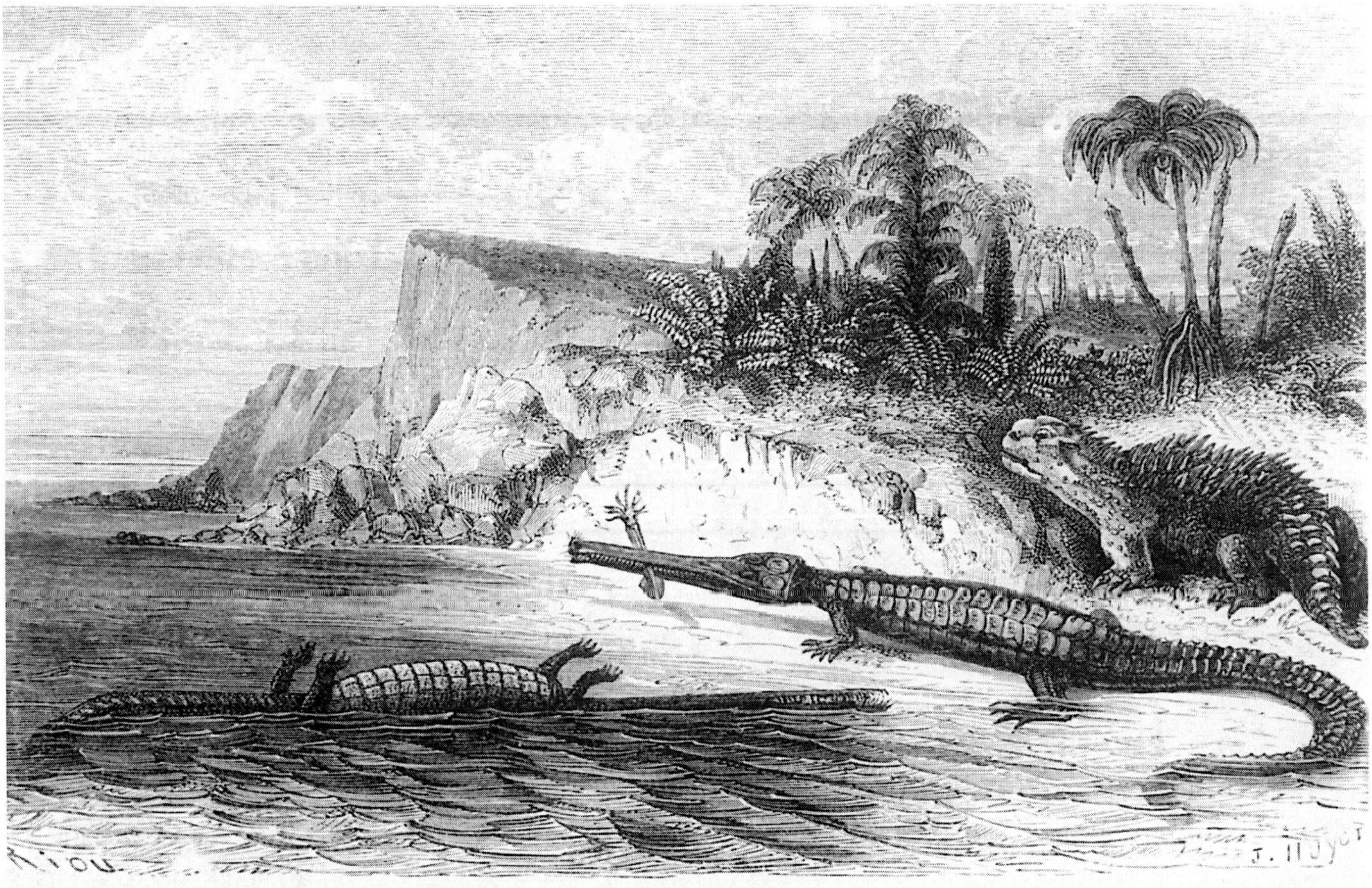

XVIII.—Ideal landscape of the Middle Oolitic Period.

Figure 33-2: Edouard Riou's Early Cretaceous scene for *The Earth Before the Deluge*. *Hylaeosaurus* stands at right of two battling teleosaurs. (Authors' collection.)

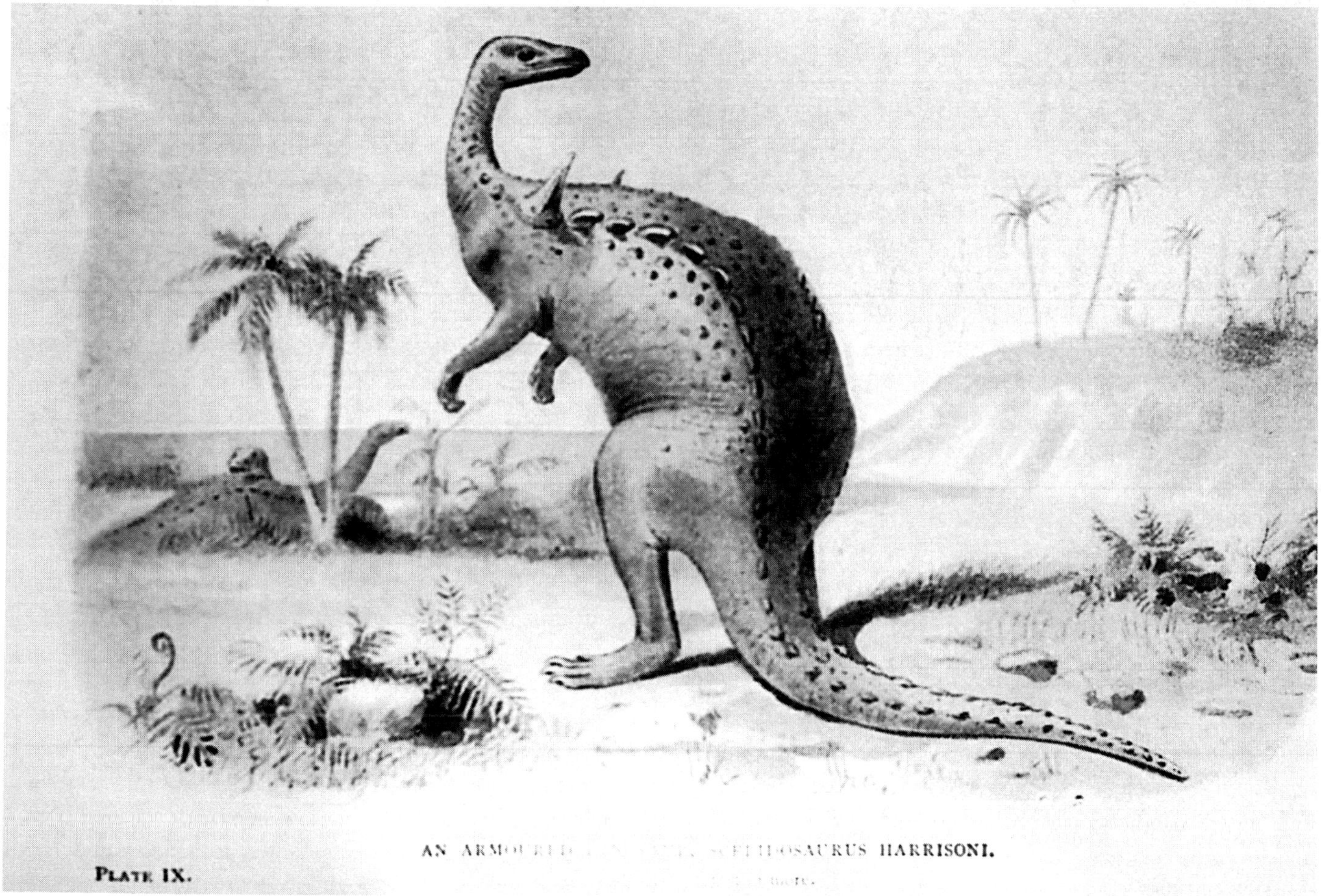

Figure 33-3: Joseph Smit's restoration of *Scelidosaurus*, a poorly known British armored dinosaur. The bipedal, Kangaroo-like posture mimics earlier views of Hadrosaurus. (From *Extinct Monsters*.)

to judge the accuracy of Hawkins' forty-year-old restorations.

However, another Early Cretaceous armored dinosaur, strikingly similar to *Hylaeosaurus*, had been discovered in 1865, from the Wealden of the Isle of Wight, by the Reverend William Fox (1813–1881). This was another headless specimen apparently adorned with two parallel rows of sharply pointed spines arranged along its backbone. The specimen seemed to have an armored girdle covering the hip region. More recent study has indicated that the bony pelvic mosaic may only be a geological artifact rather than a skeletal feature.[11] Fox described the new dinosaur, named *Polacanthus*, in 1866.[12]

In 1905 Caleb Barlow, working under Hungarian paleontologist Baron Franz Nopsca, restored the *Polacanthus* as a quadrupedal animal with a shielded pelvic region, two bony rows of pointed spines situated over the backbone, and armored scutes (or low plates) projecting dorsally along the tail vertebrae. This classic skeletal reconstruction was somewhat founded in Nopsca's interpretation of a possibly related dinosaur, *Struthiosaurus*.[13] By the mid–twentieth century, dinosaurs such as the *Polacanthus* and the *Hylaeosaurus* were regarded as forms intermediary between plated dinosaurs, such as *Stegosaurus*, and fully armored dinosaur "tanks," such as *Ankylosaurus*.

Barlow's reconstruction has inspired several "life" renderings of the 15-foot-long *Polacanthus*.[14] Barlow's design was popularized, especially through Neave Parker's life restoration of *Polacanthus* printed on a British Museum of Natural History souvenir postcard available in the late 1950s through early 1970s.[15]

In 1979 dinophile George Olshevsky created the first new restoration of the *Hylaeosaurus* in over a century for a wall chart featuring evolutionary relationships among the two main groups of armored dinosaurs (e.g. the ankylosaurs and the nodosaurs).[16] His influential interpretation actually followed Mantell's original description rather closely, except that some of the large dermal spines have been restored in their natural position as originally found, facing laterally outward at shoulder height. The (fictitious) head is no longer so iguana-like, and there is no longer a single-rowed crest of spines situated along the spinal column.

Olshevsky's *Hylaeosaurus* resembles the appearance of better known nodosaurid dinosaurs (such as *Sauropelta*), and mirrored his restoration of the even more puzzling dinosaur *Acanthopholis*. In 2000, however, Olshevsky claimed his "...Hylaeosaurus picture is dated ... being a synthesis of *Polacanthus* body and *Hylaeosaurus* front end, and has only historical interest.... As for *Acanthopholis*, goodness knows what it really looked like!"[17] (See Figure 18-2.)

Paleontologist Walter P. Coombs, Jr. suggested that the two dinosaurs, *Polacanthus* and *Hylaeosaurus*, may actually be the same genus, particularly since they both have been found in the Wealden Beds. Even if it had been established that *Hylaeosaurus* and *Polacanthus* represented the same genus, the life appearance of the resulting composite dinosaur (which would have been named *Hylaeosaurus* rather than *Polacanthus* on basis of priority) would have remained in doubt.

Paleontologist David Norman's *The Illustrated Encyclopedia of Dinosaurs* (1985) featured John Sibbick's striking restoration of several nodosaur genera, including *Hylaeosaurus*, styled after Olshevsky's 1979 revision, *and* a *Polacanthus*. Sibbick angled the double row of spines on his *Polacanthus* laterally as in the case of his *Hylaeosaurus* restoration. Norman wrote, "...At present the [*Hylaeosaurus*] skeleton, consisting of the front half of the animal, is embedded in a block of limestone.... This [*Hylaeosaurus*] specimen has, unfortunately, never been prepared out of the stone in which it was embedded. Nevertheless, the parts which are exposed seem to show an animal with rows of large, curved plates running down its back. Few other remains of *Hylaeosaurus* have been discovered to date, so that not a great deal is known about either its appearance or its relations with other ankylosaurs. Another early British nodosaurid is *Polacanthus* ... the skeleton which consisted of the hind part of the animal including many spines, various back and tail vertebrae, the hips and hind legs, had evidently weathered out of the cliffs on the coast of the Isle of Wight.... The frustrating thing about this specimen is that although it must have lived at about the same time as *Hylaeosaurus*, the parts of both skeletons do not overlap (one being the front half, the other the back half). It is therefore impossible to compare the two skeletons directly to prove they belonged either to the same or to different animals."[18]

Both genera are also separately portrayed in another popular reference, *The MacMillan Encyclopedia of Dinosaurs and Prehistoric Animals* (1988).[19] Although the text clearly suggests that the two genera may be *synonomous*, such restorative efforts imply visually that such genera were actually distinctive.

So, which interpretation turned out to be correct? Were the two dinosaurs joined at the hip, or are they separate genera?

In 1979 paleontologist William T. Blows recovered some new *Polacanthus foxii* material from the Isle of Wight, where Fox found his weathered specimen over a century before. Then, seventeen years later, Blows described a second species of *Polacanthus*, named *P. rudgwickensis*. The new Early Cretaceous (Wealden) specimen was estimated to have been 30 percent larger than the 15-foot-long *P. foxii*. *Polacanthus rudgwickensis* was discovered in 1985 in southeast England and was only the fourth known specimen attributed to the genus. Blows' examination of the new *Polacanthus* specimens fortified his opinion that *Hylaeosaurus* and *Polacanthus* are distinct genera.[20]

Blows also suggested that *Polacanthus* is unique from *Hylaeosaurus* on skeletal as well as geological grounds. The tibia of *P. rudgwickensis* has a fused astragalus, unlike the condition in *Hylaeosaurus*. Furthermore, *Hylaeosaurus* is known from an older Wealden biozone than the two *Polacanthus* species, which are geologically younger. Additionally, the type of dermal armor differs between *Hylaeosaurus* and *Polacanthus*, and the scapula is fused to the coracoid in *Polacanthus*. "*Polacanthus*' laterally projecting shoulder spines are of a completely different shape than *Hylaeosaurus*', and the presacral spines are of different lengths."[21]

But the definitive "blow" in the matter is instead attributed to a 1993 publication by paleontologist Javier Pereda-Suberbiola. Pereda-Suberbiola reexamined the remains of *Hylaeosaurus* and *Polacanthus*, claiming that, "*Polacanthus Owen* can be distinguished from *Hylaeosaurus Mantell* on the basis of significant differences in the pectoral girdle, hindlimb and dermal armour. A comprehensive review of the dinosaurs from the Wealden Series of England and contemporary basins of western Europe adds further evidence of a biological separation between Lower Wealden (Hastings Beds, containing *Hylaeosaurus armatus*) and Upper Wealden fanuas (Wealden Group of the Isle of Wight—containing *Polacanthus foxii*...."[22] Not only are there some distinctive differences that may be discerned between the fossilized bones of the two genera, but the geological ages in which they evidently lived are not the same.

So, despite superficial similarities, *Polacanthus* and *Hylaeosaurus* were not evidently joined at the hip after all. So while new restorations of *Polacanthus* shall continue to appear in popular publications, *Hylaeosaurus* will continue to be neglected.

A sad end for one of the first established "Dinosauria," or merely a beginning?

Chapter 34

The Big Sho-

The immense creature shifted violently, forcing its soft belly further into the slippery mud. Surf sprayed over its dorsal fin, exposing its slick, vulnerable hide. Solar radiance burned intensely, painfully. An air breather, its nostrils labored raspingly as retreating waves lapped seaward over its brow. Cast out into an ocean of air and cruelly tugging gravity, cavernous lungs were inexorably caving in. Abdominal musculature, unsuited for terrestrial breathing, collapsed. Betrayed by its ancestry, the great chimera, seemingly more fish than lizard, lay trapped in the shallows, amidst the waves pounding mercilessly. A great exhalation of despair would not avert the course of natural selection on this Late Triassic morn.

Man's collecting of fossils merely suspends weathering and erosion processes. Natural elements will eventually lay final claim to all bones, such as these scattered bones lying before us. Gold has been found in the nearby quartz veins; quartz has infiltrated these fossilized vertebrae dating from the golden age of reptiles. Here, only minutes from a ghost mining town, the spirits of 230-million-year-old, whale sized "fish lizards," ichthyosaurian reptiles known as *Shonisaurus*, linger in planetary memory.

What lures us to this place of solitude and utter desolation, a tomb for marine vertebrates seemingly in the unlikeliest of places—atop Nevada's Shoshone mountain range at 7,000 foot elevation? A question. Can we raise *Shonisaurus* from the ashes of time?

We must study the dead to raise living impressions of prehistoric animals. The most exquisitely preserved fossils reveal final reposes of pain, once transmitted along corroding nerves, without guaranteeing accurate visions of how the animals in question lived their lives. Achieving "accuracy" in a paleontological restoration is like trying to hit a moving target. Science is ever changing, while paleoart remains static, timeless. Our evolving impressions of "fish-lizards," the group of marine reptiles known as ichthyosaurs, represent a splendid historical coupling of art and science.

Origins of Ichthyosaurian Paleoart

It was young amateur fossil collector Mary Anning (1799–1847) who, with a brother, collected the first ichthyosaur to be reported in scientific literature. For, in 1814, a few years after its discovery, this specimen would be described in the *Philosophical Transactions of the Royal Society of London.*[1] A decade would pass before ichthyosaurs would become fertile ground for paleoartists.

Thomas Webster (1781–1841), formerly engraver and curator for Britain's Geological Society, is recognized for completing the first accurate skeletal reconstructions of the *Ichthyosaurus* (and *Plesiosaurus*), appearing in an 1824 scientific description by William Daniel Conybeare (1787–1856).[2] (Mary also found the first plesiosaur known to science in 1821.) Webster's remarkable paleoart provided a fundamental framework for comprehending how these animals appeared in life. Webster not only illustrated the bones, as preserved in the matrix, but also, as stated by Michael A. Taylor, "...in a conceptual and heuristic leap, reconstructed [them] as they had existed in the living animals. These are of enormous interest, being the first examples of this typical convention of vertebrate paleontological communication, apparently preceded only by Cuvier's drawings of Tertiary mammals from the Paris Basin."[3]

In 1830 paleontologist Henry De la Beche produced a cartoon watercolor, later distributed for sale by Mary Anning, titled *Durior antiquior*, a restoration of "a more ancient Dorset" in which ichthyosaurs take center stage. One chews a plesiosaur neck, others happily munch on fish and mollusk morsels. Following these gustatory exertions, they defecate on the sea bed. To our knowledge, this was the first restoration ever done of ichthyosaurs, shown here as bug-eyed, serpent-tailed, flippered forms.[4] Then, in 1834, Thomas Hawkins, an eccentric collector of marine saurians (and no relation to Benjamin Waterhouse Hawkins), notorious for "doctoring" specimens, published a book entitled *Memoirs of Ichthyosauri and Plesiosauri: Extinct Monsters of the Ancient Earth.* A frontispiece by artist John Samuelson Templeton memorialized Hawkins' love for ichthyosaur fossils. Like De la Beche, Templeton restored straight tailed ichthyosauri, one "beached" and sunning itself on a sea shore. The image of serpent tailed ichthyosaurs would persist for many decades in popular publications, as in Edouard Riou's depiction of an ichthyosaur battling a plesiosaur appearing in Loius Figuier's *The World Before the Deluge* (1863).[5] (See Figure 9-3.)

The frontispiece for Hawkins' *Book of the Great Sea Dragons* (1840) featured John Martin's remarkable portrayal of ichthyosaurs and other saurians posed in bloodthirsty combat. Such illustrations were quite popular at the time and certainly fueled the public's imagination and curiosity about the "pre–Adamite" world. To modern paleontologists, Martin's "grotesque apparitions" appear to be "...as outrageous as a Jules Verne fantasy."[6] Verne certainly made the most of such imagery in his novel *Journey to the Center of the Earth* (1864)! (See Figure 9-2.)

Several years later, the first life-sized models of ichthyosaurs and other "antediluvians" were reconstructed by another Hawkins, the naturalist/artist Benjamin Waterhouse Hawkins, for the Crystal Palace Exhibition of 1854 at Sydenham. Hawkins' restorations had been prepared under the guidance of Britain's reigning authority on fossil vertebrates, Sir Richard Owen.

Hawkins' "in-the-flesh" representations of ichthyosaurs, both for the Crystal Palace and in a mural for Princeton University completed in 1877 (and also for an 1858 edition of William Buckland's *Bridgewater Treatise Geology and Mineralogy*), have long serpent-like tails *and* a vertical caudal (tail) fin.[7] In 1975 Adrian J. Desmond commented that Owen envisaged ichthyosaurs as "...shallow water dwellers ... that possessed long whip-like tails having vertical flukes."[8] (See Figure 34-1.)

"Bending" the Rules

Hawkins' sculptures certainly "owe" their fishy features to Owen. Owen had discovered tail bones angled down-

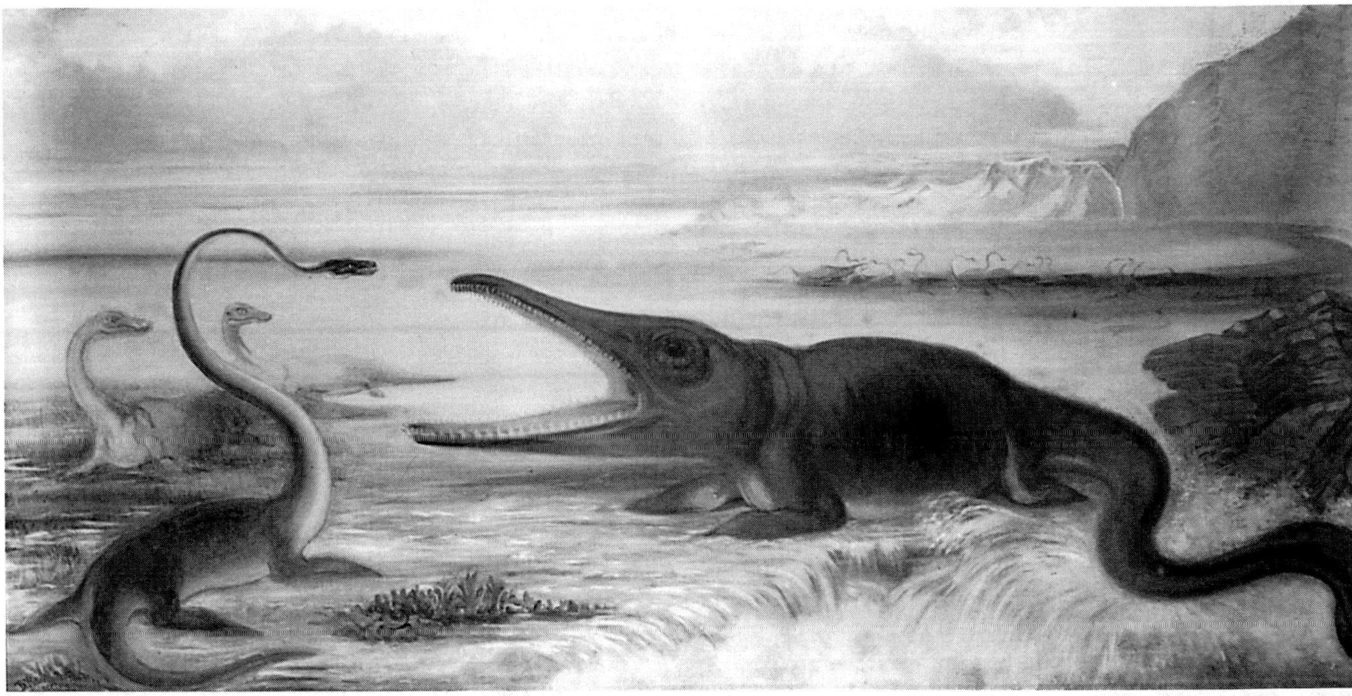

Figure 34-1: Benjamin Waterhouse Hawkins' *Early Jurassic Reptiles—Liassic Marine Reptiles of England* (1876), a restoration featuring a contest between *Plesiosaurus* (left) and *Ichthyosaurus* (right), mirroring his similarly posed life–sized sculptures displayed at Sydenham, England. This restoration suggests that both aquatic reptiles could leave the ocean depths and come ashore. (Copyright 1980, Trustees of Princeton University, reproduced with permission.)

ward in several ichthyosaurs. Owen believed these tail bone segments to have been dislocated. In other words, ichthyosaurs' tails were not fashioned after serpents—they were fishier than that.

Owen interpreted the angled tail bones as post-mortem fracturing, occurring in the tail section which in life supported a terminal, vertically oriented caudal fin. This was the tail region thought most likely to be susceptible to breakage after death. Owen's perceptive—yet erroneous—views were published in *Note on the Dislocation of the Tail at a Certain Point Observable in the Skeleton of Many Ichthyosauri* (1840).[9] Fracturing of the tail bone extremity could have resulted from phenomena such as decompositional bloating or scavenging of the carcass, and damage inflicted by wave action.

Although Hawkins' skillfully rendered restorations may be lauded for representing a (slightly misconceived) intuitive progression, relative to Webster's 1824 serpent-tailed skeletal reconstruction, the matter of tail bending in ichthyosaurs is, as we shall later see, more complex than Owen could have realized.

Contemporary scientists thought that ichthyosaurs were capable of coming onto the beaches to deposit eggs. In fact, Hawkins' Crystal Palace sculptures are posed sprawling to the shoreline. The work of German ichthyosaur researcher Bernard Hauff (1866–1950), one of the most important contributors to the study of these long extinct creatures, discredited such presumptions. Hauff, who had developed a keen fascination with the fossils of Holzmaden, proved that in many respects ichthyosaurs more greatly resembled modern day dolphins than sharks or fishes. The resemblance of the fishier ichthyosaur lot to modern sharks, porpoises and dolphins is today regarded as a marvelous example of evolutionary convergence, wherein animals sharing similar habitats will develop similar features even though they may not be closely related phylogenetically.

Scientists have speculated that the ichthyosaurs deposited in the Liassic slate quarries surrounding Holzmaden near Wurtemburg, Germany, may have perished in the marine environment and were later carried into the silty enclosed bay area. Isolated, anoxic conditions prevalent near the bottom of the bay may have poisoned any marine life encountering it (e.g. while pursuing prey or escaping enemies). These same conditions later helped to preserve the carcasses. Through his meticulous investigation of Holzmaden specimens, Hauff became the world's foremost preparator of ichthyosaurs.[10]

Hauff was an early proponent of the hypothesis that ichthyosaurs gave birth to live young, an idea first proposed by J. C. Pearce in a journal article titled "Notice of What Appears to Be the Embryo of an Ichthyosaurus Communis" (1846).[11] Numerous adult specimens had been discovered, including 14 by Hauff alone, that contained juvenile ichthyosaurs within the body cavity of the adult. Based on this fossil evidence, some paleontologists contended that ichthyosaurs were cannibalistic, sometimes devouring their young as do some modern fishes. And in order to reproduce, it was thought that female ichthyosaurs sought secluded beach areas where eggs could be laid. Hauff disagreed.

To Hauff, ichthyosaurs seemed fully adapted to a marine existence. Being stranded ashore surely meant a death sentence for fish-lizards.

The issue was resolved after Hauff identified a beautifully preserved female ichthyosaur specimen which he believed to have been fossilized in the process of giving live birth. However, it is possible that the birth happened post-mortem. Three intact embryos were also contained within the body cavity of the mother. There is little likelihood that several relatively unfragmented ichthyosaurs, all of the same size, should be contained within the body cavity of a larger individual belonging to the same genus unless these were indeed embryos. Furthermore, because ichthyosaurs were air breathers, their young would be expected to have been born tail first instead of head first, so they would not drown. And this is exactly how the fossils were found, with young emerging from the mother's uterus, tail first.

Viewers who tuned in to Tim Haines' televised BBC production *Walking with Dinosaurs* (1999) were treated to a computer created restoration showing how ichthyosaurs, in this case the 16-foot-long genus *Opthalmosaurus*, may have given birth to live young. Here an ophthalmosaur "pup" emerges from its mother's womb, thereafter quickly rising to the sea surface to catch its first breath of air.[12]

Ophthalmosaurus, a fairly common ichthyosaur genus from England's Oxford Clay, is noted for its proportionally large eye orbit. Annular bony sclerotic rings are often noted in ichthyosaur skulls. In life, these structures helped to support eye rigidity under conditions of deep sea water pressure. Benjamin Waterhouse Hawkins' Victorian era *Ichthyosaurus* sculptures show these rings as external features, encircling the eyes, when in fact they were internal, bony structures. *Opthalmosaurus* had the largest eyes of any adult vertebrate (23 centimeters in diameter) proportional to body size. Incidentally, an accurate sculptural restoration of broad-tailed, wide-eyed *Ophthalmosaurus* may be seen at the George S. Eccles Dinosaur Park in Ogden, Utah. Eccles' life-sized sculpture is dramatically poised "in flight," as if leaping over a rock like a trained Dolphin.

In 1890 Hauff made another exciting discovery. Noting a carbon-like coating on the periphery of an ichthyosaur skeleton, he was led to believe that the substance was actual tissue. About a year later he came across yet another skeleton possessing a carbonized coating that very closely followed the body contours. Other such skeletons came to light, eventually forcing Stuttgart paleontologist Eberhard Fraas to recognize the significance of Bernard's discovery.

The fossilized fleshy remains thus greatly advanced paleontologists' knowledge of ichthyosaurs. For instance, it became evident that ichthyosaurs did indeed possess a vertical, shark-like (as opposed to a horizontal, dolphin-like) tail rudder. Previously, Owen had considered the downturned tail end to be a post-mortem fracture. A century later, using computer tomography ("CAT" scans), Canadian vertebrate paleontologist Christopher McGowan demonstrated how downward tail bend flexure in Jurassic and Cretaceous ichthyosaurs results from a sequence of wedge shaped bone centra occurring in tail vertebrae. McGowan reported his findings in an article titled "The Ichthyosaurian Tailbend: A Verification Problem Facilitated by Computed Tomography (1989).[13] Now it is realized that the tailbend evident in some ichthyosaur specimens is quite natural, and opposite that of the shark's upturned tail structure. (However, McGowan has identified several cases of tailbend exaggerations evidently made during fossil preparation).[14]

Hauff's proof of a tail fin fortified the position that ichthyosaurs were not oviparous (i.e. egg laying) animals. A fishy tail fin would only have hampered excursions onto sandy or rocky terrain. Furthermore, the carbonized coating, so carefully elucidated by Hauff, bore evidence that Holzmaden ichthyosaurs *also* possessed a dorsal fin, much like the dolphins of today. (See Figure 34-2.)

Interestingly, analysis of the carbonaceous film has revealed fatty tissue and the presence of fossilized pigment cells. A fatty layer would have thermally insulated ichthyosaurs, much like blubber does in modern whales. Did some ichthyosaur species also possess fatty humps like the modern sperm whale? Even if they didn't, McGowan believes

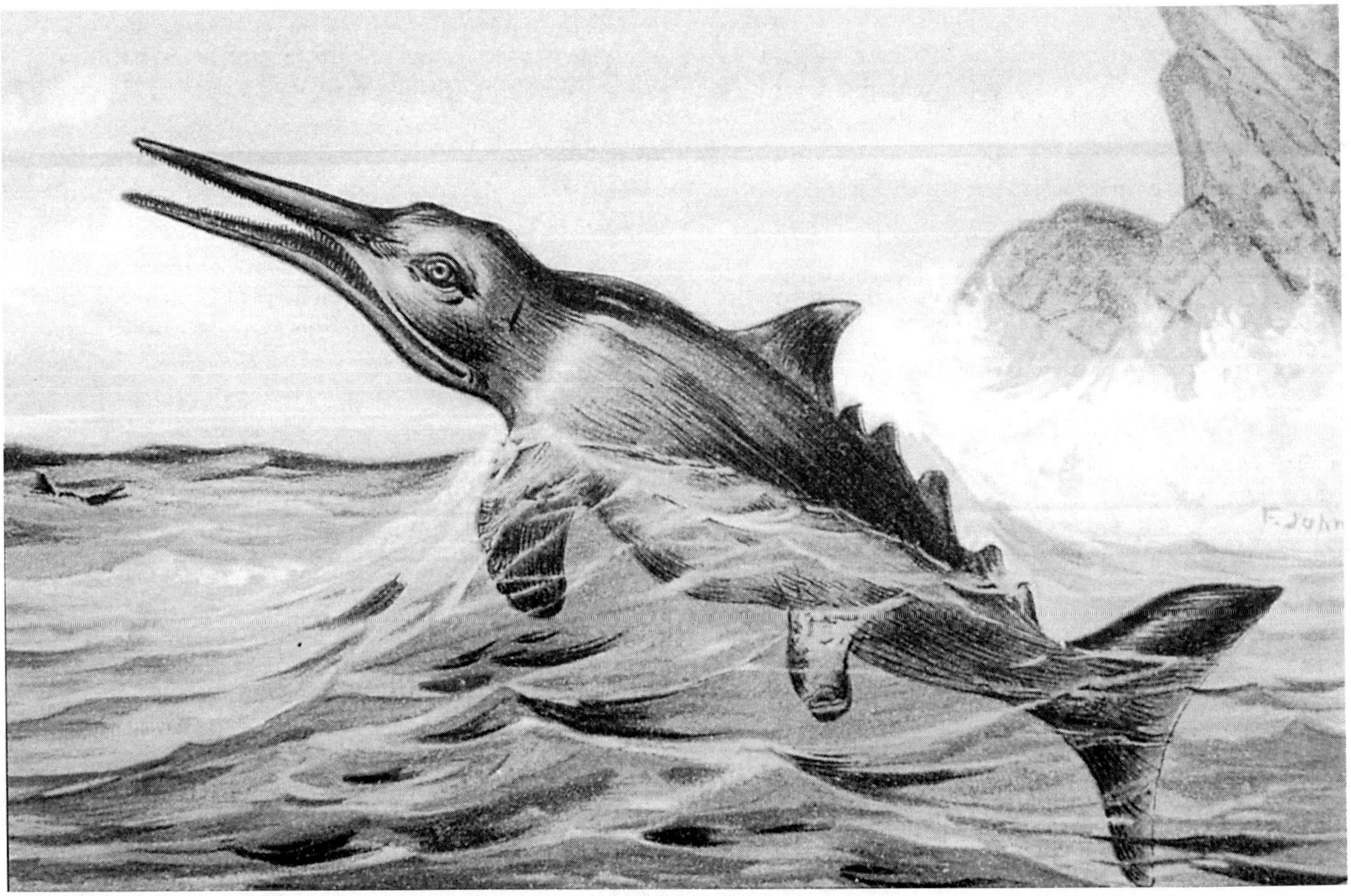

Figure 34-2: F. Long's popular restoration of *Ichthyosaurus* (circa 1905), based on Richard Lydekker's 1893 reconstruction appearing in *Extinct Monsters*, reflecting understanding that the genus possessed dorsal fin structures (although incorrectly depicted). This collectible card was originally released by the Kakao Company of Hamburg-Wandsbek, Germany, as a promotional item. (Courtesy of Allen G. Debus.)

that some, yet not necessarily all, fast swimming ichthyosaurs were, like modern tuna, endothermic (warm-blooded).[15]

Not all ichthyosaurs had significant tailbends. Owen and Hauff honed their knowledge of ichthyosaur anatomy, studying post–Triassic specimens. However, the earliest ancestral forms, dating from the Lower, Middle and Upper Triassic, actually bore closer resemblances to time-honored nineteenth century Victorian age restorations.[16]

By the Triassic, ichthyosaurs had emerged into the fossil record fully adapted to marine life. The Lower Triassic, five-foot-long Japanese genus *Utatsusaurus*, for instance, had no tailbend. One branch of ancestral ichthyosaurs is represented by a yard-long genus named *Mixosaurus*. Mixosaur skulls are narrow, resembling those of modern dolphins. Mixosaur limb bones had adapted to an aquatic lifestyle, and their tail rudders were evidently less developed than those of later ichthyosaurs. The tail-bend is evident, but with minimal flexuring. Another Middle Triassic genus, the 32-foot-long *Cymbospondylus*, also had a minor tailbend. Artist Marg Sansom's recently published *Cymbospondylus* restoration is depicted with a straight serpent tail and lacks a dorsal fin.[17] In Doug Henderson's 1992 painting, *Cymbospondylus*' tailbend supports a short vertical tail fluke.[18]

Ichthyosaurs propelled themselves underwater chiefly using their strong tail fins and powerful fin paddles. Rhythmic body movements added to the forward motion, while paddles and the triangular dorsal fin, probably supported by tough, cartilaginous tissue, steered and balanced the animal. The downturned tail of certain species provided downward thrust to overcome body buoyancy. Ancestral Triassic forms, however, with less powerfully developed tail rudders, such as the *Cymbospondylus*, may have earned a living as "ambush predators" rather than fast oceanic cruisers. Another large genus, *Leptopterygius*, dating from the Late Triassic to Lower Jurassic, also possessed reduced tailbend curvature, divulging its "sit and wait" predatory nature.

"Living Impressions" of Ichthyosaurs

Did ichthyosaurs also eat other marine reptiles? Did ichthyosaur adults battle with other monsters of the Mesozoic deep, such as plesiosaurs? Did marine reptiles snatch hovering pterodactyls out of the air? While nineteenth century paleoartists, such as De la Beche, Riou and others, depicted these speculative scenes, we have no direct evidence for primordial aquatic combat. (See Figures 12-1 and 12-2.) Yet such startling images persist to the present day.

Like modern whales and porpoises, deep diving ichthyosaurs ascended to the surface in order to breathe, possibly through a blowhole. Holzmaden ichthyosaur pigment indicates a porpoise-like dark grey coloration on the dorsal surface and a pale grey below. Skin pigment preserved in British ichthyosaurs is tortoise-shell brown. Great colors for sea "monsters" of the deep! Although in the 1860s Edouard Riou restored an *Ichthyosaurus* exhaling great volumes of water from a blowhole in whale-like fashion, that image remains conjectural.

The successes of Owen, Hauff and other paleontologists led to the earliest accurate restorations of ichthyosaurs, notably those by British paleontologist Richard Lydekker (1892) and artists Joseph Smit (1893) and Charles Knight (1902).[19] To a casual observer of the history of ichthyosaur restorations, it may have seemed that by the mid–twentieth century, all the work involved in accurately restoring ichthyosaurs had been done. Then along came *Shonisaurus*.

Reconfiguring an Ancient Sea Monster

In the late 1920s an ichthyosaur "graveyard" was discovered at a 7,000 foot elevation in the Upper Triassic (late Carnian) Luning Formation exposures in the Shoshone Mountains of Nevada. (Centuries ago, this deposit would have been offered as "proof" of the biblical Deluge, having carried the carcasses to high elevation.) Here, scientists declared, the largest ichthyosaurs on record were trapped along a muddy shoreline that later hardened into shale. It has been hypothesized that receding tides may have beached the dying creatures, as often happens to modern whales.[20] (The fictional introduction to this chapter is inspired by this interpretation.) A total of 37 specimens were found here and described in scientific papers written by American paleontologist Charles L. Camp (1893–1975).

The visitor's center building currently housing about 10 of the displayed *Shonisaurus* fossils, known as the "Ichthyosaurus Cathedral," was dedicated on August 20, 1966. Nearly a decade before then, however, the first restoration of *Shonisaurus popularis*, a masterful life-sized frieze adorning a cement wall adjacent to the building, was completed by sculptor Michael Huff in 1957, shortly after Dr. Camp began his investigations at the site. Visitors standing next to the frieze are simply dwarfed by the enormity of *Shonisaurus*' likeness. (See Figure 34-3.) So,

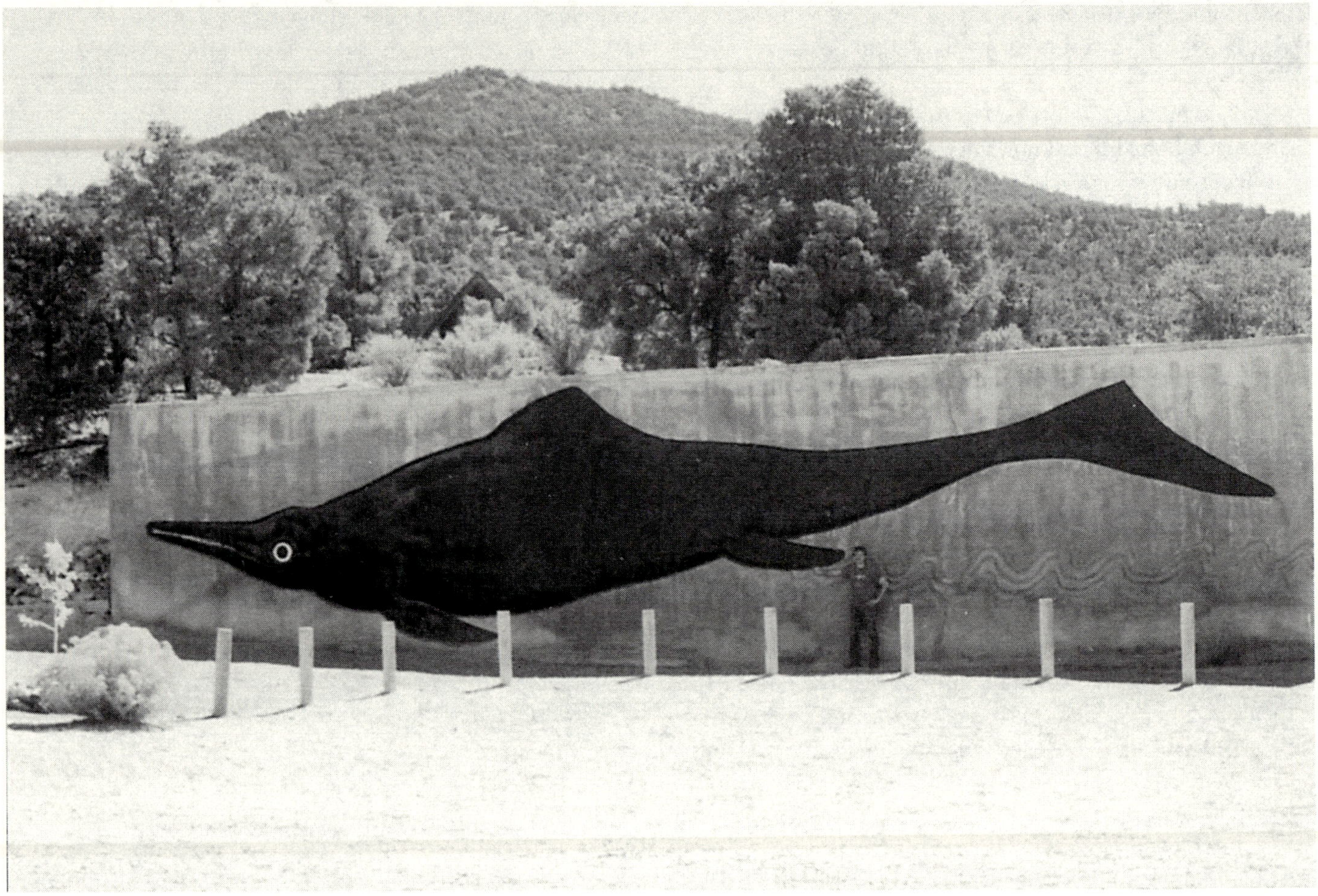

Figure 34-3: Michael Huff's life-sized sculptural *Shonisaurus* restoration displayed outside the "Ichthyosaurus Cathedral" dwarfs Allen A. Debus, standing near rear flipper. (Photograph by Diane E. Debus, 2000.)

thanks to Huff's artistry, even before entering the Cathedral one has a striking, although somewhat inaccurate, impression of how these Godzilla-sized, 50-foot-long, 40 ton, prehistoric "Flippers" appeared in life. For several hundreds of thousands—if not millions—of years, *Shonisaurus* was the largest animal ever on the planet—yes, the biggest "Sho-" on earth!

Inside the Cathedral, a mausoleum for America's grandest ichthyosaurs, the story is unveiled. The specimens on view lay heaped on one another, with bones jumbled about. The remains are not immaculately preserved, as in the case of the Holzmaden specimens. There is no evidence of carbonized tissue, or birthings captured "instamatically" on slatey film. Here lie only traces of death. In fact, it is difficult to tell which of the scattered dinner plate–sized vertebrae belong to which specimens, or even exactly how many specimens there are on display. We need artistic impressions like Huff's to solve this puzzle.

Although there is no fossil evidence of *Shonisaurus* skin impressions, paleontologists believe that, as in other ichthyosaur specimens, it may have been scaly but smooth.[21] Perhaps this is unsurprising, because the likely "reptilian" alternative—having coarsely textured, pebbled, crocodilian skin studded with nodules and scutes—would have created frictional drag in the water, reducing swimming efficiency.

Hanging on a wall inside the Cathedral, is the original Charles Camp skeletal reconstruction. A caption under the drawing claims, apologetically, "There are some notable errors in the reconstruction of this *Ichthyosaurus*, particularly in the fins, the skull and the tail. All in all it was a much more hydrodynamic predator than is obvious from this reconstruction. Dr. Camp, however, was only going by the specimens he was excavating and can be forgiven a few errors when one realizes that he had no intact skull and was working under very primitive and arduous conditions in what was, then, an extremely remote location."

Camp's reconstruction first appeared in a 1976 German journal published a year following his death.[22] However, it is likely that Camp actually prepared the reconstruction during the late 1960s. Most ichthyosaur specialists and laymen became familiar with Camp's reconstruction when it appeared, posthumously, in other publications. Since that time, the few life restorations that have been done of *Shonisaurus*, including striking renditions by artists Larry Jacox, Douglas Henderson, Cliff Green and Eli Thibodeaux, have apparently been founded on Camp's reconstruction, now known to be erroneous. Despite anatomical errors, in painted restorations *Shonisaurus* has been correctly depicted as a slow swimmer rather than a creature capable of active pursuit.

What's wrong with Camp's *Shonisaurus* and restorations founded upon it? In 1990 Bradley Kosch of the Lake Tahoe Nevada State Park reexamined Camp's field notes and fossil specimens. He discovered that the figured skull, ribs and paddles in Camp's skeletal reconstruction were too large, proportionally, and that the tail was too long relative to the overall body length. Camp also underestimated the number of presacral vertebrae. Consequently, restorations resulting from reliance on Camp's illustration appear big headed, with swollen bellies and enormous paddles. Kosch noted that Camp's reconstruction "...is a composite of the study specimens; however, its development preceded final preparation, study, and description of the fossils in the late 1960s. Although Camp refined much of his written description, ill-health prevented him from finalizing his work and revising the reconstruction.... Camp indicated in both his monograph ... and his field notes ... that he was aware of at least some of the reconstruction's deficiencies."[23]

So Camp's problematic reconstruction really was, at best, preliminary in nature. The artists who have relied on it may have jumped the gun. Then again, paleontologist Joseph T. Gregory, who took responsibility for seeing Camp's paper through to final publication, may have had no opportunity to have a revised version of the reconstruction prepared for final publication, thereby contributing to the confusion.[24] Humans tend to believe what they see, however, and the striking image of America's largest sea monster, Camp's reconstruction (which has since been reproduced in vertebrate paleontology texts and popular books), will naturally stand out from tedious technical language.

In 1990 Kosch published a revised, more conservative skeletal reconstruction of *Shonisaurus popularis*, satisfactorily reflecting the outcome of his paleontological forensics. Instead of a 12.4-foot-long skull, Kosch posited a skull length of only 9.1 feet, even though there is no complete skull available for study. Kosch's revised estimate for the dorsal length, based on a recount of dorsal vertebrae, was 16.5 to 17.5 feet, as compared to the 14.2 foot length figured in Camp's reconstruction. This revision relieved *Shonisaurus*' peculiar "pot-bellied" appearance. (See Figure 34-4.)

Additionally, the tail length was reduced from 19.1 feet, as conveyed in Camp's skeletal reconstruction, to only 15.2 feet (which is the estimated length Camp described in his paper, reflecting an inconsistency between his visual and stated interpretations). A tailbend, based on the discovery of a fossilized wedge-shaped centrum, was discussed in Camp's field notes and shown in Camp's reconstruction. Kosch incorporated a slight tailbend into his 1990 revision, while stating that, "...its existence must be considered questionable."[25] Christopher McGowan affirms, however, that the evidence for a tailbend does exist.[26]

Shonisaurus' front and rear paddles were originally illustrated at lengths of 9.6 and 10.6 feet, respectively. For his 1990 revision, Kosch relied on Camp's measured values of 6.3 and 5.9 feet for front and rear paddles, as published in the 1980 paper.

Consequently, Kosch claims that, "...reduction in the size of the head, paddles and tail, and the lengthening of the body, indicates that *Shonisaurus* was a longer-bodied ichthyosaur with proportions more typical of other Triassic ichthyosaurs...."[27] Although Camp estimated that some *Shonisaurus* individuals attained lengths of up to 50 feet, Kosch's conservative estimate for the overall length of this genus is 45 feet. Kosch believes that *Shonisaurus* may have a vertebral column that was slightly arched, more so than in his 1990 reconstruction, as in other ichthyosaur skeletons. The first revised, accurate restoration of *Shonisaurus*, appearing in a September 1991 issue of *Dinonews*, was illustrated by Australian paleontologist Dr. John Long.[28] Proportionally, however, Huff's 1957 sculpture agrees rather nicely with Kosch's revised reconstruction.

A dramatic portrayal of death appears on page 96 of *Dinosaurs: A Global View* (1990), by Sylvia J. Czerkas and Stephen A. Czerkas.[29] Here we spy Doug Henderson's restoration, completed prior to the date of Kosch's skeletal revision. Henderson restored over a dozen *Shonisaurus* bodies lying dead or dying on an ancient sandy shore, harkening images of beached whale pods. Henderson even portrayed one juvenile in the foreground, lying adjacent to its mother.

Reflecting Dr. Camp's original, popularized assessment of the shonisaur beds, in their *Global View* text Czerkas and Czerkas state, "Curiously, a telling of their depositional preservation reflects a behavioral similarity that is seen in the mysterious mass deaths of whales and other present day sea-going mammals ... the uniform parallel alignment of the numerous *Shonisaurus* bodies facing the same direction are strikingly indicative of an accidental, misguided mass beaching."[30]

According to the latest information, however, this may not be the story told by the rocks here at Berlin,

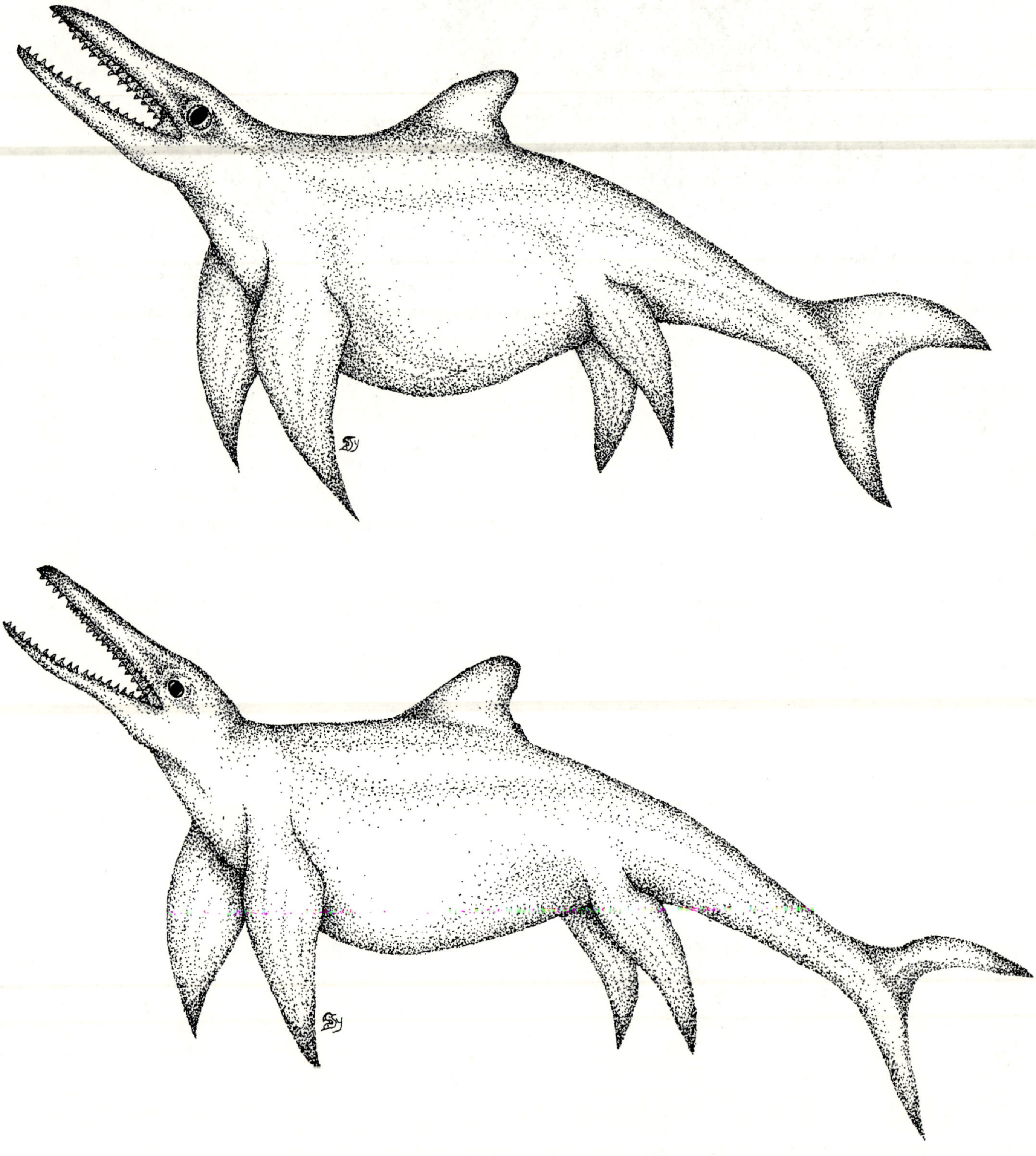

Figure 34-4: Restorations by artist Shannon S. Yeager (2001) comparing (top) Charles Camp's and (bottom) Brad Kosch's reconstructed views of *Shonisaurus popularis* to same scale. See text for details. (Reproduced with permission.)

Nevada. Although in his earlier popular 1950s writings Camp originally supported the dramatic beach stranding hypothesis, later in life he came to doubt its accuracy. Dr. Kosch explained, "In ... popular accounts Dr. Camp stated the skeletons were oriented like logs on a beach. I have never been able to see that orientation.... There really is no data to support the hypothesis that *Shonisaurus* frequented intertidal waters or was prone to stranding."[31] However tempting it may be to make behavioral analogies founded primarily on size similarities, shonisaurs were not even remotely related to modern whales. (Did we fool you with our fictional introduction?)

As Berlin Ichthyosaur State Park tour guide Jeff Morris explained during the summer of 2000, many of the shonisaur specimens have been found deposited on top of one another in poorly oxygenated deep water, not shallow intertidal sediments. So the fossils here most likely do not record an ancient beaching event. Alternatively, the current belief is that the carcasses were individually collected by the tides and buried at depth. Over time, bodies accumulated and were deposited in an overlying sequence. No juveniles have been found here, lessening the possibility that there was a social structure to any of the individuals found here.

Morris claims the assemblage represents a "localized extinction," making an analogy to waters bacterially poisoned by the "red tide." It is possible, although perhaps challenging to prove, that the Berlin shonisaurs may have been afflicted by poisoning from seawater pollutants. The fact that only large individuals have been found here is consistent with the notion that it may have taken a long time for toxins, steadily accumulating in their systems, to reach lethal levels. Dr. Kosch cautions that while this is a reasonable scenario, the red tide theory of shonisaur mortality has received more attention than is justified.

And Now for Our Really Big Sho-

Dr. Camp's visions of his tremendous *Shonisaurus* set a wave in motion among the ichthyosaurian "camp." The earliest perceptions of how ichthyosaurs appeared in life proved to be inaccurate, and as scientific knowledge accumulated about these fascinating creatures, ichthyosaur restorations steadily evolved. By the mid–nineteenth century, ichthyosaurs would symbolically represent prehistory in a broad sense and, for a time, become emblematic of the emerging science of paleontology as well. Over a century later, Camp's posthumously published review of shonisaurs rejuvenated flagging interests in Triassic ichthyosaurs. As stated by paleontologist P. Martin Sander, presently, "The study of Triassic ichthyosaurs is experiencing an unprecedented renaissance that began in the early 1980s."[32]

It is the Triassic forms which will hopefully someday reveal ancestral patterns in the ichthyosaur lineage. Now, through their redoubled efforts, paleontologists recognize that differences formerly noted between Triassic and Lower Jurassic ichthyosaur fauna were actually gradual in nature, rather than abrupt.

There were other giant ichthyosaurs besides *Shonisaurus*. The Lower Jurassic genus *Stenopterygius* had a seven-foot-long skull and, despite early erroneous estimates ranging from 40 to 65 feet, may have grown to a "mere" 20 feet in length. The German Lower Jurassic genus *Leptopterygius* grew to over 30 feet in length. Another Lower Jurassic genus, *Temnodontosaurus*, which probably preyed on other ichthyosaurs and plesiosaurs, attained a 30 foot length.[33] Christopher McGowan records a poorly known genus known only from teeth and bone scrap that may have feasted on other ichthyosaurs, from the Lower Liassic of England, which evidently grew to a 49 foot length.

McGowan has also examined new Late Triassic ichthyosaurs, important transitional fauna comprising a mosaic of Late Triassic and Lower Jurassic evolutionary features, discovered at Williston Lake in northeastern British Columbia. McGowan excitedly proclaims, "The close of the Triassic was therefore probably not marked by a major turnover in ichthyosaurian species. Instead, there was probably a more gradual period of transition that may have lasted for most of the Triassic."[34] At Williston Lake, McGowan and others have identified "very large" ichthyosaurs new to science, including one thought to have attained the size of a large baleen whale (i.e. 50 to 100 feet).

Yet, in light of recent discoveries, Jeff Morris is reluctant to yield *Shonisaurus*' heavyweight title so readily. Yes, folks, we have a "*really* Big Sho-" for you today! Pointing to an isolated specimen of extraordinarily large shonisaur ribs poorly preserved at the Berlin Ichthyosaur State Park, Morris claims the bones can be speculatively extrapolated to a record-setting individual body length of 110 feet! An apparition like that, when "Free," would give anyone ... the "Willies."

When it comes to the science and associated paleoart of slumbering giants, size *does* matter after all.[35]

Chapter 35

THUNDER in Their Footsteps

Forget that trivial debate over which carnivorous dinosaur may have been a tad bigger than *T. rex*. Sauropods had them all beat!

As any youngster knows, sauropods were those colossal dinosaurs, the ones with names like *Brachiosaurus*, *Ultrasaurus*, *Argentinosaurus*, *Sauroposeidon* and *Seismosaurus*, that could easily peer into the fourth story window of a building by virtue of their long sinuous necks.[1] Their gigantic limbs, when joined to bulky, muscular torsos, made their puny heads seem poorly proportioned. Their great tails and necks waving rhythmically in time to ponderous footfalls as they strolled in herds through the Mesozoic forests and jungles would have made an awe inspiring spectacle indeed!

To little avail, sauropods have valiantly challenged *T. rex*'s iconic supremacy. By the 1930s the Sinclair Oil Company selected a sauropod image ("Dino") instead of a tyrannosaur for their corporate logo. In the climactic scene of the silent film *The Lost World* (1925), a live brontosaur smashes its way through London, setting a precedent for many subsequent fantasy dinosaur monster movies of the 1950s and 1960s. The *Lost World*'s brontosaur plunged into the Thames River, managing an escape. Then an amphibious brontosaur claimed more human victims than all the rest of Skull Island's dino denizens in RKO's *King Kong* (1933). And the scene in which Dr. Grant and colleagues first spy the giant *Brachiosaurus* upon their entry into Michael Crichton's *Jurassic Park* (1993) is a triumph of science fiction cinema.[2]

Yet despite their majestic size, weighty industrial symbolism and commanding "live" screen presence, sauropod imagery has never made it "big." Instead, *T. rex* prevailed, relegating the intriguing history of sauropods' evolving reconstructions and restorations to relative obscurity.[3]

Perhaps because early authorities could not imagine how Nature's most immense land-living creatures could have ever survived a terrestrial existence, in most early restorations sauropods were confined to watery places. Before his death in 1832, Frenchman Baron Georges Cuvier examined some of the earliest known sauropod remains, a gigantic rib and an arm bone found near Oxford, England. Cuvier thought the fossils came from a prehistoric whale, but he might have made a better interpretation had there been more material to study.

By 1842, British anatomist Sir Richard Owen had introduced the "Dinosauria" to the annals of paleontology. However, no sauropods were included among their ranks, even though a year before Owen had described fossils now known to be of a sauropod, collected on the Isle of Wight. Owen claimed that these fossils belonged to a new genus of extinct reptiles which he named *Cetiosaurus*, or "whale-lizard." These Early Cretaceous remains included several enormous vertebrae, a gigantic fossilized rib and an arm bone (but no skull or hip bones). Judging from its apparent huge size, Owen was convinced that *Cetiosaurus* was an aquatic whale-like creature, although not a dinosaur.[4]

Owen claimed that *Cetiosaurus* was strictly aquatic, propelled in open water by a long serpentine tail terminating in a vertical fin. Owen also thought *Cetiosaurus* had webbed feet, and that it may have been carnivorous. Although no pictorial restorations of Owen's imaginary colossal creature have survived, he may have envisioned a peculiar cross between the more familiar "fish-lizard" *Ichthyosaurus* and a crocodile, which propelled itself through the water using a whip-like tail equipped with a vertical fluke and webbed toes.[5]

Then, in 1848 and 1868, *Cetiosaurus*' hind limb bones came to light, prompting Thomas H. Huxley to rank *Cetiosaurus* among the dinosaurs in 1869.[6] Huxley even caricatured *Cetiosaurus* as a *bipedal* dinosaur with diminutive

T. rex–like arms (even though *Cetiosaurus*'s arms were unknown and *rex* still awaited discovery in Montana). Owen, however, continued to classify *Cetiosaurus* as an *aquatic* dinosaur. (See Figure 3-3.)

During the last quarter of the nineteenth century, paleontologists discovered skeletons of sauropods in the western United States that were far more complete than the material Owen had access to. Some of these genera are well known today, such as "*Brontosaurus*," meaning "thunder-lizard" (or, more likely, implying that it was a "thundering big animal"[7]), *Diplodocus* and *Camarasaurus*. Edward D. Cope attempted the first visual skeletal reconstruction of a sauropod dinosaur in 1877, an essentially accurate representation.[8] (See Figure 35-1.)

Both Cope and his arch scientific rival, Othniel C. Marsh, believed that such genera (and other sauropods) were amphibious lake-dwellers that could rarely (if ever) reach the shore because their limbs were too weak. Such creatures would not have caused the earth to shake with their footsteps. They would only squish bottom mud between their toes.

Some scientists wondered how such creatures, so committed to an aquatic existence, could have ever evolved without having also developed flippers to maneuver efficiently through water, instead of massive limbs and banana-clawed toes.

Due to their extreme weight (estimated at 38 tons for an adult "Brontosaurus" in 1905), most paleontologists concurred that the limbs of sauropods were not sufficiently sturdy to prevent collapse without relying on water for buoyancy.[9] It seemed improbable that sauropods would have been capable of standing upright on land for prolonged periods, or rising into tripodal position. Presumably, sauropods came to the shore like turtles do during egg laying season, but only for brief periods.

However, others imagined sauropods as terrestrially walking "reptilian giraffes." The essence of this debate was captured in early restorations. Sauropods were depicted as bipeds rising on their hind limbs into tripodal position, either off the lake bottom or in a terrestrial setting.[10] (See Figure 35-2.)

During the early twentieth century, steel magnate and philanthropist Andrew Carnegie had his museum staff prepare casts of his dinosaur *Diplodocus*. These casts were installed in several museums around the world, helping to popularize the dinosaurs of western North America. But soon this act of goodwill erupted into paleontological controversy!

Carnegie's chief paleontologist involved with casting and exhibition, William J. Holland, had mounted *Diplodocus* in an upright, elephantine posture. But European paleontologists challenged this interpretation, claiming that *Diplodocus* must have slithered on its belly. (See Figure 35-3.) An American paleontologist, Oliver Hay, also restored *Diplodocus* as a creeping, slithering giant.[11] Hay poorly reasoned that because *Diplodocus* was a reptile, and because reptiles crawl like crocodiles, *Diplodocus* must also have been a crawling creature. Furthermore, because *Diplodocus* was *known* to have inhabited swampy, marshy terrain, a slithering locomotion would have prevented it from sinking into the ground.

Holland demonstrated that if *Diplodocus*' body had been contorted in life the way Hay and German scientists had restored its bones, sauropods would have spent their lives in "indescribable agony." Furthermore, Holland whimsically noted that in order for *Diplodocus*' limbs to have been oriented in crocodilian fashion, its belly would have necessarily been sunk into a groove running into the earth below its feet.[12]

By 1910 it was clear that sauropods did not slither. But could sauropods' limbs support their great weight on land for extended periods? Certainly Marcel Delgado's

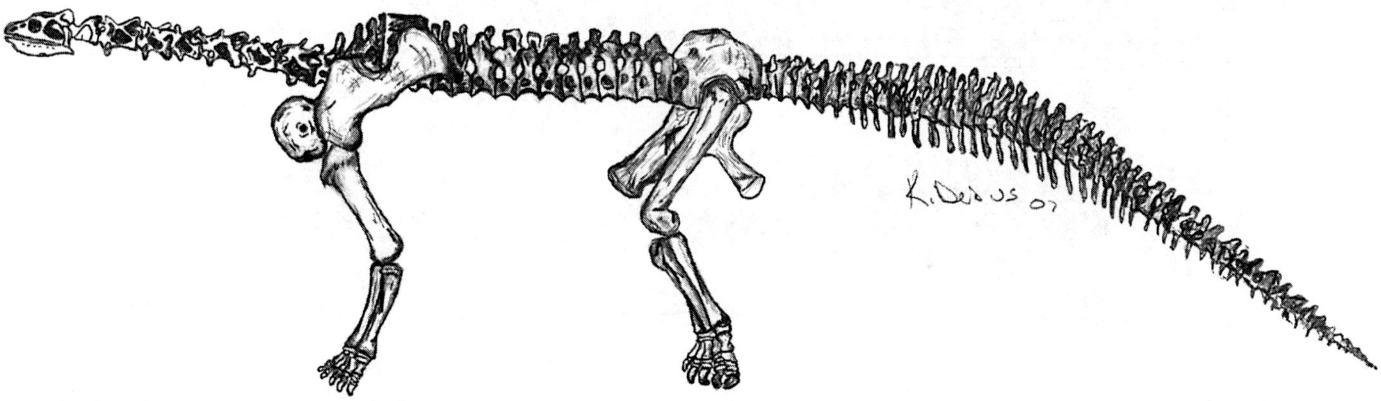

Figure 35-1: A representation of Cope's 1877 skeletal reconstruction of the *Camarasaurus*. (Illustration by Kristen Debus, 2001.)

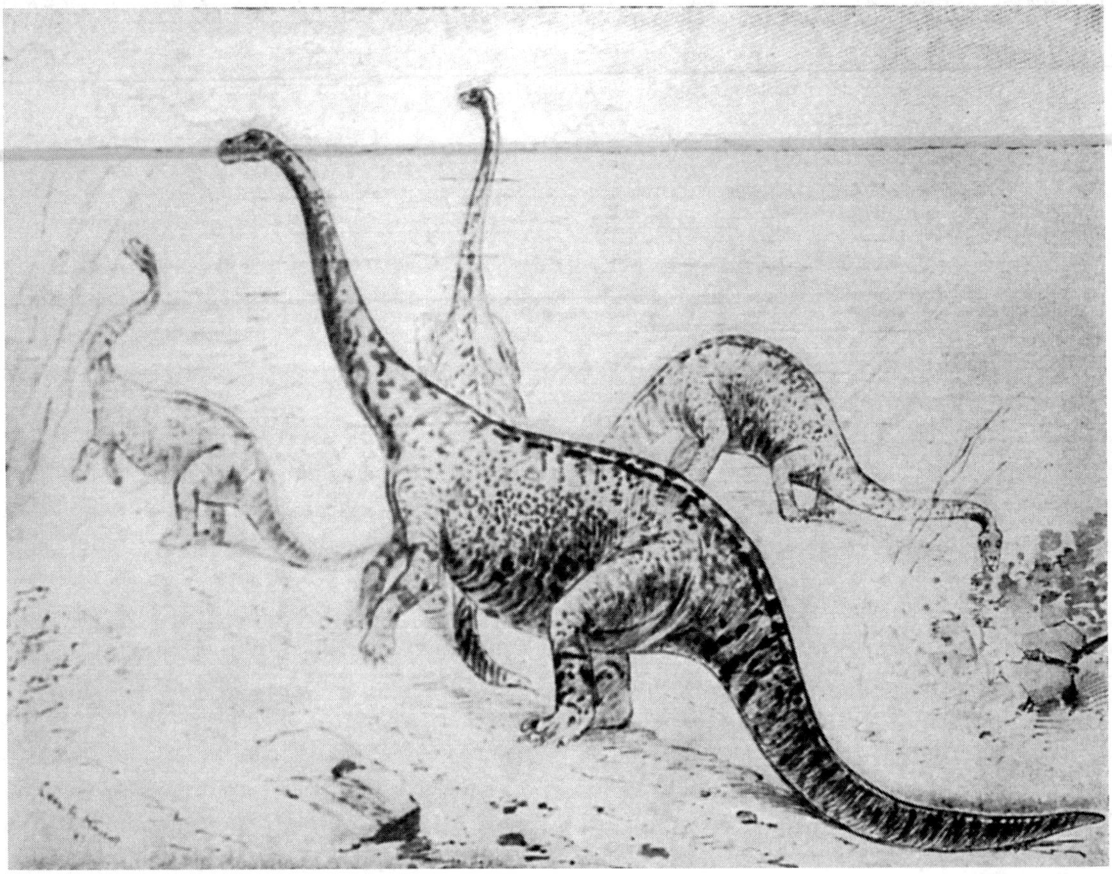

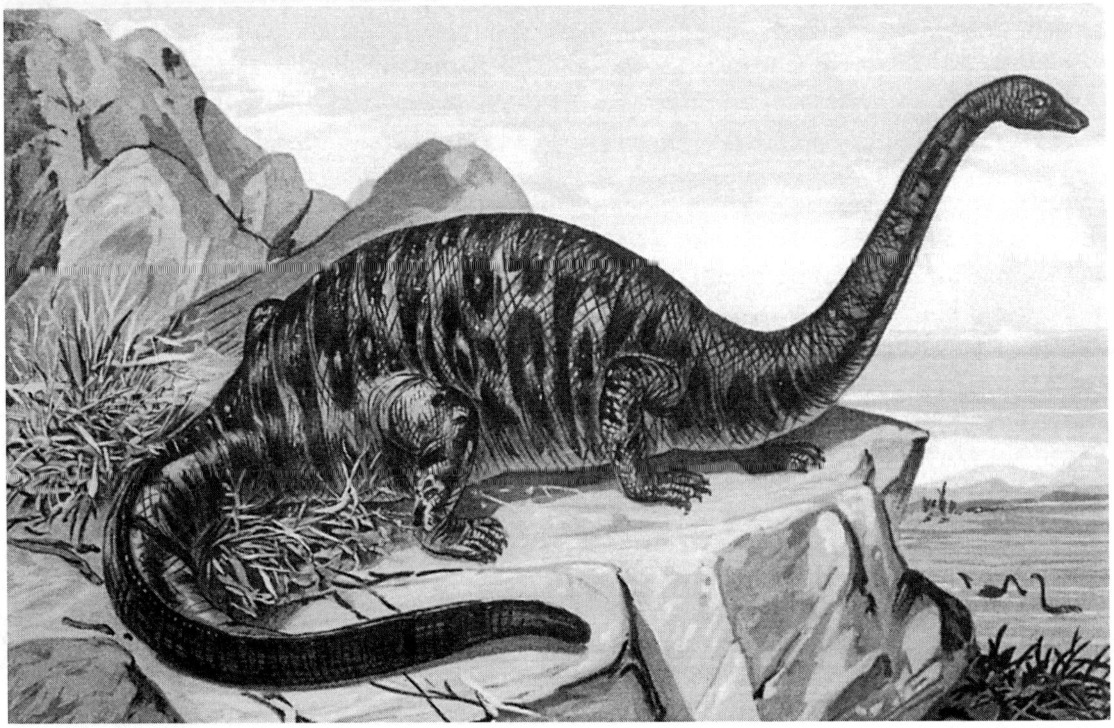

Top: Figure 35-2: Charles Knight's 1897 restoration of the poorly known enormous sauropod *Amphicoelias*, as published in *Century Magazine*. (Authors' collection.) *Bottom:* Figure 35-3: F. Long's contemporary restoration of *Diplodocus*, circa 1905, mirrors views of European paleontologists who believed sauropods slithered on their bellies. This collectible card was originally released by the Kakao Company of Hamburg-Wandsbek, Germany, as a promotional item. (Courtesy of Allen G. Debus.)

stop-motion animated, amphibious brontosaurs appearing in *The Lost World* (1925) and *King Kong* (1933) were fully capable of raising havoc while on shore leave. Then, during the 1940s, sauropod tracks discovered in Texas were interpreted as evidence that sauropods could swim, *and* that they could also walk upright over mud bars, just like Delgado's fantasy creations.[13] Case closed?

Although fossil trackways discovered during the 1940s indicated sauropods could walk on land over mudflats, through the mid-1960s they were usually restored by artists in marshy, lake or coastal settings. Their high-placed nostrils were viewed as an amphibious adaptation.[14] Was it even possible for sauropods to snorkel air while standing on the lake bed, as conventionally shown in many restorations dating from this period, or would the submerged giants have simply drowned? During the early 1960s, Edwin H. Colbert, for example, supported the view that sauropods could breathe while submerged.[15] The high-placed nostrils indicated an adaptation to life in the water.

Several years later, Robert Bakker rejected the premise that sauropods' limbs were not sturdy enough for walking on land. To Bakker, sauropods were warm-blooded, land-roving, giraffe-like creatures, as beautifully illustrated in his 1968 restoration of the *Barosaurus*.[16]

A decade later, through his studies of the distribution of *Alamosaurus* fossils in western North America, Bakker discovered that sauropods preferred dry, well drained environmental conditions. But they evidently couldn't tolerate swampy terrain.[17] When early Cretaceous climatic conditions became wetter and deltaic-forming, northernmost sauropod populations were driven southward from the forming swamplands toward drier, well drained forested areas. Sauropods were landlubbers after all! In contemporary restorations, sleek and shiny skin (formerly reflecting sauropods' supposed watery existence) was exchanged for a leathery, elephant-like hide.

During the past 25 years, the influential Bakker has dazzled us with his artistry and challenged our intellect with his "revolutionary" opinions concerning sauropods. His speculative restorations showing sauropodian internal soft tissue anatomy and a *Diplodocus* bestowed with an elephantine trunk have certainly enlivened scientific discussions concerning these amazing animals.[18] Bakker also championed Marsh's century old idea that sauropods gave live birth to their young, a hypothesis inspiring Doug Henderson's 1987 painting showing a mother *Apatosaurus* giving live birth.[19]

The live birth theory of sauropods was considerably weakened by recent discovery of sauropod embryos and egg shells found in a 71-to-89-million-year-old deposit in Argentina. The fifteen-inch-long hatchlings found here would have grown into 45-foot-long sauropods known as titanosaurs. Contrary to Doug Henderson's art showing live sauropod birth, artist Mick Ellison's restoration appearing in the December 1998 issue of *National Geographic* depicted a "good mother" titanosaur gazing fondly at her hatchling brood.[20] Paleontologist Rodolfo Coria clarified that social behavior doesn't "fossilize," and that no adult sauropod bones have been found in the deposit to date. Therefore, one can only speculate whether, or to what degree, adult titanosaurs protected their egg clutches and nurtured their young.

Although the nature of sauropod physiology remains unresolved, Bakker contends that sauropods were warm-blooded. Paleontologists believe the lung-air sac systems of theropods was analogous to that of modern birds, which are metabolically active "avian theropods." Sauropods have structurally similar trunk design to their saurischian (theropod) cousins, suggesting to some paleontologists that sauropods also had evolved a bird-like lung-air sac system as well.[21] So were sauropods also warm-blooded? Today, most paleontologists regard sauropods as cold-blooded "gigantotherms."[22]

Bakker also suggested that "brontosaurs" were "high feeders" fully capable of rearing into tripodal position to feed off the very highest branches.[23] With their small heads, could sauropods have ever eaten enough to fuel a warm-blooded metabolism, or would they have literally overheated as they digested such great quantities of vegetation?

Sauropod blood pressure must have climbed to enormous levels when lifting necks to foliage high among the tree tops. Bakker claims that even if sauropods felt faint while munching on lofty foliage, there was still sufficient oxygen fueling their brains so that they wouldn't pass out. Because their brains didn't require high blood pressures, as is the case with mammals, sauropods could lower their heads without having the suddenly elevated blood pressure blow off their heads.[24]

Today American Museum of Natural History patrons can marvel at a *Barosaurus* skeleton marvelously balanced on powerful hind limbs and tail. Museum imagetext for this display suggests that *Barosaurus* has risen into tripodal stance to fend off allosaurs attacking its young. This display pose, perhaps inspired by John Gurche's 1991 classic painting, captures the spirit of the "land-lubbing" sauropod most poignantly.[25] However, it remains conjectural whether sauropods' circulatory systems could have withstood the tremendous blood pressures involved in raising their heads skyward, let alone rearing upward in such a fashion. What sort of heart could have withstood the strain?[26] Does it really matter, as long as those necks stretch majestically ever upward toward dizzying heights in popular paleoart? (See Figure 35-4.) Bakker also thinks

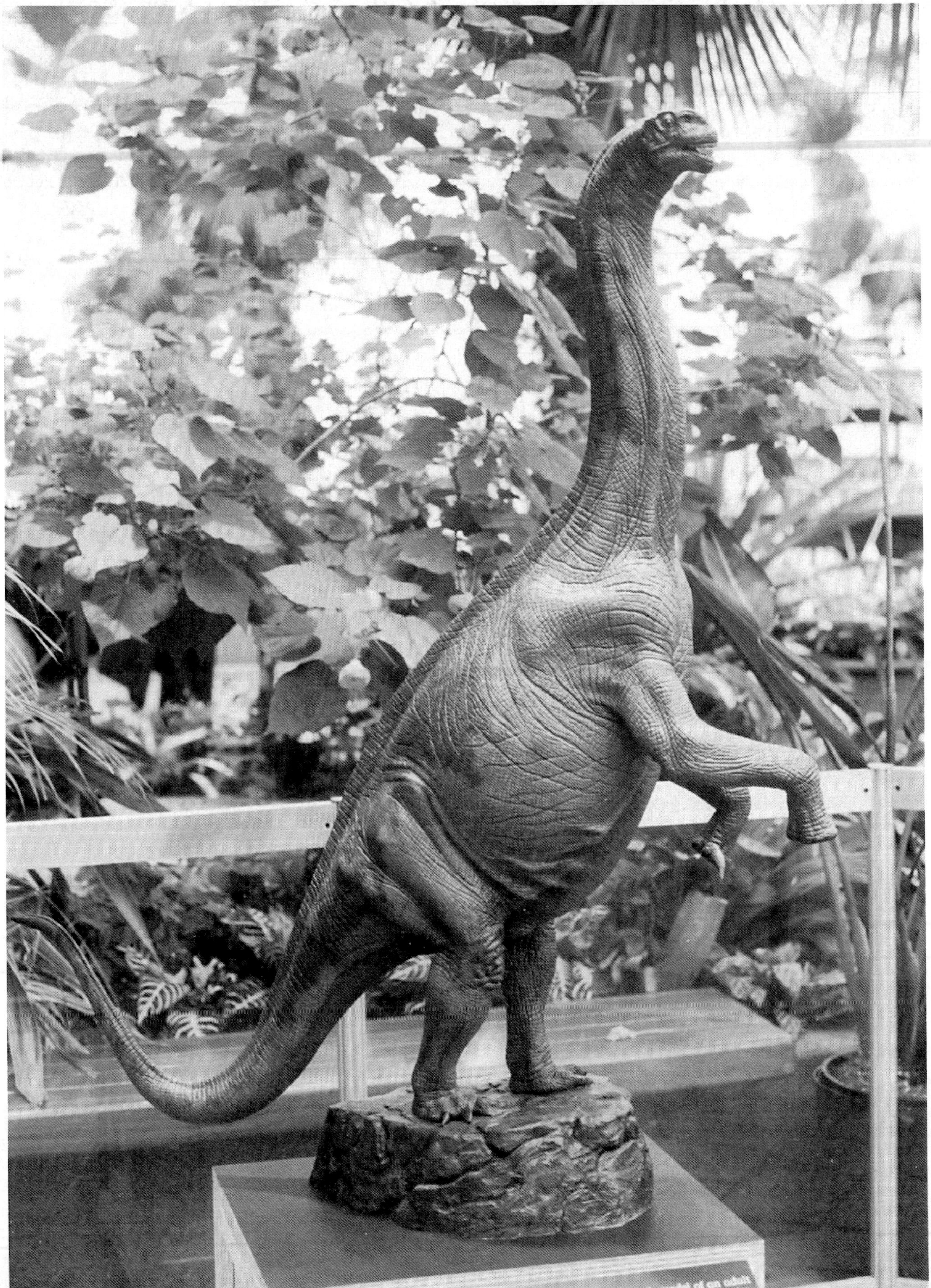
this flesh model of an adult

that in an upright position, sauropods would have passed quite a bit of gas through their posterior, a thought which tarnishes their tripodal "majesty."

Sauropods' massive fossilized remains inhibit experimental reconstruction. However, sophisticated computer programs have been designed to virtually restore their movements in cyberspace. University of Oregon computer scientist Kent Stevens, and University of Northern Illinois vertebrate paleontologist Michael Parrish, have investigated the biomechanics of sauropod neck vertebrae using computer technology.[27] As opposed to Burian's and Zallinger's classic sauropod portrayals, cyber-sauropods virtually recreated using actual bone dimensions had highly limited cervical mobility. It is as if they all suffered from stiff necks, surely an excruciating malady for any sauropod. Furthermore, Stevens and Parrish claimed, the heads of *Apatosaurus* and *Diplodocus* only rose to low elevation above ground level.

Computer simulations demonstrated that *Apatosaurus*' head could only rise as high as 17 feet above the ground or sway 13 feet from the midline. (*Diplodocus* was far less flexible, even though its neck was 25 percent longer, and *Camarasaurus* could nearly hold its neck vertically.) *Brachiosaurus* had a peculiar limitation. It paradoxically couldn't lower its neck completely to ground level for a drink of water. (Garbage in, garbage out?) Gregory Paul thinks this data is illogical.[28] Paul questions why sauropods wouldn't have been able to raise their heads to spy approaching predators or prospective mates from a "bird's eye" view.

Certainly evidence of balloon air sacs situated in between cervical bones would have lightened sauropods' necks and facilitated their movement into the tallest branches. Furthermore, Paul claims that computer simulations didn't fully account for discs and cartilage, which would have separated neck vertebrae. If sauropods had thick intervertebral discs, their necks would have been more flexible, both vertically and horizontally.[29] Stevens' and Parrish's cyber-sauropod researches did not falsify Bakker's claim that some sauropods could rear up tripodally, possibly to feed on higher tree limbs.

In most recent restorations, instead of dragging their tails limply behind on the ground, sauropods hold their whip-like tails aloft. Physicist and "cyberpaleontologist" Nathan Myrhvold has assessed sauropodian tail function in the process of testing another of Robert Bakker's speculations. In his *The Dinosaur Heresies* (1986), Bakker had suggested that when diplodocids "...swung their hugely muscled tails, the whiplash effect could inflict crippling wounds on an unwary predator." Bakker's accompanying restoration depicted a dazed allosaur, dramatically felled by a supple *Diplodocus* defending her offspring with vigorous tail swipes.[30] Myrhvold concluded on "sound" principles that the tail of *Apatosaurus* could crack like a bullwhip. Sauropods evidently were "into whips." Perhaps the tail functioned in mating rituals, Myrhvold speculated, instead of being used to ward off predators.[31]

In 1992 Stephen Czerkas presented new fossil evidence indicating that some diplodocid sauropods (i.e. *Diplodocus* and *Barosaurus*) sported a row of spines along the vertebral column.[32] The most peculiar spines, however, were sported by South American Early Cretaceous genus *Amargasaurus*. Paleoartist Brian Franczak dazzled dinophiles in 1992 with visions of *Amargasaurus*' cervically situated twin sail, supported by bony spines approaching 21 inches in length.[33] Certainly spiny structures and bony nodules such as those recently discovered in the African sauropod genus *Malawisaurus*, are not anticipated in sauropods, creatures whose integument has become characteristically restored like the leathery hide of modern elephants.[34]

Czerkas believes sauropod skin had a scaly, instead of leathery or elephantine, texture, stating, " there is often a circular rationale based on realistic portrayals that instill the belief that sauropods had a smooth, wrinkled hide like that of an elephant. The underlying effect from this misperception further exaggerates other popular beliefs that if dinosaurs look so mammalian, why should they not be considered endothermic?.... However, if dinosaurs were accurately portrayed according to their fossilized skin impressions, one could hardly dismiss the biological affinities of their reptilian appearance ... recent discoveries of sauropod skin not only reaffirm the noticeably scaly nature of their hide but also demonstrate that they looked far more reptilian than previously suspected."[35] About a year after Czerkas penned those words, paleontologists identified rows of reptilian, tubercle-like scales in titanosaur embryo skin.[36]

A century ago the sheer immensity of nature's grandest land creatures would have been startling enough. There seems to be no size limitation among the sauropods! With the discovery of a 137 foot long *Seismosaurus*, the 100 ton *Argentinosaurus* and 60 ton *Sauroposeidon*, new contenders are "weighing in" for the all-time terrestrial heavyweight title.[37] (See Figure 35-5.) Years ago, it would have been truly difficult to imagine sauropods as forest-dwelling

Opposite: Figure 35-4: Michael Trcic's bronze sculpture of the African sauropod *Jobaria*. See chapter 27 for further information on this genus. (Photograph courtesy of Michael Trcic.)

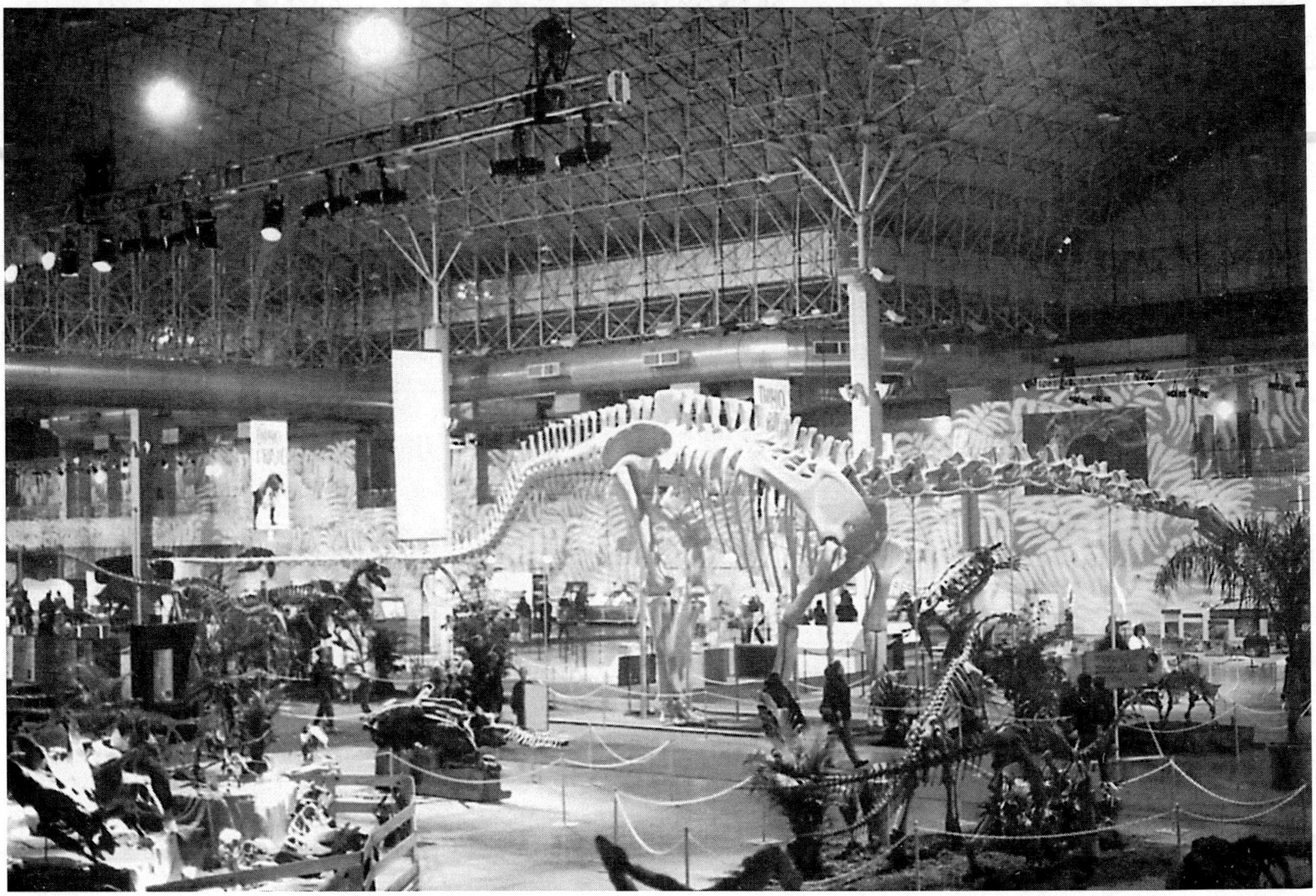

Figure 35-5: *Seismosaurus,* hypothetically reconstructed using synthetic materials—proportioned after a more familiar and completely known genus, *Diplodocus.* (Photograph by Allen A. Debus, 2000.)

"landlubbers." Now it is evident that some sauropod genera were adorned with dermal armor, crests, bony tail clubs and spines—features unknown to nineteenth and early twentieth century paleontologists. Although inconsistently shown in newer sauropod restorations, it is likely that (instead of ichthyosaurian flippers) only the first clawed toe of sauropods' hands protruded from thick hide. How different these visions seem when compared to Owen's imaginary "whale-crocodile!"

Both beauty and accuracy in paleoart are judged by the eyes of the beholder. Paleoimagery that inspires awe or becomes fully absorbing doesn't necessarily represent objective truth. In fact, the more beautiful the image, the more speculative it may be. Consider, for example, sauropod imagery broadcast in Tim Haines' BBC production, *Walking with Dinosaurs* (1999), a remarkable televised production viewed by nearly a third of the United Kingdom's population when it first aired. The pros and cons of *Walking*'s *Diplodocus* sequences, computer-animated for the "Time of the Titans" episode, became hotly contested by paleontologists, dinophiles and artists.[38]

Walking's diplodocids are the first animated sauropods to be adorned with Czerkas' vertebral spines. And the diplodocids also displayed limited range of neck motion, with necks favoring low elevations per Kent Stevens' suggestions.[39] The necks appeared relatively straight and in relaxed mode, sloping downward with heads suspended just above ground level. These were not Bakker's active "high feeders" but lumbering cyber-browsers instead. Their flesh was massive and their skin texture reptilian. And they were impressive to behold.

However, invertebrate paleontologist Norman MacLeod scorned the production, claiming it was merely "creative fantasy" and that he was "...embarrassed for the profession." The scene where a female *Diplodocus* lays her eggs through an ovipositor was considered "screenwriter's fantasy." Scenes showing diplodocid hatchlings abandoned by their parents, MacLeod said, are wrought with speculation.[40]

Conversely, *Science* reporters Andrew Watson and Erik Stokstad praised the *Diplodocus* restorations. They stated, "In one amazing shot, the camera looks up at a diplodocus walking overhead, the dinosaur's neck takes

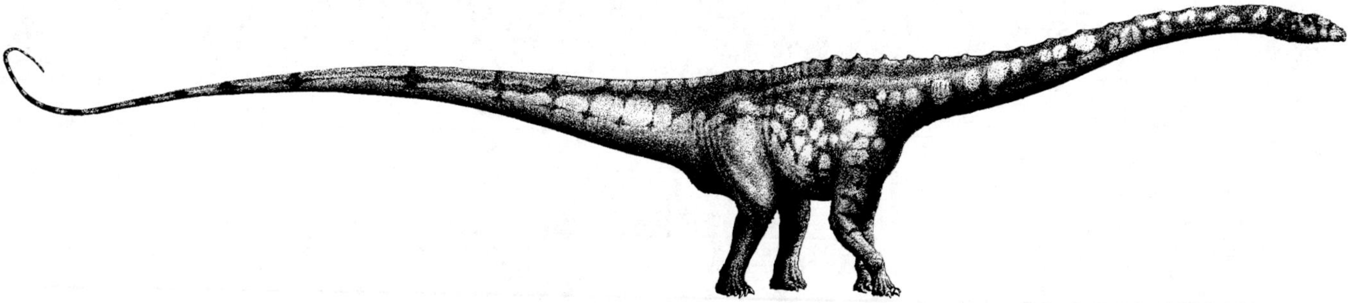

Figure 35-6: Bob Nicholls' accurate restoration of the *Diplodocus* (2000). (Reproduced with permission.)

seconds to pass over, then huge wattles on its fleshy belly fill the screen.... [The] herds look so heavy you can almost feel the ground shake." In further defense of Haines' effort, Dr. Thomas Holtz of the University of Maryland stated, "If you want an animal to actually walk across the screen you've got to start speculating."[41]

British paleoartist Bob Nicholls was not enamored of *Walking*'s diplodocids. So he decided to restore *Diplodocus* with enhanced accuracy. (See Figure 35-6.) In a popular article, Nicholls claimed, "As a palaeo artist and an ancient life fanatic I am always very critical of modern prehistoric life illustration. Every day I am frustrated by images which are clearly unresearched and based upon stereotypical perceptions of dinosaurs. In fact, some images have become so distorted and characterized that they barely resemble a dinosaur at all! I believe that to produce an accurate representation of an extinct animal, the artist must approach the project with 'new eyes.' For example, when I began to design *Diplodocus* I questioned every characteristic previously associated with the beast. How flexible was the tail? How did the muscle attach to the skeleton and effect its external forms? How were the internal organs arranged and how would this influence the dinosaur's physique? Basically, I rebuilt the animal from the inside out."[42]

The vertebral column of Nicholls' new *Diplodocus* restoration is not as densely spiked as in Czerkas 1992 reconstruction. It appears bloated from production of digestive gases, yet not as portly-looking as *Walking*'s version, and it has an "...upper lip and an uninflated air sack for courtship, over the nostrils."[43] Nicholls' new restoration evidently borrows more from Bakker than from Stephen Czerkas.

Many aspects of sauropod natural history simply seem impossible or even paradoxical. Sauropods evidently evolved special attributes ensuring their survival for 120 million years. They may have died out, comedian Jay Leno joked, because of a "terminal case of hickies."[44]

Sauropods remain one of Nature's most puzzling zoological forms. A century from now, Man's understanding of the sauropods may markedly contrast with views of the present. However, as a result of over 150 years of dedicated scientific investigation, it is no longer difficult to imagine that, indeed, wherever they went their footsteps thundered!

CHAPTER 36

Renaissance Dinosaurs

One wonders whether the only commonality between the historical Renaissance and the lauded Dinosaur Renaissance is the fact that in both cases creative individuals designed things that could fly, machines in the former case and dinosaurs in the latter. The term "Dinosaur Renaissance," labels the current era in the history of dinosaur and prehistoric animal awareness. Yet it would seem incongruous associating dinosaur restorations with a term conjuring masterpieces of the great fifteenth and sixteenth Renaissance painters and sculptors—Leonardo da Vinci, Michelango and Raphael.

Historian of science Professor Emeritus Allen G. Debus suggests that the word "...Renaissance ... [has] been employed with so many connotations that there is little hope of satisfying any two scholars with a single definition. To be sure, the Renaissance did involve a kind of 'rebirth' of knowledge—no less than it did a rebirth of art and literature.... But having granted this, it is necessary to be careful to avoid simplification."[1]

Our dinosaur renaissance is visually enhanced, fueled by paleoimagery with the art and science of dinosaurs coupled to a degree further than ever before. The term "Dinosaur Renaissance" implies an intellectual "rebirth" of thinking about dinosaurs. At the outset, a nostalgic return to Huxley's precocious Victorian views on dinosaurs and older, livelier dinosaur images (like Charles R. Knight's battling dryptosauri or his sauropod rearing on its hind legs, or Erwin S. Christman's 1921 illustration showing a gorgosaur racing after a pack of corythosaurs) was implicit.[2] So what is meant by the vague, underappreciated yet often used term "Dinosaur Renaissance," a term that is rapidly becoming synonymous with "paleoimagery"?

Although Robert Bakker inaugurated the "renaissance" in dinosaurs with his popular April 1975 *Scientific American* article titled "Dinosaur Renaissance," the scientific foundation for this movement had been mounting for a decade.[3] The essence of the self-proclaimed Dinosaur Renaissance rested on two precepts, that dinosaurs had a superior physiological edge because they had an endothermic (warm-blooded) metabolism, and that dinosaurs were the evolutionary ancestors of birds. The short "abstract" prefacing Bakker's article said it all: "Dinosaurs were not obsolescent reptiles but were a novel group of 'warm-blooded' animals. And the birds are their descendants."[4] Sarah Landry's illustrations introduced readers to "alien" creatures never before perceived: feathered dinosaurs chasing down prey, restorations of furry winged pterosaurs and hairy therapsids—mammalian ancestors of the Permian period.[5] Unconventional evolutionary diagrams showed birds branching from ancestral dinosaur stock. Amidst such dazzling imagery, Bakker supported the controversial premise of his renaissance affair expertly. (See Figure 36-1.)

Bakker had not originated the warm-blooded dinosaur theory. Canadian paleontologist Loris Russell had, in fact, explored similar terrain in 1965, a year after Bakker's college advisor, John Ostrom, had discovered *the* "renaissance" dinosaur—*Deinonychus*.[6] Ostrom described *Deinonychus* in a 1969 monograph, "Osteology of *Deinonychus antirrhopus*, an Unusual Theropod from the Lower Cretaceous of Montana."[7] Bakker had assisted in the 1964 fossil dig in Montana where the Early Cretaceous carnivore had been found. From the appearance of its skeleton, which included a weight counterbalancing tail held rigidly in life and long scythe-like talons on the second toes of the feet, Ostrom suggested that *Deinonychus* was a warm-blooded dinosaur. Bakker's restoration of a sprinting (unfeathered) *Deinonychus* (1969), an apt symbol for the Dinosaur Renaissance, appeared as the frontispiece to Ostrom's monograph. Of Bakker's restoration, Ostrom commented, "Although *Deinonychus* may not have looked

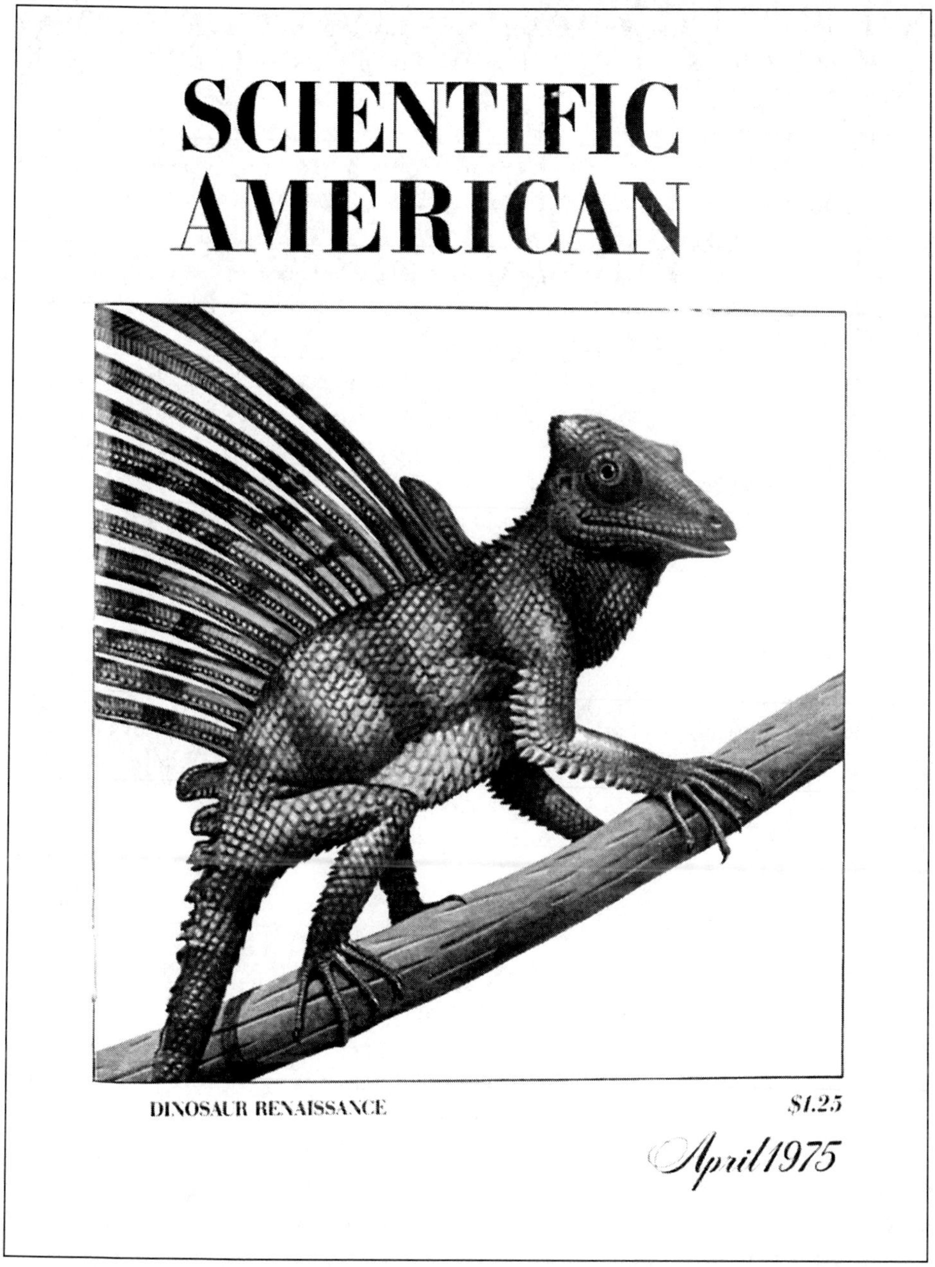

Figure 36-1: Cover of 1975 *Scientific American* issue announcing the beginning of the "Dinosaur Renaissance." (With permission of *Scientific American* and Richard Ellis.)

ing three-horned *Chasmosaurus*, restored with upright, rhinocerine limb posture, charged across the magazine cover.

Bakker's ideas became further celebrated following publication of Adrian J. Desmond's *The Hot-Blooded Dinosaurs: A Revolution in Palaeontology* (1975).[10] Desmond's new-wave book expertly captured the essence of the rapidly "evolving" dinosaur. Bakker's restorations of active, warm-blooded dinosaurs were highlighted. Desmond addressed the plausibility of warm-blooded pterosaurs, the evolution of dinosaurs from warm-blooded thecodont ancestors, the evolution of birds from dinosaurs, and the theory that dinosaurs had been exterminated by a supernova. Between Bakker's *Scientific American* article and Desmond's book, the founding principles of the dinosaur renaissance were becoming widely disseminated. With his long hippie hair and flamboyant style, Bakker's controversial stance struck a popular chord during America's 1970s historical period of social unrest and cultural upheaval. His convention-defying ideas appealed intellectually, while the new disturbing paleoimagery excited us subconsciously. Cold-blooded dinosaurs were an "establishment" to be overturned, and Bakker led a spirited protest.

exactly as Bob restored it, I think his sketch is probably quite close."[8] In retrospect, Ostrom's seminal paper didn't set a wave in motion. However, through his art and wisdom, the idiosyncratic Bakker became the chief instigator of the dinosaur renaissance movement of the 1970s and 1980s.

A year earlier, in the spring of 1968, Bakker published an article in *Discovery* titled "*The Superiority of Dinosaurs.*"[9] This issue of limited circulation featured two of his now classic dinosaur restorations. A pair of elegant giraffe-like *Barosaurus* graced one page of his article, and two gallop-

The years 1978 through 1983 became a pivotal period for the dinosaur renaissance. New discoveries were sensationalized by the media.[11] News coverage escalated. Associating astronomical catastrophes with dinosaur extinctions contributed substantially to paleontology's rising popularity. In 1980 Walter and Luis Alvarez, Helen Michel, and Frank Asaro claimed that a six mile asteroid collided with the earth in the Late Cretaceous.[12] The impact generated a sunlight-blocking dust cloud that cooled temperatures globally for a geologically brief episode, leading to ecological devastation and extinctions of many marine and terrestrial organisms. It was received with much derision from the scientific community. But the asteroid impact theory was sound and testable, offering many useful predictions, many of which have become substantiated. By early 1984, other researchers presented further evidence indicating that global mass extinctions were celestially caused, periodic events recurring once every 26 to 33 million years.[13]

And who can forget the announcement of little "Maia," the dinosaur embryo discovered in 1983 in a fossilized egg by paleontologist John Horner? Paleontologists had discovered the now famous dinosaur nest egg site near Bozeman, Montana, in 1978. Like no other dinosaur genus, the "good mother lizard," *Maiasaura peeblesorum*, symbolized that rollicking 1980s decade of dinosaur frenzy and love.[14]

Images of prehistoric animals were popularized during this interval to an unprecedented level, both in books and magazines. Paleoimagery was making a new insurgence, further popularizing the science. Books by Donald F. Glut, such as *The Dinosaur Scrapbook* (1980), *The New Illustrated Dinosaur Dictionary* (1982) and *Dinosaurs, Mammoths and Cavemen: The Art of Charles R. Knight* (with Sylvia Massey Czerkas; 1982), and Edwin H. Colbert's *Dinosaurs: An Illustrated History* (1983) celebrated prehistoric animal restorations, underscoring their significance in timeless imagery.

While such books favored conventional restorations and dinosaurs depicted in customary poses, Stephen J. Gould's essay collection *The Panda's Thumb* (1980) included two chapters titled "*The Telltale Wishbone*" and "*Were Dinosaurs Dumb?*" featuring dinosaur restorations by Gregory S. Paul, an artist then rapidly rising to paleoartistic prominence.[15] Another bestseller, *The Dinosaurs: A Fantastic New View of a Lost Era* (1981), introduced us to the Mesozoic visions of artist William Stout. Two decades later, Stout's book is revered by dinophiles for having catalyzed the dinosaur renaissance movement. Stout claimed that, "...my book was the first to collect all of the 'new' dinosaur theories in one place and present them to the public.... [It] is generally considered to be the 'spark' that set off the dinosaur explosion (for me it was mainly *The Hot-blooded Dinosaurs*)."[16]

During a segment of an early 1980s televised dinosaur program, host Donald F. Glut recalls artist Mark Hallett explaining that he should be introduced on the show as an "artist-paleontologist." Until that time most animal artists referred to themselves as "wildlife artists," even if they specialized in dinosaur restorations. By 1986, Hallett was referring to himself as a "paleoartist," and "paleoart" was all the rage among dinophiles.[17]

On February 15, 1986, an exhibition *Dinosaurs Past and Present* opened at the Natural History Museum of Los Angeles County. For this thought-provoking display, Sylvia Czerkas selected various paintings, illustrations and sculptures showing dinosaurs as envisioned by artists of the past (such as Benjamin Waterhouse Hawkins, Charles R. Knight, Margaret Colbert, Ernest Untermann, Rudolph Zallinger, Charles Whitney Gilmore and Richard Swann Lull) placed beside the restorations of the Dinosaur Renaissance's leading lights. This became a traveling exhibition, touring many North American and Canadian cities. The display underscored how "...the most striking change to occur in the restorations is in the depiction of body posture. The emerging new look is one of anatomically correct, sleeker, and more varied animals that successfully adapted to almost every ecologic niche now occupied by modern mammals and birds. New information about dinosaur anatomy and physiology has indicated that some dinosaurs were capable of rapid locomotion and may have been warm-blooded."[18] Paleoimagery was transforming under its own "renaissance." While paleoart of the old masters could still be celebrated, the new generation of artists preferred to restore dinosaurs as lively, social creatures. The visual image of dinosaurs was shifting rapidly, fueling the renaissance movement.

Almost overnight, leading paleoartists were becoming as celebrated as their contemporary scientific beacons, paleontologists Horner, Gould and Phil Currie. By the late 1990s there were hundreds of artists producing high quality paleoimagery worldwide. While many gifted artists passionately painted and sculpted a host of prehistoric animals, the names of several were becoming increasingly familiar to dinophiles, who had savored their restorations in museum displays, books and magazines. These few movers and shakers of the paleoart industry lovingly recreated the Mesozoic world more accurately and in finer detail, with keener insights and armed with more exact knowledge, than at any prior episode in the history of prehistoric animal image-making.

Accordingly, no "A team" listing of dinosaur renaissance painters and illustrators would be complete without mentioning Bakker, Ely Kish, Gregory Paul, Mark Hallett, William Stout, Douglas Henderson, John Sibbick, vertebrate paleontologist Kenneth Carpenter, Brian Franczak, John Gurche and Michael Skrepnick.[19] American dinophiles were most attracted to the sculptural creations of Stephen Czerkas, David Thomas (1928–1999), Michael Trcic, Tony McVey, Bob Morales and Shane Foulkes.[20] (See Figure 36-2.) While the startling productions of the renaissance "masters" bolstered the lore, hype and popularity of dinosaur *science*, paleoimagery had become a prominent commercial force of its own. In particular, Paul sounded rather disenchanted by having to face such worthy competition.[21]

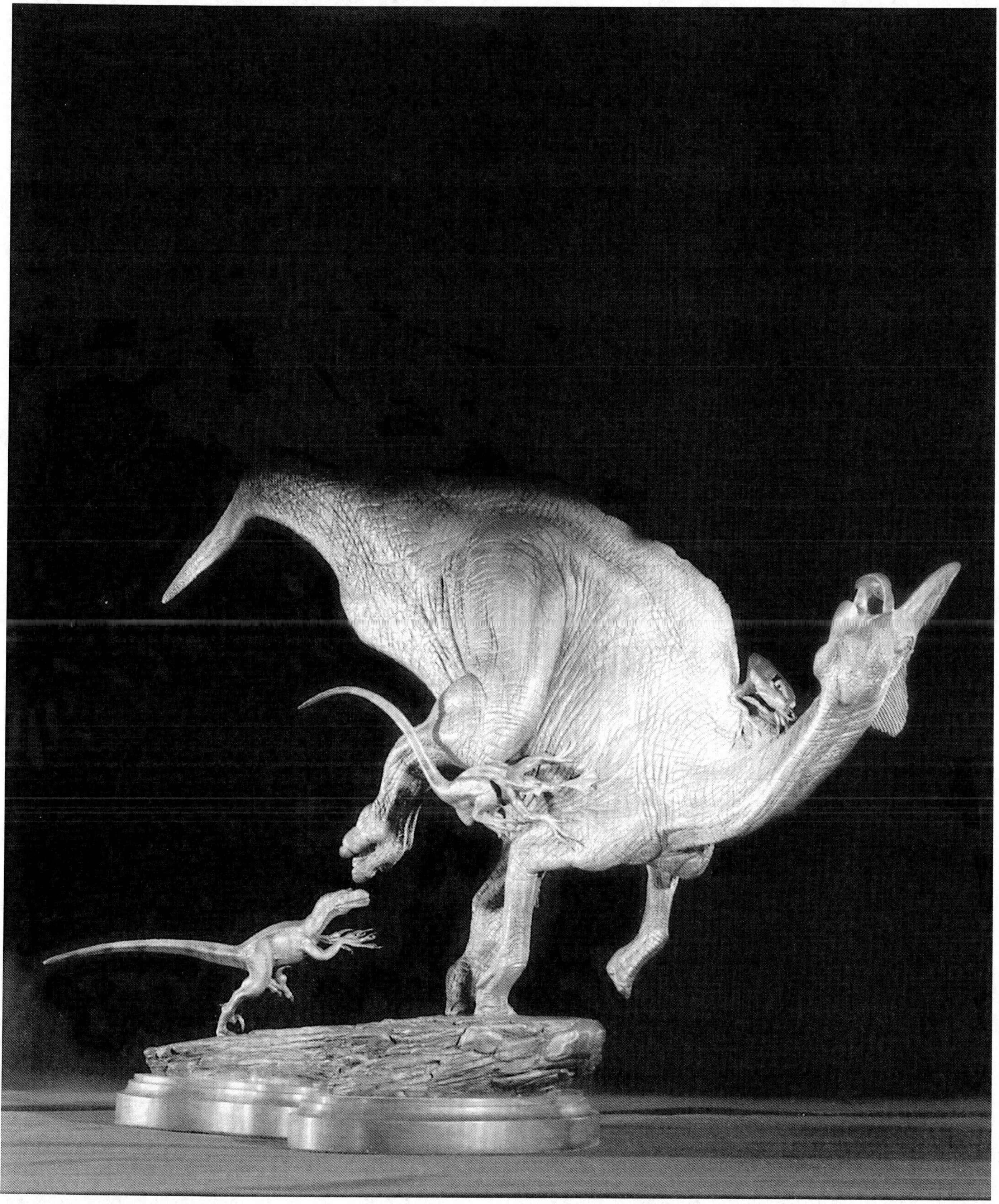

Figure 36-2: Michael Trcic's fantastic ⅛ scale sculpture—*Lambeosaurus and Dromaeosaurs* (1996). Through his art, Trcic skillfully captures the essence of the idea of prehistoric "savagery," a theme originating with scientific studies of the American Mastodon. (Reproduced with permission.)

During the early 1990s, while two traveling paleoart exhibits toured the country and abroad, the Dinosaur Renaissance exploded with the publication of Michael Crichton's novel *Jurassic Park* (1990).[22] Fans anxiously awaited Stephen Speilberg's film adaptation, which premiered on June 11, 1993. The movie *Jurassic Park* set records as the highest grossing movie in history, no doubt due to the riveting imagery of its fantastic dinosaurs, created through use of computerized special effects, then the most realistic and frightening ever filmed. The film's real stars were the demonic "raptors," unfeathered intelligent harbingers of chaos and mayhem, thinly disguised versions of Ostrom's *Deinonychus* incarnate. *Jurassic Park* kindled public interest in dinosaurs like no other science news story, fossil discovery, book or museum display had ever before. The movie's use of paleoimagery rejuvenated older dinophiles' waning taste for dinosaurs and whetted appetites of millions of youngsters.[23]

While the *Jurassic Park* phenomenon was in full swing, two new organizations, the U.S. Dinosaur Society (founded in 1992) and the UK Dinosaur Society (founded in 1994) aimed to support dinosaur research and educate the public, especially children, about dinosaurs. A news publication issued by the U.S. Dinosaur Society stated, "...billions are made on dinosaur products. Few of these products succeed in accurately communicating the discoveries and the excitement of dinosaur science. Meanwhile, sustained youthful interest in all of science is in jeopardy."[24]

The U.S. Dinosaur Society inaugurated an "artistic review committee" comprised four paleoartists. This panel was charged with the task of reviewing proposals for artist's projects to illustrate educational or scientific projects. Furthermore, the committee evaluated and certified dinosaur products and toys for "consonance with scientific research and informed speculation on dinosaur appearance."[25] The U.S. Dinosaur Society also maintained a listing of active dinosaur artists and published a guide to help artists receive adequate compensation for their work. Across the Atlantic, the UK Dinosaur Society also placed emphasis and value on the nature and history of paleoart, and helped sponsor the efforts of practicing paleoartists.

By the time the U.S. Dinosaur Society went bankrupt in 1997, other venues were accessible to dinophiles. These included several dinosaur fanzines, club newsletters and publications dedicated to dinosaurs—paleomag "candy" especially for those who enjoyed reading articles about paleoart and paleoartists' careers, or on how to become a paleoartist. Chief among those emphasizing dinosaurs was Mike Fredericks' bimonthly publication, *Prehistoric Times*, with others (including Lynne Clos' *Fossil News*, *Dinosaur World*, the *Archosaurian Archive* and *Dinosaurus Magazine*) quickly following suit. Each of these presented interesting angles on paleoimagery, offered interviews with notable "renaissance" artists and featured restorations and reconstructions of every variety of dinosaur and prehistoric animal known to science.

Prehistoric Times, inspired by *Jurassic Park*'s successes, promoted dinosaurabilia collecting, especially dinosaur toy and model imagery. It was treasured by dinosaur modelers and beginning sculptors, and became a common "watering hole" for paleoartists worldwide. (See Figure 36-3.) And it greatly facilitated John Lanzendorf's efforts to gather the world's foremost privately held collection of "dinosaur fine art." John collaborated on a lavish coffee table–style book featuring photographs of his collection, and he also lent his name to the "Lanzy," a set of awards presented annually to paleoartists by the Society of Vertebrate Paleontology (for the first time in October 2000).[26]

By the turn of the millennium, paleoimagery had proliferated beyond Osborn's wildest expectations, with websites devoted to dinosaur art and images, a number of popular publications offering guidance on how to restore dinosaurs through illustration or sculpture, biannual Dinofest conventions emphasizing paleoimagery and the contributions of paleoartists, and more broadcast dinosaur imagery than at any prior episode in television history. Indeed, the Dinosaur Renaissance seemed boundless and all encompassing.[27]

A quarter of a century ago, the dinosaur "renaissance" was perceived as a "revolution in palaeontology," which was hyperbole because the movement was just underway. What is meant by scientific "revolution?" Historian of science I. Bernard Cohen defined his meaning of the term in *Revolution in Science* (1985) by identifying four stages of progression in the pattern of development leading toward a revolution in science.[28] The pattern is initiated when a scientist or team of scientists conceives a new solution to a significant problem, resulting in a new theory or knowledge that alters perception of how the problem can be examined. Secondly, the innovative idea or knowledge is *recorded*, thereby establishing the originator's *commitment* to the new theory, concept or method. Thirdly, the innovation must be *publically disseminated* through correspondence or conversation. Finally, unless a significant proportion of other scientists *concur* with the innovation, *revolution* cannot occur. According to Cohen, few "revolutionary ideas" survive this final but critical hurdle. Cohen does not cite papers by Bakker or Ostrom in his book, nor is there any reference to a dinosaur renaissance. If there has been no recognized *revolution* in dinosaur science, does this imply the *renaissance* in dinosaurs is not genuine? Not necessarily.

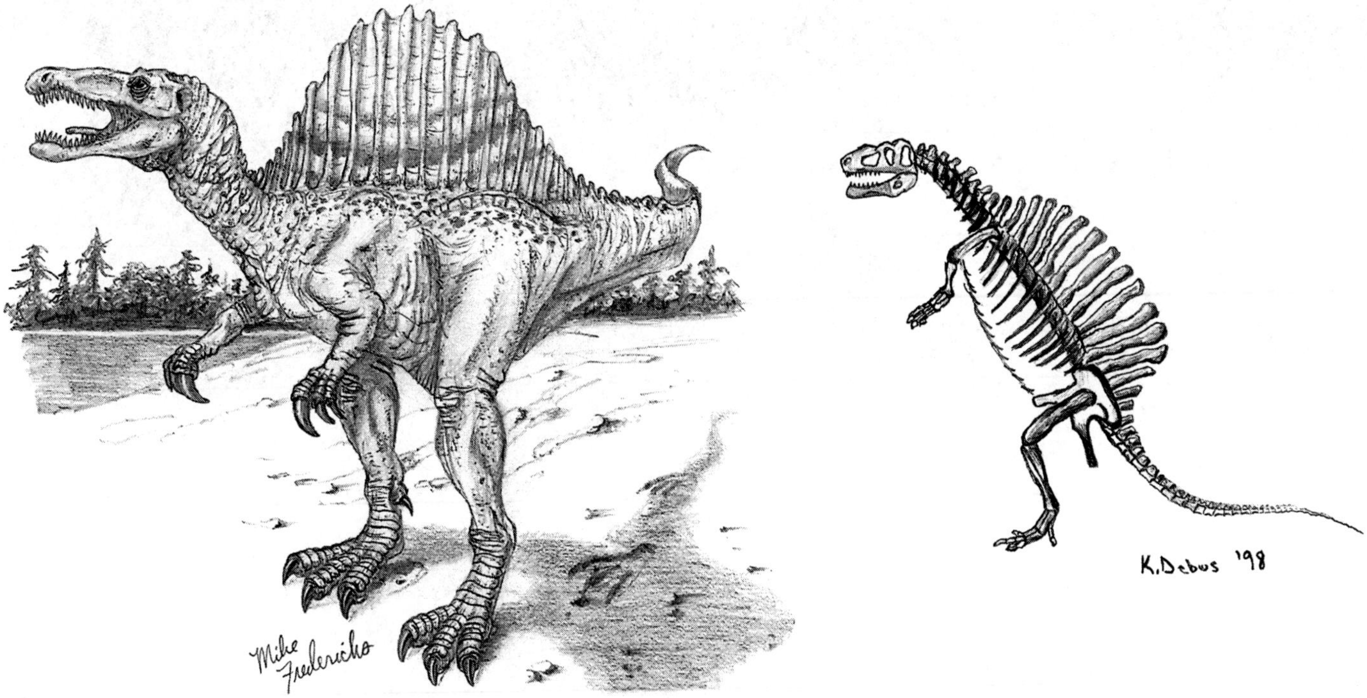

Figure 36-3: At left, Michael Fredericks' illustration of the "twilight zone" dinosaur—*Spinosaurus* (2001). Fredericks' interest in paleoart did much to promote the field of paleoart during the 1990s through his globally popular magazine, *Prehistoric Times*. At right, Kristen Debus' 1998 illustration, after a reconstruction by A. Lapparent, showing how *Spinosaurus* was formerly reconstructed before discovery of related dinosaurs *Baryonyx* and *Suchomimus*. (Reproduced with permission.)

Allen G. Debus claims, "Use of [the terms]'Renaissance' and 'Scientific Revolution' can be somewhat different and complicated. Historians most frequently use Renaissance in relation to the recovery of ancient knowledge, with an emphasis on accuracy of translations from both Greek and Latin, to the flowering of art and literature, and the move away from the philosophical and theological subjects that dominated studies in the medieval universities. Part of this move away from the past involved a greater interest in nature, but it was not until the late fifteenth century that the recovery of ancient scientific and medical texts became a part of this movement.... During this time, ancient scientific texts were translated and criticized. The first reaction was to treat them as 'gospel,' and only slowly did people see flaws in the work of the major ancient authorities.... As the criticism increased, some figures, such as Paracelsus, began to attack the ancient learning, and there was a move toward an emphasis on new observations and experiments, on mathematical analysis and on quantification as a basis for a new science ... as to a rebirth in dinosaurs after years of less interest ... and this is because of specific discoveries and new theories (extinction views, warm-blooded theory, evolution of birds, etc., as well as the spectacular new fossil discoveries), I think that the use of the term 'renaissance' is appropriate."[29]

The striking manner in which paleoimagery has transformed over two centuries may not be apparent to all. Historian of science and paleontologist Martin J. S. Rudwick claims that the genre of "scenes from deep time" had become established with the publication of 1860s editions of Figuier's popular *Earth Before the Deluge*.[30] Furthermore, by the 1860s most of the major themes prevalent in paleoart today had already been expressed. Thereafter, from the 1860s onward, "it is no accident that scenes of dinosaurs are those that best exemplify the modern genre."[31] How shall we regard this "modern genre"? Can we identify a "renaissance" of image-making therein?

In *The Last Dinosaur Book* (1998) William J. T. Mitchell presents a "Schematic History of Dinosaur Images," wherein three historical periods are recognized.[32] These include the American Revolution through Civil War, the Gilded Age to the Depression, and from World War II to the end of the Cold War. While not denying the present condition of a renaissance in dinosaurs, Mitchell excluded specific mention of a "Bakkerian" renaissance period from his schematic tabulation. This is curious because Mitchell is a cultural iconologist, and dinosaur images experienced their most dramatic wave of replication, novelty and diversification as a direct consequence of the scientific renaissance in dinosaurs.[33]

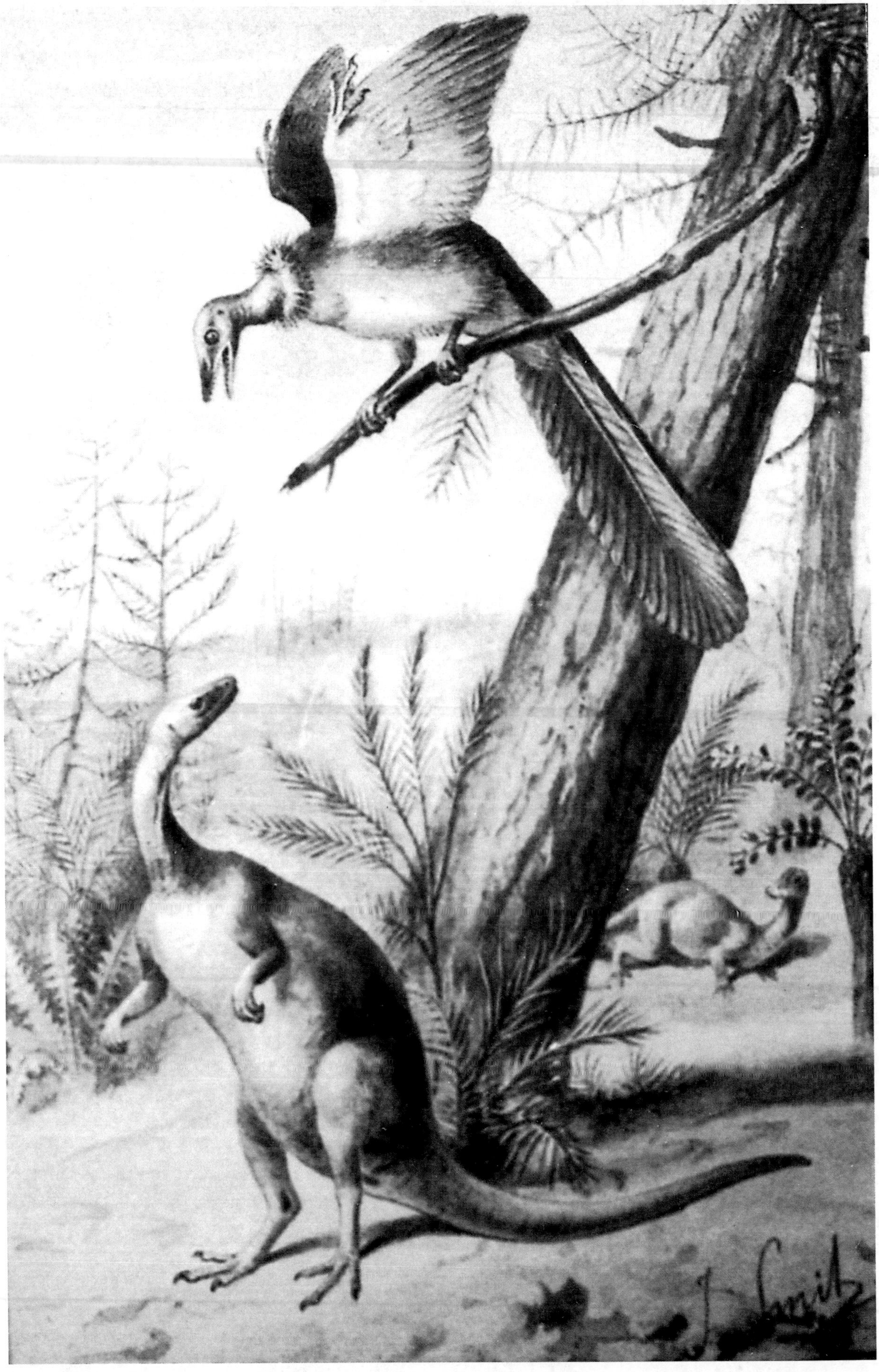

There would appear to be several paleoart phases or "eras." The first phase, in which the earliest portrayals of prehistoric life were made (1800 to circa 1900) was punctuated by Osborn's "Reformation" in paleoart of the early 1900s. Next, in the dinosaur dominated phase, 1900 through the early 1970s, portrayals of prehistoric life and other paleoimagery became refined yet conventionalized, that is until Bakker's "Dinosaur Renaissance" enticed paleoartists to show us the "new dinosaurs" and the world in which they lived. But Bakker's "renaissance" can be further subdivided into three periods. In the first, 1965 to 1978, paleontologists labored over the warm-blooded question. In the next, 1979 to 1989, *Maiasaura* and mass extinction hype shared the spotlight. In the current phase, dawning with the publication of *Jurassic Park* (1990), paleontologists have been relying on computer technology to create the most lifelike paleoimagery of all—cyber-dinosaurs—both for sheer entertainment and scientific purposes, while relentlessly pursuing paleontology's "holy grail"—finding the evolutionary fossil link between birds and dinosaurs. With the discovery of China's feathered dinosaurs, this last phase may be whimsically dubbed the "Feather Revolution."[34] (See Figure 36-4.)

Throughout its enduring two century history, paleoimagery has supported many themes and ideals. From a scientific perspective, paleoimagery's utility has been realized most keenly through scientific illustration of vertebrate fossils, reconstruction of fossil vertebrates, restoration of prehistoric animals, depiction of scenes from "deep time," portrayals of sequences of scenes from geological history and "violent prehistory," and the projection of theoretical geology and evolutionary iconography.[35]

From the earliest glimpses of America's *Mastodons* and Victorian visions of primeval ichthyosaurs, through the dinosaur's "Reformation" phase and into our "dinosaur renaissance," perceptions of the "prehistoric world" have become increasingly accurate; yet they are far from crystal clear. This is a direct reflection of how reliant the art of paleontology is on the science. The scientific techniques and devices available to today's breed of earth scientists far exceed in sophistication the methods and instruments used by late nineteenth century scientists. Thus, data applied to our paleoworld reconstructions is more accurate and precise than ever before, allowing more definitive conclusions to be drawn about the fossil organisms that lived in those times. New data has championed new theoretical notions and falsified older conventional ideas. Statistical reasoning is more cogently allied to our comprehension of phylogeny than during Darwin's day, when Hawkins' restorations were all the rage.[36]

Yet beyond the strictly scientific, there are three fundamental "iconic themes" recurrent throughout the history of prehistoric animal image-making. These appeal to human psychological, intellectual and ecological senses. They are shadowed in the idea of the dinosaur's name

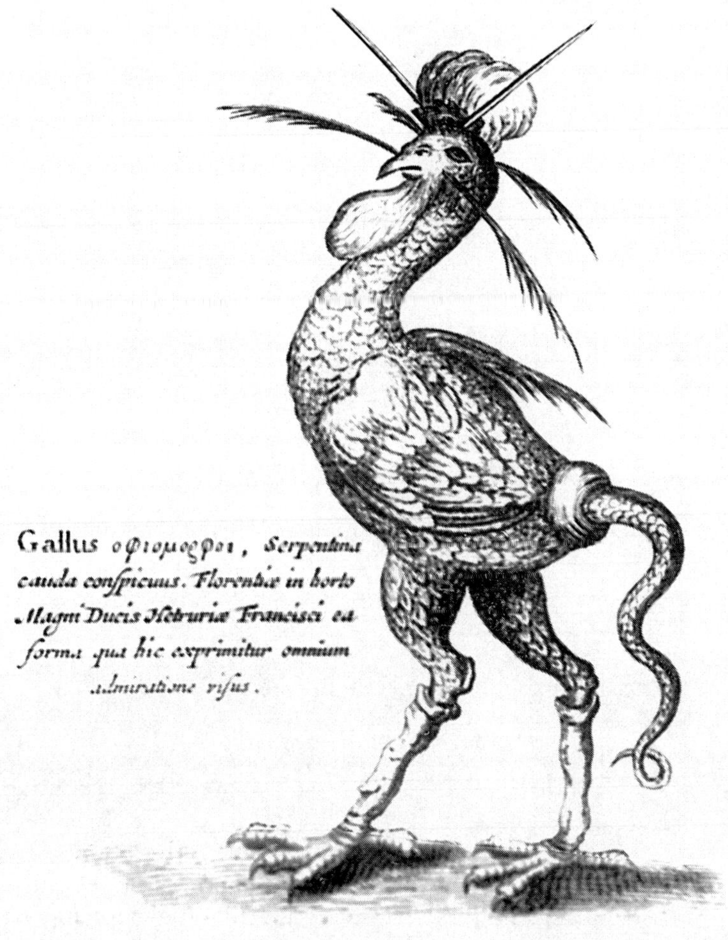

Opposite: Figure 36-4: "Holy grail" incarnate! Joseph Smit's 1905 restoration showing two related Bavarian species, *Archaeopteryx* and *Compsognathus*. Modern paleoartists would perhaps have restored the latter with feathers. From *Nebula to Man* (New York: J. M. Dent and Company.)

Above: Figure 36-5: Medieval illustration showing fantasy creature, the "⅔Serpentine Cock," inspired by occurrences of fossilized bones and thought by scholars to have inhabited the earth's interior. Note how closely this ancient "restoration" also captures the flavor of modern evolutionary ideas of a dinosaur ancestry for birds. From the German scholar Athanasius Kircher's 1665 edition of *Mundus Subterraneus*. (Photograph by Allen A. Debus, courtesy of Allen G. Debus.)

("fearfully great reptiles"), the *quest* for the chimerical feathered bird-dinosaur "link" (perhaps forever missing),[37] and our innate, prehistoric fascination with animal *forms*, first cultivated by our Cro-Magnon ancestors who painted mammoths, horses and woolly rhinos on the walls of caves.[38] These themes are manifested today in perceived rivalry between "savagery" and giants (i.e. *T. rex* vs. all contending sauropod icons), the search for a feathered dinosaur "holy grail" bird ancestor, and in our need to *accurately* portray prehistoric organisms as they really lived, not as monstrous distortions. (See Figure 36-5.)

Yet, in the ever popular image-making arena, there is still room for artistic license. A century hence, today's "accurately" depicted "prehistoria" will likely seem as improbable to us then as those of yesteryear do now.

Notes

Chapter 1

1. William J. T. Mitchell, *The Last Dinosaur Book: The Life and Times of a Cultural Icon*, Chicago: The University of Chicago Press, 1998, pp. 52, 287.

2. Mark Hallett, "Bringing Dinosaurs to Life," in *Dinosaurs Past and Present*, Volume 1, Seattle: Natural Museum of Los Angeles County in association with University of Washington Press, 1986. In a December 23, 1999, personal letter, Allen identified Mr. Hallett as the originator of the term "paleoartist."

3. William E. Swinton, *Dinosaurs*, London: British Museum—Natural History, publication no. 542, 1969, pp. 38–39. Paleoartist Michael Skrepnick further clarified the relation between restorations and reconstructions in a 1999 *Dinosaur World* interview. "When you go to a museum and you see a (i.e. resin or plaster) cast skeleton of a dinosaur, that is in fact a replica produced of bones excavated from the field, you have creativity injected by the artist to either reproduce a life restoration of this animal, or some other interpretation, perhaps a technical rendering of what the artist observes of the bones. So, in either case the artist is taking it a step further, whereas the replicas in the museum are reproduced down to finite, almost molecular detail to accurately reproduce the material removed from the field. Casting bones creates a record of the actual fossil, while the artist goes beyond that level by infusing his own interpretations and artistry. The skeletal mount is not just a recreation of the articulated bones of a specimen. Now, the person who has created that mount is trying to interpret behavior through a skeletal mount and that is where the artistic nature of it is 'in play.' So, the bones themselves as a standing mount is one thing. But when someone takes it a step further to show you something outside the dimension of that animal, then you are taking artistic license. To put flesh on the animal is an artistic interpretation. How good an artistic interpretation it is may vary, and different models are often meant for different purposes. It's strictly put together to build an image of the living animal, and you want to stop at that (i.e. functional) point, that's fine. Such studies provide a good deal of information. But I think the real "meat" of the artwork is to show us more. Don't just show us the animal, show us how it moved, show us some behavior, some interaction with other individuals, give us a sense we are living in its world. When more of that happens, the viewer is more captivated. It engages them. They want to know more. The more you offer them, the more they want." Skrepnick interviewed by Allen A. Debus, quoted in *Dinosaur World*, no. 7, Winter 1999/2000, pp. 64–65.

4. Personal e-mail communication from paleoartist Dan Varner, July 9, 1999.

5. This definition appears on the Society of Vertebrate Paleontology website at: http://www.museum.state.il/us/svp/awards/lanzendorf/html

6. Allen A. Debus, "Fine Paleontology vs. Fine Art: The Emergence of American Paleoart" (Part 3)," in *Prehistoric Times*, Oct./Nov. 2000, no. 44, pp. 17–20.

7. Martin J. S. Rudwick, *Scenes from Deep Time: Early Pictorial Representations of the Prehistoric World*, Chicago: The University of Chicago Press, 1992.

8. Stephen J. Gould, "The Lying Stones of Marakech," in Gould's essay collection, *The Lying Stones of Marakech*, New York: Harmony Books, 2000, p. 25.

9. Personal communication with Mr. Lanzendorf, May 2000.

10. An idea first expressed in, Allen A. Debus, "Paleoart Passion and Modern Meaning," in *Prehistoric Times*, no. 45, Dec./Jan. 2001, pp. 17–18.

11. William J. T. Mitchell, *op. cit.*, p. 272.

12. *Dinosaur Imagery: The Lanzendorf Collection*, San Diego: Academic Press, 2000, Photography by Michael Tropea, Foreword by Phil Currie, p. xiv.

13. Personal correspondence from Donald F. Glut, dated Feb. 2, 1999.

14. Michael Skrepnick interviewed by Allen A. Debus (June 1999) in *Dinosaur World*, Spring, 2000, no. 8, pp. 47–48.

15. During the early 20th century, restoration of Tertiary "titanotheres" reflected H. F. Osborn's views of "orthogenesis."

16. Recently discussed and reflected in Peter Dodson, *The Horned Dinosaurs: A Natural History*, Princeton: Princeton University Press, 1996, p. 250.

17. Jeff Liston, "Archaeopteryx and the Evolution of Feathered Flight: The Hidden Story," in *The Quarterly Journal* (UK Dinosaur Society), 2000, vol. 4, no.1, p. 14.

18. See, for example, Dougal Dixon, *After Man: A Zoology of the Future*, New York: St. Martin's Press, 1981; and George Olshevsky, "The Origin and Evolution of the Stegosaurus," in *Kyoryugaku Saizenzen* (*Frontiers of Dinoaurology*), Gakken Mook, 1993, no. 4, pp. 65–103, in Japanese

(English translation provided by Mr. Olshevsky).

19. Paul Sereno, "The Evolution of Dinosaurs," in *Science*, June 25, 1999, vol. 284, no. 5423, pp. 2137–2147.

20. It would seem that dinosaurs should rightfully have promoted evolutionary ideas, much as Michael Jordan became a "good guy" public relations figure for pro sports. Until bird-dinosaur evolutionary relationships became publicized during the "dinosaur renaissance," however, this wasn't the case—which seems counterintuitive. Certainly, with the discovery of feathered dinosaurs (see chapter 20), dinosaurs have been increasingly representative of the ideas inherent in evolution. Perhaps the most significant recent contribution to evolutionary understanding, Stephen J. Gould and Niles Eldredge's "punctuated equilibria" theory—"Punctuated Equilibria: An Alternative to Phyletic Gradualism," in *Models in Paleobiology*, San Francisco: Freeman, Cooper and Company, 1972, pp. 82–115, T. J. M. Schopf (ed.)—was published just prior to the onset of the "dinosaur renaissance" phenomenon. Yet Gould and Eldredge relied on trilobites and other invertebrate fossils as a data base, not dinosaurs. At the outset, "punctuated equilibria" did little to popularize either dinosaurs or trilobites. Of trilobites as an evolutionary symbol, paleontologist Allison Palmer stated, "*...trilobites surpass the dinosaurs in their contribution to evolutionary theory*" ("A Fondness for 'Beetles of the Paleozoic,'" in *Science*, October 6, 2000, vol. 290, p. 59).

Chapter 2

1. Robert Bakker, "Jurassic Sea Monsters," in *Discover*, September 1993, pp. 78–85; and Bakker, "Plesiosaur Extinction Cycles—Events That Mark the Beginning, Middle and End of the Cretaceous," in W. G. E. Caldwell and E. G. Kauffman (eds.), *Evolution of the Western Interior Basin* (Geological Association of Canada Special Paper 39), 1993, pp. 641–664.

2. In personal communication dated August 6, 2000, Kenneth Carpenter noted that the idea about plesiosaur relationships sprang from his initial undergraduate work on plesiosaurs conducted in the late 1970s and early 1980s, long before he (or Bakker) published those ideas. Carpenter published three papers on this matter, one of which is "Comparative Cranial Anatomy of Two North American Cretaceous Plesiosaurs," in *Ancient Marine Reptiles*, J. M. Callaway and E. L. Nicholls (eds.) in *Ancient Marine Reptiles*, San Diego: Academic Press, 1997, pp. 191–216.

3. Josef Augusta, *Prehistoric Sea Monsters*, London: Paul Hamlyn, Ltd., 1964, p. 28; for more recent information concerning *Kronosaurus*, see John A. Long, *Dinosaurs of Australia and New Zealand: And Other Animals of the Mesozoic Era*, Cambridge: Harvard University Press, 1998; and a website posting, "Giant Pliosaurs—Real and Imaginary," by Ben Creisler (Dino-Dispatches no. 1, 11/08/98); and Allen A. Debus, "Kronosaurus: An Australian Sea Monster That 'Got Away,'" in *The Earth Science News* (news bulletin of the Earth Science Club of Northern Illinois), December 1988, vol. 39, pp. 17–23.

4. Robert Bakker, "Jurassic Sea Monsters," in *Discover*, September, 1993, p. 85.

5. Heber A. Longman, "Some Queensland Fossil Vertebrates," in *Memoirs of the Queensland Museum*, vol. 8, 1924, p. 26; also see Susan Turner's article "Vertebrate Paleontology in Australia" for further details on the discovery of Australia's *Kronosaurus*. Turner's article was published in *Earth Sciences History* (Journal of the History of the Earth Sciences Society), 1986, vol. 5, no. 1, pp. 50–65.

6. Personal communication from Ralph Molnar, then of the Queensland Museum, dated April 5, 1988.

7. Heber A. Longman, *op. cit.*, 1924, p. 28.

8. R. Hall and M. Hall, *About the Exhibits: Museum of Comparative Anatomy*, Cambridge: Harvard University, 1975, p. 10.

9. Molnar, *op. cit.*

10. The phrase," *Monsters from the Id*," is borrowed from the classic 1956 sci-fi movie, *Forbidden Planet*.

11. This popular plastic Marx figure was a favorite of the author's prehistoric play menagerie during the early 1960s. A photo of this figure appears on page 77 of *Dinosaur Collectibles* (Norfolk, VA: Antique Trader Books, 1999), by Dana Cain and Mike Fredericks.

12. Personal communication from Kenneth Carpenter, dated August 6, 2000. Also, a 1967 reconstruction of *Liopleurodon* bearing a vertical tail fluke is reproduced as Figure 12–13 on page 248 of Robert L. Carroll's *Vertebrate Paleontology and Evolution*, New York: W. H. Freeman and Company, 1988. William E. Swinton reproduced a skeletal reconstruction of *Plesiosaurus macrocephalus*, outlined with a vertical tail rudder on page 64 of his *Fossil Amphibians and Reptiles*, London: British Museum (Natural History), 1965.

13. Carl Sagan, *The Dragons of Eden*, New York: Ballantine Books, 1977.

Chapter 3

1. For the significance of these "cartoons," see Martin J. S. Rudwick, *Scenes from Deep Time: Early Pictorial Representations of the Prehistoric World*, Chicago: The University of Chicago Press, 1992 and Stephen K. Gould, *Time's Arrow, Time's Cycle: Myth and Metaphor in the Discovery of Geological Time*, Cambridge: Harvard University Press, 1987.

2. Simply put, Cuvier's "principle" stated that if one understood the relationships of animals' organ morphologies, soft tissue features, skeletal apparatus, etc., analogies could be made to comparable extinct species. Given current understanding of the range and pattern of skeletal and morphologies noted in living species, missing fossil bone components of extinct species could be inferred by analogy from the living species. According to historian Paul Semonin, however, in 1728, decades before Cuvier's celebrated contributions, Sir Hans Sloan of Britain's Royal Society "...inaugurated a new era of comparative anatomy..." that led to the understanding that bones of ancient mammals (i.e. Mammoths) were not remains of human giants, but more closely related to modern elephants. Paul Semonin, *American Monster: How the Nation's First Prehistoric Creature Became a Symbol of National Identify*, New York: New York University Press, 2000, pp. 76–77.

3. For more on the relationships between "geognosy" and "geology," see Rachel Laudan's *From Mineralogy to Geology: The Foundations of a Science 1650–1830*, Chicago: The University of Chicago Press, 1987; and Mott T. Greene's *Geology in the Nineteenth Century*, Ithaca: Cornell University Press, 1982.

4. See chapter three, "Life's Revolutions," in Martin J. S. Rudwick's *The Meaning of Fossils: Episodes in the History of Paleaontology*, Chicago: The University of Chicago Press, 1985.

5. Martin J. S. Rudwick, *Georges Cuvier, Fossil Bones and Geological Catastrophes: New Translations and Interpretations of the Primary Texts*, The University of Chicago Press, 1997.

6. *Ibid.*

7. Martin J. S. Rudwick, *op. cit.*, 1985, chapter three.

8. Stephen J. Gould, *op. cit.*, 1987.

9. See chapter seven, "A Pouched Marsupial from Paris," in Martin J. S. Rudwick, *op. cit.*, 1997.

10. Martin J. S. Rudwick, *op. cit.*, 1985, p. 87.

11. Martin J. S. Rudwick, *op. cit.*, 1997, chapter 15, "The Revolutions of the Globe."

12. Martin J. S. Rudwick, *op. cit*, 1992, p. 35.

13. *Ibid.*, pp. 31–33; "Cuvier's" Mastodon reconstruction may have been drawn by British anatomist Everard Home. See caption to Mastodon figure in John C. Greene, *The Death of Adam*, Ames: Iowa State University Press, 1959, p. 122.

14. Martin J. S. Rudwick, *op. cit*, 1992, p. 34.

15. *Ibid.*, pp. 36–37.

16. For more on the Marsh-Cope vendetta, see Mark Jaffe's *The Gilded Dinosaur: The Fossil War Between E. D. Cope and O. C. Marsh and the Rise of American Science*, New York: Crown Publishers, 2000.

17. The story of *Iguanodon*'s discovery is expertly recounted by Dennis Dean in *Gideon Mantell and the Discovery of Dinosaurs*, Cambridge: Cambridge University Press, 1999. See, especially, Figure 4.2.

18. *Ibid.*, pp. 95–96.

19. *Ibid.*, p. 138.

20. Richard Owen, "Report on British Fossil Reptiles," in *Report of the British Association for the Advancement of Science II*, 1842, pp. 60–204.

21. Dennis Dean, *op. cit.*, 1999, p. 190.

22. *Ibid.*, and Edwin H. Colbert, *Men and Dinosaurs: The Search in Field and Laboratory*, New York: E. P. Dutton, 1968.

23. Dennis Dean, *op. cit*, 1999, p. 113.

24. Edwin H. Colbert, *op. cit.*, 1968, p. 19.

25. Dennis Dean, *op. cit*, 1999, p. 121.

26. *Ibid.*, p. 123.

27. *Ibid.*, p. 131; also see pp. 107, 122 and 243. As Dean suggests on page 235, while Owen retained the visage of the rhinocerine horn in his collaborations with Benjamin Waterhouse Hawkins, by the 1850s Mantell no longer attributed a "horn" to his *Iguanodon*.

28. *Ibid.*, p. 261.

Chapter 4

1. Martin J. S. Rudwick, *Scenes from Deep Time: Early Representations of the Prehistoric World*, Cambridge: The University of Chicago Press, 1992, chapters 1 and 3; and Allen A. Debus, "Palaeoartists of the Past—John Martin and the Element of Paleoart Fantasy," in *Quarterly* (The Dinosaur Society UK), 1998, vol. 3, no. 1, pp. 10–11.

2. Martin J. S. Rudwick, *op. cit.*, 1992, pp. 23–26.

3. *Ibid.*, p. 78.

4. Gideon Mantell, *Wonders of Geology*, London, 1848 ed.; also see Dennis Dean, *Gideon Mantell and the Discovery of Dinosaurs*, Cambridge: Cambridge University Press, 1999, pp. 166–170. Dean claims that, "The first book to feature such a frontispiece was Thomas Hawkins' *Memoirs of Ichthyosauri* and *Plesiosauri* (1834)." Thomas Hawkins' frontispiece was illustrated by artist John Samuelson Templeton.

5. From Mantell's "Description of the Frontispiece" in *Wonders of Geology*, 1848 ed.

6. The theme of savage nature in wildlife art can be traced to 1760s paintings by British artist George Stubbs (1724–1806). One celebrated example is Stubbs' painting of a stately-looking horse being ferociously bitten by a lion that has leaped upon its back. According to historian Paul Semonin, this work "...foreshadow[s] the later arrival of warring species and the struggle for existence in the era of Darwin and the dinosaurs." See Semonin's *American Monster: How the Nation's First Prehistoric Creature Became a Symbol of National Identity*, New York: New York University Press, 2000, pp. 259–262.

7. This restoration has been recently published in Martin J. S. Rudwick, *op. cit.*, 1992, p. 85.

8. William Feaver, *The Art of John Martin*, Oxford: Clarendon Press, 1975.

Chapter 5

1. Roswell Ward, *Henry A. Ward: Museum Builder to America* (The Rochester Historical Society Publications—XXIV, 1948), pp. 263–264; also see entry for Henry A. Ward in *The National Cyclopaedia of National Biography*, 1940, New York: James T. White & Co., vol. 28, pp. 169–170.

2. Roswell Ward, *op. cit.*, p. 90.

3. Henry A. Ward, *Catalogue of Casts of Fossils from the Principal Museums of Europe and America with short Descriptions and Illustrations*, Rochester, New York, 1866, pp. 11–14, 16.

4. Roswell Ward, *op. cit.*, p. 130.

5. *Ibid.*, p. 137.

6. Frederic Holder, "The California Academy of Sciences," in *The California Illustrated Magazine*, vol. III, December 1892 to May 1893, pp. 229–244. See J. Aronniz' illustration of the bipedal *Megatherium*.

7. Barbara Franco, *The Cardiff Giant*, Cooperstown, New York: The New York State Historical Association,1990.

8. Roswell Ward, *op. cit.*, pp. 185–186; America may have been predisposed to receive Ward's giant ground sloth cordially, for, according to Todd Womack, William Hodgson's book *Memoir on the Megatherium* (Bartlett & Welford, 1846) "...had an impact on American science and the American imagination, and a 'sloth craze' swept the country for almost the next two decades." See Todd Womack, "Plentifully Charged with Fossils: The 1822 Discovery of the *Eremotherium* at Skidaway," in *Fossil News: Journal of Avocational Paleontology*, vol. 6, no. 7, July 2000, pp. 14–16. According to Bernard F. Reilly of the Chicago Historical Society, "The 1875 exposition referred to in Roswell Ward was probably the Chicago Inter-State Industrial Exposition, otherwise known as the Chicago Exposition. This exhibition was first opened in September 1873 and was repeated annually until 1877. Unfortunately we have no published catalogs of the contents of the exposition from year to year, so we can't confirm for certain that the Ice Age sloth was exhibited there. But this is the most likely venue for the specimen." Personal communication received on September 7, 2000.

9. Roswell Ward, *op. cit.*, p. 186.

10. Letter—E. J. Koestner, the Dayton Museum of Natural History, dated October 12, 1977, addressed to W. C. Gamble, President of Ward's Natural Science Establishment. Also see editorial in *Ward's Bulletin*, April 1966, pp. 1–2, titled "The Mighty Megatherium."

11. Letter—William A. Fischer, Pro-

fessor of Geology, dated October 5, 1977, addressed to W. C. Gamble, President of Ward's Natural Science Establishment.

12. Henry A. Ward, *The Mammoth: Being Articles Concerning This Celebrated Specimen Which Have Appeared in Various American and European Newspapers to Which Is Added "The Last of the Mammoths"* (Stevenson, Foster & Co., 1879).

13. Roswell Ward, *op. cit.*, 1948, chapter XVII, titled "The Mammoth Comes to Rochester"; Martin may have constructed the Mammoth as early as (circa) 1860. Rare photographs show the colossus on display in Stuttgart and in St. Petersburg.

14. Roswell Ward, *op. cit.*, p. 197.

15. *Ibid.*, p. 198.

16. Winchell quoted in Henry A. Ward, *op. cit.*, 1879, chapter titled "Mammoth and Mastodon" by Alexander Winchell.

17. William Gunning quoted in Henry A. Ward, *op. cit.*, 1879, p. 14.

18. Alexander Winchell quoted in Henry A. Ward, *op. cit.*, 1879, p. 7.

19. Quotation from Jeffrey Hantman, "Brooks Hall at the University of Virginia: Unraveling the Mystery," in *The Magazine of Albemarle County History*, 1989, vol. 47.

20. Quotation from Jeffrey Hantman, *op. cit.*, 1989.

21. *Ibid.*, 1989.

22. Personal communication from Jeffrey Hantman, July 12, 2000.

23. Roswell Ward, *op. cit.*, see figure opposite page 176.

24. Personal communication from Jean DeMouthe, February 23, 2000.

25. This was the first of two Chicago's World fair events that featured prehistoric animal restorations. The second took place in 1933/34.

26. For more information on prehistoric exhibits at the 1893 Chicago World's Fair, see Donald F. Glut, *Carbon Dates*, Jefferson, NC: McFarland Publishers, 1999; and Donald F. Glut, *The Dinosaur Scrapbook*, Secausus, NJ: The Citadel Press, 1980.

27. Micheal Reed and Jean Sellar, "'Ziggy' Finds New Home at Field Museum," in *Field Museum of Natural History Bulletin*, November 1982, vol. 53, no. 10, p. 8.

28. In personal communication dated July 19, 2000, Don Glut stated that the Field Museum's *Megatherium* isn't the Ward cast. This was confirmed by Bill Simpson of the Field Museum on October 18, 2000. However, the Field Museum's Irish Elk skeleton *is* the Ward cast originally displayed during the Chicago World's Fair of 1893. A photograph of the Ward *Megatherium* displayed in bipedal pose at the Field-Columbian Museum was recently published on page 103 of Donald F. Glut's *Carbon Dates* (Jefferson, NC: McFarland & Company Inc., Publishers, 1999). Evidently, as Don Glut discovered in a 1935 issue of the *Field Museum News* (vol. 6, no. 3), the Ward *Megatherium* was offered for sale or trade in 1935, about the time that the genuine skeletal mount was erected. In 1999 Simpson also stated that the Field Museum does not have Benjamin Waterhouse Hawkins' cast *Hadrosaurus* in storage.

29. Ward's company memorandum (undated). According to this document, "Most of Ward's early records were lost in the fire of September 30, 1930." Some records have been retained by the University of Rochester Library, however.

Chapter 6

1. Steve McCarthy and Mick Gilbert, *The Crystal Palace Dinosaurs: The Story of the World's First Prehistoric Sculptures*, London: The Crystal Palace Foundation, 1994.

2. See entry for Benjamin Waterhouse Hawkins in *Appleton's Cyclopaedia of American Biography*, New York: J. G. Wilson & John Fiske (eds.),1888, vol. 3, p. 119.

3. Adrian J. Desmond, *The Hot-Blooded Dinosaurs: A Revolution in Palaeontology* (Warner Books, 1975), pp. 24–25; Adrian J. Desmond, *Archetypes and Ancestors: Palaeontology in Victorian England 1850–1875*, Chicago: The University of Chicago Press, 1982, pp. 70–72.

4. As Steve McCarthy states in "Owen's Visions: The Crystal Palace Dinosaurs," in *Dinosaur World* , February 1997, vol. 1, no. 1, p. 41, the dinosaur was "invented" by Owen not in 1841 but in 1842, when he published his views on British Fossil Reptiles. See also Hugh Torrens, "When Did the Dinosaur Get Its Name?" in *New Scientist*, April 4, 1992. For more on Mantell's involvement in designing dinosaurs for the Crystal Palace exhibition, see Dennis Dean, *Gideon Mantell and the Discovery of Dinosaurs*, Cambridge: Cambridge University Press, 1999, p. 261.

5. Hawkins, quoted in Steve McCarthy's article, "Owen's Visions: The Crystal Palace Dinosaurs," in *op. cit.*, 1997, pp. 40–41.

6. Quotation from Desmond, *op. cit.*, 1975, p. 29.

7. See Desmond, *op. cit.*, 1975, pp. 312–313; and William J. T. Mitchell, *The Last Dinosaur Book: The Life and Times of a Cultural Icon*, Chicago: The University of Chicago Press, 1998, pp. 126, 292.

8. See chapter 5, "Domesticating the Monsters," in Martin J. S. Rudwick, *Scenes from Deep Time: Early Representations of the Prehistoric World*, Chicago: The University of Chicago Press, 1992, for more on Hawkins' educational illustrations and wall charts.

9. Letters from Andrew H. Green, and Hawkins' reply, are quoted in Edwin H. Colbert, *Men and Dinosaurs: The Search in Field and Laboratory*, New York: E. P. Dutton, 1968, pp. 48–50.

10. Leonard Warren, *Joseph Leidy: The Last Man Who Knew Everything*, New Haven: Yale University Press, 1998, p. 87; Mark Jaffe, *The Gilded Dinosaur: The Fossil War Between E. D. Cope and O. C. Marsh and the Rise of American Science*, New York: Crown Publishers, 2000, p. 75.

11. Item no. 35 in the recently discovered Waterhouse Hawkins Album held in the Academy of Natural Sciences, Ewell Sale Stewart Library, includes an old photograph showing the reconstruction of a bipedal *Laelaps* made of bent iron armature and plaster leg bones. Donald Baird suggests this photograph may have been taken in the Central Park Arsenal in May 1869, or possibly later.

12. Allen A. Debus and Steve McCarthy, "A Scene from American Deep Time," in *The Mosasaur* (The Journal of the Delaware Valley Paleontological Society), May 1999, vol. VI, pp. 105–115; Richard C. Ryder, "Dusting Off America's First Dinosaur," in *American Heritage*, March 1988, vol. 39, no. 2, pp. 68–73.

13. Donald F. Glut, *The Dinosaur Scrapbook*, Secaucus, New Jersey: The Citadel Press, 1980, p. 28.

14. According to Bill Simpson of Chicago's Field Museum of Natural Museum, they are not in storage. (Personal communication dated October 18, 2000.) Might the hadrosaur and possibly pieces of Ward's Great Siberian Mammoth be retained in the bowels of Chicago's Museum of Science and Industry? In personal correspondence from Donald Baird,

dated October 6, 1981, Baird claimed, "My friends in Chicago say they have no idea what became of the Hadrosaurus, in fact they never knew it existed."

15. Donald F. Glut, *op. cit.*, p. 28.

16. Charles Schuchert and Clara LeVene, *O. C. Marsh: Pioneer Paleontologist*, New Haven: Yale University Press, 1940, pp. 295–296.

17. Personal communication from Steve McCarthy dated March 3, 2001.

18. O. C. Marsh is quoted in Desmond, *op. cit.*, 1975, p. 57.

19. Steve McCarthy, "Restoring the Vision: Saving the Crystal Palace Dinosaurs," in *Dinosaur World*, June 1997, vol. 1, no. 2, pp. 19–22.

20. Allen A. Debus' miniature 1995 sculpture recreating the prehistoric Cretaceous and Cenozoic scenes that Hawkins would have completed in sculpture for the Palaeozoic Museum, based on surviving design sketches, has been on display at the Burpee Museum of Natural History in Rockford, Illinois, since the spring of 1999.

Chapter 7

1. See Gregory S. Paul, "The Art of Charles R. Knight," in *The Scientific American Book of Dinosaurs*, Gregory S. Paul (ed.), New York: St. Martin's Press, 2000, p. 114.

2. *The Century Magazine*, November 1897, vol. LV, no. 1, pp. 10–23.

3. Sylvia Massey Czerkas and Donald F. Glut, *Dinosaurs, Mammoths and Cavemen: The Art of Charles R. Knight*, New York: E. P. Dutton, Inc., 1982, pp. 13–19.

4. See Nicholas Hotton III's editorial in *Earth Sciences History* (Journal of the History of the Earth Sciences), 1988, vol. 7, no. 1, pp. 62–63.

5. See Gregory S. Paul, *op. cit.*, 2000, p. 116–117. True Knight aficionados might find this statement "a-Pauling"!

6. Knight's ideas are quoted in Czerkas and Glut, *op. cit.*, 1982, pp. 33–34.

7. Gregory S. Paul, *op. cit.* 2000, p. 116. Paul suggests that Knight "...typically painted misty scenes, possibly because of his poor long-range vision...." However, in many existing photographic portraits of Knight, the artist is wearing vision-correcting glasses.

8. See Paul, *op. cit.*, 2000, p. 117, for further insights on the sense of "realism" in Knight's paintings.

9. Charles R. Knight, *Prehistoric Man: The Great Adventurer*, New York: Appleton-Century-Crofts, Inc., 1949, p. 133.

10. "Dinosaurs, Mammoths, and Cavemen—The Art of Charles R. Knight," an exhibit of paintings, drawings and sculpture of prehistoric life. Natural History Museum of Los Angeles County, 5 March through 30 May 1982. (Souvenir Brochure.) Interestingly, a less heralded exhibit of Knight's work took place about two decades earlier, between January 15 and April 9, 1961, at the American Museum of Natural History. See *Park Avenue Social Review*, "The Career of Charles R. Knight," 1961. (Xerox of tear sheet in personal collection.) While Knight didn't produce how own life-sized sculptural restorations of prehistoric animals, many other sculptors have based their work on Knight's. Thus is it particularly interesting that Knight did in fact plan to build a display of life-sized dinosaurs with fellow paleosculptor, Louis Paul Jonas. (See chapter 22.) Furthermore, a preliminary rough sketch for a remarkable dinosaur fountain, showing sauropods (one standing tripodally), tyrannosaurs, and parasaurolophuses intended for Manhattan's Central Park was recently published on page 50 of *Prehistoric Times*, no. 48 June/July 2001. (See "Interview with Dean Hannotte," by Tony Campagna.)

Chapter 8

1. Adrienne Mayor, *The First Fossil Hunters: Paleontology in Greek and Roman Times*, Princeton: Princeton University Press, 2000, p. 80.

2. Jack Horner quoted in Adrienne Mayor, *op. cit.*, p. 46.

3. William J. T. Mitchell, *The Last Dinosaur Book: The Life and Times of a Cultural Icon*, Chicago: The University of Chicago Press, 1998, pp. 55–56.

4. Leonard Warren, *Joseph Leidy: The Last Man Who Knew Everything*, New Haven: Yale University Press, 1998, p. 58; In the vertebrate fossil arena, America embarrassed itself twice during the mid-nineteenth century when Albert Koch gathered first the bones of an American Mastodon, and then later the bones of the fossil whale *Basilosaurus*, reconstructing a single individual of each but including the vertebrae and ribs of additional specimens to make his reconstructions appear impossibly huge and therefore more impressive. These false reconstructions were exhibited in London, where British anatomist Sir Richard Owen debunked them. For more on Koch's *Basilosaurus* (i.e. "Hydrachos") see Carl Zimmer's *At the Water's Edge: Macroevolution and the Transformation of Life*, New York: Simon & Schuster, Inc., 1998, pp. 141–144. For more on Koch's Mastodon (i.e. "Missourium") see Paul Semonin's *American Monster: How the Nation's First Prehistoric Creature Became a Symbol of National Identity*, New York: New York University Press, 2000, pp. 382–386.

5. Leonard Warren, *op. cit.*, p. 183.

6. Kevin Padian, "The Case of the Bat-Winged Pterosaur: Typological Taxonomy and the Influence of Pictorial Representation on Scientific Perception," in *Dinosaurs Past and Present*, vol. II, Seattle: Natural History Museum of Los Angeles County in Association with University of Washington Press, 1987 (Sylvia J. Czerkas and Everett C. Olson, eds.), pp. 77–78.

7. Leidy quoted in Edwin H. Colbert, *Men and Dinosaurs: The Search in Field and Laboratory*, New York: E. P. Dutton & Company, Inc., 1968, p. 47.

8. By 1886 artists began utilizing the earliest images of *bipedally* postured British dinosaurs, *Megalosaurus* and *Iguanodon*, for popular books, as in Camille Flammarion's *Le Monde Avant la Creation de l'Homme*, Paris: C. Marpon and E. Flammarion, 1886.

9. See Chapter 12 in this book for a detailed discussion of this point. Some of the modern paleoartists find themselves critiquing the work of early "masters," such as Charles R. Knight. For example, Gregory S. Paul has stated, "Knight used his vast knowledge of anatomy to make extinct forms appear so real that his viewers could easily believe he had seen them. This ability no doubt explains why his pictures continue to look plausible today. But this seeming realism was in some ways superficial. Although Knight sketched detailed musculoskeletal studies of living animals, he did not produce similar studies of dinosaurs—in part because skeletons reveal limited information about an animal's musculature." Paul, "The Art of Charles R. Knight," in *The Scientific American Book of Dinosaurs*, New York: St. Martin's Press, 2000, p. 117.

10. Thomas S. Kuhn, *The Structure of Scientific Revolutions*, 2nd ed., Chicago: University of Chicago Press, 1970.

11. For more on Pulcher, see Adrienne Mayor, *op. cit*, pp. 144–146. For more on medieval views, hippos and unicorns,

see Paul Semonin's *American Monster*, New York: New York University Press, 2000, pp. 72–75. According to Semonin, by the late seventeenth century, "Prehistoric nature was still an unknown, enigmatic realm just beginning to emerge from the Christian unconscious, and its lack of definition made it an imaginary place, often shaped as much by fantasy as fact." (p. 73) Also, amateur paleontologist Steve Brusatte has researched the origins of a peculiar painting by artist Hecht Maik (1963). This painting may offer clues to Native American myths concerning their understanding of fossil bones. Maik's painting may indirectly "restore" the animals perceived by Native Americans in the myth of Unktehi. Steve Brusatte, "Native American Fossil Myths," in *Dinosaur World*, Summer 2001, no. 10 (in press).

12. Leonard Warren, *op. cit.*, p. 188.

13. Rolf E. Johnson and John H. Ostrom, "The Forelimb of *Torosaurus* and an Analysis of the Posture and Gait of Ceratopsian Dinosaurs," in *Functional Morphology in Vertebrate Paleontology*, J. Thomason, ed. Cambridge: Cambridge University Press, 1995, pp. 205–218.

14. Carl Zimmer, "Dinosaurs in Motion," in *Discover*, November 1997, pp. 96–109.

15. David Krentz quoted in "The PT Interview," *Prehistoric Times*, Dec./Jan. 2001, no. 45, p. 13.

Chapter 9

1. William Schoell, "The Making of Jules Verne's Journey to the Center of the Earth," in *Filmfax: The Magazine of Unusual Film and Television*, March/April 1996, no. 55, pp. 51–57, 74.

2. Jules Verne, *Journey to the Center of the Earth: A New Translation by William Butcher*, Osford: Oxford University Press, 1988.

3. This more familiar quotation from the standard 1871 English translation.

4. John Cleves Symmes, A *Symzonia: Voyage of Discovery*, New York: J. Seymour, 1820 (reprinted by New York: Arno Press, 1975). For more on Symmes, see Peter Costello, *Jules Verne: Inventor of Science Fiction* (Charles Scribner's Sons, 1978); and William H. Goetzmann, *New Lands, New Men: America and the Second Great Age of Discovery* (Viking Penguin, Inc., 1986).

5. See Butcher's "Introduction" to his 1988 translation of Verne's novel, *op. cit.*

6. Martin J. S. Rudwick, *Scenes from Deep Time: Early Representations of the Prehistoric World*, Chicago: The University of Chicago Press, 1992, pp. 166–170.

7. Allen G. Debus, *Man and Nature in the Renaissance*, Cambridge University Press, 1978, p. 26. For further reflections on this point, see Allen A. Debus, "At the Center of Verne's Paleontology," in *Dinosaur World*, Summer 2001, no. 10 (in press).

8. Although there is no specific mention that he ever did visit Hawkins' models at Sydenham in Peter Costello's biography of Verne, *op. cit.*, 1978.

9. Edmondo Marcucci cited in Arthur B. Evans, "The Illustrators of Jules Verne's *Voyages Extraordinaires*," *Science-Fiction Studies*, XXV: 2, July 1998, pp. 241–270.

10. This quotation excerpted from Butcher's 1988 translation of Verne's *Journey*, p.153.

11. This quotation is excerpted from the standard 1871 English translation of Verne's *Journey*, p. 194.

12. These restorations are discussed in Martin J. S. Rudwick, *op. cit.*, 1992, chapters 1 to 3.

13. Erik Trinkaus and Pat Shipman, *The Neandertals: Changing the Image of Mankind*, New York: Alfred A. Knopf, 1993, chapter 3 titled "L'Affaire Moulin Quignon 1865–1885."

14. However, Don Glut claims, "In 1962 French writer Jean Boullet suggested that *King Kong*, the 'prehistoric' ape, might have been strongly influenced by Johnathan Swift's novel of Gulliver's Travels. Illustrations from various editions of Gulliver's Travels depicting an enormous monkey or ape tenderly caressing Gulliver, reaching for him through a window, carrying him across the rooftops and finally holding him while surmounting a tall towerlike building, though published long before *King Kong*, bear striking resemblances to scenes in [*King Kong*]." *Classic Movie Monsters*, Metuchen, NJ: The Scarecrow Press, Inc., 1978, p. 295.

Chapter 10

1. Nicolaas A. Rupke, "The Picturing of Geological Time," in *History of Science*, March 1998, vol. 36, Part 1, no. 111, pp. 61–90. For more on the subject of "visual language," see Martin J. S. Rudwick, "The Emergence of a Visual Language for Geological Sciences," in *History of Science*, 1976, vol. 14, pp. 149–195.

2. See Figure 2 in Rupke, *op. cit.*, 1998, p. 64.

3. Martin J. S. Rudwick, *Scenes from Deep Time: Early Representations of the Prehistoric World*, Cambridge: The University of Chicago Press, 1992, pp. 54–55; Rudwick claims that "...Buckland [urged] De la Beche to add some more scenes without delay, before the Germans stole all the best ideas."

4. William Buckland, *The Bridgewater Treatises on the Power Wisdom and Goodness of God as Manifested in the Creation*, Treatise VI, "Geology and Mineralogy Considered with Reference to Natural Theology," London: William Pickering, 1837, 2nd ed., Plate 1.

5. See Rudwick, *op. cit.*, 1992, pp. 86–87.

6. *Ibid.*, pp. 92–93.

7. *Ibid.*, chapter 4, "A First Sequence of Scenes."

8. C. C. Albritton, Jr., *The Abyss of Time: Changing Conceptions of the Earth's Antiquity*, Los Angeles: Jeremy P. Tarcher, Inc., 1986, p. 175.

9. Gary R. Williams, "Prehistoric Journeys of Karel Zeman," in *Dinosaur World*, October 1997, vol. 1, no. 3, pp. 49–52; Neil Pettigrew, *The Stop-Motion Filmography*, Jefferson, NC: McFarland & Company, Inc., Publishers, 1999, pp. 117–124.

10. Another early example is a sauropod dinosaur associated with the age of the earth using imagetext in an illustration by V. T. Hamlin, *Alley Oop*'s creator. This figure was drawn for a Sunday edition of the now defunct, *Chicago Times*, circa 1935, and was one installment in a series titled *Dinny's Family Album*. (From Prehistoric Animals' scrapbook #1, compiled by Allen G. Debus.)

Chapter 11

1. Allen A. Debus, "Stegosaurus: A Paleontological Puzzle" (Parts 1 to 4), in *The Earth Science News* (news bulletin of the Earth Science Club of Northern Illinois), April to June and July/August, 1986, vol. 37 (and unpublished notes in Debus' "Stegosaurus Scrapbook").

2. Edwin H. Colbert, *Dinosaurs: Their Discovery and Their World*, New York: E. P. Dutton, 1961, chapter 9, "Plated Dinosaurs"; Othenio Abel, *Geschichte und Methode der Rekonstruction Vorzeitlicher*

Wirbeltiere (Jena: Gustav Fischer, 1925), pp. 191–213; Stephen A. Czerkas, "A Reevaluation of the Plate Arrangement on Stegosaurus Stenops," in *Dinosaurs Past and Present* (Sylvia J. Czerkas and Everett C. Olson, eds.), Seattle: Natural History Museum of Los Angeles County in association with University of Washington Press, 1987, pp. 83–99; Charles Whitney Gimore, "Osteology of the Armored Dinosauria in the United States National Museum, with Special References to the Genus *Stegosaurus*," 1914, in *Memoirs of the United States National Museum*, no. 89; Allen A. Debus, "Historical Dinosaurs: Episodes in Discovery and Restoration," in *Earth Sciences History* (Journal of the History of the Earth Sciences Society), 1993, vol. 12, no. 1, pp. 60–69; Allen A. Debus, "Stegosaurus: A Plated Puzzle," in *The Dinosaur Report* (news bulletin of the Dinosaur Society), Winter 1994, pp. 8–9.

3. Stephen A. Czerkas, *op. cit.*, 1987, p. 86; Interestingly, it may be the case that Arthur Lakes may have completed the earliest *Stegosaurus* restoration possibly in the late 1870s or early 1880s. This often neglected restoration published as Plate XXVII (top) in *O.C. Marsh: Pioneer Paleontologist*, New Haven: Yale University Press, 1940, by Charles Schuchert and Clara LeVene, has a paired double row of plates along the entire vertebral column, as Marsh originally may have speculated. According to John S. McIntosh and Ronald Rainger (personal communications—Summer 2001), Lakes painted his dinosaur restorations circa 1880.

4. James Erwin Culver, "Some Extinct Giants," in *The California Illustrated Magazine*, April 1892, pp. 501–507; H. N. Hutchinson, *Extinct Monsters: A Popular Account of Some of the Larger Forms of Ancient Animal Life*, London: Chapman and Hall, 1893, 2nd ed.

5. See "Preface to Second Edition," in H. N. Hutchinson, *op. cit.*, 1983, and Plate X.

6. Edwin H. Colbert, *op. cit.*, 1961, p. 148.

7. See Knight's restoration recently reproduced in Sylvia M. Czerkas and Donald F. Glut, *Dinosaurs, Mammoths and Cavemen: The Art of Charles R. Knight*, New York: E. P. Dutton, 1982, p. 6.

8. William H. Ballou, "Strange Creatures of the Past," in *The Century Magazine*, November 1897, vol. LV, pp. 21

9. William H. Ballou, *op. cit.*, 1987, p. 22.

10. E. Ray Lankester, *Extinct Animals*, New York: H. Holt, 1905, p. 208.

11. Edwin H. Colbert, *op. cit.*, 1961, p. 149.

12. Sir Arthur Conan Doyle, *The Lost World*, New York: A. L. Burt Company Publishers, 1912, p. 44; Edgar Rice Burroughs' fictional description appears on pp. 23–25 of *Out of Time's Abyss* (recently reprinted by New York: Ace Books).

13. For more on Delgado's animated *Stegosaurus* scenes see Jeff Rovin, *From the Land Beyond Beyond: The Making of the Movie Monsters You've Known and Loved*, New York: Berkley Publishing Corporation, 1977, chapters 2 and 3; Donald F. Glut, *The Dinosaur Scrapbook*, Secaucus, New Jersey: The Citadel Press, 1980, chapter 5; Neil Pettigrew, *The Stop-Motion Filmography*, Jefferson, NC: McFarland and Company, Inc., Publishers, 1999, pp. 391–409, 425–438; Donald F. Glut, *Classic Movie Monsters*, Metuchen, NJ: The Scarecrow Press, 1978, pp. 303–304. In this last reference, Glut claims, "The King Kong Stegosaurus, showing the dorsal plates paired and with eight deadly spikes on the tail, was actually in error, apparently based upon restorations of the animal by E. Ray Lankester in 1905 and Richard Swann Lull in 1910."

14. See Plate 68 in Edwin H. Colbert, *op. cit.*, 1961.

15. See Stephen A. Czerkas, *op. cit.*, 1987, p. 85.

16. See Figure 88, page 70, in Sylvia M. Czerkas and Donald F. Glut, *op. cit*, 1982.

17. See Figure 9 in Stephen A. Czerkas, *op. cit.*, 1987, p. 88.

18. William J. T. Mitchell, *The Last Dinosaur Book: The Life and Times of a Cultural Icon*, Chicago: The University of Chicago Press, 1998, chapter 3 "A *Stegosaurus* made of Money"; Donald F. Glut, *The Dinosaur Scrapbook*, Secaucus, New Jersey: The Citadel Press, 1980, p. 275.

19. The *Chicago Tribune* poem is printed in Donald F. Glut *op. cit.*, 1980, p. 276. Also see Henry R. Knipe, *Nebula to Man*, New York: J. M. Dent and Company, 1905, pp. 70–71; the stegosaur portion of Knipe's poem reads:

"Whilst through Europa's land these monsters range
Upon Columbia's scenes are forms as strange
Here lumbers Stegosaurus on his fours,
With high-arched back, a kind of dinosaurs,
Big is his bulk, with solid bones secure,
And great his length—a score of feet and more.
Upon his back are bony plates implanted,
That look like rows of tombstones, small and slanted
These oft may save him, when attacked by foes
And clear his way, when through thick woods he goes
And e'en may serve, when he on love is bent
To act effective, as a blandishment
Small head the brute possesses for his size,
And taken up well high with jaws and eyes.
But though his brain is lodged in humble space,
Down by his haunches, in a larger case,
Some brainy tissue lies, as if in store,
Against a time his skull might harbor more.
Much must he eat, for herbage is his food;
And save in fight his jaws are free from blood."

20. Henry R. Knipe, *Evolution in the Past*, New York: J. B. Lippincott Company, 1912, plate facing page 90; also see Allen A. Debus, "Plated Dragons: Unearthing New Facts About *Stegosaurus*," in *Crystal Cluster*, March 1993 (Part 2 of 4).

21. See Figures 140 and 149 in Othenio Abel, *op. cit.*, 1925.

22. The paired double row plate arrangement idea was highly influential throughout this period. Pallenburg's *Stegosaurus* sculpture seems to have been inspired by Arthur Lakes' watercolor painting (see Note 3).

23. Francis B. Messmore, "The History of the 'World a Million Years Ago'" in *Dinosaur World*, October 1997, vol. 1, no. 3, pp. 27–32; also see two contemporary Sinclair Fair souvenirs bearing headlines "Sinclair Exhibits Weird Dinosaurs" and "Millions See Weird Sinclair Dinosaurs." The contents of these newsletters show a staggered arrangement in Sinclair's *Stegosaurus*. Donald F. Glut confirmed that the "World a Million Years Ago" stegosaur was "*...definitely staggered, two rows.*" (Personal communication, October 21, 2000.)

24. This is shown in Stephen A. Czerkas, *op. cit.*, 1987, Figure 20, p. 95.

25. Othenio Abel, *op. cit.*, 1925, Figure 150, p. 212.

26. Stephen A. Czerkas, *op. cit.*, 1987, p. 98.

27. Stephen A. Czerkas quoted in "Stephen and Sylvia Czerkas Interview with Cliff Green," in *Prehistoric Times*, Aug./Sept., no. 43, 2000, p. 12.

28. One of these is displayed in Salt Lake City's Utah Museum of Natural History. The other is displayed in a museum that opened in the summer of 2000, the Museum of Prehistoric Life, situated about 30 miles south of Salt Lake City.

29. See entry for *Stegosaurus* in Donald F. Glut's *Dinosaurs—The Encyclopedia*, Jefferson, NC: McFarland and Company, Inc., Publishers, 1997.

30. Kenneth Carpenter quoted in *Dinosaur World*, Spring 1998, no. 4, p. 14.

31. Kenneth Carpenter interviewed

in *Dinosaur World*, Spring 1998, no. 4, p. 14. The new genus, *Hesperosaurus*, was described in chapter 3, "New Primitive Stegosaur from the Morrison Formation, Wyoming," in *The Armored Dinosaurs*, Bloomington: Indiana University Press, 2001, Kenneth Carpenter and Clifford Miles (eds.).

32. See page 157 of David Norman's *The Illustrated Encyclopedia of Dinosaurs*, New York: Crescent Books, 1985; also see Robert Bakker's *The Dinosaur Heresies*, New York: William Morrow and Company, Inc., 1986, pp. 231–233. Bakker claimed that *Stegosaurus* was "...a grand performer under attack—a five-ton ballet dancer with an armor-plated tutu of flipping bony triangles and a swinging war club." (p. 233).

33. John C. McLoughlin, *Archosauria: A New Look at the Old Dinosaur*, London: Allen Lane, 1979, p. 94.

34. Robert Bakker, *op. cit.*, p. 188.

35. James Farlow, Carl V. Thompson and Daniel Rosner, "Plates of the dinosaur *Stegosaurus*: forced convection heat loss fins?" in *Science*, June 11, 1976, vol. 192, no. 4244, pp. 1123–1125; Daniel Rosner's restoration of *Stegosaurus* appeared on the cover of this *Science* issue.

36. Kenneth Carpenter, "A New Look at *Stegosaurus*," in *Dinosaur Discoveries*, Sept./Oct. 1996, no. 1, pp. 6–7; also see pp. 73–74 in Kenneth Carpenter's *Eggs, Nests, and Baby Dinosaurs: A look at Dinosaur Reproduction*, Bloomington: Indiana University Press, 1999

37. See Luis Rey's illustration on page 77 of Kenneth Carpenter, *op. cit.*, 1999.

38. Robert Bakker quoted in "Tyrannosaurus Sex: A Love Tail," by Sandy Fritz in *Omni*, February 1988, p. 78.

39. British paleontologist Beverly Halstead is one of the earliest scientists to give due consideration to the provocative question of "how" dinosaurs did it.

40. Othenio Abel, *op. cit.*, 1925, Figure 149, p. 212.

41. See entry for *Stegosaurus* in Donald F. Glut's, *Dinosaurs—The Encyclopedia*, Jefferson, NC: McFarland and Company, Inc., Publishers, 1997.

42. Tracy Lee Ford, *How to Draw Dinosaurs: Book 1* (Dino Hunter, 1999), p. 5.

43. George Olshevsky and Tracy L. Ford, "The Origin and Evolution of the Stegosaurs," in *Gakken Mook*, no. 4, pp. 92–119; English translation supplied by Mr. Olshevsky on November 13, 1993.

44. Tracy Lee Ford, *op. cit.*, 1999.

45. A high percentage of stegosaur restorations include either an attacking allosaur or ceratosaur. There are too many restorations devoted to this popular theme to mention here.

46. Robert Bakker, *op. cit.*, 1986, p. 227.

47. Emily Giffin quoted in Jeffrey Kluger's interview in *Discover*, February 1992, p. 70.

48. Stephen A. Czerkas, "The Beaked Jaws of *Stegosaurus* and Their Implications for Other Ornithischians," in *Vertebrate Paleoantology in Utah* (Utah Geological Survey, Miscellaneous Publication 99–1), pp. 143–149.

Chapter 12

1. While paleontology's defining ideas (i.e. evolution, extinction and paleobiology) remain steadfast, paleontology's "visual" nature has long been appreciated by scientists. Of all the myriad forms of prehistoric life iconography, the restorations of extinct fauna, recreated as living organisms, have proved most appealing. Regardless, there have been few attempts to discuss the historical period during which the practice of making scientific restorations became fully established. Allen A. Debus, "Fine Paleontology vs. Fine Art: The Emergence of American Paleoart" (Parts 1 through 3), in *Prehistoric Times*, June/July 2000, no. 42, pp. 17–23; Aug./Sept. 2000, no. 43, pp. 17–20; Oct./Nov. 2000, no. 44, pp. 17–20.

2. Stephen J. Gould, "Reconstructing (and Deconstructing) the Past," preface to *The Book of Life: An Illustrated History of the Evolution of Life on Earth*, New York: W. W. Norton and Company, 1993, Stephen J. Gould (General Editor), pp. 6–21; Allen A. Debus, "Sorting Fossil Vertebrate Iconography" in *Fossil News: Journal of Amateur Paleontology* (now the *Journal of Avocational Paleontology*), November 1997, vol. 3, nos. 10 and 11, pp. 2–5; Allen A. Debus, "The Art of Paleocatastrophe: Portraying Survivorship at the Quantum Crossroads," in *Fossil News: Journal of Avocational Paleontology*, vol. 4, January 1998, pp. 14–16; Kenneth Carpenter, *Eggs, Nests, and Baby Dinosaurs: A Look at Dinosaur Reproduction*, Bloomington: Indiana University Press, 1999.

3. Although many attribute first use of the term "Dinosaur Renaissance" to a *Scientific American* publication, April 1975, including an article by Robert Bakker on pp. 58–78, the term was actually coined in 1958. See article on dinosaur toys and collectibles in July 27, 1958 issue of *This Week Magazine*, pp. 8–9, by Leslie Lieber. Charles Schuchert and Clara LeVene coined the "Reformation" term in *O. C. Marsh: Pioneer Paleontologist*, New Haven: Yale University Press, 1940, p. 91.

4. Martin J. S. Rudwick, *Scenes from Deep Time: Early Pictorial Representations of the Prehistoric World*, Chicago: University of Chicago Press, 1992, pp. 98 and 131.

5. The tradition of America's Mastodon as a vital national symbol, incorporating Peale's contributions to our understanding of its early imagery and natural history, is skillfully addressed in Paul Semonin's *American Monster: How the Nation's First Prehistoric Creature Became a Symbol of National Identity*, New York: New York University Press, 2000. Also see William J. T. Mitchell, *The Last Dinosaur Book: The Life and Times of a Cultural Icon*, Chicago: University of Chicago Press, 1998, pp.111–123. Cuvier had already conceived the nature of a "prehistoric world" by 1797, however.

6. Steve McCarthy and Mick Gilbert, *The Crystal Palace Dinosaurs: The Story of the World's First Prehistoric Sculptures*, London: The Crystal Palace Foundation, 1994.

7. Martin J. S. Rudwick, *Georges Cuvier, Fossil Bones and Geological Catastrophes: New Translations and Interpretations of the Primary Texts*, Chicago: University of Chicago Press, 1997.

8. Dennis R. Dean, *Gideon Mantell and the Discovery of Dinosaurs*, Cambridge: Cambridge University Press, 1999, pp. 197–200, 225, 233, 246–247

9. Edward D. Cope, "The Fossil Reptiles of New Jersey," in *American Naturalist*, 1869, vol. III, pp. 84–91.

10. Donald F. Glut, *Carbon Dates: A Day by Day Almanac of Paleo Anniversaries and Dino Events*, Jefferson, NC: McFarland and Company, Inc., Publishers, 1999, pp. 170–171.

11. John A. Ryder, "Gigantic Extinct Armadillos and Their Peculiarities With a Restoration," in *Popular Science Monthly*, May to October 1878, vol. XIII, pp. 139–145. Ryder states that, "The restoration is ... based on the figures in H. Burmeister's work. It is believed to be approximately correct, since nothing was needed to make the originals assume the appearance of life except to clothe the skull and neck with flesh, and furnish the extremities with claws and hoofs, muscles and tendons." (p. 145)

12. Henry Fairfield Osborn, *Cope: Master Naturalist*, Princeton: Princeton University Press, 1931, pp. 191–193.

13. Edward D. Cope, *The Penn Monthly*, August 1873.

14. O. C. Marsh, "Additional Observations on the Dinocerata," in *American Journal of Science*, April 1873, vol. V, no. 28, pp. 292–296.

15. Guillaume Louis Figuier, *The World Before the Deluge*, London: Cassell and Company, Ltd, 1867; Note that Rudwick, *op. cit.*, chapter 6, 1992, suggests the proper translation for the title is "The Earth Before the Deluge".

16. Arthur B. Evans, "The Illustrators of Jules Verne's Voyages Extraordinaires," in *Science Fiction Studies*, July 1993, vol. XXV, no. 2, pp. 241–270; Allen A. Debus, "At the Center of Verne's Paleontology," in *Dinosaur World*, no. 10, Spring 2000 (in press).

17. D. Alexander Winchell, *Sketches of Creation: A Popular View of Some of the Grand Conclusions of the Sciences in Reference to the History of Matter and of Life Together with a Statement of the Intimation of Science Respecting the Primordial Condition and the Ultimate Destiny of the Earth and the Solar System*, New York: Harper and Brothers, 1874; for more on Winchell, see "Sketch of D. Alexander Winchell" in *Popular Science Monthly*, October 1892, vol. 41, pp. 837–843.

18. Martin J. S. Rudwick, *op. cit.*, 1992, pp. 216–218.

19. D. Alexander Winchell, *op. cit.*, 1874, p. 175.

20. Allen A. Debus and Steve McCarthy, "A Scene from American Deep Time: New York's 'Palaeozoic Museum—Revisited,' in *The Mosasaur* (Journal of the Delaware Valley Paleontological Society, May 1999, vol. 6, pp. 105–115.

21. Steve McCarthy and Mick Gilbert, *op. cit.*, 1994, p. 41.

22. O. C. Marsh quoted in Charles Schuchert and Clara LeVene, *op. cit.*, 1940, p. 295.

23. Charles Schuchert and Clara LeVene, *O. C. Marsh: Pioneer Paleontologist*, Yale University Press, 1940, pp. 295–296.

24. O. C. Marsh is quoted in Charles Schuchert and Clara LeVene, *op. cit.*, 1940, p. 385.

25. H. N. Hutchinson, *Extinct Monsters: A Popular Account of Some of the Larger Forms of Ancient Animal Life*, London: Chapman and Hall, 1893, 2nd ed., pp. V–VI; Schuchert and LeVene, *op. cit.*, 1940, p. 385.

26. Leonard Warren, *Joseph Leidy: The Last Man Who Knew Everything*, New Haven: Yale University Press, 1998, p. 188.

27. Clayton Hoaglund, "They Gave Life to Bones," in *Scientific Monthly*, Feb. 1943, p. 129; considering how little is known of the genus, possibly a more accurate "schematic representation" of *Dryptosaurus* was recently published in *Journal of Vertebrate Paleontology*, Sept. 3, 1997, vol. 17, no. 3, p. 562.

28. Richard C. Ryder, "Dusting Off America's First Dinosaur," in *American Heritage*, vol. 39, no.2, March 1988, pp. 68–73.

29. Charles Schuchert and Clara LeVene, *op. cit.*, 1940, p. 91.

30. *Ibid.*, p. 91.

31. Allen A. Debus, *op. cit.*, 1997, pp. 2–5. The term "paleoart" is used here anachronistically for all cases prior to Mark Hallett's first use of the term in 1986.

32. Martin J. S. Rudwick, *op. cit.*, 1992, pp. 32–36.

33. Richard Owen, *A History of Fossil Mammals and Birds*, London: J. Van Voorst, 1846.

34. James W. Hulke, "An Attempt at a Complete Osteology of *Hypsilophodon foxii*," in *Philosophical Transactions of the Royal Society of London*, 1882, vol. 173, pp. 1035–1062.

35. Louis Dollo, "Note sur les Restes de Dinosauriens reconstres dans le Cretace Superieur de la Belgique" (Bulletin du Musee Royale d'Histoire Naturelle de Belgique, 1883), vol. 2, pp. 205–221.

36. Josef Augusta, *Prehistoric Sea Monsters*, London: Paul Hamlyn, 1964, pp. 16–19; Stephen J. Gould, "Bent Out of Shape," in Gould's *Eight Little Piggies*, New York: W. W. Norton, 1993, pp. 79–94; H. N. Hutchinson, *op. cit.*, 1893, Plate II, and pp. 264–266.

37. H. N. Hutchinson, *op. cit.*, 1893, Plate II, and pp. 264–266.

38. H. G. Seeley, *Dragons of the Air*, London: Methuen, 1901.

39. Martin J. S. Rudwick, *op. cit.*, 1992, pp. 236–237.

40. Richard Owen, *op. cit.*, 1846, pp. 324, 439.

41. William Gunning, *Life History of our Planet*, New York: Worthington Co., 1st ed., 1879 (2nd ed. 1886); William Guyot, *Creation, or the Biblical Cosmogony in the Light of Modern Science*, New York: Charles Scribner's Sons, 1887; H. N. Hutchinson, *op. cit.*, 1893.

42. James Erwin Culver, "Some Extinct Giants," in *The California Illustrated Magazine*, April 1892, pp. 501–507.

43. William Ballou, "The Serpent-like Sea Saurians," in *Popular Science Monthly*, June 1898, vol. 53, pp. 209–225.

44. Henry F. Osborn, "Prehistoric Quadrupeds of the Rockies," in *The Century Magazine*, vol. LII, Sept. 1896, pp. 705–715.

45. *Ibid.*, p. 715.

46. Henry F. Osborn, "A Great Naturalist: Edward Drinker Cope," in *The Century Magazine*, 1897, vol. LV, pp. 10–-15; William H. Ballou, "Strange Creatures of the Past," in *The Century Magazine*, 1897, vol. LV, pp. 15–23.

47. Frederic Holder, "The California Academy of Sciences," in *The California Illustrated Magazine*, vol. III, Dec. 1892 to May 1893, pp. 229–244 (illustrations by J. Aronniz and Harold Herkiner).

48. Personal correspondence from Jean F. Demouthe of the California Academy of Sciences, dated Feb. 23, 2000.

49. For more on Henry A. Ward, see Roswell Ward's *Henry A. Ward: Museum Builder to America* (The Rochester Historical Society—Publication XXIV, 1948).

50. Ronald Rainger, *An Agenda for Antiquity: Henry Fairfield Osborn and Vertebrate Paleontology at the American Museum of Natural History, 1890–1935*, Tuscaloosa: The University of Alabama Press, 1991.

51. Henry F. Osborn, "Fossil Wonders of the West," in *The Century Magazine*, 1904, vol. 68, p. 689.

52. Charles Schuchert and Clara LeVene, *op. cit.*, 1940, pp. 383–385

53. *Ibid.*, p. 91.

54. For a complete listing of sources, see Charles Schuchert and Clara LeVene, *op. cit.*, 1940, "Bibliography of Othniel Charles Marsh," years 1880 to 1895.

55. Charles Schuchert and Clara LeVene, *op. cit.*, 1940, pp. 383–385; Marsh's "restorations of extinct animals," Plates 1 and 2, were privately printed in November 1895.

56. Richard Lydekker, *Nature*, July 27, 1893, pp. 302–304; Othniel C. Marsh, *Nature*, September 7, 1893, pp. 438–439.

57. Charles Schuchert and Clara LeVene, *op. cit.*, 1940, pp. 296–297.

58. *Ibid.*, p. 296.

59. *Ibid.*, p. 297.

60. William J. T. Mitchell, *op. cit.*, 1998, pp. 15–19.

61. Charles Schuchert and Clara LeVene, *op. cit.*, 1940, pp. 1, 383.

62. Charles Schuchert and Clara LeVene, *op. cit.*, 1940, p. 421; O. C. Marsh, "The Wings of Pterodactyles," in *American Journal of Science* (1882) (3), vol. 23, pp. 251–256.

63. Allen A. Debus, "Bully for Laelaps," in *The Dinosaur Society UK Quarterly Magazine*, 1996, vol. 2, nos. 1 and 2, pp. 7–8.

64. Personal correspondence from Dr. Carl Klaus Hagenbeck, dated July 10, 1990; Donald F. Glut, *op. cit.*, 1999, p. 158.

65. A silent film, *Evolution* (1923), produced by Max Fleischer, suggests that a *Tyrannosaurus rex* sculpture had been completed by that time, although on closer inspection this appears to be the *Iguanodon*. A few of the statues built for this display may have been destroyed during World War II. Interestingly, Carl Hagenbeck (1844–1913) financed hunting expeditions to the African Congo in pursuit of what he believed were *living* dinosaurs, circa 1910. See "Carl Hagenbeck and the Rhodesian Dinosaurs," in *Strange Magazine*, 1990, no. 5, pp. 50–52. The authors thank Gary R. Williams for supplying a copy of this "strange" article and other information related in this note. According to Dr. Carl Claus Hagenbeck (personal communication dated July 10, 1990), "...the dinosaurs originally designed in concrete by Joseph Pallenberg 1909 ... [were] reshaped and covered with fiberglass in the 1960s."

66. Frederic A. Lucas, "The Restoration of Extinct Animals," *Annual Report of the Board of Regents of the Smithsonian Institution: Showing the Operations, Expenditures, and Condition of the Institution for the Year ending June 30, 1900*, Washington D. C.: Smithsonian, 1901, pp. 490, 492.

67. Ronald Rainger, *op. cit.*, 1991, p. 166.

68. William K. Gregory, "Erwin S. Christman, 1885–1921: Draughtsman, Artist, Sculptor," in *Natural History*, 1921, vol. 21, p. 623.

69. William H. Goetzmann, *New Lands, New Men: America and the Second Great Age of Discovery*, New York: Viking Penguin Inc., 1986, p. 56. Thanks to Ronald Rainger for elucidating when Osborn, who as a young boy, had taken painting lessons from Church.

70. William K. Gregory, "Obituary for William D. Matthew," in *Natural History*, Nov./Dec. 1930, pp. 665.

71. William K. Gregory, "The Weight of the Brontosaurus," in *Science*, Sept. 30, 1905, vol. XXII, no. 566, p. 572.

72. Walter L. Beasley, "A Carnivorous Dinosaur: A Reconstructed Skeleton of a Huge Saurian," in *Scientific American*, Dec. 14, 1907, pp. 446–447.

73. Adrian J. Desmond, *The Hot-Blooded Dinosaurs: A Revolution in Palaeontology*, New York: Warner Books, 1975, p. 166.

74. *Ibid.*, p. 169.

75. William J. Holland, "A Review of Some Recent Criticisms of the Restorations of Sauropod Dinosaurs Existing in the Museums of the United States, with Special Reference to that of *Diplodocus carnegiei* in the Carnegie Museum," in *The American Naturalist*, May 1910, vol. XLIV, p. 264.

76. George Gaylord Simpson, "Charles W. Gilmore (1874–1945)," in *Society of Vertebrate Paleontology Newsbulletin*, 1945; Richard Swann Lull, *The National Cyclopaedia of American Biography*, New York: James T. White and Co., 1926, vol. A, p. 419; C. Bertrand Schultz, "Memorial to Erwin Hinckley Barbour," in *Proceedings Volume of the Geological Society of America—Annual Report for 1947*, 1948, pp. 109–117; William B. Scott, *The National Cyclopaedia of American Biography*, New York: James T. White and Co., 1950, vol. 36, pp. 426–427.

77. Richard Swann Lull, "Restoration of the Horned Dinosaur, Diceratops," in *American Journal of Science*, series 4, vol. 20, 1905, pp. 420–422.

78. Charles Whitney Gilmore, "Osteology of the Armored Dinosauria in the United States National Museum with Special Reference to the Genus, Stegosaurus," in *U.S. National Museum Bulletin* no. 89, 1914.

79. Charles Whitney Gilmore, *The National Cyclopaedia of American Biography* (James T. White and Co., 1948), vol. 34, pp. 362–363; here it is stated that, "Creating dinosaur models was his [Gilmore's] hobby...." A series of Gilmore's miniature plaster sculptures of North American dinosaurs were exhibited in many museums during the twentieth century.

80. William B. Scott, *A History of Land Mammals in the Western Hemisphere*, New York: The Macmillan Company, 1913, pp. 44; William B. Scott, *op. cit.*, 1937 ed., pp. 61–72.

81. C. Bertrand Schultz, *op. cit.*, 1948, p. 110; Dorothy C. Meade, "Morrill Hall ... Looking Back," *Museum Notes—University of Nebraska State Museum*, Feb. 1991, no. 78.

82. Samuel W. Williston, "Restorations of Some American Permocarboniferous Amphibians and Reptiles," in *Journal of Geology*, 1914, vol. 22, no. 1, pp. 57–70. Williston stated, p.70, "...I may add that, whatever may be the merits of these restorations as works of art, they have been drawn with most scrupulous accuracy so far as form and proportions are concerned, the musculature derived from the study of living reptiles. And ... they are all based upon practically complete skeletons...."

83. Samuel W. Williston, *Water Reptiles of the Past*, Chicago: University of Chicago Press, 1914; Two additional examples of Williston's early paleoart appeared in his articles "The Wing-Finger of Pterodactyls, with Restoration of *Nyctosaurus*" in *Journal of Geology*, Nov.–Dec. 1911, vol. XIX, no. 8, pp. 696–705 and "Restoration of *Limnoscelis*, a Cotylosaur Reptile from New Mexico" (in *American Journal of Science*, Nov. 1912, vol. XXXIV, 4th series, pp. 457–468).

84. "A Gentle Collector Caught up in a Dinosaur War," (author uncredited) in *Life Before Man* Time Life Books, 1972, pp. 81–83; Charles Schuchert and Clara Le Vene, *op. cit.*, 1940, p.409, Plate XXVII, and book dust jacket—which features Lakes' *Stegosaurus* watercolor. While it is generally believed that Lakes' dinosaur restorations were finished between 1877 and 1880, several of these do appear rather precocious for their time.

85. Allen A. Debus (with Diane E. Debus), "Say 'Yes' to Michigan's Paleo-Treasures," in *Dinosaur World*, Spring/Summer 1999, no. 6, p. 27.

86. Linda VanAller Hernick, *The Gilboa Fossils*, Rensselaerville: Givetian Press, 1996, pp. 57–64.

87. William J. T. Mitchell, *op. cit.*, 1998, p. 104; Stephen J. Gould, general editor, *The Book of Life: An Illustrated History of the Evolution of Life on Earth*, New York: W. W. Norton and Company, 1993, see essay by Gould, "Reconstructing and Deconstructing the Past," pp. 6–21.

88. Ann Shelby Blum, "A Better Style of Art: The Illustrations of the Paleontology of New York," in *Earth Sciences History: Journal of the Earth Sciences Society* (Special James Hall issue), vol. 6, no.1, 1987, pp. 72–85; Leonard Warren, *op. cit.*, 1998, pp. 233–235.

89. David Goldman, "Half Tones and Early Reconstruction Artists," in

Prehistoric Times, June/July 2000, no. 42, pp. 34–36.

90. Based on a paleontological misinterpretation, an artist restored some fossil pig teeth as a prehistoric human. This restoration appeared in the *Illustrated London News*, June 24, 1922, pp. 942–943. Also see Francis Hitching's *The Neck of the Giraffe* (New York: Mentor Books, 1982), pp. 181–183. During the late 1910s, American Museum sculptor J. H. McGregor sculpted a head bust of the Piltdown Man, "Eoanthropus," for Osborn's Hall of the Age of Man.

91. Leonard Warren, *op. cit.*, 1998.

92. Gerhard Heilmann, "Our Present Knowledge About the Origin of Birds," vols. 7–10, 1913–1916; Finn Salomonsen, "Gerhard Heilman 25 June 1859–26 March 1946," *Dansk Ornithologisk Forenings Tidsskrift*, 1946, vol. 40, pp. 146–149; Ilja Nieuwland, "Gerhard Heilmann and the Artist's Eye in Science, 1912–1927," posted on webpage—http://odur.let.rug.nl/nieuwlnd/Artisteye.html; Adrian J. Desmond, *op. cit.*, 1975.

93. Othenio Abel, *Geschichte und Methode der Rekonstruction Vorzeitlicher Wirbeltiere*, Jena, Verlag: Gustav Fischer, 1925; Allen A. Debus, "Profile—Othenio Abel: Artistic Founder of Paleobiology," in *The Dinosaur Society UK Quarterly Magazine*, Autumn 1995, vol. 1, no. 3, pp. 4–5.

94. Allen A. Debus, "Zdenek Burian's Global Visions of Prehistory," in *The Dinosaur Society UK Quarterly Magazine*, 1999, vol. 3, no. 3, pp. 10–12.

95. William J. T. Mitchell, *op. cit.*, 1998; Allen A. Debus and Diane Debus, "Domke's Dinos: Exploring Ancient Gardens," in *Dinosaur World*, Summer/Fall 1998, no. 5, pp. 7–9

96. A century after Frederic A. Lucas wrote his article for the Smithsonian's *Annual Report* (see note 66), a collection of *T. rex* original fine art, privately owned by John Lanzendorf, invaded Chicago's Field Museum of Natural History to augment its exhibition of the *Tyrannosaurus* specimen known as "Sue," marking for the first time in America that such paleoimagery has been exhibited in a major museum institution.

Chapter 13

1. James A. Effinger, "Entelodontidae," in *Evolution of Tertiary Mammals of North America: Terrestrial Carnivores, Ungulates and Ungulate Mammals*, Volume 1, Christine M. Janis, Kathleen M. Scott and Louis Jacobs (eds.), Cambridge, England: Cambridge University Press, 1998, p. 380.

2. Todd Runestead, "The Making of the Terminator Pig," in *Denver Museum of Natural History*, Museum Quarterly Notes, 1994.

3. A photograph of this model may be seen in a more recent publication, Cleophas E. O'Hara, *The White River Badlands*, Bulletin no. 13, November 1920, reprinted 1976, South Dakota School of Mines, Rapid City, South Dakota: Department of Geology, Plate no. 37-B.

4. Charles R. Knight, *Before the Dawn of History*, New York: McGraw-Hill Book Company, 1935, p. 82.

5. William E. Scheele, *The First Mammals*, New York: The World Publishing Company, 1955, pp. 72–73.

6. Cleophas E. O'Hara, *op. cit.*, Plate 37-A.

7. Todd Runestead, *op. cit.*

8. Edwin H. Colbert, *A Fossil-Hunter's Notebook*, New York: E. P. Dutton, 1980, p. 56.

9. Cleophas C. O'Hara, *op. cit.*, pp. 96–98.

10. R. J. G. Savage and M. R. Long, *Mammal Evolution: An Illustrated Guide*, New York: Facts on File Publications, 1986, p. 198; another restoration of *Chalicotherium* appears on p. 108 of Michael Benton, *The Rise of the Mammals*, London: Eagle Remainders, Ltd., 1998.

11. Margery C. Coombs, "Chalicotheroidea," in *Evolution of Tertiary Mammals of North America*, Volume 1, *op. cit.*, pp. 560–568.

Chapter 14

1. For more on Abel see entry in *Dictionary of Scientific Biography*, vol. I, New York: Charles Scribners' Sons Publishers, 1970, Charles Coulton Gillespie (ed.), pp. 17–18; Kurt Ehrenberg, "Othenio Abels Werden und Wirken," in *Mitt. Ges. Geol. Bergbaustad. Osterr.*, 1978, vol. 25, Wien, pp. 271–295.

2. Allen A. Debus, "Othenio Abel—Artistic Founder of Palaeobiology," in *Quarterly* (The Dinosaur Society of the UK Quarterly Magazine), Autumn 1995, vol. 1, no. 3, pp. 4–5.

3. Othenio Abel, *Lebensbilder aus der Tierwelt der Vorzeit*, Jena: Gustav Fischer, 1927 ed.

4. Othenio Abel, *Geschichte und Methode der Rekonstruktion Vorzeitlicher Wirbeltiere*, Jena: Gustav Fischer, 1925.

5. Other paleontologists similarly endowed with artistic abilities to express their opinions of how dinosaurs and other fossil vertebrates appeared through restoration are Edward D. Cope, Henry F. Osborn, Robert Bakker, Paul Sereno, Phil Currie, Kenneth Carpenter, George Olshevsky, Angel Cabrera, Henry De la Beche and Gerhard Heilmann. While not to be properly regarded as a "paleoartist" the artistic vertebrate paleontologist George Gaylord Simpson (1902–1984) has been lauded for his innovative use and refinement of "visual language" in his many publications on evolution and paleomammalogy. See Chapter 10, "The Mind's Eye," in Leo F. Laporte's *George Gaylord Simpson: Paleontologist and Evolutionist*, New York: Columbia University Press, 2000.

6. For instance, see Neave Parker's restoration in William E. Swinton's *Dinosaurs* (Trustees of the British Museum, 4th ed., 1969). Plate 9.

7. For an overview of this historical episode, see Adrian J. Desmond's *The Hot-Blooded Dinosaurs: A Revolution in Palaeontology*, New York: Warner Books, 1975, pp. 66–67, 114–115.

8. Abel was critical of early nineteenth century artistic efforts to restore pterodactyls.

9. Othenio Abel, *op. cit.*, 1925, pp. 1–12; Josef Augusta, *The Age of Monsters*, London: Paul Hamlyn, Ltd., 1966, pp. 29–31.

10. Stephen J. Gould, *Bully for Brontosaurus: Reflections in Natural History*, New York: W. W. Norton & Company, 1991, p. 442.

11. A listing of his publications may be found in Kurt Ehrenberg, *op. cit.*, 1978.

12. For more on Osborn, see Ronald Rainger, *An Agenda for Antiquity: Henry Fairfield Osborn and Vertebrate Paleontology at the American Museum of Natural History, 1890–1935*, Tuscaloosa: The University of Alabama Press, 1991.

Chapter 15

1. Personal correspondence from Dr. Vratislav Mazak, dated June 24, 1981; personal correspondence dated February 23, 1993, from Eva Hochmanove (Burian); Allen A. Debus, "They Painted the Dinosaurs," in *Crystal Cluster* (News bulletin of the Des Plaines Valley Paleontological

Society), September 1984 to April 1985; Richard Milner, *The Encyclopedia of Evolution* (Facts on File—Book News, Inc., Portland, Oregon), 1991 pp. 62–63; Allen A. Debus, "Palaeoartists of the Past—Zdenek Burian," in *Quarterly* (The Dinosaur Society UK Quarterly Magazine), vol. 3, no. 3, 1999, pp. 10–12. According to Petr Sadecky, Burian worked as a construction laborer to pay his art school expenses. "In his spare time, he painted pictures of his boyhood dreams of brave men and adventure—usually of the old American West—and tried using different techniques until he developed his own style. But times were extremely hard, and he returned to nature and lived with his friends as a backwoodsman—hunting for food, sleeping in the woods and washing in rivers and streams, even in the winter. He loved the outdoors and nature. In 1927, he finally became an illustrator with Villimek Publishers, painting pictures for articles in *World's of Adventure* and *Young Readers*. Since childhood he had a passion for accurate details, and purposefully collected photographs and books to properly research his subjects." Sadecky quoted in "The Art of Zdenek Burian," by Camille Cazedessus, Jr., in *ERB-dom and the Fantasy Collector*, March 1973, no. 68, pp. 2–4.

2. Josef Augusta, *Prehistoric Reptiles and Birds*, London: Paul Hamlyn, 1961; Josef Augusta, *The Age of Monsters: Prehistoric and Legendary*, London: Paul Hamlyn, 1966; Josef Augusta, *Prehistoric Man*, London: Paul Hamlyn, 1960; Josef Augusta, *Prehistoric Sea Monsters*, London: Paul Hamlyn, 1964; Josef Augusta, *Prehistoric Animals*, London: Spring Books, 1956.

3. Thomas Kovacs' letter was printed in *Dinosaur World*, Summer/Fall, 1998, no. 5, pp. 3–5.

4. Zdenek Spinar, *Life Before Man: The Fascinating Story of the Evolution of Living Things on Our Earth*, New York: Crescent Books, 1981. Spinar's book also appeared in Japanese and Finnish editions. Joseph Wolf, *The Dawn of Man*, New York: Henry N. Abrams, Inc., Publishers, 1978; Spinar studied paleontology under Augusta. He also collaborated with Burian from 1970 for a decade until Burian's death. Reflecting on the success of *Life Before Man* and Spinar's collaboration with Burian, Zybnec Rocek wrote, "....many people who are interested in paleontology as amateurs see our remote past through Spinar's and Burian's eyes." *Society of Vertebrate Paleontology Newsbulletin*, vol. 166, February 1996, p. 134.

5. Personal correspondence from Eva Hochmanova (Burian) dated February 23, 1993.

6. Gregory S. Paul's statement about Burian's *T. rex* may be found on p. 345 of his *Predatory Dinosaurs of the World: A Complete Illustrated Guide*, New York: Simon and Schuster, 1988. Paul also commented on Burian's "impressionistic" style in *The Scientific American Book of Dinosaurs*, New York: St. Martin's Press, 2000, Gregory S. Paul (ed.), p. 110, claiming "The primary challenge to [Rudolph F.] Zallinger came from across the ocean.... Burian also followed other artists of his time in not executing multiview skeletal restorations, so his dinosaurs' forms were often more impressionistic than accurate. Burian had a particular and rare affinity for creatureless prehistoric landscapes, which are among his best evocations of Mesozoic times." Burian admirer John Lavas claims, "Gregory Paul says the T. rex head of the 1938 depiction is too small, yet if you measure the drawing, it is about right. Don Lessem claims that Burian often painted his dinosaurs in swamps. But in all his work, I can only find five paintings (of four species—three sauropods, one hadrosaur) in water, and four of those have the main animal on land! Of course, large dinosaurs did utilize this part of their environment, just as elephants, rhinos and hippos do in wetter parts of Africa and Asia. It would be most unusual if they didn't....The reconstruction of [his] animals ... are very convincing and superior..." (personal communication dated November 1, 1998).

7. Personal correspondence from Dr. Vratislav Mazak, dated June 24, 1981.

8. Thomas Kovacs' letter was printed in *Dinosaur World*, Summer/Fall, 1998, no. 5, pp. 3–5; Sadecky claimed that, "Mr. Burian is the illustrator of 500 books, mostly adventurous literature; 15,000 illustrations for books by Verne, Maye, London, Stevenson, Ferry, Defoe, Ried, Kipling, Altsheller, White, Bret Harte, Assollant, Cooper, Zane Grey, Curwood, Frick, Storch and others." He also painted Tarzan for editions of Edgar Rice Burroughs' books.

9. Burian's artistry has been celebrated in a number of ways through the passing decades. Notably, his work has been celebrated through the issuance of a set of Polish postage stamps (1965) that were inspired by Burian's paintings. The artwork for the stamps was done by artist Andrzej (Andrew) Heidrich. (Personal correspondence from Bronislaw Przystas, dated September 8, 1981. Also see pp. 84–85 of *Dinosaur Stamps of the World*, by Stuart Baldwin and Beverly Halstead, Witham, Essex, England: Baldwin's Books, 1991.) Burian's restorations of prehistoric life also may have made an impact on Czech movie producer Karel Zeman. See Gary R. Williams, "Prehistoric Journeys of Karel Zeman," in *Dinosaur World*, October 1997, vol. 1, no. 3, pp. 49–52, and Thomas Kovacs' letter printed in *Dinosaur World*, Summer/Fall, 1998, no. 5, pp. 3–5.

10. Personal communication from Zdenek Spinar dated November 10, 1993. Zdenek V. Spinar, "Development of Vertebrate Paleontology in Czechoslovakia," in *Acta Musei Reginaehradecensis S.A.: Scientiae Naturales*, XXII, 1989, pp. 11–14; Jiri Brezina, "Geology in the Czech Republic," in *Geotimes*, August 1993, pp. 9–10.

Chapter 16

1. Vincent Scully, Rudolph F. Zallinger, Leo J. Hickey, John Ostrom, *The Age of Reptiles: The Great Dinosaur Mural at Yale*, New Haven: Peabody Museum of Natural History, 1990.

2. *The World We Live In*, New York: Time Incorporated, 1955, by the editorial staff of *Life* and Lincoln Barnett, pp. 98–103. The *Age of Reptiles* painting appears on pages 88 to 106. Zallinger's aquatic reptiles painting appears on pp. 104–105.

3. Zallinger quoted in "An Interview with Rudolph F. Zallinger," by Lee Grimes, in *Discovery* (Magazine of the Peabody Museum of Natural History), Fall 1975, vol. 11, no. 1, pp. 33–35.

4. Martin J. S. Rudwick discusses the "aquarium" perspective in *Scenes from Deep Time: Early Representations of the Prehistoric World*, Chicago: The University of Chicago Press, 1992.

5. Lincoln Barnett, "Strange Creatures of a Lost World," in *Life*, January 26, 1959, pp. 46–64. The cover of this issue features Rudolph Zallinger's spectacular painting showing how a "...sabertooth cat, an invader from North America attacks terrified guanaco in Bolivia valley some half million years ago." Zallinger's original *Age of Mammals* painting, contrasting considerably with the 1967 mural, appears on pp. 110–121, chapter 6, in *The World We Live In* (1955). The

Age of Mammals mural took 10 years to complete. See *Discovery* (Magazine of the Peabody Museum of Natural History), Spring 1968, vol. 3, no. 2, p. 3. Here it is stated that, "The project was initiated in 1957 and sponsored by Time-Life Inc., Peabody Museum Association and Heath M. Robinson. Physically, the completed mural consists of over a ton of angle iron, wire, mesh, plaster and casein-base water paint. All of these materials are scientifically arranged and bolted to the south wall in the Hall of Mammals."

6. Jane Werner Watson, *The Giant Golden Book of Dinosaurs and Other Prehistoric Animals*, New York: Golden Press, 1965.

7. Peter Zallinger, *Prehistoric Animals*, New York: Random House, 1978.

8. Dana Cain and Mike Fredericks, *Dinosaur Collectibles*, Norfolk, VA: Antique Trader Books, 1999, pp. 68–79, 22; for more on Chialu, see "Chialu—The Composition of Prehistorics," by Joe DeMarco, in *Prehistoric Times*, February/March 2001, no. 46, pp. 28–29, 35. For more information concerning Rudolph F. Zallinger's paleoart, see Allen A. Debus, "Palaoeartists of the Past: Rudolph Zallinger—the Ages of Zallinger," in, *Quarterly Review* (The UK Dinosaur Society Quarterly Magazine), 2000, vol. 3, no. 5, pp. 10–13; and "Prehistoric Scenes," by Allen and Diane Debus, in *Dinosaur Movies: The Complete History*, 1993, Starlog Telecommunications, Inc., pp. 81–86.

Chapter 17

1. Personal correspondence received from Maurice Burton, dated February 18, 1982, including unpublished essay titled "Toots and his dogs." Also see Allen A. Debus, "Palaeoartists of the Past—Neave Parker's Prehistoric World," in *Quarterly* (The Dinosaur Society UK Quarterly Magazine), 2000, vol. 3, no. 4, pp. 16–18.

2. Neave Parker died on May 16, 1961. An obituary notice was published in *The Illustrated London News*, May 27, 1961, p. 897.

3. Maurice Burton, "Toots and his dogs," February 18, 1982 (unpublished m.s.).

4. *Ibid.*

5. *Ibid.*

6. William E. Swinton, *The Dinosaurs*, New York: Wiley Interscience, 1970; Swinton's book was also published by George Allen and Unwin, 1970.

7. Edwin H. Colbert, *Dinosaurs: Their Discovery and Their World*, New York: E.P. Dutton and Company, 1961.

8. The series included restorations of *Triceratops, T. rex, Cetiosaurus, Polacanthus, Scelidosaurus, Megalosaurus, Stegosaurus, Macroplata, Pteranodon, Ichthyosaurus, Acanthopholis, Apatosaurus, Pterodactylus, Schlermochus, Hypsilophodon, Iguanodon* and *Ornithosuchus.*

9. For Peter Snowball's work, see Peter Charig's *A New Look at the Dinosaurs*, London: Heinemann in association with the British Museum (Natural History), 1979; Mark Hallett, "Bringing Dinosaurs to Life," in *Dinosaurs Past and Present* Volume 1, Seattle: Natural History Museum of Los Angeles County in association with University of Washington Press, 1987, Sylvia J. Czerkas and Everett C. Olson (eds.), pp. 96–113. Artist Nova Young's restorations of past global fauna were published in Edwin H. Colbert's *Wandering Lands and Animals: The Story of Continental Drift*, New York: E. P. Dutton and Company, 1973. Artist Peter Trusler's painting of Australian dinosaur fauna was discussed by Ken McNamara in "Aussie Dinosaur Stamps," in *Dinonews*, no. 6, July 1993, pp. 12–13. Trusler's painting also appeared in *Scientific American*, July 1993, illustrating an article, "Australia's Polar Dinosaurs," by Patricia Vickers-Rich and Thomas Hewitt Rich.

10. Personal correspondence from William E. Swinton dated January 13, 1982.

11. Personal correspondence received from Peter Crowther dated August 13, 1999.

12. Thanks to Peter Crowther, Philip Doughty and Ken James of the Ulster Museum, Belfast, for supplying information concerning the Ulster Museum's acquisition of the ILN illustrations and their 1993 exhibit.

13. Maurice Burton, "Toots and his Dogs," February 18, 1982.

14. Personal correspondence from William E. Swinton dated February 13, 1982.

15. Maurice Burton, "Toots and his Dogs," February 18, 1982. (unpublished m.s.).

16. Dale Russell, "Models and Paintings of North American Dinosaurs," in *Dinosaurs Past and Present* Volume 1, *op. cit.*, p. 117. In *The Last Dinosaur Book: The Life and Times of a Cultural Icon*, Chicago: The University of Chicago Press, 1998, p. 108, however, William J. T. Mitchell claims that, "This clearly cannot be right. Artists, no matter how obedient, are not simply 'eyes,' nor are paleontologists simply 'brains' waiting to be wired up to a cooperative retina ... both scientists and artists are human beings, participants in cultures that impinge on their 'pure' scientific pursuits, guaranteeing that their work will never be pure or pristinely objective. And even if their images were 'windows through which nonspecialists can see the dinosaurian world' (which they clearly are not), those nonspecialists would also be coming to those windows not with pure, innocent eyes, but with preconceptions, fantasies, and prejudices much like those shared by the scientists and artists."

Chapter 18

1. Other examples of "fringe dinosaurs" include the Mongolian Cretaceous dinosaur *Deinocheirus*, the African genus *Spinosaurus*, and the Wealden armored dinosaur *Polacanthus*. Also see, Allen A. Debus, "Sculpting Twilight Zone Dinosaurs," in *Modeler's Resource*, Dec./Jan. 2000, no. 31, pp. 58–61.

2. Donald F. Glut, *The New Dinosaur Dictionary*, Secaucus, NJ: The Citadel Press, 1982, p. 39.

3. Thomas Huxley, "On *Acanthopholis horridus*, a New Reptile from the Chalk Marl," in *The Geological Magazine* (vol. iv, pp. 65–67) (This later reappeared in Volume III of *The Scientific Memoirs of Thomas Henry Huxley*, London, Macmillan, edited by M. Foster and E. R. Lankester [1898–1902], which will be easier to access through your local reference librarian.)

4. Edwin H. Colbert's *Dinosaurs: Their Discovery and Their World*, New York: E. P. Dutton & Co., Inc., 1961; and a British Museum publication by William E. Swinton, *Dinosaurs* (British Museum of Natural History, 1969), which included Parker's fanciful restoration, and where Swinton mentions that *Acanthopholis* is also known from slightly older Cambridge Upper Greensand deposits.

5. George Olshevsky, *The Ankylosaur Family Tree*, 1979.

6. Only one plastic toy figure has been produced, and there is one rather amorphous looking sculpture labeled *Acanthopholis* displayed at a dinosaur theme park in the American south.

7. "Paleontology's 'Jurassic' Windfall," in *Science*, December 18, 1992, vol.

258, p. 1879, discusses *Jurassosaurus*. Another Jurassic ankylosaur, *Gargoyleosaurus*, was described in 1998. Kenneth Carpenter, Clifford Miles and Karen Cloward, "Skull of a Jurassic ankylosaur (Dinosauria)," in *Nature*, June 25, 1998, vol. 393, pp. 782–783.

8. From an Associated Press internet website posting, dated Friday, July 10, 1998—"New Dinosaur Species Discovered" (pasted into Debus' "dinosaur scrapbook" no. 15). On May 18, 2001, three years after "Tails—You Lose!" had been completed, I learned that the mystery of the Isle of Wight "raptor" had been resolved. The fossil was described as the 132 million year old ancestral tyrannosaurid—*Eotyrannus*. This is viewed as a "tyranno-raptor" transitional form linking the tyrannosaur and smaller coelurosaur lineages, the latter including such genera as *Velociraptor*. An early published restoration and reconstruction of this genus, by Darren Naish and Terry Smith, respectively, reveals a 5 meter long, raptor-like creature with long grasping forelimbs but without sickle-shaped toe claws. Erik Stokstad, "Early Tyrannosaur was small but well armed," *Science*, May 18, 2001, vol. 292, pp. 1278–1279. Dr. James Kirkland's notion that "*...it seems likely that Utahraptor would have been found on both* [author's note—North American and European] *continents*" was expressed in the 1999 BBC broadcast *Walking With Dinosaurs*. The cited quote was published in *Walking With Dinosaurs: A Natural History*, by Tim Haines, New York: Dorling-Kindersley Publishing, Inc., p. 159.

9. Personal e-mail communications dated October 17, 1998. Also see Peter M. Galton, "British Plated Dinosaurs (Orhithischia, Stegosauridae)" in *Journal of Vertebrate Paleontology*, September 1985, vol. 5, no. 3 , pp. 211–254.

10. Gerard Gierlinski, "The Furry Dino," in *Dinosaur World*, Spring 1998, no. 4, pp. 3–5.

11. Nomen dubium means "doubtful name" or a "*...taxon founded upon material of questionable diagnostic value*," as defined by Donald F. Glut, *Dinosaurs: The Encyclopedia*, Jefferson, NC: McFarland & Company, Publishers, 1997, p. 1019.

Chapter 19

1. Edward D. Cope quoted in *Dinosaurs, Mammoths and Cavemen: The Art of Charles R. Knight*, New York: E. P. Dutton, Inc., 1982, by Sylvia M. Czerkas and Donald F. Glut, p. 16.

2. Tracy Ford's skeletal reconstruction, recently published on page 165 of *Dinosaur Encyclopedia* (Random House of Canada, Ltd., 1993) by Don Lessem and Donald F. Glut, graphically shows how incomplete the specimen really is.

3. Edward D. Cope, "The Fossil Reptiles of New Jersey," in *American Naturalist*, 1869, vol. III, pp. 84–91. One of Cope's restorations, drawn on the back of a letter, was recently published on page 194 of *The Complete Idiot's Guide to Dinosaurs* (New York: Alpha Books, 1998) by Jay Stevenson and George R. McGhee.

4. A photograph of Hawkin's partially completed "*Laelaps*" reconstruction was recently published on page 154 of *The Complete Dinosaur* (James O. Farlow and Michael Brett-Surman, eds.), Bloomington: Indiana University Press, 1997.

5. Knight quoted from his unpublished autobiography, *Autobiography of an Artist*, Book 1, circa 1950?, p. IV–16.

6. Henry F. Osborn, "A Great Naturalist: Edward Drinker Cope," in *The Century Magazine* 1897, vol. LV, pp. 10–15; William H. Ballou, "Strange Creatures of the Past," in *The Century Magazine*, 1897, vol. LV, pp. 15–23.

7. Photographs of many of Knight's preliminary sculptural studies are shown in Czerkas and Glut, *op. cit.*, 1982. Several of these, including Knight's fighting *Laelaps* pair, were on display at the Academy of Natural Sciences in Philadelphia during the late 1980s and early 1990s.

8. David Rains Wallace, *The Bonehunters' Revenge: Dinosaurs, Greed, and the Greatest Scientific Feud of the Gilded Age*, Boston: Houghton Mifflin Company, 1999, pp. 273–274.

9. Henry R. Knipe, *Nebula to Man*, New York: J. M. Dent and Company, 1905.

10. Gregory S. Paul, *Predatory Dinosaurs of the World*, New York: Simon and Schuster, 1988, p. 347.

11. Dale A. Russell, *An Odyssey in Time: The Dinosaurs of North America*, Toronto: University of Toronto Press in Association with the National Museum of Canada, 1989, p. 190.

12. Kenneth Carpenter, Dale Russell, Donald Baird and Robert Denton, "Redescription of the holotype of *Dryptosaurus aquilunguis* (Dinosauria: Theropoda) from the Upper Cretaceous of New Jersey," in *Journal of Vertebrate Paleontology*, vol. 17, no. 3, September 1997, pp. 561–573.

13. Kenneth Carpenter's illustration was originally published as Figure 1 in Carpenter, et. al., *op. cit.*, 1997, p. 562.

14. This phrase is inspired by the title of one of Stephen J. Gould's essays, "Bully for Brontosaurus," appearing in his essay collection, *Bully for Brontosaurus: Reflections in Natural History*, New York: W. W. Norton and Company, 1991, chapter 5. In historical accounts, Cope is targeted for foolishly creating one of America's earliest, yet erroneous, skeletal reconstructions—an *Elasmosaurus* specimen cleaned and prepared by Benjamin Waterhouse Hawkins, yet mounted under Cope's direction. Cope fitted the skull to the tail end instead of to the anterior end of the cervical vertebrae, publishing his description of the reconstruction. To Cope's chagrin, Othniel Marsh identified the error and later reveled in the circumstances. Is this one reason for why relatively few reconstructions and restorations were condoned by paleontologists prior to Marsh's death in 1899? For a popular account of this incident, see chapter two of Mark Jaffe's *The Gilded Dinosaur* (New York: Crown Publishers, 2000). Also, thanks to Donald Baird for suggesting and recommending in 1991 (personal correspondence) that the "evolution" of interpretations concerning *Laelaps/Dryptosaurus* be commemorated in sculptural representation. This was accomplished by Allen in 1994.

Chapter 20

1. William J. T. Mitchell, *The Last Dinosaur Book*, Chicago: The University of Chicago Press, 1998, p. 25.

2. Allen A. Debus, "A Gilded Cage of Significant Slabs," in *Fossil News*, February 1999, vol. 5, no. 2, pp. 6–8.

3. Pei-je Chen, Zhi-ming Dong, and Shuo-nan Zhen, "An exceptionally Well-Preserved Theropod Dinosaur from the Yixian Formation of China," in *Nature*, Jan. 8, 1998, vol. 391, pp. 147–152; Michael Lemonick, "Dinosaurs of a Feather," in *Time*, July 6, 1998, pp. 82–3; Ji Quiang, Phillip Currie, Mark Norell and Ji Shu-An, "Two Feathered Dinosaurs from Northeastern China," in *Nature*, June 25, 1998, vol. 393, pp. 753–761; Jennifer Ackerman, "Dinosaurs Taking Wing," in *National Geographic*, July 1998, pp. 74–99; Justin Wang, "Scientists Flock to Explore China's 'Site of the Century,'" in *Science*, March 13, 1998, vol. 279, pp.

1626–1627; Ann Gibbons, "Dinosaur Fossils, in Fine Feather, Show Link to Birds," in *Science*, June 26, 1998, vol. 280, p. 2051.

4. There remains a possibility that the most bird-like theropods were secondarily flightless. The feathered dinosaur *Caudipteryx*, having a center of mass shifted anteriorly as in modern birds yet unlike the condition noted in the bird-like theropods, lends support to this theory. Andrzej Elzanowski, "The Flying Dinosaurs," in *The Scientific American Book of Dinosaurs*, Gregory S. Paul (ed.), New York: St. Martin's Press, 2000, pp. 171–172; Terry D. Jones, James O. Farlow, John A. Ruben, Donald M. Henderson, William J. Hillenius, "Cursoriality in Bipedal Archosaurs," in *Nature*, August 17, 2000, vol. 406, pp. 716–718; George Olshevsky, "The Birds First? A Theory to Fit the Facts," in *Omni*, July 1994, pp. 34–43, 80–86; Gregory S. Paul, *Predatory Dinosaurs of the World: A Complete Illustrated Guide*, New York: A New York Academy of Sciences Book, 1988.

5. Michael Skrepnick's *Sinosauropteryx* restoration first appeared in the *New York Times*, Saturday, October 19, 1996, pp. 1, 10, and illustrated Malcolm W. Browne's article, "Feathery Fossil Hints Dinosaur-Bird Link." This *Sinosauropteryx* restoration (1996) was recently published on page 274 of Pat Shipman's *Taking Wing: Archaepteryx and the Evolution of Bird Flight*, New York: Simon and Schuster, 1998. Michael Skrepnick was among the first western investigators to see the famous *Sinosauropteryx* slab. Skrepnick wrote about his experiences for *Dinosaur World*, February 1997, vol. 1, no. 1, pp. 4–7, "Dreaming About Dinosaurs." Skrepnick's more accurate, less hurried painted restoration of *Sinosauropteryx* appeared on the dust jacket of the *Encyclopedia of Dinosaurs*, Philip J. Currie and Kevin Padian (eds.), San Diego: Academic Press, 1997.

6. According to paleontologist Phil Currie, "Related animals more derived or advanced than Archaeopteryx are birds. But species more primitive are not.... Feathers separate birds from flying insects, flying reptiles and bats, but they are not what makes a bird unique." Quotation from *The Scientific American Book of Dinosaurs*, *op. cit.*, p. 188.

7. John H. Ostrom, "*Archaeopteryx* and the origin of flight," in *Quarterly Review of Biology*, 1974, vol. 49, pp. 27–47.

8. Gregory S. Paul also restored a less reptilian, more bird-like looking *Archaeopteryx* in 1988. See Gregory S. Paul, *op. cit.*, 1988, p. 212.

9. Darren Naish, "Tree-Climbing Dinosaurs," in *Quarterly*, "Flight Special Issue" (The Quarterly Journal of the Dinosaur Society), 2000, vol. 4, no. 1, pp. 20–24. Naish makes an interesting case for "scansoriality," tree-climbing ability in certain Mesozoic reptiles, suggesting that birds may have evolved from the trees down. He also provides much original imagery of tree-climbing dinosaurs.

10. *ZooBooks*, Wildlife Education Ltd., June 1985, vol. 1, no. 9. Hallett's painting illustrates discussion of a topic, "Did Dinosaurs Really Die Out?"

11. Richard Thulborn and Tim Hamley quoted in Pat Shipman, *op. cit.*, 1998, p. 253. An illustration of the *Archaeopteryx* wading as a predator also appears on this page. Luis Rey's restoration of an aquatic, swimming *Archaeopteryx* appears on page 16 of *Quarterly* (The Quarterly Journal of the Dinosaur Society), vol. 4, no. 1, 2000.

12. *Science*, February 5, 1993, vol. 259; inside the issue appeared ornithologist Alan Feduccia's article, "Evidence from Claw Geometry Indicating Arboreal Habits of *Archaeopteryx*." According to Andrzej Elzanowski, *op. cit.*, 2000, p. 175, "The closest we can currently get in the fossil record to the origin of flight is Archaeopteryx, which retains its position as the most primitive of all known unquestionable birds. Archaeopteryx is beyond the initial stage of the evolution of avian flight but seems to be primitive enough to cast light on the lifestyle and locomotion of the avian ancestors. That is why most current reconstructions of Archaeopteryx are heavily biased toward author's preconceptions about the origin of birds and their flight." Recent restorations of *Archaeopteryx*' wings by Derik Yalden and Siegfried Rietchel, respectively, offer contrasting opinions on its flying abilities. Yalden's restoration contains less surface area compared to Rietchel's, and is therefore a poorer flyer commanding less take off capability. See chapter 11 in Pat Shipman, *op. cit*, 1998.

13. Gregory S. Paul, *op. cit.*, 1988.

14. An intriguing theoretical idea for how the elongated (feathered) arms of an oviraptor may have facilitated egg brooding is described on p. 18 of *Quarterly* (The Quarterly Journal of the Dinosaur Society), 2000, vol. 4, no. 1. Here authors David Marjanovic and Hugo Meisl-Weg claim, "The arms of the famous brooding Oviraptor and Ingenia skeletons all circle the eggs in a way that if they could have borne wing feathers, would have covered the nests and shielded them from the sun and the rain. Apparently, there was a strictly Darwinian advantage in lengthening the arms and the feathers: longer arms = more wing area = bigger nests = room for more eggs = more offspring." Eventually these preadaptations would have led to flight.

15. There is also a "pouncing proavis" model for origin of flight. See pp. 178–179 of Andrzej Elzanowski, *op. cit.*, 2000. According to Elzanowski, early birds may have "...evolved from predators that specialize in ambush from elevated sites." These pouncing predators would have descended from tree limbs like modern paratroopers. For an early 1990s perspective on the evolution of feathers, see Carl Zimmer's article, "Ruffled Feathers," in *Discover*, May 1992, pp. 44–54.

16. Pat Shipman, *op. cit.*, 1998, p. 272; interestingly, one researcher suggests that *Archaeopteryx*'s feathers and "ground up" flight capabilities could have evolved before the evolution of endothermy in the avian lineage. (p. 265).

17. Sano Kazuhiko's restoration of the *Confucisornis* appears on the cover of the February 1998 issue of *Scientific American*, vol. 278, to advertise the featured article, "The Origin of Birds and Their Flight," by Kevin Padian and Luis M. Chiappe. The words "Both a bird and a dinosaur" appear above Mr. Kazuhiko's cover painting. A photograph of Brian Cooley's sculpture of the *Caudipteryx* graced the cover of *National Geographic*'s July 1998 issue, vol. 194, no. 1, featuring an article titled, "Dinosaurs Take Wing—The Origin of Birds." William M. Nickell's restoration of *Protoavis texensis* graced the cover of Sankar Chatterjee's *The Rise of Birds: 225 Million Years of Evolution*, Baltimore: Johns Hopkins, 1997. Also see Luis M. Chiappe, "The First 85 Million Years of Avian Evolution," in *Nature*, November 23, 1995, vol. 378, pp. 349–355; and Allen A. Debus, "New Wings on the Paleo-Perch," in *Dinosaur World*, June 1997, vol. 1, no. 2, pp. 10–14, 42.

18. Christopher Sloan, "Feathers for T. rex: New Birdlike Fossils are Missing Links in Dinosaur Evolution?" in *National Geographic*, November 1999, vol. 196, no. 5, pp. 98–107.

19. Christopher Sloan, *op. cit.*, 1999, p. 105.

20. *Ibid.*, pp. 98–99.

21. *Ibid.*, p. 100.

22. A new oviraptorid dinosaur, *Nowingia*, has been shown to have a fossil pygostyle, indicative of feathered modern birds. See p. 16 of *Quarterly* (The Quarterly Journal of the Dinosaur Society), 2000, vol. 4, no. 1.

23. Christopher Sloan, *op. cit.*, 1999; Stephen Czerkas' defensive note, "Is Flying Dinosaur a Composite?" appeared in *Prehistoric Times*, April/May 2000, no. 41, p. 5. Also see, Constance Holden, "Florida Meeting Shows Perils, Promise of Dealing for Dinos," in *Science*, April 4, 2000, vol. 288, pp. 238–239; Xu Xing's letter to *National Geographic*, March 2000, titled "Feathers for T. rex?"; Jeff Hecht, "Piltdown Bird," in *New Scientist*, January 29, 2000, p. 12; and Lewis M. Simons, "Archaeoraptor Fossil Trail," in *National Geographic*, October 2000, vol. 198, no. 4, pp. 128–133. Undaunted by the devastating experience, by the summer of2000, Stephen had opened a "Hall of Feathered Dinosaurs" in their museum located in Blanding, Utah. Meanwhile, the front end of Archaeoraptor, which *may* be a bird, hasn't been described yet. We are a visually-oriented species, and tend to believe what we see, even if the image is false or too beautiful to disbelieve.

24. Perle Altangerel, Mark Norell, Luis M. Chiappe and James Clark, "Flightless Bird from the Cretaceous of Mongolia," in *Nature*, April 15, 1993, vol. 362, pp. 623–626; Mark Norell, Luis Chiappe and James Clark, "New Limb on the Avian Family Tree," in *Natural History*, September 1993, pp. 38–42. Much as paleontologists now must consider a variety of possible origins for the different categories of winged, flying Mesozoic avians, so must equally perplexing matters concerning the *flightless* nature of other Mesozoic avians be settled. Did the mononkysaurs evolve directly from flightless "protobird" dinosaurs closely related to "ostrich dinosaurs?" Or, just as winged avian flight may have developed a multiple number of times during the Mesozoic from different theropod lineages, in turn, some branches derived from these winged descendants may have split off and converged, independently, toward a secondary flightless condition.

25. Sculptor Matt Manit described a peculiar experiment he performed in which he attempted to elucidate the function of *T. rex*'s arms for digging nests in *Dinosaur World*, June 1997, vol. 1, no. 2. He alluded to analogies to the *Mononkyus*.

26. Fernando E. Novas and Pablo F. Puerta, "New Evidence Concerning Avian Origins from the Late Cretaceous of Patagonia," in *Nature*, May 22, 1997, vol. 387, pp. 390–397.

27. Fernando E. Novas and Pablo F. Puerta, *op. cit.*, p. 392.

28. Steve Kirk's restoration illustrated Carl Zimmer's article, "A Sickle in the Clouds," *Discover*, June 1998, p. 32. T. Michael Keesey's striking restoration of the *Rahonavis* is posted on the internet at "http://dinosauricon.com/images/rahonavis-tmk.html."

29. Catherine Forster, Scott D. Sampson, Luis M. Chiappe and D. W. Krause, "The Theropod Ancestry of Birds: New Evidence from the Late Cretaceous of Madagascar," in *Science*, March 20, 1998, vol. 279, pp. 1915–1919; Kevin Padian (letters), *Science*, May 15, 1998, vol. 280, pp. 986–987.

30. Xing Xu, Xiao-Lin Wang and Xiao-Chun Wu, "A Dromaeosaurid Dinosaur with a Filamentous Integument from the Yixian of China," in *Nature*, vol. 401, pp. 262–266, September 16, 1999. Recent morphological analysis of the feathery filamental structures observed in *Sinornithosaurus* "...support[s] the hypothesis that feathers evolved and initially diversified in terrestrial dinosaurs before the origin of birds and flight." Xing Xu, Zhong-he Zhou, Richard O. Prum, "Branched Integumental Structures in *Sinornithosaurus* and the Origin of Feathers," in *Nature*, March 8, 2001, vol. 410. However, *Sinornithosaurus* did not possess avian, bipinnate flight feathers. Therefore, most likely *Sinornithosaurus* was not a flying dinosaur. Another recent discovery is the 115-million-year-old *Eoalulavis* from Lay Hoyas Spain, a little bird with a 17 centimeter wingspan found in a remarkable state of preservation. Scientists could readily discern feather impressions arranged in a natural position and a special wing apparatus evolved to enhance low speed flight maneuverability (known as the alula, or "bastard wing"), as well as the fossilized contents of its final meal. Apparently, *Eoalulavis* fed on aquatic organisms and could expertly control its landings and take offs due to the presence of an alula on the wings' leading edge. This is the oldest known species for which there is positive evidence of such a sophisticated structure. Despite the fact that *Eoalulavis*' hand structure differed from that of modern avians (i.e. hand and wrist bones rigidly fused into a structure known as a carpometacarpus), the species had evolved a means of controlled flight far beyond the capacity of *Archaeopteryx*. While Nature can provide alternative adaptational solutions to address similar ecological challenges, evolutionary "forces" must build on the existing structures found in ancestral species.

31. Some interesting examples of paleoimagery include Gerhard Heilmann's "restoration" of the hypothetical bird "Proavis" (1926), more recently reproduced in Adrian J. Desmond's *The Hot-Blooded Dinosaurs: A Revolution in Palaeontology*, New York: Warner Books, 1975, p. 202; and Franz Nopsca's previous version of "Proavis" recently published in Pat Shipman, *op. cit.*, 1998, p. 111. The imagetext between Heilmann's and Nopsca's illustrations contrasts considerably. Heilmann's protobird is a tree-climber, whereas Nopsca's is a "ground up" cursorial. Decades later, following Sankar Chatterjee's 1983 discovery of *Protoavis*, the alleged Late Triassic flying bird, paleoartist John Sibbick provided a restoration first seen by the authors in the *Chicago Tribune*, June 12, 1989. Also see Carl Zimmer, *op. cit.*, 1992.

32. Xing Xu, Zhonghe Zhou and Xiaolin Wang, "The Smallest Known Non-Avian Theropod Dinosaur," in *Nature*, December 7, 2000, vol. 408, pp. 705–708; Erik Stokstad, "Tiny, Feathered Dino Is Most Birdlike Yet," in *Science*, December 8, 2000, vol. 290, pp. 1871–1872. According to paleoartist Tracy L. Ford, the name *Archaeoraptor* remains valid for the specimen described by Xu, Zhou and Wang as *Microraptor*, with the latter name being a junior synonym for the former. Tracy Ford, "How to Draw Dinosaurs—Microraptor," in *Prehistoric Times*, February/March 2001, no. 46, p. 15.

33. Article "Amazing New Feathered Dinosaur Found in China," and "News—*Microraptor*: The Smallest Adult Dinosaur Yet is Unearthed, and It's a Lot Like a Bird," were posted on the *New Scientist* website in December 2000, along with a restoration reminiscent of Heilmann's 1926 "Proavis."

34. Jeff Liston, "The Hidden Story: *Archaeopteryx* and the Evolution of Feathered Flight," in *Quarterly*, "Flight Special Issue" (The Quarterly Journal of the

Dinosaur Society), 2000, vol. 4, no. 1, p. 12.

35. Luis Rey, "International Paleoartist—A "Dinosaur Renaissance?," in *Dinosaur World*, June 1997, vol. 1, no. 2, pp. 26–27.

36. This chapter was written in December 2000, but it is our speculation that forthcoming *Jurassic Park III*'s "raptor" will be feathered! You'll all know if we were right by the time this book is published.

37. Dougal Dixon, *The New Dinosaurs: An Alternative Evolution*, Topsfield, Mass: Salem House Publishers, 1988, p. 11.

38. Sarah Landry's restoration of *Syntarsus* appears on pp. 58–59 of Robert Bakker, "Dinosaur Renaissance" in *Scientific American*, April 1975. Although key sources elected not to respond, we suspect the term "Dinosaur Renaissance," as published in April 1975, may have been coined by *Scientific American*'s editorial staff. Reminiscing, Bakker claimed in an interview with Tony Campagna that, "I was the first to put feathers on dinosaurs. That was back in 1975 in a *Scientific American* article. I remember people hooting and hollering about, 'Oh! There can't be feathers on dinosaurs! Blahdiddy blahdiddy dah....' John Ostrom believed that birds descended from dinosaurs and wrote an article and said dinosaurs didn't have feathers. Now we know dinosaurs had feathery insulation." *Prehistoric Times*, February/March 2001, no. 46, p. 51.

39. Robert Bakker, *The Dinosaur Heresies*, New York: William Morrow and Company, Inc., 1986, p. 310; Gregory S. Paul, *op. cit.*, 1988, p. 172.

40. *Ibid.*, p. 310.

41. Personal communication from John Long dated October 29, 2000. John Long restoration appeared in his article, "Fur Coats for Flesh-Eaters: Polar Dinosaur Discoveries from Victoria," in *Dinonews*, June 1990, no. 1, p. 7.

42. Erik Stokstad, "Feathers, or Flight of Fancy?" in *Science*, vol. 288, June 23, 2000, pp. 2124–2125; Terry D. Jones, John A. Ruben, Larry D. Martin, Evgeny N. Kurochkin, Alan Feduccia, Paul F. A. Maderson, William J. Hillenius, Nicholas R. Geist, Vladimir Alifanov, "Nonavian Feathers in a Late Triassic Archosaur," in *Science*, June 23, 2000, vol. 288, pp. 2202–2205.

43. *Longisquama* made an anachronistic Late Cretaceous appearance when it was restored in Disney's *Dinosaur* film (2000). Jeff Kurtti, *Dinosaur: The Evolution of an Animated Feature*, Disney Enterprises Inc., 2000, pp. 50, 119.

44. Robert Bakker, *op. cit.*, 1975; Terry D. Jones, John A. Ruben, Larry D. Martin, Evgeny N. Kurochkin, Alan Feduccia, Paul F. A. Maderson, William J. Hillenius, Nicholas R. Geist, Vladimir Alifanov, *op. cit.*, 2000.

45. Not without contention, the presence of feathers in Triassic species has been historically hypothesized or inferred by a number of workers. See Gerard Gierlinski's article, "The Furry Dino," in *Dinosaur World*, Spring 1998, no. 4, pp. 3–5. Based on Gierlinski's interpretations, Polish artist Marta Szubert created a striking life-sized sculptural restoration of a "furry" *Dilophosaurus*, nicknamed "Dyzio," shown on page 4 of this citation. Yet, understandably, stone cold evidence of feathers will prove most persuasive to the paleontological jury.

46. Or, as Andrzej Elzanowski put it, "There are several reasons to believe that birds not only are flying dinosaurs, but baby flying dinosaurs." Elzanowski, *op. cit.*, 2000, p. 179; Phil Currie has stated, "The presence of feathers on dinosaurs does not prove that birds came from dinosaurs." Phil Currie, *op. cit.*, p. 187.

47. George Olshevsky, *op. cit.*, 1994; Andrzej Elzanowski, *op. cit.*, 2000, pp. 171–172. Elzanowski claims, "If flightlessness had evolved at a stage of avian evolution close to Archaeopteryx, it would be extremely difficult to distinguish it from the primary flightlessness."

48. Pat Shipman, *op. cit.*, p. 242; By comparison, little interest lingers over the discovery of a "furry" pterosaur, *Sordes pilosus* (the "hairy devil"). And the oldest known bat, 50-million-year-old *Icaronycteris*, has no documented fossil ancestors. The more mysterious matter of how bats adapted to flight is never sensationalized by the media or in paleoart. Yet news of feather-sprouting dinosaurs always seizes headlines.

49. In the fall of 2000 a new book for children titled *Feather Dinosaurs*, written by Christopher Sloan, was published by *National Geographic*. Doubtless there will be many more books of this nature to come. Herein, we are merely chronicling early phases of the feathered dinosaur "revolution." Such publications will appeal to younger readers, impressionistic dinophiles in the making, who will embrace the feathered dinosaur idea and imagetext.

Chapter 21

1. J. Alan Holman, *Ancient Life of the Great Lakes Basin*, Ann Arbor: The University of Michigan Press, 1995, pp. 209–222; John A. Dorr, Jr., and Donald Eschman, *Geology of Michigan*, Ann Arbor: The University of Michigan Press, 1970.

2. Personal communication, Richard Rush, January 18, 2001.

3. John C. Greene, *The Death of Adam: Evolution and Its Impact on Western Thought*, Ames: Iowa State University Press, 1959, chapter 4; Paul Semonin, *American Monster: How the Nation's First Prehistoric Creature Became a Symbol of National Identity*, New York: New York University Press, 2000, chapters 4 and 5.

4. Noah T. Clarke, "Method of Restoring the Cohoes Mastodon," in *New York State Bulletin Eighteenth Report to the Director of the State Museum and Science Department, 1922*, March 1924, no. 251, pp. 131–133; Judith Drumm, *Mammoths and Mastodons: Ice Age Elephants of New York*, educational leaflet no. 13, Albany: The University of the State of New York, 1963, pp. 25–26.

5. John Noetzel, "Damaged by Fire in 1981, the Famous Cohoes Mammal Is Restored" (undated xerox of newspaper clipping sent courtesy of Caroline Edwards); Douglas R. Butler, "Mastodont Makeover," in *Hudson Valley*, August 1996, pp. 26–29.

6. Personal communication from Caroline Edwards, August 2000.

7. Unsigned page titled, "Setting the Stage for Santa," in *The Express Messenger* (Railway Express Monthly Magazine), December 1927, vol. X, no. 6.

8. The Louis Paul Jonas Studios is perhaps best known to dinophiles for their expert renditions of dinosaurs, displayed by Sinclair during the 1964/65 New York World's Fair.

9. Personal communication from Katherine Hamilton-Smith dated July 12, 2000.

10. Jean Moore, "Richard Rush: World-Renowned Artist," in *The Daily Journal* (DuPage Sesquicentennial profile), November 2, 1988, p. 27; Allen and Diane Debus, "An Interview with Richard Rush," in *Prehistoric Times*, July/Aug 1995, no. 13, pp. 14–15.

11. Donald Baird, note in *Society of Vertebrate Paleontology News Bulletin*, Feb. 1988, pp. 44–45.

12. Lull's mount was completed by

the early 1930s. Personal communication from John Ostrom dated October 28, 1991. According to Ostrom, this was one of the earliest exhibits in the Peabody Museum.

13. Conversation with Richard Rush on January 18, 2001.

14. Paul Semonin, *op. cit.*

15. *Ibid.*; Peale's son, Rembrandt Peale, stated, "Nothing in either skeleton is imaginary" (p. 327).

16. *Ibid.*, pp. 370–371. Fittingly, an image of this painting appeared on a U.S. postage stamp to commemorate the 150th anniversary of the Pennsylvania Academy of Fine Arts. This three cent stamp, issued in 1955, may be regarded as America's first "dinosaur" stamp.

17. Paul Semonin, *op. cit*, p. 333.

18. *Ibid.*, p. 364.

19. *Ibid.*, pp. 386–387, 403–404.

20. *Ibid.*, pp. 382–386; advertising for the Missourium boasted, "This unparalleled Gigantic remains, when its huge frame was clad with its peculiar fibrous integuments, and when moved by its appropriate muscles, was Monarch over all the Animal Creation; the Mammoth, and even the mighty *Iguanodon* may easily have crept between his legs" (p. 382). Semonin refers to this symbolic confrontation, "...the mastodon's first encounter with the saurian species," as a metaphorical disaster for America's Mastodon.

21. *Ibid.*, pp. 400–401. Semonin claims, "Hawkins's [1868] drawing of his tableau depicts the eclipsing of the mastodon by the dinosaurs. The American monster, with its trunk uplifted, appears forlornly in the background, upstaged by the immense, kangaroo-like figure of *Hadrosaurus*, turning its hind side to the former rulers of the antediluvian world.... Drawn to scale, the lithograph shows the intimidating size of the new saurian monsters, which tower over the spectators viewing the prehistoric scene from behind a railing." In 1852 Hawkins looked forward principally to the opportunity of sculpting the American Mastodon for the Crystal Palace exhibition at Sydenham, although this never came to fruition (*Ibid.*, p. 394). Ultimately, first at Crystal Palace and eventually elsewhere, Owen's Dinosauria "triumphed" over the American Mastodon.

22. *Ibid.*, p. 402; Allen A. Debus, "Why Not the Mammals?" Chapter 5, pp. 85–110 in *Dinosaur Memories*. Lincoln: iUniverse, 2002.

Chapter 22

1. Coloman Jonas, *From "Stuffing" to Sculpturdermy: The Jonas Story*, Denver: Jonas Brothers, Inc., 1983, pp. 21–22; for more on Louis Paul Jonas, Jr., see "The World of Louis Paul Jonas," by Zella Jonas Merritt, in *American Artist*, January 1967 (reprint).

2. For information on Akeley, see Douglas Preston's *Dinosaurs in the Attic: An Excursion into the American Museum of Natural History*, New York: Ballantine Books, 1988, chapter 7. In 1911 Lanteri published a book available as a Dover reprint titled *Modeling and Sculpting Animals* (Dover reprint, 1985).

3. Stages in the process of building dinosaurs like Jonas did are illustrated in photographs published in *Dinosaurs Return: Elbert H. Porter—Dinosaur Sculptor*, by Elbert H. Porter (Davis Printing Co., 1982).

4. When Jonas' dinosaurs "walked" the earth once again, they were standing within a few miles of the site where the famous "Giant of Claverack" was found in the seventeenth century. This "giant" turned out to be the bones of an American *Mastodon*, the genus which symbolized American "virility" during the late eighteenth and early nineteenth centuries. (In this context, we have borrowed the term "virility" from Gary Williams.) See *American Monster: How the Nation's First Prehistoric Creature Became a Symbol of National Identity*, New York: New York University Press, 2000, by Paul Semonin, chapter 1.

5. Donald F. Glut, *The Dinosaur Scrapbook*, Secaucus, NJ: The Citadel Press, 1980, p. 42; Ed Lathrop, "The Jonas Studios," in *The Prehistoric Times*, March/April 1994, no. 5, pp. 10–11.

6. A rival dinosaur exhibit, created by WED Enterprises for Walt Disney, also graced the New York World's Fair grounds. These masterful, animatronic restorations were still on display during our last visit to Disneyland's "Primeval World" in Anaheim, California, during the early 1980s.

7. They arrived in Chicagoland during the summer of 1966.

8. These and other Jonas dinosaur memorabilia were on display at the Berkshire Museum in Pittsfield, Massachusetts, during the summer of 2000.

Chapter 23

1. Dougal Dixon quoted in "Dinosaur World Interview with Dougal Dixon," in *Dinosaur World*, no. 9, Winter 2000/2001, p. 50.

2. Dougal Dixon, *After Man: A Zoology of the Future*, New York: St. Martin's Press, 1981.

3. *Ibid.*, p. 32.

4. *Ibid.*, Desmond Morris's "Introduction," p. 9.

5. Dougal Dixon "Interview," *op. cit.*

6. *Ibid.*, p. 51.

7. *Ibid.*, pp. 50–51; Dougal Dixon, *The New Dinosaurs: An Alternative Evolution*, New York: Salem House Publishers, 1988; Dale A. Russell, "Models and Paintings of North American Dinosaurs," in *Dinosaurs Past and Present*, vol. 1, Seattle: Natural History Museum of Los Angeles County in association with University of Washington Press, 1986, Sylvia J. Czerkas and Everett C. Olson (eds.), pp. 114–131.

8. H. G. Wells, "The Time Machine," in The Science Fiction Hall of Fame: The Greatest Science Fiction Novellas of All Time, vol. IIA, New York: Avon Books, 1973.

Chapter 24

1. (1) Ice Age catastrophes signaling Life's annihilation are reflected both in Isabella Duncan's *Pre-Adamite Man* (1860) and in one of Edouard Riou's (1833–1900) prehistoric scenes for 1860s editions of Guillaume Louis Figuier's (1819–1894) *The Earth Before the Deluge*, entitled "Deluge of the North of Europe." Such scenes portrayed times of ecological adversity, episodes that were unique in geologic time. Therefore, the *sequence* of continuously graded scenes included in became punctuated by these depicted events. In a large folding tinted lithograph of prehistoric animals by Charles Moody (after W. R. Woods), published in Duncan's book, a lifeless Ice Age boundary separates Tertiary life from that of the modern historical period. In Figuier's book, Agassiz's Ice Age catastrophe is rationalized with three deluges. The first two occur during an episode of global cooling. Directly afterward the full onslaught of cold leads to "*annihilation of organic life*" in northern and central Europe. Louis Figuier, *The World Before the Deluge*, London: Cassell and Company, Ltd., 1867, p. 439. Figuier exploited the

evidence of diluvial deposits found in association with other Ice Age markers. Riou illustrated a violent flooding caused by uplifting of mountains in Norway. The waters descending southward from partially frozen seas are seen to cause widespread devastation on a broad regional scale. A second (unillustrated) deluge was caused by the uprising of the Alps. Following the "glacial period," and the creation of Man, another (illustrated) deluge occurred, this time the one recorded in Biblical records. Only after early editions of the book appeared was it understood that prehistoric Man did coexist with some of the fauna which Figuier had exterminated with his deluge/Ice Age combination. Robert H. Silliman places the dilivium-drift problem in proper context in "Agassiz vs. Lyell: Authority in the Assessment of the Diluvium-Drift Problem by North American Geologists with Particular Reference to Edward Hitchcock," in *Earth Sciences History*, 1994, vol. 13, no. 2. (2) Riou also provided an illustration, "Condensation and Rainfall on the Primitive Globe," included in a chapter entitled "The Beginning." The illustration shows the young steaming earth *devoid* of life. This scene from deep time, presaging the early "space art" genre of the twentieth century, symbolizes the union of astronomy and geophysics with respect to merging contemporary interests in the earth's mode of formation and age of origin. (3) The Crystal Palace dinosaurs dating from the early 1850s have been cited for their innovative temporal (and visual) perspective *through* ages of prehistory. The outdoor exhibition on the Crystal Palace grounds at Sydenham, England (sculpted during the early 1850s by Benjamin Waterhouse Hawkins under the care of Richard Owen [1804–1892], was of theoretical importance in illustrating Owen's ideas of evolutionary progress (conveying an anti-Lamarckian message). However, in delivering his biological statement (i.e. imagetext), Owen was forced to invoke an (ad hoc) theory of dinosaurian extinction it being caused by *atmospheric* changes. Obviously, Hawkins wasn't able to *sculpt* the mechanism of Owen's postulated extinction out of thin air (i.e. an increasing oxygen concentration). Due to its inherently (anti) evolutionary theme, the Hawkins/Owen portrayal may also be viewed as a pre–Darwinian (ancestral) example of a closely related paleontological art form, "evolutionary iconography."

2. Luis W. Alvarez, Walter Alvarez, Frank Asaro, Helen V. Michel, "Extraterrestrial Cause for the Cretaceous-Tertiary Extinction," in *Science*, June 6, 1980, vol. 208, no. 4448, pp. 1095–1108; Walter Alvarez, *T. rex and the Crater of Doom*, Princeton: Princeton University Press, 1997; Vincent Courtillot, *Evolutionary Catastrophes: The Science of Mass Extinctions*, Cambridge: Cambridge University Press, 1999; Richard Muller, *Nemesis: The Death Star*, New York: Weidenfeld & Nicholson, 1988; James Lawrence Powell, *Night Comes to the Cretaceous*, New York: W.H. Freeman and Company, 1998; David M. Raup, *Extinction: Bad Genes or Bad Luck?*, New York: W.W. Norton & Company, 1991.

3. One such image showing dinosaurs shivering in the cold appears in a popular children's book, W. Maxwell Read's *The Earth for Sam*, New York: Harcourt, Brace and Co., 1930.

4. Allen A. Debus, "Before Alvarez—The Early Dinosaur Killers," in *Fossil News*, December 1999, vol. 5, no. 12, pp. 10–15.

5. Neave Parker's illustration was first published in *The Illustrated News*, March 5, 1960, pp. 402–403.

6. Kenneth Carpenter, *Eggs, Nests, and Baby Dinosaurs: A Look at Dinosaur Reproduction*, Bloomington: Indiana University Press, 1999, p. 256.

7. Robert Bakker, *The Dinosaur Heresies*, New York: William Morrow and Company, 1986, p. 96.

8. *Journal of Vertebrate Paleontology*, December 27, 1995, vol. 15, no. 4, p. 718.

9. Dale Russell and Wallace Tucker stated that the Late Cretaceous extinctions were "...of unusual magnitude and geologically brief duration." *Nature*, February 19, 1971, vol. 229, p. 554.

10. Kish's 1974 painting was published in Edwin H. Colbert's *Dinosaurs: An Illustrated History*, Maplewood, NJ: Hammond, Inc., 1983, pp. 208–209.

11. *Ibid.*, p. 207.

12. Carl Sagan, *Cosmos*, New York: Ballantine Productions, 1980, p. 285. Even though the Alvarez theory was not discussed in *Cosmos*, Davis depicted *both* supernova and asteroid impact extinctions imagery in the same painting. The caption to Davis' painting reads, "The death of the dinosaurs. In one astronomical hypothesis, it was due to a nearby supernova explosion, seen in this painting by Don Davis in the sky at top right. In another hypothesis a large asteroid strikes the Earth; the fine debris from the impact persists in the stratosphere, reduces sunlight available for plants the dinosaurs eat, and cools the earth. Over hundreds of millions of years, both events must have happened at least once. The extinction of bipedal intelligent reptiles cleared the stage for the evolution of mammals and humans."

13. For statistical considerations concerning the K-T boundary, see Michael Benton, "Scientific Methodologies in Collision: The History of the Study of the Extinction of the Dinosaurs," in *Evolutionary Biology*, 1989, vol. 24, pp. 371–400.

14. *Time*, May 6, 1985—cover illustration by Braldt Bralds; the words "Did Comets Kill the Dinosaurs—A New Theory About Mass Extinctions" were placed above a comet-caused eruption and to the right of a portly looking *T. rex* eyeing the catastrophe. *T. rex* coupled with mass extinction imagery has proven popular ever since. Natalie Angier's feature story, "Did Comets Kill the Dinosaurs," appeared on pp. 72–83.

15. There were exceptions, however, including an animated segment of a late 1980s televised documentary concerning the asteroid impact occurring during the Upper Triassic, forming Canada's 210-million-year-old Manicougan crater. Also, a series of paintings by paleoartist Jay H. Matternes portrayed a local (decidedly not cosmic) event, the one entombing many species of prehistoric mammals 10 million years ago at Ashfall, Nebraska. See "Dwarfing the St. Helens Eruption—Ancient Ashfall Creates Pompeii of Prehistoric Animals," by Michael Voorhies, in *National Geographic*, January 1981, vol. 159, no. 1, pp. 66–75.

16. Carl Sagan and Ann Druyan, *Comet*, New York: Random House, 1985, p. 293; artists John Gurche, David Hardy, Bob Walters, Brian Franczak, Jack Arata and others have also produced striking images of the impending event.

17. John Dawson's restoration appeared on the dust jacket of the 1986 hard cover edition. Kenneth J. Hsu, *The Great Dying: Cosmic Catastrophe, Dinosaurs, and the Theory of Evolution*, New York: Harcourt Brace Jovanovich Publishers, 1986. This painting is privately owned by Dr. Hsu.

18. Dale A. Russell, "Models and Paintings of North American Dinosaurs," in *Dinosaurs Past and Present*, Seattle: Natural History Museum of Los Angeles County in association with University of

Washington Press, 1987, Sylvia J. Czerkas and Everett C. Olson (eds.), pp. 114–131; here Russell wrote, "An evening impact was chosen for a site located four hundred kilometers off the west coast of North America [the actual impact occurred off the coast of Yucatan].... Two seconds before impact the asteroid would resemble a disk of light four times the diameter of the sun and fourteen times as brilliant. Ten seconds after impact the ejecta plume would be one hundred kilometers high, but because of the curvature of the earth at that distance it would appear only twenty-eight moon diameters wide. The artistic problem was greater. How do you portray on canvas an object that is fourteen times brighter than the sun? After several preliminary studies, Kish effectively solved the problem by depicting not the brightness but the strain that is produced in the eye by an intensely bright object."

19. Dale A. Russell, *An Odyssey in Time: The Dinosaurs of North America*, Toronto: University of Toronto Press in Association with the National Museum of Canada, 1989, p. 205; Ely Kish's precedent setting paintings vividly showing night scenes at the end of the Cretaceous, one second before and ten seconds after impact, were published on pages 192 to 193.

20. Carl Sagan and Ann Druyan, *op. cit.*, 1985.

21. *Zoobooks*, June 1985, vol. 1, no. 9 (Wildlife Education, Ltd.); Hallett's spectacular painting illustrates the topic, "How Did the Dinosaurs Die Out?" Four conventional alternatives to the showcased asteroid impact theory are also illustrated. Besides Hallett's tsunami restoration, artist Karen Carr also splendidly portrayed the K-T asteroid–caused tsunami in Louis Jacobs' *Lonestar Dinosaurs* (Merrick Natural Environment series—no. 22, 1995, p. 126). Also see Don Davis' dramatic depiction of the K-T tsunami wave appearing in *Sky & Telescope*, March 1990, illustrating articles by Paul R. Weissman, "Are Periodic Bombardments Real?" and David Morrison & Clarke R. Chapman, "Target Earth: It Will Happen." Davis depicted two horrific scenarios, a land strike and an oceanic strike. The caption to Davis' paintings reads, "Regardless of whether the object struck land or ocean, a huge fraction of Earth's inhabitants were wiped out by the cataclysmic consequences."

22. Sylvia J. Czerkas and Stephen A. Czerkas, *Dinosaurs: A Global View*, New York: Mallard Press, 1990, p. 235. Rarely are competing theories to the K-T impact theory ever portrayed. However, some of you may recall a children's instructional book on an unusual theme, titled *Time Machine 22: Last of the Dinosaurs*, New York: Byron Preiss, 1988, by Peter Lerangis. This book thoughtfully challenges readers to consider a *variety* of factors possibly leading to extinction of the dinosaurs in the late Cretaceous. This book was skillfully illustrated by Douglas Henderson.

23. Joseph Wallace, *The Rise and Fall of the Dinosaur*, New York: Gallery Books, 1987, pp. 118–199.

24. Joseph Wallace, *op. cit.*, p. 119; recently, however, geologist Michael Rampino has suggested that the last of Jack Horner's 74-million-year-old nesting dinosaur fauna in Montana (e.g. *Maiasaura*) may have been exterminated by the impact of the bolide creating the 22 mile diameter crater near Manson, Iowa, situated some 900 miles to the east. This impact event has *also* been dated to 74 million years ago. (See *The Planetary Report*, January/February, 1998, pp. 6–11.) Thus, the imagetext of Henderson's painting as published in Wallace's book may be accurate after all.

25. Douglas Henderson, *Asteroid Impact*, New York: Dial Books for Young Readers, 2000, pp. 22–25; Henderson's paintings will doubtless seem stirring to a younger generation of dinophiles (as they are to adults!). Henderson feels the magnitude of the impact has been underestimated by prior artists and paleontologists. During a February 5, 2001, telephone discussion, he claimed the expanding shell of debris and incandescence would have rounder, hemispheric shape. Henderson believes that the predominantly vertical ejecta trajectory, as portrayed in Kish's 1985 composite restoration, isn't quite accurate. The appearance of a six mile asteroid or comet blazing through the earth's atmosphere, Henderson claimed, would be far more dramatic (and blinding) than portrayed by any artist previously.

26. For more on Bonestell, see Ron Miller and Frederick C. Durant III, *Worlds Beyond: The Art of Chesley Bonestell*, Norfolk/Virginia Beach: Donning, 1983.

Chapter 25

1. John's special art exhibit featuring paleoart from his private collection was on display for six months at Chicago's Field Museum of Natural History. For more on John's collection, see "Realm of the Dinosaur—the incredible Dinosaur Art Collection of John Lanzendorf," by Allen A. Debus and Gary R. Williams, in *Prehistoric Times*, 1997, no. 23, pp. 40–41.

2. John R. Horner and Don Lessem, *The Complete T. rex*, New York: Simon & Schuster, 1993, p. 11.

3. A startling painting, "Impending Doom" (1999), by artist Bob Nicholls blends the idea of the dinosaurs' demise with the idea of a comet or asteroid collision. Nicholls' trio of "rexes" seems influenced by discovery of "Sue," as if the fate of an individual dinosaur, "Sue," was no less a momentous event in time than the end of an era. Such was the media impact of Sue's discovery and symbolism.

4. Dougal Dixon, *The New Dinosaurs: An Alternative Evolution*, Topsfield, Mass.: Salem House Publishers, 1988, p. 75.

5. Robert J. Sawyer, "Just Like Old Times," in *Dinosaur Fantastic*, New York: Daw Books, 1993, Mike Resnick & Martin H. Greenberg (eds.), pp. 13–26.

6. Besides Michael Crichton's *Jurassic Park* (1990), for example, there is O'Neil De Nous' "Tyrannous and Strong," in *Asimov's Science Fiction*, February 2000, pp. 66–81.

7. Another aspect to the matter of accurately restoring the nonavian theropod, *T. rex*, is the body posture. Gregory S. Paul states, "Today the debate is whether [bipedal dinosaurs] walked and ran with their bodies nearly horizontal like birds, or tilted up about 20 degrees.... Most theropods ... did not have retroverted hips.... Most bipedal dinosaurs, therefore walked and ran with, at most, a slight upward tilt to the body." Gregory S. Paul, "Restoring the Life Appearances of Dinosaurs," in *The Scientific American Book of Dinosaurs*, New York: St. Martin's Press, 2000, p. 82.

8. Chris Brochu lecture, "The Rex Files," at the Field Museum of Natural History on May 30, 1999, chronicled in *Dinosaur World*, Winter 1999/2000, no. 7, p. 16. Also see "*T. Rex* was Fierce, Yes, but Feathered, Too," by Tim Appenzeller, in *Science*, September 24, 1999, vol. 285, pp. 2052–2053. Paleoartist Michael Skrepnick painted the first restoration of a (juvenile) feathered *T. rex*, published in *National Geographic*, November 1999, vol. 196, no. 5, pp. 98–99.

9. The first time we saw this illustration, dated 1979, in a popular magazine, there was one tyrannosaur (*Albertosaurus*) and one depicted styracosaur (*Monoclonius*). But Paul continually adds to his illustrations, and in a more recent 1985 rendition there were two tyrannosaurs and more horned dinosaurs. The 1985 version appears on pp. 16–17 of "The Science and Art of Restoring the Life Appearance of Dinosaurs and their Relatives—A Rigorous How-to Guide," by Gregory S. Paul, in *Dinosaurs Past and Present*, Vol. II, Seattle: Natural History Museum of Los Angeles County in association with University of Washington Press, 1987, Sylvia J. Czerkas and Everett C. Olson (eds.). Paul added further details to this illustration in 1988.

10. Paul's influential running *T. Rex* was published on page 26 of Gregory S. Paul, *op. cit.*, 1987.

11. Kish's 1985 painting was recently published in *Encyclopedia of Dinosaurs: A Visual Guide to Prehistoric Life with Over 140 Dinosaur Profiles*, New York: Beekman House, 1990, p. 236.

12. A recent topic of debate regarding tyrannosaurids concerns whether they had lips. William L. Abler claims, "If the armamentarium of tyrannosaurs did include septic oral flora, we can postulate other characteristics of its anatomy. To help maintain a moist environment for its single-celled guests, tyrannosaurs probably had lips that closed tightly, as well as thick, spongy gums that covered the teeth. When tyrannosaurs ate, pressure between teeth and gums might have cut the latter, causing them to bleed. The blood in turn may have been a source of nourishment for the septic dental bacteria." "The Teeth of the Tyrannosaurs," in *The Scientific American Book of Dinosaurs*, *op. cit.*, 2000, p. 276. While *T. Rexes* are ordinarily restored with their mouths open to make them appear scarier, when their mouths are closed, traditionally their dagger-like teeth jut from their jaws. Stephen Czerkas has sculpted one of the first closed-mouth *T. Rexes* shown with lips that hide the teeth. A photograph of his creation appears on page. 8 of *Dinosaur Imagery: The Lanzendorf Collection*, San Diego: Academic Press, 2000.

13. Gregory S. Paul, *Predatory Dinosaurs of the World*, New York: Simon & Schuster, 1988, p. 28.

14. Gregory S. Paul's painting of *Quetzalcoatlus* and *Albertosaurus* first appeared in Michael Parrish's *Smithsonian* article, "The Big Guy's Back, Flying High Again and Creating a Flap," March 1986, pp. 73–81. A decade later this scene was memorialized in a model kit diorama by sculptor Bob Morales. Bakker's rendition on the "Rodan/Godzilla-esque" theme appeared in *The Dinosaur Heresies*, New York: William Morrow and Company, Inc., 1986, p. 274. David Peters' 1996 painting was published on page 142 of *Dinosaur Imagery: The Lanzendorf Collection*, San Diego: Academic Press, 2000. *Quetzalcoatlus* must be regarded as one of the principal characters of the "Dinosaur Renaissance," as its discovery, interpretation and restoration coincided with the 1970s and early 1980s media frenzy concerning dinosaur paleontology.

15. See chapter 4, "The Image of *T. rex*: The Making of a Monster," in John R. Horner and Don Lessem, *op. cit.*, 1993), pp. 79–97. The quotation is from p. 97.

16. James O. Farlow, Matt B. Smith and John M. Robinson, "Body Mass, Bone 'Strength Indicator,' and Cursorial Potential of *Tyrannosaurus Rex*," in *Journal of Vertebrate Paleontology*, vol. 15, no. 4, December 27, 1995, pp. 713–725; Paleontologist Per Christiansen believes, based on his studies of limb proportions in tyrannosaurs, that *T. rex* could sprint at speeds approaching 30 miles per hour. Gregory M. Erickson, "Breathing Life Into *Tyrannosaurus rex*," in *The Scientific American Book of Dinosaurs*, *op. cit.*, 2000, p. 274.

17. Skrepnick's painting was recently published in Louis Psihoyos' (with John Knoebber) *Hunting Dinosaurs*, New York: Random House, 1994, pp. 234–235. Rebecca R. Hanna, however, believes that *T. rex* Sue lived a long while after suffering injuries now evident in its skeletal abnormalities. Rebecca R. Hanna, "Dinosaurs Got Hurt Too," in *The Scientific American Book of Dinosaurs*, *op. cit.*, 2000, p. 122. There is particular fascination with "Sue." Paleoartist William Stout has also recently painted "Sue." Stout claims his painting, "...is the very first picture ... of a specific dinosaur.... It is not just a picture of a Tyrannosaurus; it is a depiction of that particular T. rex. I consulted with the Field Museum of Natural History paleontologists and got a list of Sue's pathologies, her injuries, and incorporated them into my reconstruction...." Stout quoted from *Prehistoric Times*, Dec. 2000/Jan. 2001, no. 45, p. 28.

18. Jeff Hecht, "Common as Muck: Yesterday's Rock Stars are Now a Bunch of Has-Beens," in *New Scientist*, October 14, 2000, p. 18; as discussed by Hecht, John Horner recently found a "graveyard" of five *T. rexes* in Montana.

19. Steve Fiffer, *Tyrannosaurus Sue*, New York: W. H. Freeman and Company, 2000; Allen A. Debus and Diane E. Debus, "Like When Sabertooths Fight—South Dakota's Dinosaurs: From the Days of Custer to the Reign of 'Sue'," in *Dino saur World*, Winter 1999/2000, no. 7, pp. 7–17.

20. Chris Brochu quoted from his May 30, 1999, Field Museum lecture, "The Rex Files." A revision to *T. rex's* olfactory condition was announced in August 2001, when Lawrence Witmer declared that the nostrils, when correctly positioned in restorations, should be situated in front, lower and closer to the front lip, over the rostral area, and not as traditionally portrayed further up on the snout. Michael Skrepnick's restorations illustrated Witmer's article, "Nostril position in Dinosaurs and other vertebrates and its significance for nasal function," *Science*, August 3, 2001, vol. 293, pp. 850–852.

21. Rainer Zangerl, "Dinosaurs, Predator and Prey—The *Gorgosaurus* and *Lambeosaurus* Exhibit in Chicago Natural History Museum," Chicago: Chicago Natural History Museum Press, 1956; Interestingly, in Zangerl's booklet it is stated that the pose adapted for "Gorgy" was, "...while the predator was looking it over, something disturbed him; he has reared up startled." This is the scenario and inspiration captured in John Gurche's 2000 painting of "Sue." The Field Museum's *Gorgosaurus* specimen, once renamed *Albertosaurus* (1970) and then subsequently *Daspletosaurus* (1999), has been (more accurately) reassembled with the spinal column mounted horizontally, instead of vertically with tail waving rather than dragging limply behind.

22. For half a year, two brachiosaurs adorned the Field Museum premises, one standing inside and the other outdoors. This second fiberglass skeleton, fortified with an ultraviolet-resistant coating, was first displayed in August of 1999 on the Museum's eastern grounds to lure tourists in from Lake Shore Drive. In late 2000, an impressive 137-foot-long *Seismosaurus* skeleton (based on *Diplodocus*) invaded Chicago's Dino Fest with little fanfare. Nevertheless, the emblematic dinosaur for Chicago's grand Dino Fest event was, you guessed it, a *T. rex*!

23. Allen A. Debus and Steve Brusatte, "Long Live the Queen! *T. rex* Reigns at the Field," in *Fossil News: Journal of Avocational Paleontology*, September 2000, vol. 6, no. 9, pp. 14–21.

24. It is another Black Hills Institute discovery, the *T. rex* known as "Stan," which is purportedly larger than "Sue." According to Lessem, there were at least six genera of theropod dinosaurs which measure longer than *T. rex*, including the longest of all—at 65 feet—enigmatic *Bruthkayosaurus*. See *Bigger Than T. rex: The Discovery of the Biggest Meat-Eating Dinosaur Ever Found*, New York: Crown Publishers, Inc., 1997, by Don Lessem.

25. Mike Milbourne, "Giganotosaurus Not Rex Enough," in *Prehistoric Times*, August/September 1997, no. 25, pp. 40–41; also see Letters from Don Lessem, Phil Currie and Paul Sereno in *Prehistoric Times*, September/October 1997, no. 26, p. 6, and letter from Allen A. Debus in *Prehistoric Times*, December/January 1997/98, no. 27, p. 3. Our predilection with the *size* and *ferocity* of *T. rex* may be traced to the founding fathers' erection of the American Mastodon as a huge, savage and fierce symbol of national identity. The Mastodon was originally viewed by many as a savage carnivore, the *largest* fossil species known! Echoing the themes of yesteryear, *T. rex* has replaced the Mastodon as our favored prehistoric symbol of national identity. See Paul Semonin's *American Monster: How the Nation's First Prehistoric Creature Became a Symbol of National Identity*, New York: New York University Press, 2000.

26. "Bigger Than Rex," in *Daily Herald*, Saturday, March 11, 2000, pp. 1–3; also see Tony Campagna's interview with Phil Currie in *Prehistoric Times*, Oct./Nov. 2001, no. 44, p. 57. Currie states, "We have only one single bone which indicates there may be an animal in there that is bigger than Giganotosaurus, and I emphasize just ONE bone. So the press picks up on these things of course and makes it into a big deal but until we get some more bones from that animal we can't be 100 percent sure that it is a bigger animal...." Theropods are the most celebrated of dinosaurs both in the media and in scientific circles. Humans, like theropods, have predatory instincts, so we can relate to the theropod "persona" more easily than we can to the "cow-like" nature of dinosaurian herbivores. Yet beyond their blood-thirsty appeal, theropods were an evolutionary diverse group of animals, which is why many (but not all) scientists tend to find them more scientifically intriguing than herbivorous dinosaurs. Their apparent evolutionary diversity increases the likelihood that newly discovered dinosaurs will be theropods. It is also interesting how *T. rex* has become both a cult dinosaur and a cultural icon. It's sort of like in baseball when Roger Maris and, later Hank Aaron challenged Babe Ruth's homerun records. The "Babe" was the all time favorite because he was *the* baseball icon. Rex and Ruth supporters grumble loudly when history is about to be made. Yet years later after Ruth's record fell, Roger Maris still hasn't been admitted to the Baseball Hall of Fame. Today, many *T. rex*-loving children aren't even aware of *Giganotosaurus*.

Chapter 26

1. Personal correspondence from Greg Van Wie dated November 17, 1992, with illustrated newsclipping (dated "circa 1958") titled, "Dells Builds Prehistoric Animal Park."

2. Information supplied by Greg Van Wie.

3. Personal communication from Carol Witt-Smith dated August 1999.

4. Klaus Westphal interviewed in *Dinosaur World*, Spring/Summer 2000, no. 8, p. 91.

5. Klaus tells an amusing tale concerning the Geology Museum's *Edmontosaurus*. "Tour guides tell the story about how the whole herd of edmontosaurs vanished in South Dakota so many millions of years ago, and were drowned and their bones were pulled apart over a two mile wide area. And then the guide would say, 'Okay, incidentally, do you think the dinosaur is a male or female?' And every parent, teacher and student has to really think about this. The guide asks for a show of hands but usually the teachers do not have an opinion. So the guide asks the teacher, say, 'Mrs. Johnson, what do you think?' And the teacher, under pressure, guesses, 'I think it is a "she."' Then the guide reveals that they're all wrong, because it is both 'she' and 'he,' because it is a composite of both sexes. The tour guide always says something like, 'Even Mrs. Johnson was wrong.' And the grade school students are amazed that their teacher could be wrong. But they understand why, and it's like a high wattage light bulb going on in their minds. You know those kids 'get it.' But one year, I was taking a confirmation class here from our church, comprising a pastor leading a group of teenagers. And I asked the same questions about the *Edmontosaurus*. And you know how teenagers are about these sorts of things at that age. They weren't easily participating. So you know the question of whether it's a 'he' or a 'she' is dangerous territory. When I asked it, the pastor doesn't say anything. The kids are looking around everywhere, quite embarrassed, at all the other exhibits. But there was one 11-year-old who replied, 'It's a "he," right?' And I couldn't say either way, because it's neither, but I didn't want to give away the answer yet. So instead I asked, 'Why do you think that?' And he walked up slowly to the skeleton and pointed to the ribcage and said, 'Is he missing a rib?' And the pastor smiled." Quote from *Ibid*, p. 95.

6. Rolf E. Johnson and John H. Ostrom, "The Forelimb of *Torosaurus* and an Analysis of the Posture and Gait of Ceratopsian Dinosaurs," in *Functional Morphology in Vertebrate Paleontology*, J. Thomason, ed., Cambridge: Cambridge University Press, 1995, pp. 205–218.

7. Peter Sheehan quoted from interview in *Dinosaur World*, Spring/Summer 2000, no. 8, p. 79.

8. According to Richard Rush, personal communication dated January 19, 2000, the first *T. rex* cast was delivered to The Children's Museum of Indianapolis in 1976. A photograph of this model, displayed outdoors, can be seen in *Where to Find Dinosaurs Today*, New York: A Puffin Unicorn, 1992, p. 83. The following year a second cast was made for the Natural Science Center in Greensboro, North Carolina. Milwaukee's cast was the third.

9. Rev. H. N. Hutchinson, *Extinct Monsters and Creatures of Other Days: A Popular Account of Some of the Larger Forms of Ancient Animal Life*, New York: D. Appleton and Company, 1917 (see Plate XXIX); Henry R. Knipe, *Nebula to Man*, New York: J. B. Lippincott Company, 1905.

10. Personal correspondence from Don Glut dated October 20, 1999.

11. Personal correspondence from Mark Zimmerman dated December 17, 1999.

Chapter 27

1. Sereno quoted in *Chicago Tribune*, August 11, 2000, section 2, p. 3;

Sereno's standing as the "Indiana Jones" among paleontologists has been perpetuated through articles such as Don Lessem's "Dinosaur Man—No Bones About It: The 'Indiana Jones' of Paleontology Is the University of Chicago's Paul Sereno," in *Chicago Tribune Magazine*, June 14, 1992, p. 34. Fellow paleoartist Gary Staab has said of Sereno, "The great thing about Paul is that he realizes the importance of the artistic impact of a restoration, as well as the need for it to be anatomically correct. He also ... can stand far enough back from a project to allow me to do the things which let me declare ownership of composition of the sculpture and call it mine, while giving critical analysis of its structure." Staab quoted in "Dinosaur World Interviews Paleosculptor Gary Staab," in *Dinosaur World*, Winter 2000/2001, no. 9, p. 98.

2. One of Sereno's first professional dinosaur restorations (1992) depicted a lithe *Eoraptor* chasing down a small protomammal, *Probelesodon*.

3. Paul Sereno, "Early Cretaceous Dinosaurs from the Sahara," in *Science*, October 14, 1994, vol. 226, pp. 267–271; surprisingly, skeletal anatomy evident in the new African dinosaur fauna reflects characteristics of northern Laurasian dinosaurs rather than other dinosaurs thought to be distinctive of Gondwanaland. It is possible that during the Late Jurassic, as Pangaea separated, land connections extended between Gondwanaland and Laurasia, allowing migrations of dinosaurs through Europe.

4. In an interview, Sereno claimed, "It was fun working with Garfield. He has a real knack to bring out the details in the skin. He is an artist that works well with a scientist." *Dinosaur World*, Spring 1998, no. 4, p. 12.

5. Paul Sereno, "Dinosaurs and Drifting Continents," in *Natural History*, January 1995, pp. 40–47.

6. *Dinosaur Imagery: The Lanzendorf Collection*, San Diego: Academic Press, 2000, Foreword by Philip J. Currie, pp. 72–73.

7. Paul Sereno, "Africa's Dinosaur Castaways," in *National Geographic*, June 1996, pp. 106–119.

8. Paul Sereno, et. al., "Predatory Dinosaurs from the Sahara and Late Cretaceous Faunal Differentiation," in *Science*, May 17, 1996, vol. 272, pp. 986–991; *Carcharodontosaurus*' skull profile appeared more highly arched toward the middle than in the case of its more famous North American cousin, *T. rex*. An expansive antorbital fenestra, the opening in the skull in front of the eye socket, comprises nearly one third of the skull's overall length.

9. Paul Sereno, "Africa's Dinosaur Castaways," in *National Geographic*, June 1996, pp. 106–119.

10. Sereno quoted in "The PT Interview: Paul Sereno," by Tony Campagna, in *Prehistoric Times*, Aug/Sept. 1998, no. 31, p. 38.

11. Allen A. Debus, "Dinosaurs Come to Life at Chicago's Navy Pier," in *Fossil News*, March 1999, vol. 5, no. 3, pp. 16–18.

12. *Colossal Claw* was originally broadcast on November 15, 1998, three days after the discovery was formally announced with great fanfare.

13. Paul Sereno, et. al. "A Long-Snouted Predatory Dinosaur from Africa and the Evolution of Spinosaurids," in *Science*, November 13, 1998, vol. 282, pp. 1298–1302.

14. Jack Bowman Bailey, "Neural Spine Elongation in Dinosaurs: Sailbacks or Buffalobacks?" in *Journal of Paleontology*, vol. 71, no. 6, 1997, pp. 1124–1146. Unfortunately, remains of *Spinosaurus* were destroyed during World War II. Although, until recently, restored with a *Tyrannosaurus* sort of head, *Spinosaurus*' skull may have more closely resembled that of the British *Baryonyx*, which had a head described (superficially in appearance) as "crocodolian." Also see entry for *Spinosaurus*, pp. 329–333, in Donald F. Glut's *Dinosaurs. The Encyclopedia* (Supplement 1), Jefferson, NC: McFarland & Company, Inc., Publishers, 2000.

15. Bailey quoted in "Dinosaur World Interviews—Dr. Jack Bowman Bailey," in *Dinosaur World*, Summer/Fall, 1998, no. 5, p. 71.

16. William Stout, *The New Dinosaurs*, New York: A Byron Preiss Book, 2000; Stout's new version of *Spinosaurus*, illustrating "Prehistoric Times Interview: William Stout," first appeared on page 37 of *Prehistoric Times*, Oct./Nov. 2000, no. 44.

17. Gregory S. Paul, "Restoring the Life Appearances of Dinosaurs," in *The Scientific American Book of Dinosaurs*, New York: St. Martin's Press, 2000 (Gregory S. Paul ed.), p. 82.

18. Kenneth Carpenter, *Eggs, Nests, and Baby Dinosaurs: A Look at Dinosaur Reproduction*, Bloomington: Indiana University Press, 1999, p. 60.

19. When this chapter was written, in December 2000, rumors began to circulate that a *Spinosaurus* would augment the cast of *Jurassic Park III*, fighting a bloody battle to the death with *T. rex* in a summer 2001 scheduled release. This indeed proved to be the case.

20. Another hundred million years later and nature spawned the modern gavial crocodile, *Gavialis*, now an endangered species. When that last hundred or so remaining *Gavialis* pass on into obscurity, that toothy reptilian countenance so well adapted to spearing fish will have faded into Time's mists. It will be forever beyond our capacities to revivify ancient spirits. Yet, shapes are shifting, ephemeral and recyclable. *Gavialis* may fade into extinction, but *we* will not witness the resurrection of the *Rutiodon* or *Suchomimus* in its next convergent evolutionary reincarnation.

21. Allen A. Debus and Diane E. Debus, "Joe Who? *Jobaria*," in *Fossil News: Journal of Avocational Paleontology*, June 2000, vol. 6, no. 6, pp. 5–8; a credits sign mentions that the skeletal designs are Paul Sereno's creation and that the skeletal mountings were prepared by P.A.S.T. Educational materials and outreach are a product of Gabrielle Lyon's and Dr. Sereno's efforts. The exhibit was created by Project Exploration.

22. Paul Sereno, et. al., "Cretaceous Sauropods from the Sahara and the Uneven Rate of Skeletal Evolution Among Dinosaurs," in *Science*, November 12, 1999, vol. 286, pp. 1342–1347.

23. Michael Trcic's *Jobaria* statue won the first "Lanze" (pronounced "Lanzee") prize at the Society of Vertebrate Paleontology 2000 meeting in Mexico City.

24. See Phil Currie's note in *Prehistoric Times*, Oct./Nov. 2000, no. 44, p. 57. A new South American Carcharodontosaurid may be larger than *Giganotosaurus*.

25. Paul Sereno, et. al, "Early Cretaceous Dinosaurs from the Sahara," in *Science*, October 14, 1994, vol. 266, p. 269.

26. Paul C. Sereno, "The Evolution of Dinosaurs," in *Science*, June 25, 1999, vol. 284, pp. 2137–2147. For a more popular perspective on the nature of cladistics as a phylogenetic tool in classifying dinosaur species, see "Classification and Evolution of the Dinosaur Groups" by Thomas R. Holtz, Jr., in *The Scientific American Book of Dinosaurs*, New York: St. Martin's Press, 2000 (Gregory S. Paul, ed.), pp. 140–168.

27. Paul Sereno, *op. cit.*, November 12, 1999.

Chapter 28

1. Thomas H. Huxley, "On *Hypsilophodon foxii*, a New Dinosaurian from the Wealden of the Isle of Wight," in *The Quarterly Journal of the Geological Society of London*, 1870—read November 10, 1869, vol. xxvi, pp. 3–12.

2. Thomas H. Huxley, "Further Evidence of the Affinity Between the Dinosaurian Reptiles and Birds," *Ibid.*, p. 472.

3. James W. Hulke, "An Attempt at a Complete Osteology of *Hypsilophodon foxii*," in *Philosophical Transactions of the Royal Society of London*, 1883, vol. 173, pp. 1035–1062, with Plates 71–82. Barlow's reconstruction appears as Plate 82 in this publication.

4. Rev. H. N. Hutchinson, *Extinct Monsters and Creatures of Other Days: A Popular Account of Some of the Larger Forms of Ancient Animal Life*, 1917 ed., New York: D. Appleton and Company, p. 172; Darren Naish, "Tree-Climbing Dinosaurs," in *Quarterly* (The Quarterly Journal of the Dinosaur Society), 2000, vol. 4, no. 1, pp. 20–23.

5. Francis Baron Nopsca, "Notes on British Dinosaurs: Part 1—*Hypsilophodon*," in *Geological Magazine*, May 1905, vol. 5, no. 2 p. 205.

6. Adrian J. Desmond, *The Hot-Blooded Dinosaurs: A Revolution in Palaeontology*, New York: Warner Books, 1975, pp. 66–67.

7. Othenio Abel, *Geschichte und Methode der Rekonstruktion Verzeitlicher Wirbeltiere*, Jena: Gustave Fischer, 1925, pp. 183–191; Alfred Brehm, *Die Saugtiere*, Leipzig and Wien: Bibliographilches Institut, 1912.

8. Gerhard Heilmann, *The Origin of Birds*, London: Witherby, 1926.

9. William E. Swinton, "Notes on the Osteology of *Hypsilophodon*, and on the family Hypsilophontidae," in *Proceedings of the General Meetings for Scientific Business of the Zoological Society of London*, 1936, pp. 555–578; sculptor Vernon Edwards' restoration of tree-climbing *Hypsilophodon* is discussed in William E. Swinton's article, "A New Exhibit of *Hypsilophodon*," in *Natural History Magazine* (The Trustees of the British Museum), 1936, vol. 5, pp. 331–336.

10. Peter M. Galton, "The Mode of Life of *Hypsilophodon*, the Supposedly Arboreal Ornithopod Dinosaur," in *Lethaia*, 1971, vol. 4, 453–464.

11. Adrian J. Desmond, *op. cit.*, 1975, figure 30, p. 114.

12. John R. Horner, "Ecologic and Behavioral Implications Derived from a Dinosaur Nesting Site," in *Dinosaurs Past and Present*, Volume II, Seattle: Natural History Museum of Los Angeles County in association with University of Washington Press, 1997, p. 62, figure 14.

13. J. R. Horner and D. B. Weishampel, "Letters to Nature—A Comparative Embryological Study of Two Ornithischian Dinosaurs," in *Nature*, September 5, 1996, vol. 383, p. 103.

14. Personal communication from Douglas Henderson dated December 7, 2000.

15. Patricia Vickers-Rich and Thomas Hewitt Rich, "Australia's Polar Dinosaurs," in *Scientific American*, July 1993, pp. 50–55.

16. Paul E. Fischer, Dale A. Russell, et. al., "Cardiovascular Evidence for an Intermediate or Higher Metabolic Rate in an Ornithischian Dinosaur," in *Science*, April 21, 2000, vol. 288, pp. 503–505; According to Erik Stokstad's comments in *Science*, however, there is suspicion that the fossil heart is instead an ironstone concretion. Erik Stokstad, "Doubts Raised About Dinosaur Heart," in *Science*, February 2, 2001, vol. 291, p. 811. Is this specimen a genuine heart fossil? The jury is still out.

17. Vertebrate paleontologist John A. Long illustrated one of the earliest restorations of a feathered polar hypsilophodont, *Leaellynasaura*, appearing in *Dinonews*, November 1990, no. 2, p. 7.

18. Kenneth Carpenter, *Eggs, Nests and Baby Dinosaurs: A Look at Dinosaur Reproduction*, Bloomington: Indiana University Press, 1999, p. 65.

Chapter 29

1. A schematic illustration of the brontosaur's innards was recently published in Francis B. Messmore's article, "The History of the 'World a Million Years Ago,'" in *Dinosaur World*, October 1997, no. 3, pp. 27–32. Also see p. 40 in Donald F. Glut's *The Dinosaur Scrapbook*, Secaucus, NJ: The Citadel Press, 1980; "Just Mechanics and Brains," in *Opportunity*, 1923, p. 31, 51, 55–56; "Lifelike Dinosaur," in *Science and Invention*, October 1925, p.510; "Moving Model of a Brontosaur," in *The Literary Digest*, November 21, 1925, vol. 87, no. 8, pp. 22–23; "Hidden Motors Give Life to Prehistoric Monsters," in *Popular Science Monthly*, June 1933, vol. 122, no. 6, p. 38; and "Ancient Monster Lives Again in Moving Model," in *Popular Mechanics*, vol. 45, January 1926, p. 43.

2. Francis B. Messmore, *op. cit.*, 1997.

3. "Mechanical Monster 'Eats' Girl on Movie Stage," in *Popular Science Monthly*, October 1931, vol. 119, no. 10; "Life-like Ancient Monsters That Move and Roar," in *American Weekly*, 1935; Edwin W. Teale, "Giant Metal 'Earth' Shows Life Forms of All Ages," in *Popular Science Monthly*, June 1932, vol. 120, no. 6, pp. 16–17; H. W. Magee, "The World a Million Years Ago," in *Popular Mechanics*, 1932, vol. 58, pp. 824–828.

4. George H. Messmore, "Head Movement for a Device Simulating a Dinosaur or Other Animal," *United States Patent Office*, # 1,898,587, patented February 21, 1933, Serial no. 552,885.

5. A photo of the Nazi brontosaur "apparition" was recently printed in Francis B. Messmore, *op. cit.*, 1997.

6. Donald F. Glut, *Carbon Dates*, Jefferson, NC: McFarland & Company Inc., Publishers, 1999, p. 216.

7. Today some of the robots reside in storehouses in New York, Indianapolis and Japan. Donald F. Glut happens to own a *World a Million Years Ago* pterodactyl and cave bear that still work!

8. Marilyn Gillespie quoted in "High Stakes Dinosaur: The Las Vegas 'Robotosaurs,'" by Allen A. Debus and Diane E. Debus, in *Dinosaur World*, Spring/Summer 1999, no. 6, pp. 9–11.

Chapter 30

1. *Dinosaurs: Next Exit*, Cleval Video, Ltd., 1994.

2. Riff Smith, "A Trip Back in Time: A Journey to Prehistoric Gardens," in *Prehistoric Times*, Dec./Jan. 98/99, no. 33, pp. 6–7.

3. Molly Walker, "Celebrating 40 Years of Building a Boyhood Dream: Ernie Nelson's Love of Prehistoric Gardens," in *Rogue Coast*, August 25, 1993, pp. 13–18.

4. We suspect the inspirational publication may have been Barnum Brown's article, "Hunting Big Game of

Other Days," in *The National Geographic Magazine*, May 1919, vol. XXXV, no. 5, pp. 407–429. Ernie claimed to have been influenced by American Museum paleontologists and displays.

5. Personal correspondence from Riff Smith dated December 26, 1998.

6. Donald F. Glut, *The Dinosaur Scrapbook*, Secaucus, NJ: The Citadel Press, 1980, p. 262.

7. E-mail communication from Jack Arata dated October 2, 1998. Personal communication with Kiki Halcumb, November 1999.

8. Paleoartist Ernie V. Nelson passed away on January 17, 1999, at the age of 91. In late 1999 came rumors that Prehistoric Gardens might be forced into closure, and that the sculptures would be removed or demolished. A campaign was inaugurated by Kiki to save the Gardens for posterity.

Chapter 31

1. Donald F. Glut wrote a science fiction novel, *Spawn* (New York: Laser Books, no. 43, 1976), involving a dinosaur theme park for dinosaurs discovered on another planet. As in *Jurassic Park*, the park's owners lose control of the dinosaurs. Don autographed our copy with the inscription, "To Al and Diane—Here's the 'original' 'Jurassic Park' ... sort of, Don Glut, 10/7/97." William J. T. Mitchell discusses the theme of "eating" in the context of Hawkins' 1854 dinosaur park statues and in relation to the dinosaurs of *Jurassic Park* in *The Last Dinosaur Book: The Life and Times of a Cultural Icon*, Chicago: The University of Chicago Press, 1998. A "real" *Jurassic Park* theme park now exists at Universal Studios in Los Angeles. See "Jurassic Park—The Ride," by James Van Hise, in *Prehistoric Times*, April/May 1999, no. 35, pp. 49–51.

2. Josef Augusta, *The Age of Monsters*, London: Paul Hamlyn, Ltd., 1966, p. 31.

3. Adrienne Mayor, *The First Fossil Hunters: Paleontology in Greek and Roman Times*, Princeton: Princeton University Press, 2000.

4. Allan D. Griesemer, Roger A. Vandiver and Michael Voorhies, "A Dinosaur Revival," in *University of Nebraska–Lincoln Museum Notes*, September 22, 1976, vol. 56, no. 6.

5. Allan D. Griesemer, Roger A. Vandiver and Michael Voorhies, *op. cit.*, September 22, 1976, vol. 56, no. 6.

6. Personal communication from Doug Dunn, March 9, 2000. Also see *The Explorer*, vol. 3, no. 1, January 1961; and "Museum Notes" (The Cleveland Museum of Natural History), August/September 1960, vol. 2, nos. 8–9, pp. 103–106. A picture of the ceratosaur appears in each of these publications, as well as in Donald F. Glut's *The Dinosaur Scrapbook*, Secaucus, NJ: The Citadel Press, 1980, p. 271.

7. See chapter 13 in William J. T. Mitchell, *op. cit.*, 1998. Interestingly, there is also visual and written documentation indicating that Benjamin Waterhouse Hawkins envisioned pterodactyles as genuine "dragons."

8. The authors borrowed this phrase from the title of a popular video produced by Cleval Video, Ltd., 1994.

9. John Paul Gries, *Roadside Geology of South Dakota*, Missoula, MT: Mountain Press Publishing Company, 1996, p. 314.

10. This is the fourth such full-sized sculptural display of these battling legends that we are aware of. Another striking pairing may be found in Domke's Dinosaur Gardens, while Douglas Van Howd's bronze *T. rex/Triceratops* sculpture is displayed outside the Los Angeles County Museum of Natural History.

11. South Dakota's world famous drugstore now comprising much of Wall's city commerce first opened in 1931. However, not until its owners, Dorothy and Ted Hustead, realized they could attract motorists to their doorstep with billboard signage advertising free glasses of ice water did the Wall Drug Store business really take off. For us, though, the paraphernalia available in the Wall Drug was of lesser interest than the famous dinosaur. So, in the late 1950s Sullivan was called upon again to design and guide the construction of the dinosaur, along with assistance from the Brady Engineering company and another sculptor named Adrian Forette. Equipped with electric bulb eyes, which glow ominously in the dark, few passing motorists have missed this spectacle. A decade later, in 1968, a planned extension of I-90 threatened the dinosaur. So, rather than demolishing it, the Wall Drug owners decided to move the cement dinosaur to its current position at a cost of $6,000. The big sauropod, standing on fully erect, columnar legs conceived before Bakker's time, has a chain link fence built around it. Good thing, for kids could seriously injure themselves if they climbed up its tail and tumbled to the ground. Sullivan built a number of other dinosaur statues as well, but none so famous or as influential as his signature Dinosaur Park grouping.

12. Allen A. Debus and Diane E. Debus, "Like When Sabertooths Fight ... South Dakota's Dinosaurs: From the Days of Custer to the Reign of 'Sue,'" in *Dinosaur World*, Winter 1999/2000, no. 7.

13. For more information on Rapid City's "Dinosaur Park," see Donald F. Glut's *The Dinosaur Scrapbook*, Secaucus, NJ: The Citadel Press, 1980.

14. A relatively new dinosaur theme park, "Dinosaur World," near Plant City and Tampa, Florida, has about 100 life-sized prehistoric animal figures on display. See "Introducing ... Dinosaur World, the Park," by Daniel Searfoss, in *Dinosaur World*, Spring/Summer 1999, no. 6, pp. 88–89.

15. Stuart Baldwin and Beverly Halstead, *Dinosaur Stamps of the World*, Witham, Essex: Baldwin's Books, 1991.

16. Allen A. Debus, "Standing Against the Hurricane—Cuba's Dinosaurs!" in *Fossil News: Journal of Avocational Paleontology*, December 2000, vol. 6, no. 12, pp. 8–9; Gary R. Williams and Allen A. Debus, "An Island of Dinosaurs: Cuba's Valle de la Prehistoria," in *Dinosaur World*, Winter 2000/2001, no. 9, pp. 12–14; Personal correspondence from Rogelio Oran Fontaine dated July 20, 1998, translated by Brunilda Debus.

17. Allen A. Debus and Diane E. Debus, "Domke's Dinos: Exploring Ancient Gardens," in *Dinosaur World*, Summer/Fall 1998, no. 5, pp. 7–9; Carolyn Kraus, "The Art of Extinction," in *Metropolitan Detroit*, October 1997, pp. 110–113, 167.

18. The brochure does, however, fail to mention a dreaded element lurking around every bend in the trail past the park's 25 paleo-scenes. Considering visitors are hiking through beautiful, lush woodlands, they will tangle with a variety of characters, some made of concrete and others of a living nature. None so relentless as the vicious mosquitoes, not yet hermetically sealed in amber, which welcome us in full strength. One is thrilled to the core and, yes, bitten to the bone!

19. Another religiously themed dinosaur park is being built at the Answers in Genesis museum under construction near Louisville, Kentucky. See "Attack of

the Missionary Lizards," by Christopher Kemp, in *Louisville Eccentric Observer*, September 20, 2000, pp. 14–16.

20. This description is based on Allen's visit with Gary R. Williams in August 2000.

21. Gary Williams, "The Canadian Connection: Canada's 'Hidden' Prehistoric World," in *The Prehistoric Times*, Nov./Dec. 1995, no. 9, p. 18.

22. Zofia Kielan-Jaworowska, *Hunting for Dinosaurs*, MIT Press, 1969.

23. Edwin H. Colbert, *Dinosaurs: An Illustrated History*, New York: Hammond Inc., 1983, p.133

24. Karol Sabath, "Poland's 'Valley of the Dinosaurs,'" in *Dinosaur World*, Spring/Summer 1999, no. 6, p. 35.

25. See Society of Vertebrate Paleontology *News Bulletin*, June 1988, no. 143, pp. 25–26; also, personal communication from Franz Gruss dated October 31, 1992; Allen A. Debus, "Sculpting Dinosaurs," in *Dinonews* (Western Australian Museum), December 1992, no. 5, pp. 14–16.

26. "Franz Gruss—One of Germany's Premier Sculptors of Prehistoric Life, an Appreciation and Interview," in *Dinosaur World*, Winter 1999/2000, no. 7, pp. 35–38.

27. The largest dinosaur statues, relative to their natural size, would be sculptor Claude Bell's *Tyrannosaurus* and "Dinny" (i.e. *Apatosaurus*), near Cabazon, California. See Daniel Cohen and Susan Cohen, *Where to Find Dinosaurs Today*, New York: A Puffin Unicorn, 1992), pp. 177–178.

28. "Take a Walk on the Wild Side in Ogden's Eccles Dinosaur Park and Museum" (Management and Training Corp., 1996).

29. This modern scene would seem to be inspired by one of John Martin's 1830s fantasies or Benjamin Waterhouse Hawkins' Crystal Palace sculptures.

30. So Dinosaurierpark may be viewed as a much grander version of Connecticut's Dinosaur State Park, which has also adapted the enclosed structure for preserving dinosaur trackways and incorporated several life restorations (e.g. *Anchisaurus* and *Dilophosaurus*) to inform visitors about the animals that walked in ancient sediments.

31. Barbara A. Slater, "A Visit to Munchehagen, Germany's Dinosaurierpark," in *Dinosaur World*, Summer/Fall 1998, no. 5, pp. 72–73.

32. Personal communication to Gary Williams from Mr. Kuchnio dated January 20, 1999.

33. "Building Life-Size Creatures from Past Times—Interview with Polish Sculptor Kryzystof Kuchnio," in *Dinosaur World*, Spring/Summer 1999, no. 6, pp. 37–40.

34. *Dinosaur World*, no. 6 interview, *op. cit.*, 1999.

35. Jane Stern and Michael Stern, *The Encyclopedia of Bad Taste*, New York: Harper Collins, 1990, pp. 97–98.

36. Eventually, newer dinosaur theme park displays will doubtless reflect the dinosaur "feather revolution" already in motion, even though feathers are trickier to sculpt in cement than are scales. There are numerous dinosaur theme parks situated globally, far too many to describe here in sufficient detail.

Chapter 32

1. Michael A. Taylor, "Before the Dinosaur: The Historical Significance of the Fossil Marine Reptiles," in *Ancient Marine Reptiles*, San Diego: Academic Press, 1997, Jack M. Calloway (ed.) , p. xxviii.

2. Scale models of the "Pterodactyle"—9" by 9", *Megalosaurus*—23" by 12", *Iguanodon*—19" by 14", *Labyrinthodon*—13" by 8", and *Ichthyosaurus with Plesiosauri*—22" by 14" were advertised on pp. 80 to 81 of the 1866 edition of Ward's Catalog.

3. Knight's miniature sculptures include *Dimetrodon* (two styles), Mastodon (two styles), *Agathaumas*, *Naosaurus*, *Allosaurus*, *Stegosaurus*, *Edaphosaurus*, *T. rex*, *Triceratops*, *Diplodocus*, *Apatosaurus*, *Hadrosaurus*, Cave Bear, *Smilodon*, Irish Elk, Mammoth and a Neandertal Man.

4. Charles Whitney Gilmore, *The National Cyclopaedia of American Biography*, New York: James T. White and Co., 1948, vol. 34, pp 362–363.

5. Barbara L. Narendra, "From the Archives: The Early Years of the Great Hall," in *Discovery* (Magazine of the Yale Peabody Museum of Natural History), 1990, vol. 22, no. 1, pp. 18–23.

6. Mike Fredericks and Riff Smith, "Messmore and Damon's 'World a Million Years Ago Toys,'" in *Prehistoric Times*, Dec./Jan. 2001, no. 45, pp. 48–49.

7. *This Week Magazine* (author not credited) July 27, 1958, p.9; for more on the history of prehistoric animal figures popular during this period, see "J. H. Miller—Revisited," by Jeff Quinn, Ron Tanner, Jack Arata, Rick Koch, et. al., in *Prehistoric Times*, September/October 1997, no. 26, pp. 28–30; and Dana Cain's and Mike Fredericks' *Dinosaur Collectibles*, Dubuque: Antique Trader Books, 1999, chapter titled "Marx Plastic Figures and Prehistoric Playsets," by Joe DeMarco, pp. 68–79.

8. Eric Stokstad, "Fossils Made to Order," in *Science*, June 9, 2000, vol. 288, p. 1729. Reflecting on the future state of the dinosaur art and sculptural industry, paleosculptor Gary Staab predicts that, "I see programs like *Walking with Dinosaurs* and see the digital realm expanding as the technology becomes more approachable. In a way these programs may make people like me, who push clay around, go extinct." Staab interviewed in *Dinosaur World*, Winter 2000/2001, no. 9, p. 97.

Chapter 33

1. Mantell quoted in Dennis Dean, *Gideon Mantell and the Discovery of Dinosaurs*, Cambridge, England: Cambridge University Press, 1999, p. 111.

2. Gideon Mantell, *Geology of the South East of England*, London, 1833; also see chapter six in Dennis Dean, *Ibid.*, for more information about Mantell's early discovery and description of the *Hylaeosaurus*.

3. Gideon Mantell, *Petrifactions and Their Teachings; or, a Hand-Book to the Gallery of Organic Remains of the British Museum*, London: Bohn, 1851, p. 323.

4. *Ibid.*, p. 321.

5. Adrian J. Desmond, *The Hot-Blooded Dinosaurs: A Revolution in Palaeontology*, New York: Warner Books, 1975, pp. 20–25.

6. In 1833 Mantell had also considered alternatives, that the spines were situated on *Hylaeosaurus*' tail, or as they had been found naturally, in situ. See Dennis Dean, *op. cit.*, p. 122. Writing in his "Petrifactions" of 1851, Mantell stated, "Hylaeosaurus' ... body was broader than high, and terminated in a long slender, flexible tail; the limbs were relatively short; the skin was studded with scutes and tubercles; and a row of very large, angular spines extended down the back and formed a serrated dermal crest." (pp. 327–328).

7. Dennis Dean, *op. cit.*, p. 123.

8. Gideon Mantell, *Wonders of*

Geology: or, a Familiar Exposition of Geological Phenomena. Being the Substance of a Course of Lectures Delivered at Brighton, from Notes Taken by G. F. Richardson, Curator of the Mantellian Museum (2 volumes), London: Relfe and Fletcher, 1838, p. 369.

9. Rev. H. N. Hutchinson, *Extinct Monsters: A Popular Account of Some of the Larger Forms of Ancient Animal Life*, London: Chapman & Hall, Ltd., 1893 ed., pp. v–vi; Besides sculpting this elusive dinosaur, Benjamin Waterhouse Hawkins also illustrated the *Hylaeosaurus*, as in his painting for Princeton University titled *Jurassic Life of Europe* (undated, yet circa 1875–77).

10. *Ibid.*, pp. 103–104, Plate IX.

11. Donald F. Glut, *The New Dinosaur Dictionary*, Secaucus, New Jersey: The Citadel Press, 1982, pp. 148–149, 201.

12. George Olshevsky, *The Ankylosaur Family Tree* (wall chart by George Olshevsky), 1979. Interestingly, George Olshevsky claimed that Fox's manuscript was lost and it was Thomas Henry Huxley who authored the first published account of the *Polacanthus* in 1867, with J. Hulke naming the type specimen *Polacanthus foxii* in 1882. Evidently, however, Richard Owen now gains credit for having named *P. foxii* through two *anonymously* written1865 articles, one published in the *Illustrated London News* (vol. 47, September 16, 1865, p. 270) and the second in *Geological Magazine* (vol. 2, p. 432). Fox's paper, in which he named what amounted to the posterior end of his dinosaur, titled "On a New Wealden Saurian Named *Polacanthus*," appeared in *Report of the British Association for the Advancement of Science*, 1866, vol. 35, p. 56.

13. Baron Franz Nopsca, "Notes on British Dinosaurs—Part II: *Polacanthus*," in *Geological Magazine*, June 1905, vol. II, no. 6, pp. 241–250; Barlow's *Polacanthus* reconstruction, appearing as Plate XII in Nopsca's article, was the first published skeletal reconstruction of the *Polacanthus*.

14. Many of the early restorations were published in Othenio Abel's *Geschichte und Methode der Rekonstruktion Verzeitlicher Wirbeltiere*, Jena: Gustav Fischer, 1925, pp. 213–216. Two more recent striking and contrasting restorations of *Polacanthus* are those by Maurice Wilson, published as Plate 17 in W. E. Swinton's, *Fossil Amphibians and Reptiles* (London: Trustees of the British Museum [Natural History], 1965 ed.), and by Luis Rey (1994), recently appearing on the cover of *Quarterly* (The Dinosaur Society of the UK Quarterly Magazine), Winter 1995/96, vol. 1, no. 4. While Wilson's *Polacanthus* derived from Caleb Barlow's reconstruction, is posed in a static position, Rey's restoration shows a spiny-looking fellow sporting lateral spines (as well as the traditional array) being attacked by four charging megalosaurs.

15. Edwin H. Colbert, *Dinosaurs: Their Discovery and Their World*, New York: E. P. Dutton & Co., Inc., Plate 76.

16. George Olshevsky, *op. cit.*

17. Personal communication from George Olshevsky, November 2000.

18. David Norman, *The Illustrated Encyclopedia of Dinosaurs*, New York: Crescent Books, 1985, pp. 160–161. In 1990 Coombs and Maryanska echoed this opinion, stating, "there are no homologous elements in the type specimens of *H. armatus* and '*P. foxii*,' so the question of synonymy will never be fully resolved." W. P. Coombs, Jr. and T. Maryanska, "Ankylosauria," in *The Dinosauria*, D. B. Weishampel, P. Dodson and H. Osmolska (eds.), Berkeley: University of California Press, pp. 456–483.

19. Dougal Dixon, et. al., *The Macmillan Illustrated Encyclopedia of Dinosaurs and Prehistoric Animals: A Visual Who's Who of Prehistoric Life*, New York: Macmillan Publishing Company, 1988, pp. 158–161.

20. William T. Blows, "The Armoured Dinosaur *Polacanthus foxii* from the Lower Cretaceous of the Isle of Wight," in *Paleontology*, 1987, vol. 30, pp. 557–580; William Blows, *Geological Magazine*, 1996, vol. 133, no. 6, pp. 671–682. The dermal armor of the new species described in 1996 has characteristically ankylosaurian roof shaped scutes with tall dorsal keels, and scapular spines were situated near the dorsal border and are not hollow. The morphology of caudal vertebrae differs considerably, as well, between the two *Polacanthus* species.

21. Personal communication received from William Blows, 1995.

22. Javier Pereda-Suberbiola, "*Hylaeosaurus, Polacanthus*, and the Systematics and Stratigraphy of Wealden Armoured Dinosaurs," in *Geological Magazine*, 1993, vol. 130, pp. 767–781.

Chapter 34

1. This specimen was described by Sir Everard Home. A valuable reference concerning the early history of ichthyosaur discoveries is S. R. Howe, T. Sharpe and H. S. Torrens, *Ichthyosaurs: A History of Fossil 'Sea-dragons'* (National Museum of Wales, 1981), 32 pp. For more on Mary Anning, see "The Three Mary Annings," by William A. S. Sarjeant, in *Wat on Earth* (A Newsletter for the Earth Sciences—University of Waterloo), Spring 2000, vol. 13, no. 2, pp. 16–17.

2. William D. Conybeare, "On the Discovery of an Almost Perfect Skeleton of the *Plesiosaurus*," in *Transactions of the Geological Society of London*, 1824, vol. 2, no. 1, pp. 382–389.

3. Michael A. Taylor, "Before the Dinosaur: The Historical Significance of the Fossil Marine Reptiles," in *Ancient Marine Reptiles*, San Diego: Academic Press, 1997, Jack M. Calloway (ed.), p. xxviii.

4. Martin J. S. Rudwick, *Scenes from Deep Time: Early Pictorial Representations of the Prehistoric World*, Chicago: The University of Chicago Press, 1992, pp. 44–48.

5. Louis Figuier, *The World Before the Deluge*, London: Cassell and Company, Ltd., 1867 ed., Plate XV.

6. Adrian J. Desmond, *The Hot-Blooded Dinosaurs: A Revolution in Palaeontology*, New York: Warner Books, 1975, p. 229.

7. Martin J. S. Rudwick, *op. cit.*, 1992, p. 153.

8. Adrian J. Desmond, *op. cit*, 1975, p. 143.

9. Stephen J. Gould, "Bent Out of Shape: The Odd Tail Bend of the Marine Fossil Ichthyosaur Is a Clue to Its Reptilian Roots," in *Natural History*, May 1990, pp. 12–25.

10. Josef Augusta, *Prehistoric Sea Monsters*, London: Paul Hamlyn, Ltd., 1964, pp. 16–24.

11. This topic is discussed further by Christopher McGowan in *Dinosaurs, Spitfires and Sea Dragons*, Cambridge: Harvard University Press, 1991.

12. *Walking with Dinosaurs* (BBC Video, 1999); also see Tim Haines, *Walking with Dinosaurs: A Natural History*, New York: Dorling Kindersley Publishing, Inc., 1999, pp. 123–126.

13. Christopher McGowan, "The Ichthyosaurian Tailbend: A Verification Problem Facilitated by Computed Tomography," in *Paleobiology*, 1989, vol. 15, no. 4, pp. 429–436.

14. Christopher McGowan, "Problematic Ichthyosaurs from Southwestern England: A Question of Authenticity," in

Journal of Vertebrate Paleontology, vol. 10, no. 1, March 1990, pp. 72–79.

15. Christopher McGowan, *op. cit.*, 1991.

16. The earliest forms may have evolved an undulatory swimming style. The biomechanics of swimming, and ichthyosaurian body shape and dorsal fin morphologies, are discussed by Ryosuke Motani, "Rulers of the Jurassic Seas," in *Scientific American*, December 2000, pp. 56–57.

17. Samson's restoration was published in Christopher McGowan, *op. cit.*, 1991.

18. Judy A. Massare, "Ancient Mariners," in *Natural History*, September, 1992, pp. 49–52.

19. Henry N. Hutchinson, *Extinct Monsters: A Popular Account of Some of the Larger Forms of Ancient Animal Life*, London: Chapman and Hall, 1893 ed., pp. 264–266 and Plate II; Sylvia M. Czerkas and Donald F. Glut, *Dinosaurs, Mammoths and Cavemen: The Art of Charles R. Knight*, New York: E. P. Dutton, Inc., 1982, p. 47.

20. For instance, this view continues to be popularized in Camp's posthumously published booklet, *Child of the Rocks: The Story of the Berlin-Ichthyosaur State Park*, Special Publication 5 (Nevada Bureau of Mines and Geology in cooperation with Nevada Division of State Parks and Nevada Natural History Association, 1981). According to paleontologist Brad Kosch, "Almost all popular information ... was based on popular articles written by Dr. Camp in the 1950s during the first year or two of the excavations, and have persisted ever since." Personal correspondence dated December 12, 2000.

21. Personal communication from Brad Kosch dated September 27, 2000.

22. Charles L. Camp, "Vorlaufige Mitteilung uber Grosse Ichthyosaurier aus der Oberen Trias von Nevada," in *Osterr Akad. Wiss., Sitzber, Math.-Naturwiss. Kl., Abt. I.*, vol. 185, no. 57 (1976), pp. 125–134; Camp's reconstruction appeared on p. 130.

23. Bradley F. Kosch, "A Revision of the Skeletal Reconstruction of *Shonisaurus Popularis* (*Reptilia Ichthyosauria*)," in *Journal of Vertebrate Paleontology*, December 1990, vol. 10, no. 4, pp. 512–514; Kosch's revised reconstruction appears on p. 513.

24. Bradley F. Kosch, *op. cit.*, 1990; Camp's more widely known paper is "Large Ichthyosaurs from the Upper Triassic of Nevada," in *Paleontographica*, 1980, Abteilung A., vol. 170, pp. 130–201. However, Camp's reconstruction did not appear in this publication.

25. Bradley F. Kosch, *op. cit.*, 1990.

26. Christopher McGowan, *op. cit.*, 1991.

27. Bradley F. Kosch, *op. cit.*, 1990.

28. John Long, "Sea Monsters from the Age of Dinosaurs: The Largest Marine Reptiles That Ever Lived," in *Dinonews* (Western Australian Museum), no. 3, September 1991, pp. 6–7.

29. Sylvia J. Czerkas and Stephen A. Czerkas, *Dinosaurs: A Global View*, New York: Mallard Press, 1990, pp. 96–97.

30. Sylvia J. Czerkas and Stephen A. Czerkas, *op. cit.*, p. 95.

31. Personal correspondence from Brad Kosch dated December12, 2000.

32. P. Martin Sander, "The Paleobiogeography of *Shastasaurus*," in *Ancient Marine Reptiles*, San Diego: Academic Press, 1987, Jack M. Calloway (ed.), p. 17.

33. Chris McGowan, "Giant Ichthyosaurs of the Early Jurassic," in *Canadian Journal of Earth Sciences*, 1996, vol. 33, pp. 1011–1021; incidentally, *Temnodontosaurus platyodon* is the formal name of the first ichthyosaur specimen found (by Mary Anning).

34. Chris McGowan, "Transitional Ichthyosaur Fauna," in *Ancient Marine Reptiles*, San Diego: Academic Press, 1987, Jack M. Calloway (ed.), p. 77.

35. To a certain degree, America's predilection with *size* of fossil species representing our prehistoric heritage may be viewed as a vestige of the eighteenth century founding fathers' concern with Georges Louis Leclerc Buffon's (1707–1788) theory of American *degeneration* of wildlife. Buffon's idea proved insulting to America's nationalistic pride. Even though Buffon's theory was refuted well over a century ago, people can still get caught up in the fervor over which country's dinosaurs are bigger. For more on Buffon and species degeneration fueling our early quest for big fossils (such as our Mastodon), see Paul Semonin, *American Monster: How the Nation's First Prehistoric Creature Became a Symbol of National Identity*, New York: New York University Press, 2000.

Chapter 35

1. When standing in tripodal position, as some paleontologists and artists have imaginatively suggested would be possible, *Diplodocus* could peer into the eleventh floor of a high rise building. Such a hypothetical image appeared on the first page of a November 1898 edition of the *New York Journal*. This image was recently reproduced in William J. T. Mitchell, *The Last Dinosaur Book: The Life and Times of a Cultural Icon*, Chicago: The University of Chicago Press, 1998, p. 158.

2. Paleoartist Gregory S. Paul has referred to *Jurassic Park*'s brachiosaurs as "fat-limbed." "This is a mistake. Slow-moving sauropods and stegosaurs should have had long, lightly muscled legs, like those of elephants." Gregory S. Paul, "Restoring the Life Appearances of Dinosaurs," in *The Scientific American Book of Dinosaurs*, New York: St. Martin's Press, 2000, Gregory S. Paul (ed.), p. 87.

3. Thomas J. Holtz claims that, "...the creatures that most immediately come to mind when the word 'dinosaur' is mentioned are the giant sauropods." Thomas J. Holtz, "Classification and Evolution of the Dinosaur Groups," in *The Scientific American Book of Dinosaurs*, *op. cit.*, p. 158. A symbolic "face-off" between the biggest of the theropods vs. the most titanic of sauropods (based on relatively complete and studied fossil material) will be displayed at the Fernback Museum of Natural History in Atlanta, Georgia, where visitors will see mounted skeletons of *Giganotosaurus* and *Argentinosaurus*. See Michael J. Porter's interview, "Giganotosaurus at the Fernback Museum in Atlanta," in *Prehistoric Times*, Dec./Jan. 2001, vol. 45, pp. 30–31.

4. Adrian J. Desmond, *The Hot-Blooded Dinosaurs: A Revolution in Palaeontology*, New York: Warner Books, 1975, pp. 142–150, 181; Desmond states that, "Owen was unable to make the break from traditional ideas. Even though his monster was endowed with massive limbs to support its weight, Owen never really allowed the creature to leave the water that had been its home for so long." (p. 150) A modern representation of the *Cetiosaurus*, based on the skeletal reconstruction of a new reasonably complete specimen from Morocco, may be found on p. 133 of *Dinosaurs: A Global View*, New York: Mallard Press, 1991.

5. One of the earliest pictorial representations of the whale-like cetiosaur was in Allen Debus' small sculpture, "Historical Sauropods" (1993). A photograph of this creature was published in Allen's article, "Mysterious Giants: Historical Sauropods," in *The Dinosaur Report* (a

quarterly publication of the Dinosaur Society), Spring 1994, pp. 8–9.

6. Desmond, *op. cit.*, pp. 148–149; Thomas H. Huxley, "Further Evidence of the Affinity Between the Dinosaurian Reptiles and Birds," in *Quarterly Journal of the Geological Society of London*, 1870, vol. 26, pp. 3–12.

7. Roy Chapman Andrews, *On the Trail of Ancient Man*, New York: G. P. Putnam & Sons, 1926, p. 242.

8. David Norman, *Dinosaur!*, New York: Prentice Hall, 1991, pp. 68–69. Of his new dinosaur, Cope claimed, as opposed to Owen's vision of the aquatic *Cetiosaurus*, that *Camarasaurus* "suggests a resemblance in form and habits between those huge reptiles and the giraffe." Cope quoted in Mark Jaffe's, *The Gilded Dinosaur: The Fossil War Between E. D. Cope and O. C. Marsh and the Rise of American Science*, New York: Crown Publishers, 2000, p. 192.

9. The use of miniature scale models for determining brachiosaur weights is discussed in Per Christiansen, "Dinosaur Biomechanics," in *The Scientific American Book of Dinosaurs*, New York: St. Martin's Press, 2000, Gregory S. Paul (ed.), p. 65. The merit of using toy models versus more accurate miniature representations for determining weights is discussed on p. 70 of Christiansen's article. Here Christiansen claims, "Because the commercial [museum gift shop] model is too bulky, it produces an excessive mass estimate." In order to judge the relative strength of sauropod limbs, weights of the "living" dinosaurs must be accurately known.

10. The earliest known example is Cope's sketch reproduced on page 302 of Henry F. Osborn's biography, *Cope: Master Naturalist*, New Jersey: Princeton University Press, 1931. Of this, Desmond, *op. cit* (p. 324), claims, "Osborn's date for this sketch [1878] seems too late, since the sauropods are depicted with diminutive front limbs, and were obviously thought to be bipeds. In 1877, Cope was already aware of the massive limbs of *Camarasaurus*." Other early restorations of interest include Charles R. Knight's restorations of the *Amphicoelias altus*, illustrating William H. Ballou's article, "Strange Creatures of Other Days," in *The Century Magazine*, 1897, vol. LV, p. 20; and an unidentified genus recently reproduced in Sylvia M. Czerkas' and Donald F. Glut's *Dinosaurs, Mammoths and Cavemen: The Art of Charles R. Knight*, New York: E. P. Dutton, Inc., 1982, p. 42.

11. Hay wrote, "It is difficult to understand why an animal whose immediate ancestors must have walked about in a crocodile manner ... an animal which could by means of its long neck reach up from the bottom many feet to the surface and from the surface many feet to the bottom—why such a reptile should need ... the ability to walk along the river bottoms like a mammal." Oliver Hay quoted in Desmond, *op. cit.*, p. 175.

12. W. J. Holland, "A Review of Some Recent Criticisms of the Restorations of Sauropod Dinosaurs Existing in the Museums of the United States, with Special Reference to that of *Diplodocus carnegiei* in the Carnegie Museum," in *The American Naturalist*, May 1910, vol. XLIV, no. 521, pp. 259–283. Holland's use of reconstructions and whimsical discussion did much to diminish the accuracy of the sprawling, crocodilian sauropod image.

13. Roland T. Bird, "Did Brontosaurus Ever Walk on Land?" in *Natural History*, 1944, vol. 53, pp. 61–67; Roland T. Bird, "Thunder in His Footsteps," in *Natural History*, May 1939, vol. 48, pp. 255–261, 302.

14. Paleoartist Zdenek Burian's 1941 *Brachiosaurus* restoration is the classic example. See Josef Augusta, *Prehistoric Animals*, London: Spring Books, 1956, Plate 26.

15. Edwin H. Colbert, *Dinosaurs: Their Discovery and Their World*, New York: E. P. Dutton & Co., Inc., 1961, p. 104.

16. Robert T. Bakker, "The Superiority of Dinosaurs," in *Discovery* (Magazine of the Yale Peabody Museum of Natural History), Spring 1968, vol. 3, no. 2, p. 15.

17. Robert Bakker, *The Dinosaur Heresies*, New York: William Morrow and Company, Inc., pp. 115–118.

18. *Ibid.*, p. 141. In a recently finished sculpture, forensic anatomist Richard Neave has restored *Diplodocus*' head with a walrus-like, fleshy upper lip. See pp. 8–9 of "Dinosaur Society Conference Report," in *Quarterly Review* (The UK Dinosaur Society Quarterly Magazine), 2000, vol. 3, no. 5.

19. See Henderson's restoration in *Discover*, March 1987, p. 27.

20. Ellison's restoration was published in *National Geographic*, December 1998, vol. 194, no. 6, p. 41.

21. Guy Leahy, "Noses, Lungs and Guts," in *The Scientific American Book of Dinosaurs*, New York: St. Martin's Press, 2000, *op. cit.*, pp. 52–63.

22. Edwin H. Colbert, *Dinosaurs: An Illustrated History*, Maplewood, New Jersey: Hammond Incorporated, 1983, pp. 156, 182.

23. Robert Bakker, *op. cit.*, 1986, chapter 9.

24. Diane Debus' recounting of Bakker's October 5, 1996, lecture at "Raptors to Rex" exhibition. Diane Debus, "Enraptored with Bakker: Previewing Raptors to Rex," in *Dinosaur World*, October 1997, vol. 1, no. 3, pp. 6–7.

25. Gurche's restoration was recently reproduced on p. 71 of Louis Psihoyos' (with John Knoebber) *Hunting Dinosaurs*, New York: Random House, 1994.

26. Per Christiansen, "Dinosaur Biomechanics," in *The Scientific American Book of Dinosaurs*, New York: St. Martin's Press, 2000, *op. cit.*, pp. 73–74.

27. Carl Zimmer, "Dinosaurs in Motion," in *Discover*, November 1997, pp. 96–109; Kent A. Stevens and J. Michael Parrish, "Neck Posture and Feeding Habits of Two Jurassic Sauropod Dinosaurs," in *Science*, April 30, 1999, vol. 284, pp. 798–799.

28. Gregory S. Paul, "Restoring the Life Appearances of Dinosaurs," in *The Scientific American Book of Dinosaurs*, New York: St. Martin's Press, 2000, *op. cit.*, pp. 92–93.

29. *Ibid.*

30. Bakker, *op. cit.*, 1986, p. 267; Robert Bakker, "Bite of the Bronto: Revenge of the Herbivores," in *Earth Magazine*, November 1994.

31. N. P. Myhrvold and P. J. Currie, "Supersonic Sauropods? Tail Dynamics in the Diplodocids," in *Paleobiology*, 1997, vol. 23, pp. 393–409.

32. Stephen A. Czerkas, "Discovery of Dermal Spines Reveals a New Look for Sauropod Dinosaurs," in *Geology*, December 1992, vol. 20, pp. 1068–1070.

33. George Olshevsky, *Amargasaurus*, in *The Dinosaur Report* (A quarterly publication of the Dinosaur Society), Summer 1992, pp. 2–3; personal communication from paleoartist Brian Franczak dated September 18, 1993.

34. Louis Jacobs, *Quest for the African Dinosaurs*, New York: Villard Books, 1993, chapter 4.

35. Stephen A. Czerkas, "Skin," in *Encyclopedia of Dinosaurs*, Phillip J. Currie and Kevin Padian (eds.), San Diego, California: Academic Press, p. 670.

36. Luis M. Chiappe, Rodolfo Coria, Lowell Dingus, et. al., "Sauropod Dinosaur Embryos from the Late Cretaceous

of Patagonia," in *Nature*, November 19, 1998, vol. 396, pp. 258–261.

37. Gregory S. Paul, "Art and Restoration: Big Sauropods—Really, Really Big Sauropods," in *The Dinosaur Report* (a quarterly publication of the Dinosaur Society), Fall 1994, pp. 12–13. On the basis of a single complete trunk vertebrae found in 1878 in Late Jurassic deposits, Paul makes his case that the largest sauropod of all was *Amphicoelias fragillimus*. He estimated the genus to have weighed 100 to 150 tons and measured up to 60 meters (over 180 feet) in length!

38. *Walking with Dinosaurs*, BBC Video, 1999.

39. Kent Stevens, "I Walked with a Dinosaur!" in *Quarterly* (The UK Dinosaur Society Quarterly Magazine), 2000, vol. 3, no. 4, pp. 10–11.

40. Quotations from "Dinosaur Docudrama Blends Fact, Fantasy," by Andrew Watson and Erik Stokstad, in *Science*, April 7, 2000, vol. 288, pp. 29–31.

41. *Ibid.*

42. Robert Nicholls, "*Diplodocus longus*—'The Real Double Beam,'" in *Down to Earth*, May 2000, vol. 31, p. 5.

43. *Ibid.*

44. Jay Leno, February 3, 2000.

Chapter 36

1. Allen G. Debus, *Man and Nature in the Renaissance*, Cambridge: Cambridge University Press, 1978, p. 2.

2. This restoration is held in the collection of the American Museum of Natural History.

3. Robert T. Bakker, "Dinosaur Renaissance," in *Scientific American*, April 1975, pp. 58–79.

4. *Ibid.*, p. 58.

5. In *The Scientific American Book of Dinosaurs* New York: St. Martin's Press, Gregory S. Paul, ed., 2000, p. 3, paleoartist Gregory S. Paul has confessed how great was the influence of seeing Sarah Landry's restorations: "But perhaps packing the most punch of all in that article—to a degree that is difficult to comprehend today—was the speculative sketch of a feathered Syntarsus dashing down the slope of a dune in hot pursuit of a gliding lizard, only one of its feet barely touching the sand."

6. Loris Russell, "Body Temperature of Dinosaurs and Its Relationship to Their Extinction," *Journal of Paleontology*, May 1965, vol. 39, no. 3, pp. 497–501; Russell wrote, "We might be justified, therefore, in exploring the possibility that dinosaurs differed physiologically in various ways from modern reptiles and that one of the results was a very different reaction to environmental temperature. Such would have been the case if the dinosaurs had been ... [warm blooded] ... even if only imperfectly so." (p. 497).

7. John H. Ostrom, "Osteology of *Deinonychus antirrhopus*, an Unusual Theropod from the Lower Cretaceous of Montana" (Bulletin 30), New Haven: Peabody Museum of Natural History—Yale University, July 1969.

8. John H. Ostrom, "The Supporting Chain," in *Discovery* (Magazine of the Peabody Museum of Natural History), Fall 1969, vol. 5, no. 1, p. 14. Of Bakker's restoration, paleontologist Michael Benton claimed in 2000, "Bakker's vision of *Deinonychus* in life was so startling, that the news spread. Pictures were published widely in popular science magazines and newspapers, generating new excitement." Benton, "A Brief History of Dinosaur Paleontology," in *The Scientific American Book of Dinosaurs*, New York: St. Martin's Press, *op. cit.*, p. 35.

9. Robert T. Bakker, "The Superiority of Dinosaurs," in *Discovery* (Magazine of the Peabody Museum of Natural History), Spring 1968, vol. 3, no. 2, pp. 11–22.

10. Adrian J. Desmond, *The Hot-Blooded Dinosaurs: A Revolution in Palaeontology*, New York: Warner Books, 1975.

11. In particular, the story of engineer Paul MacCready's attempt to create a fully functioning life-size flying restoration of the 40-foot-wing-span pterosaur, *Quetzalcoatlus*, was sensationalized by the media during the mid–1980s. Patrick Cooke, "The Man Who Launched a Dinosaur," in *Science 86*, April 1986, pp. 27–35; Michael Parrish, "The Big Guy's Back, Flying High Again and Creating a Flap," in *Smithsonian*, March 1986, pp. 73–81. Here MacCready claimed, "I'm trying to combine the good features of a museum and a zoo. The zoo has only the present day creatures. The museum shows old ones, but they're dead.... You put a giant pterodactyl on the TV screen or in the newspaper, and it's just so exciting that people can't help but be interested in finding out more." (p. 74); Wann Langston, Jr., "Pterosaurs," in *Scientific American*, February 1981, pp. 122–136.

12. Luis W. Alvarez, Walter Alvarez, Frank Asaro, Helen V. Michel, "Extraterrestrial Cause for the Cretaceous-Tertiary Extinction," in *Science*, vol. 208, no. 4448, June 6, 1980, pp. 1095–1108.

13. Richard Muller, *Nemesis—The Death Star*, New York: Weidenfeld & Nicholson, 1988.

14. The discovery inspired a very popular children's book, beautifully illustrated by Douglas Henderson, titled *Maia: A Dinosaur Grows Up* (Philadelphia: Running Press, 1985), by John R. Horner and James Gorman. This book traced the life of a real female *Maiasaura* from hatchling stage through harrowing experiences (surviving attacks from predators such as *Troodon* and *Albertosaurus*, and through a migration phase). Eventually "Maia," perhaps the dinosaurian equivalent of Walt Disney's "Bambi," must learn to defend *her* own nest from marauding predators.

15. Stephen J. Gould, *The Panda's Thumb: More Reflections in Natural History*, New York: W. W. Norton & Company, 1980, pp. 261–262, 269.

16. Personal communication from William Stout dated December 2000. William Stout's restoration of mating *Parasaurolophus* appearing in a 1981 book, *The Dinosaurs: A Fantastic New View of a Lost Era* (New York: Mallard Press), reflected the sexual frenzy of America's 1970s "sexual revolution."

17. Personal communication from Don Glut December 2000.

18. John M. Harris, "Introduction" to Volume I of *Dinosaurs Past and Present*, Seattle: Natural History Museum of Los Angeles County in association with University of Washington Press, Sylvia J. Czerkas and Everett C. Olson (eds.), 1986, pp. 1–3.

19. During the 1980s and 1990s it was common for the most celebrated paleoartists to participate in interview sessions with popular paleo-mag contributors. For Bakker, see "Up Close with Dr. Robert Bakker," interviewed by Steve Brusatte in *Dinosaur World*, Winter 1999/2000, no. 7, pp. 68–70. For Kish, see "The PT Interview—Ely Kish," interviewed by Tony Campagna in *Prehistoric Times*, Oct./Nov. 1999, no. 38, pp. 9–10. For Paul, see "The PT Interview—Gregory S. Paul, Parts 1 and 2," interviewed by Bob Morales in *Prehistoric Times*, April/May 1999, no. 35, pp. 9–11 and June/July 1999, no. 36, pp. 9–11. For Hallett, see "Interview—Mark Hallett," interviewed by Mike Fredericks in *Prehistoric Times*, Oct./Nov. 1998, no. 32, p. 7. For Stout, see "The PT Interview—

William Stout, Parts 1 and 2," interviewed by Mike Fredericks in *Prehistoric Times*, Dec./Jan. 1997–98, no. 27, pp. 29–31, and April/May 1998, no. 28/29, pp. 45–47; and "Prehistoric Times Interview—William Stout, Parts 1 and 2," interviewed by Rick Savage in *Prehistoric Times*, Oct./Nov. 2000, no. 44, pp. 34–37, and Dec 2000/January 2001, no. 45, pp. 28–29. For Henderson, "The PT Interview—Douglas Henderson," interviewed by Mike Fredericks in *Prehistoric Times*, August/September 1997, no. 25, pp. 32–35; and "The PT Interview—Douglas Henderson," in *Prehistoric Times*, Dec./Jan. 1998/1999, no. 33, pp. 8–9. For Sibbick, see "John Sibbick—The PT Interview," in *Prehistoric Times*, September/October 1997, no. 26, pp. 22– 25; also see Sibbick's article, "Experiencing Paleoworlds," in *Dinosaur World*, October 1997, vol. 1, no. 3, pp. 8–11. (Incidentally, in 1995 John Sibbick was selected as the head of the UK Dinosaur Society's artist guild.) For Franczak, see "Artist of the Month—Brian Franczak," in *The Dinoaur Times*, Summer 1993, vol. 1, no.1, pp. 21–24. For Michael Skrepnick, see "Feature Interview with Michael Skrepnick and John Lanzendorf—Parts 1 and 2," interviewed by Allen A. Debus in *Dinosaur World*, Winter 1999/2000, no. 7, pp. 63–67, and Spring/Summer 2000, no. 8, pp. 40–49.

20. For Stephen Czerkas, see "The PT Interview—Stephen and Sylvia Czerkas," interviewed by Cliff Green in *Prehistoric Times*, Aug./Sept. 2000, no. 43, pp. 9–13; and Don Lessem, "Skinning the Dinosaur," in *Discover*, March 1989, pp. 39–42; and Caroline Arnold, "Dino-Mite Dinos," in *Dinosaurus Magazine*, Sept./Oct. 1997, pp. 27–30. Czerkas also had a stellar career as a stop-motion animator for dinosaur science fiction films. Czerkas' dinosaur animations are expertly recounted in Neil Pettigrew's *The Stop-Motion Filmography*, Jefferson, NC: McFarland & Company, Inc., Publishers, 1999. For Thomas, see "David A. Thomas: Tracking the Past in Art," interviewed by Allen A. Debus in *Dinosaur World*, Summer/Fall 1998, no. 5, pp. 49–53; and obituary notice by David Gillette in *Society of Vertebrate Paleontology News Bulletin*, October 1999, no. 177, pp. 54–55. For Trcic, see "Dinosaur Sculptor Supreme: An Interview with Artist Michael Trcic," interviewed by Allen A. Debus in *Dinosaur World*, Spring/Summer 1999, no. 6, pp. 29–32. Bob Morales owns a successful business named "Dragon Attack!," while Shane Foulkes founded "Cretaceous Creations."

21. Gregory S. Paul claimed, "Today, dinosaur restoration is a minor industry, practiced around the world. Necessarily, much of the romance that [Charles R.] Knight enjoyed—having been almost alone in the field—has disappeared." Paul, "The Art of Charles R. Knight," in *The Scientific American Book of Dinosaurs*, New York: St. Martin's Press, Gregory S. Paul (ed.), 2000, p. 116.

22. Michael Crichton, *Jurassic Park*, New York: Alfred A. Knopf, 1990.

23. See *The Making of Jurassic Park*, Ballantine Trade, 1993, by Don Shay and Jody Duncan. Besides Crichton, artist and author James Gurney also became celebrated during the 1990s for his creation of an original fantasy dinosaur menagerie. Gurney was the author of two children's books, *Dinotopia—A Land Apart from Time* (Atlanta: Turner Publishing, 1992) and *Dinotopia—The World Beneath* (Atlanta: Turner Publishing, 1995), concerning the imaginary land of "Dinotopia," with attendant images of realistic looking yet fantasy dinosaurs. See "Dinosaur World interviews Dinotopia's James Gurney," interview by Allen A. Debus, in *Dinosaur World*, October 1997, vol. 1, no. 3, pp. 21–23. Gurney also restored two Mesozoic scenes—Cretaceous and Jurassic—featuring North America's dinosaur genera on a sheet of U.S. Postage stamps in 1996; see James Gurney, "How I Painted the Dinosaur Stamps," in *Prehistoric Times*, March/April 1997, no. 23, pp. 8–9.

24. "The Crisis in Dinosaur Science," in *The Dinosaur Report*, 1992, p. 1.

25. *Ibid.*, p. 2. By 1994, the Artist's Committee had prepared a "Directory" of 81 "dinosaur artists" affiliated with the U.S. Dinosaur Society (even though many of these lived outside the U.S.). Two years later, Tess Kissinger compiled a 90-page Dinosaur Society publication, the first of its kind, "Copyrights, Contracts, Pricing and Ethical Guidelines for Dinosaur Artists and Paleontologists" (1996), offering guidance on how to conduct the business of paleoart. Paleontologist Peter Dodson, Vice President of the Dinosaur society, wrote a Foreword to this publication in which he stated, "If the eyes are the window to the soul, so dinosaur art is the window to the lost world. My professional life is immeasurably richer because of the activities of dinosaur artists who bring to life a world my eyes are too dim to see. Their skills begin precisely where mine end. It is not my professional activity that has drawn the interest of the public, but rather it is my work made accessible through the gifts of the artist. The partnership between the paleontologist and the artist is of long standing and of great importance."

26. *Dinosaur Imagery: The Lanzendorf Collection*, Photography by Michael Tropea, San Diego: Academic Press, 2000. The impact of John's apartment is described in "Realm of the Dinosaur: The Incredible Dinosaur Art Collection of John Lanzendorf," by Allen A. Debus and Gary Williams, in *Prehistoric Times*, March/April 1997, no. 23, pp. 40–42. The first-place winner of the first "Lanze" award—three dimensional art category—was sculptor Michael Trcic for his bronze statue of *Jobaria*.

27. One arena into which dinosaur imagery diversified during this period was "fantasy dinosaur" art. For instance, besides Gurney's "Dinotopians" (which are depicted as "domesticated" dinosaurs), for the first time ever, there was much activity in the art of Godzilla. One particularly striking painting by Jean-Pierre Normand, showing the ice-covered fin of a swimming Godzilla cutting into the ill-fated Titanic's hull, appeared on the cover of *G-Fan*, Jan./Feb. 1998, no. 31. Another imaginative example is Dr. Kenneth Carpenter's illustrated article, "A Dinosaur Paleontologist's View of Godzilla," on pp. 102–106 in *The Official Godzilla Compendium*, by J. D. Lees and Marc Cerasini, New York: Random House, 1998. Also of great interest are popular publications of the 1990s written by artists offering instruction and guidance on how to draw and sculpt prehistoric animals. Some examples include: Sylvia J. Czerkas, " Reconstruction and Restoration," in *Encyclopedia of Dinosaurs*, Philip J. Currie and Kevin Padian (eds.), San Diego: Academic Press, 1997, pp. 626–629; Stephen A. Czerkas, "Skin," in *Encyclopedia of Dinosaurs, op. cit.*, pp. 669–675; Douglas Henderson, "Restoring Dinosaurs as Living Animals," in *The Complete Dinosaur*, James O. Farlow and M. K. Brett-Surman (eds.), Bloomington: Indiana University Press, 1997, pp. 165–172; Allen A. Debus, Bob Morales, and Diane E. Debus, *Dinosaur Sculpting: A Complete Beginners' Guide*, Hanover Park, IL: Hell Creek Creations, 1995;

Tracy Lee Ford, *How to Draw Dinosaurs—Part 1*, DinoHunter, 1999; Gregory S. Paul, "Restoring the Life Appearances of Dinosaurs," in *The Scientific American Book of Dinosaurs, op. cit.*, pp. 78–106; Michael Anderson, "Fleshing out the Past: Reconstructing Fossil Faces," in *Discovery* (Magazine of the Yale Peabody Museum of Natural History), 1990, vol. 22, no. 1, pp. 11–15; Brian Cooley, "Sculpting Dinosaurs," in *Alberta* (Studies in the Arts and Sciences), 1988, vol. 1, no. 1, pp. 119–130.

28. I. Bernard Cohen, *Revolution in Science*, Cambridge: The Belknap Press of Harvard University Press, 1985, pp. 42–47.

29. Personal communication from "Dad," dated January 7, 2001.

30. Martin J. S. Rudwick, *Scenes from Deep Time: Early Pictorial Representations of the Prehistoric World*, Chicago: The University of Chicago Press, 1992, chapter 6—"The Genre Established," and pp. 224–225.

31. *Ibid.*, p. 261.

32. William J. T. Mitchell, *The Last Dinosaur Book: The Life and Times of a Cultural Icon*, Chicago: The University of Chicago Press, 1998, p. 101

33. Stephen J. Gould takes a more conventional, if not commonsensical, perspective than Mitchell in his introduction, "Reconstructing (and Deconstructing) the Past," in *Life: An Illustrated History of the Evolution of Life on Earth*, Stephen J. Gould (General Ed.), New York: W. W. Norton & Company, pp. 6–21.

34. Paleontologist Phil Currie recorded his observation that, following the announcement of the discovery of a feathered dinosaur—*Sinosauropteryx*—in 1996, "*...controversies on the ancestry of birds and warm-bloodedness erupted on a scale that surpassed even public interest in dinosaur extinction!" The Scientific American Book of Dinosaurs, op. cit.*, p. 184.

35. Allen A. Debus, "Sorting Fossil Vertebrate Iconography," in *Fossil News*, November 1997, vol. 3, nos. 10 & 11, pp. 2–5.

36. For more on the power of cladistics in elucidating the phylogeny (i.e. evolutionary family trees) of dinosaur, see "Phylogenetic System," by Kevin Padian, pp. 543–545, in *Encyclopedia of Dinosaurs, op. cit.*

37. On the heels of new discoveries, an entire "flock" of new restorations of Mesozoic birds and feathered dinosaurs resulted during the 1990s. See, for example, Allen A. Debus, "New Wings on the Paleo-Perch," in *Dinosaur World*, June 1997, vol. 1, no. 2, pp. 10–14.

38. See Charles R. Knight's 1920 American Museum of Natural History mural portraying Cro-Magnon Man painting mammoths on cave walls at France's Font-de-Gaume, recently published in Sylvia M. Czerkas and Donald F. Glut, *Dinosaurs, Mammoths, and Cavemen: The Art of Charles R. Knight*, New York: E. P. Dutton, Inc, 1982, pp. 68–69. Reflecting on the inspiration for Knight's painting, Stephen J. Gould wrote about Cro-Magnon cave art in "Up Against a Wall," recently published in *Leonardo's Mountain of Clams and the Diet of Worms*, New York: Three Rivers Press, 1998, pp. 161–178.

Bibliography: Principal Sources for Paleoimagery

Abel, Othenio. *Geschichte und Methode der Rekonstruction Vorzeitlicher Wirbeltiere.* Jena: Gustav Fischer, 1925.

_____. *Lebensbilder aus der Tierwelt der Vorzeit.* Jena: Gustav Fischer, 1927.

Ackerman, Jennifer. "Dinosaurs Taking Wing." *National Geographic* (July 1998): 74–99.

Albritton, Claude C. *The Abyss of Time: Unraveling the Mystery of the Earth's Age.* Los Angeles: Jeremy P. Tarcher, Inc., 1986.

Altangerel, Perle, Mark Norell, Luis M. Chiappe and James Clark. "Flightless Bird from the Cretaceous of Mongolia." *Nature* 362 (April 1993): 623–626.

Alvarez, Luis W., Walter Alvarez, Frank Asaro and Helen V. Michel. "Extraterrestrial Cause for the Cretaceous-Tertiary Extinction." *Science* 208, no. 4448 (June 1980): 1095–1108.

Alvarez, Walter. *T. rex and the Crater of Doom.* Princeton: Princeton University Press, 1997.

Augusta, Josef. *Prehistoric Animals.* London: Spring Books, 1956.

_____. *Prehistoric Man.* London: Paul Hamlyn, Ltd., 1960.

Augusta, Josef. *Prehistoric Reptiles and Birds.* London: Paul Hamlyn, Ltd., 1961.

_____. *Prehistoric Sea Monsters.* London: Paul Hamlyn, Ltd., 1964.

_____. *The Age of Monsters: Prehistoric and Legendary.* London: Paul Hamlyn, Ltd., 1966.

Bailey, Jack Bowman. "Neural Spine Elongation in Dinosaurs: Sailbacks or Buffalobacks?" *Journal of Paleontology* 71, no. 6 (1997): 1124–1146.

_____. (Interviewed by Allen A. Debus) *Dinosaur World* no. 5 (Summer/Fall 1998): 65–71.

Bakker, Robert T. "Dinosaur Renaissance." *Scientific American* (April 1975): 58–78.

_____. *The Dinosaur Heresies.* New York: William Morrow and Company, Inc., 1986.

_____. "Jurassic Sea Monsters." *Discover* (September 1993): 78–85.

_____. "Plesiosaur Extinction Cycles—Events That Mark the Beginning, Middle and End of the Cretaceous." *Evolution of the Western Interior Basin (Geological Association of Canada Special Paper 39)* (1993): 641–664.

_____. (Interviewed by Steve Brusatte) *Dinosaur World* no. 7 (Winter 1999/2000): 68–70.

Baldwin, Stuart, and Beverly Halstead. *Dinosaur Stamps of the World.* Witham, Essex: Baldwin's Books, 1991.

Ballou, William H. "Strange Creatures of the Past." *The Century Magazine* LV (November 1897): 15–23.

_____. "The Serpentlike Sea Saurians." *Popular Science Monthly* 53 (June 1898): 209–225.

Barnett, Lincoln (and the editorial staff of *Life*). *The World We Live In.* New York: Time Incorporated, 1955.

Benton, Michael. "Scientific Methodologies in Collision: The History of the Study of the Extinction of the Dinosaurs." *Evolutionary Biology* 24 (1989): 371–400.

Bird, Roland T. "Thunder in His Footsteps." *Natural History* 48 (May 1939): 255–261, 302.

_____. "Did Brontosaurus Ever Walk on Land?" *Natural History* 53 (1944): 61–67.

Blows, William T. "The Armoured Dinosaur *Polacanthus foxii* from the Lower Cretaceous of the Isle of Wight." *Paleontology* 30 (1987): 557–580.

Brezina, Jiri. "Geology in the Czech Republic." *Geotimes* (August, 1993): 9–10.

Buckland, William. *The Bridgewater Treatises on the Power, Wisdom and Goodness of God as Manifested in the Creation. Treatise VI, Geology and Mineralogy Considered with Reference to Natural Theology.* London: William Pickering, 1837.

Burroughs, Edgar Rice. *Out of Time's Abyss.* New York: Ace Books, 1918.

Cain, Dana, and Mike Fredericks. *Dinosaur Collectibles.* Norfolk, VA: Antique Trader Books, 1999.

Camp, Charles L. "Vorlaufige Mitteilung uber Grosse Ichthyosaurier aus der Oberen Trias von Nevada," *Osterr. Akad. Wiss., Sitzber., Math.-Naturwiss. Kl., Abt. I.*, vol. 185, no. 57 (1976): 125–134.

_____. *Child of the Rocks: The Story of the Berlin-Ichthyosaur State Park.* Special Publication 5, Nevada Bureau of Mines and Geology in cooperation with Nevada Division of state Parks and Nevada natural History Association, 1981.

Carpenter, Kenneth. "A New Look at *Stegosaurus*." *Dinosaur Discoveries* no. 1 (September/October 1996): 6–7. Carpenter, Kenneth, Dale Russell, Donald Baird and Robert Denton. "Redescription of the Holotype of *Dryptosaurus acquilunguis* (Dinosauria: Theropoda) from the Upper Cretaceous of New Jersey." *Journal of Vertebrate Paleontology* 17, no. 3 (September 1997): 561–573.

____. "Comparative Cranial Anatomy of Two North American Cretaceous Plesiosaurs." *Ancient Marine Reptiles*. Eds. J. M. Calloway and E. L. Nicholls, San Diego: Academic Press 1997: 191–216.

____. (Interviewed by Allen A. Debus) *Dinosaur World* no. 4 (Spring 1998): 13–14.

____. "A Dinosaur Paleontologist's View of Godzilla." *The Official Godzilla Compendium*. (Eds.) J. D. Lees and Marc Cerasini, New York: Random House, 1998.

____. *Eggs, Nests, and Baby Dinosaurs: A New Look at Dinosaur Reproduction*. Bloomington: Indiana University Press, 1999.

Carroll, Robert L. *Vertebrate Paleontology and Evolution*. New York: W. H. Freeman and Company, 1988.

Charig, Peter. *A New Look at the Dinosaurs*. London: Heinemann in association with the British Museum (Natural History), 1979.

Chen, Pei-je, & Zhi-ming Dong and Shuo-nan Zhen. "An Exceptionally Well-Preserved Theropod Dinosaur from the Yixian Formation of China." *Nature* 391 (January 1998): 147–152.

Clarke, Noah T. "Method of Restoring the Cohoes Mastodon." *New York State Bulletin Eighteenth Report to the Director of the State Museum and Science Department, 1922*, no. 251 (March 1924): 131–133.

Cohen, I. Bernard. *Revolution in Science*. Cambridge: The Belknap Press of Harvard University Press, 1985.

Colbert, Edwin H. *Dinosaurs: Their Discovery and Their World*. New York: E. P. Dutton & Company, 1961.

____. *Men and Dinosaurs: The Search in Field and Laboratory*. New York: E. P. Dutton, 1968.

____. *Wandering Lands and Animals: The Story of Continental Drift*. New York: E. P. Dutton & Company, 1973.

____. *A Fossil Hunter's Notebook: My Life with Dinosaurs and Other Friends*. New York: E. P. Dutton, 1980.

____. *Dinosaurs: An Illustrated History*. Maplewood, NJ: Hammond, Inc., 1983.

Colbert, Margaret M. (Interviewed by Allen A. Debus) *Dinosaur World* no. 8 (Spring/Summer 2000): 50–52.

Conan Doyle, Arthur. *The Lost World*. New York: A. L. Burt Company Publishers, 1912.

Conybeare, William D. "On the Discovery of an Almost Perfect Skeleton of the *Plesiosaurus*." *Transactions of the Geological Society of London* 2, no. 1 (1824): 382–389.

Cooke, Patrick. "The Man Who Launched a Dinosaur." *Science 86* (April 1986): 27–35.

Cooley, Brian. "Sculpting Dinosaurs." *Alberta* 1, no. 1, Studies in the Arts and Sciences (1988): 119–130.

Coombs, Margery C. "Chalicotheroidea." *Evolution of Tertiary Mammals of North America: Terrestrial Carnivores, Ungulates and Ungulate Mammals*. Volume 1 (eds.) Christine M. Janis, Kathleen M. Scott and Louis Jacobs, Cambridge, England: Cambridge University Press, 1998.

Cope, Edward D. "The Fossil Reptiles of New Jersey." *American Naturalist* III (1869): 84–91.

Costello, Peter. *Jules Verne: Inventor of Science Fiction*. New York: Charles Scribner's Sons, 1978.

Crichton, Michael. *Jurassic Park*. New York: Alfred A. Knopf, 1990.

Culver, James Erwin. "Some Extinct Giants." *The California Illustrated Magazine* 2 (April 1892): 501–507.

Currie, Phillip, and Kevin Padian (eds.). *Encyclopedia of Dinosaurs*. San Diego: Academic Press, 1997.

Currie, Phillip, Mark Norrell and Ji Shu-An. "Two Feathered Dinosaurs from Northeastern China." *Nature* 393 (June 1998): 753–761.

Czerkas, Stephen A. "Reevaluation of the Plate Arrangement on *Stegosaurus stenops*." *Dinosaurs Past and Present* (Vol. 2). Eds. Sylvia Czerkas and Everett C. Olson, Seattle: Natural Museum of Los Angeles County in association with University of Washington Press, 1987.

____. "Skin." *Encyclopedia of Dinosaurs*. (Eds.) Phillip J. Currie and Kevin Padian, San Diego: Academic Press, 1997.

____. "The Beaked Jaws of *Stegosaurus* and Their Implications for Other Ornithischians." *Vertebrate Paleontology in Utah*. Utah Geological Survey, Miscellaneous Publication 99-1 (1999): 143–149.

____. "Is Flying Dinosaur a Composite?" *Prehistoric Times* no. 41 (April/May 2000): 5.

____. (Interviewed by Cliff Green) *Prehistoric Times* no. 43 (August/September 2000): 9–13.

____, and Sylvia J. Czerkas. *Dinosaurs: A Global View*. New York: Mallard Press, 1990.

Czerkas, Sylvia. "Reconstruction and Restoration." Phillip Currie and Kevin Padian (eds.), *Encyclopedia of Dinosaurs*. San Diego: Academic Press, 1997.

____, and Donald F. Glut. *Dinosaurs, Mammoths and Cavemen: The Art of Charles R. Knight*. New York: E. P. Dutton, 1982.

Dean, Dennis. *Gideon Mantell and the Discovery of Dinosaurs*. Cambridge: Cambridge University Press, 1999.

Debus, Allen A. "Historical Dinosaurs: Episodes in Discovery and Restoration." *Earth Sciences History* 12, no. 1 (1993): 60–69.

____. "Stegosaurus: A Plated Puzzle." *The Dinosaur Report*. (Winter 1994): 8–9.

____. "Bully for Laelaps." *The Dinosaur Society UK Quarterly Magazine* 2, nos. 1–2 (1996): 7–8.

____. "Sorting Fossil Vertebrate Iconography." *Fossil News: Journal of Amateur Paleontology* 3, nos. 10–11 (November 1997): 2–5.

____. "Dinosaurs Come to Life at Chicago's Navy Pier." *Fossil News: Journal of Amateur Paleontology* 5, no. 3 (March 1999): 16–18.

____. "Palaeoartists of the Past—Neave Parker's Prehistoric World." *The Dinosaur Society UK Quarterly Magazine* 3, no. 4 (2000): 16–18.

____. "Sculpting Twilight Zone Dinosaurs." *Modeler's Resource* no. 31 (December/January 2000): 58–61.

____. "Paleoart, Passion and Modern Meaning." *Prehistoric Times* no. 45 (December/January 2001): 17–18.

____. "Fin-tastic Mammals—A Quick Look at "Naosaurus." *Prehistoric Times* nos. 54-55, June/Jul.–Aug./Sept. 2002.

____, and Diane E. Debus. "Prehistoric Scenes." *Dinosaur Movies: The Complete History*, New York: Starlog Telecommunications, Inc. (1993): 81–86.

____, and ____. "An Interview with

Richard Rush." *Prehistoric Times* no. 13 (July/August 1995): 14–15.
_____, and _____. "Joe Who? *Jobaria*." *Fossil News: Journal of Avocational Paleontology* 6, no. 6 (June 2000): 5–8.
_____, and _____. *Dinosaur Memories: Dinotrekking for Beasts of Thunder and Fantastic Saurians.* Lincoln: iUniverse, 2002. ISBN 0-595-22988-3.
Debus, Allen A., and Steve McCarthy. "A Scene from American Deep Time." *The Mosasaur*, vol. 6 (May 1999): 105–115.
Debus, Allen A., Bob Morales and Diane E. Debus. *Dinosaur Sculpting: A Complete Beginners' Guide.* Bartlett, IL: Hell Creek Creations, 1995.
Debus, Allen A., and Gary R. Williams. "Realm of the Dinosaur—the Incredible Art Collection of John Lanzendorf." *Prehistoric Times* no. 23 (March/April 1997): 40–41.
Debus, Allen G. *Man and Nature in the Renaissance.* Cambridge: Cambridge University Press, 1978.
Desmond, Adrian J. *The Hot-Blooded Dinosaurs: A Revolution in Palaeontology.* New York: Warner Books, 1975.
_____. *Archetypes and Ancestors: Paleontology in Victorian England.* Chicago: The University of Chicago Press, 1982.
Dingus, Lowell, and Timothy Rowe. *The Mistaken Extinction: Dinosaur Evolution and the Origin of Birds.* New York: W. H. Freeman and Company, 1998.
Dixon, Dougal. *After Man: A Zoology of the Future.* New York: St. Martin's Press, 1981.
_____. *The Macmillan Illustrated Encyclopedia of Dinosaurs and Prehistoric Animals: a Visual Who's Who of Prehistoric Life.* New York: Macmillan Publishing Company, 1988.
_____. *The New Dinosaurs: An Alternative Evolution.* Topsfield, Mass.: Salem House Publishers, 1988.
_____. (Interviewed by Allen A. Debus) *Dinosaur World* no. 9 (Fall/Winter 2000/2001): 50–52.
Dodson, Peter. *The Horned Dinosaurs: A Natural History.* Princeton: Princeton University Press, 1996.
Dorr, John A., and Donald Eschman. *Geology of Michigan.* Ann Arbor: The University of Michigan Press, 1970.
Effinger, James A. "Entelodontidae." *Evolution of Tertiary Mammals of North America: Terrestrial Carnivores, Ungulates and Ungulate Mammals.* Volume 1 (eds.) Christine M. Janis, Kathleen M. Scott, and Louis Jacobs, Cambridge, England: Cambridge University Press, 1998.
Ehrenberg, Kurt. "Othenio Abels Werden und Wirken." *Mitt. Ges. Geol. Bergbaustad. Osterr.* 25 (1978): 271–295.
Evans, Arthur B. "The Illustrators of Jules Verne's *Voyages Extraordinaires*." *Science Fiction Studies* XXV: 2 (July 1998): 241–270.
Farlow, James, Carl V. Thompson, and Daniel Rosner. "Plates of the Dinosaur *Stegosaurus*: Forced Convection Heat Loss Fins?" *Science* 192, no. 4244 (June 1976): 1123–1125.
Farlow, James O., Matt B. Smith, and John M. Robinson. "Body Mass, Bone 'Strength Indicator,' and Cursorial Potential of *Tyrannosaurus rex*." *Journal of Vertebrate Paleontology* 15, no. 4 (December 1995): 713–725.
Farlow, James O., and Michael Brett-Surman (eds.). *The Complete Dinosaur.* Bloomington: Indiana University Press, 1997.
Feaver, William. *The Art of John Martin.* Oxford: Clarendon Press, 1975.
Fiffer, Steve. *Tyrannosaurus Sue.* New York: W. H. Freeman and Company, 2000.
Figuier, Guillaume Louis. *The World Before the Deluge.* London: Cassell & Company, Ltd., 1867.
Fischer, Paul E., and Dale A. Russell, et. al. "Cardiovascular Evidence for an Intermediate or Higher Metabolic Rate in an Ornithischian Dinosaur." *Science* 288 (April 2000): 503–505.
Flammarion, Camille. *Le Monde Avant la Creation de l' Homme.* Paris: C. Marpon and E. Flammarion, 1886.
Ford, Tracy L. *How to Draw Dinosaurs 1.* Dino Hunter, 1999.
France, Barbara. *The Cardiff Giant.* Cooperstown: The New York State Historical Association, 1990.
Franczak Brian. "Artist of the Month." *The Dinosaur Times* 1, no. 1 (Summer 1993): 21–24.
Fredericks, Michael, and Riff Smith. (Interviewed by Allen and Diane Debus) *Dinosaur World* no. 7 (Winter 1999/2000): 71–74.
_____, and _____. "Messmore and Damon's 'World a Million Years Ago' Toys." *Prehistoric Times* no. 45 (December/January 2001): 48–49.
Galton, Peter M. "The Mode of Life of *Hypsilophodon*, the Supposedly Arboreal Ornithopod Dinosaur." *Lethaia* 4 (1971): 453–464.
_____. "British Plated Dinosaurs (Ornithischia, Stegosauridae)." *Journal of Vertebrate Paleontology* 5, no. 3 (September 1985): 211–254.
Gibbons, Ann. "Dinosaur Fossils, in Fine Feather, Show Link to Birds." *Science* 280 (June 1998): 2051.
Gierlinski, Gerard. "The Furry Dino." *Dinosaur World* no. 4 (Spring 1998): 3–5.
Gillette, David. Obituary Notice for David Thomas. *Society of Vertebrate Paleontology News Bulletin* no. 177 (October 1999): 54–55.
Gilmore, Charles Whitney. "Osteology of the Armored Dinosauria in the United States Museum, with Special References to the Genus *Stegosaurus*." *Memoirs of the United States National Museum* no. 89, 1914.
Glut, Donald F. *Classic Movie Monsters.* Metuchen, NJ: The Scarecrow Press, Inc., 1978.
_____. *The Dinosaur Scrapbook.* Secaucus, NJ: Citadel Press, 1980.
_____. *The New Dinosaur Dictionary.* Secaucus, NJ: Citadel Press, 1982.
_____. *Dinosaurs: The Encyclopedia.* Jefferson, NC: McFarland & Company, Inc., Publishers, 1997.
_____. *Carbon Dates: A Day by Day Almanac of Paleo Anniversaries and Dino Events.* Jefferson, NC: McFarland & Company, Inc., Publishers, 1999.
_____. *Dinosaurs: The Encyclopedia (Supplement 1).* Jefferson, NC: McFarland & Company, Inc., Publishers, 2000.
Goetzmann, William H. *New Lands, New Men: America and the Second Great Age of Discovery.* New York: Viking Penguin, Inc., 1986.
Gould, Stephen, J. *Time's Arrow, Time's Cycle: Myth and Metaphor in the Discovery of Geological Time.* Cambridge: Harvard University Press, 1987.
_____. *Bully for Brontosaurus.* New York: W. W. Norton & Company, 1991.
_____. "Reconstructing (and Deconstructing) the Past." Preface to *The Book of Life: An Illustrated History of the Evolution of Life on Earth.* New York: W. W. Norton & Company, 1993.
_____. "Bent Out of Shape." *Eight Little Piggies.* New York: W. W. Norton, 1993.
_____. "The Lying Stones of Marakech." *The Lying Stones of Lying Marakech.* New York: Harmony Books, 2000.
Greene, John C. *The Death of Adam.* Ames: Iowa State University Press, 1959.
Gregory, William K. "The Weight of the Brontosaurus." *Science* XXII, no. 566 (September 1905): 572.

_____. "Erwin S. Christman, 1885–1921." *Natural History* 21 (1921): 620–625.

_____. "Obituary for William D. Matthew." *Natural History* (November/December 1930): 665.

Griesemer, Allan D., Roger A. Vandiver, and Michael Voorhies. "A Dinosaur Revival." *University of Nebraska–Lincoln Museum Notes* 56, no. 6 (September 22, 1976).

Grimes, Lee. "An Interview with Rudolph F. Zallinger." *Discovery*. 11 (Magazine of the Peabody Museum of Natural History). (Fall 1975): 33–35.

Gruss, Franz. "One of Germany's Premier Sculptors of Prehistoric Life, an Appreciation and Interview." *Dinosaur World* 7 (Winter 1999/2000): 35–38.

Gunning, William. *Life History of Our Planet*. New York: Worthington Co., 1879.

Gurney, James. *Dinotopia: A Land Apart from Time*. Atlanta: Turner Publishing, 1992.

_____. *Dinotopia: The World Beneath*. Atlanta: Turner Publishing, 1995.

_____. "How I Made the Stamps." *Prehistoric Times* no. 23 (March/April 1997): 8–9.

_____. (Interviewed by Allen A. Debus) *Dinosaur World* 1 no. 3 (October 1997): 21–23.

Guyot, William. *Creation or the Biblical Cosmogony in the Light of Modern Science*. New York: Charles Scribner's Sons, 1887.

Hall, R., and M. Hall. *About the Exhibits: Museum of Comparative Anatomy*. Cambridge: Harvard University, 1975.

Hallett, Mark. "Bringing Dinosaurs to Life." *Dinosaurs Past and Present*. (Vol. 1). Eds. Sylvia Czerkas and Everett C. Olson, Seattle: Natural Museum of Los Angeles County in association with University of Washington Press, 1987.

_____. (Interviewed by Mike Fredericks) *Prehistoric Times* no. 32 (October/November 1998): 7.

Hantman, Jeffrey. "Brooks Hall at the University of Virginia: Unraveling the Mystery." *The Magazine of Albemarle County History* 47 (1989).

Hecht, Jeff. "Piltdown Bird." *New Scientist* (January 29, 2000): 12.

_____. "Common as Much. Yesterday's Rock Stars Are Now a Bunch of Has-Beens." *New Scientist* (October 14, 2000): 18.

Heilmann, Gerhard. *The Origin of Birds*. London: Witherby, 1926.

Henderson, Douglas. (Interviewed by Mike Fredericks) *Prehistoric Times* no. 25 (August/September 1997): 32–35.

_____. "Restoring Dinosaurs as Living Animals." Phillip Currie and Kevin Padian (eds.), *Encyclopedia of Dinosaurs*. San Diego: Academic Press, 1997.

_____. *Asteroid Impact*. New York: Dial Books for Young Readers, 2000.

Hoaglund, Clayton. "They Gave Life to Bones." *Scientific Monthly* (February 1943): 114–133.

Holden, Constance. "Florida Meeting Shows Perils, Promise of Dealing for Dinos." *Science* 288 (April 2000): 238–239.

Holder, Frederic. "The California Academy of Sciences." *California Illustrated Magazine* 3 (December 1892–May 1893): 229–244.

Holland, William J. "A Review of Some Recent Criticisms of the Restorations of Sauropod Dinosaurs Existing in the Museums of the United States, with Special Reference to that of *Diplodocus carnegii*." *The American Naturalist* XLIV, no. 521 (May 1910): 259–283.

Holman, J. Alan. *Ancient Life of the Great Lakes Basin*. Ann Arbor: The University of Michigan Press, 1995.

Horner, John R. "Ecologic and Behavioral Implications Derived from a Dinosaur Nesting Site." *Dinosaurs Past and Present* (Vol. 2) Eds. Sylvia Czerkas and Everett C. Olson, Seattle: Natural Museum of Los Angeles County in association with University of Washington Press, 1987.

_____, and D. B. Weishampel. "Letters to Nature—A Comparative Embryological Study of Two Ornithischian Dinosaurs." *Nature* 383 (September 1996): 103.

Howe, S. R., T. Sharpe, and H. S. Torrens. *Icthyosaurs: A History of Fossil "Sea-Dragons."* National Museum of Wales, 1981.

Hsu, Kenneth, J. *The Great Dying*. New York: Harcourt Brace Jovanovich Publishers, 1986.

Hulke, James W. "An Attempt at a Complete Osteology of *Hypsilophodon foxii*." *Philosophical Transactions of the Royal Society of London* 173 (1883): 1035–1062.

Hutchinson, Henry Neville. *Extinct Monsters: A Popular Account of Some of the Larger Forms of Ancient Animal Life* 2nd ed. London: Chapman & Hall, 1893.

_____. *Extinct Monsters and Creatures of Other Days: A Popular Account of Some of the Larger Forms of Ancient Animal Life*. New York: D. Appleton and Company, 1917.

Huxley, Thomas H. "On *Hypsilophodon foxii*, a New Dinosaurian from the Wealden of the Isle of Wight." *The Quarterly Journal of the Geological Society of London* xxvi (1870—read November 10th 1869): 3–12.

_____. "On *Acanthopholis horridus*, a New Reptile from the Chalk Marl." *The Geological Magazine* IV in *The Scientific Memoirs of Thomas Henry Huxley* London: Macmillan, eds. M. Foster, and E. R. Lankester, E. R. (1898–1902).

Jacobs, Louis. *Quest for the African Dinosaurs*. New York: Villard Books, 1993.

Jaffe, Mark. *The Gilded Dinosaur: The Fossil War Between E. D. Cope and O. C. Marsh and the Rise of American Science*. New York: Crown Publishers, 2000.

Johnson, Rolf E., and John H. Ostrom. "The Forelimb of *Torosaurus* and an Analysis of the Posture and Gait of Ceratopsian Dinosaurs." *Functional Morphology in Vertebrate Paleontology*. Ed. J. Thomason, Cambridge: Cambridge University Press, 1995.

Jonas, Coloman. *From "Stuffing" to Sculpturdermy: The Jonas Story*. Denver: Jonas Brothers, Inc., 1983.

Jones, Terry D., James O. Farlow, John A. Ruben, Donald M. Henderson, and William J. Hillenius. "Cursioriality in Bipedal Archosaurs." *Nature* 406 (August 2000): 716–718.

Kemp, Christopher. "Attack of the Missionary Lizards." *Louisville Eccentric Observer* (September 20, 2000): 14–16.

Kielan-Jaworowska Zofia. *Hunting for Dinosaurs*. MIT Press, 1969.

Kish, Ely. (Interviewed by Tony Campagna) *Prehistoric Times* no. 38 (October/November 1999): 9–10.

Knight, Charles R. *Before the Dawn of History*. New York: McGraw-Hill Book Company, 1935.

_____. *Prehistoric Man: The Great Adventurer*. New York: Appleton-Century-Crofts, Inc., 1949.

Knipe, Henry R. *Nebula to Man*. New York: J. M. Dent & Company, 1905.

_____. *Evolution in the Past*. New York: J. B. Lippincott Company, 1912.

Kosch, Bradley F. "A Revision of the Skeletal Reconstruction of *Shonisaurus popularis (Reptilia* Ichthyosauria)," *Journal of Vertebrate Paleontology* 10, no. 4 (December 1990): 512–514.

Kraus, Caroline. "The Art of Extinction."

Metropolitan Detroit (October 1997): 110–113, 167.

Kuchnio, Krzysztof. (Interviewed by Allen A. Debus) *Dinosaur World* 6 (Spring/Summer 1999): 37–40.

Kuhn, Thomas. *The Structure of Scientific Revolutions*. Chicago: University of Chicago Press, 1970.

Lankester, E. Ray. *Extinct Animals*. New York: H. Holt, 1905.

Lanzendorf, John, and Michael Skrepnick. (Interviewed by Allen A. Debus) *Dinosaur World* nos. 7–8 (Winter/1999/2000): 63–67, and (Spring/Summer 2000): 40–49.

Laporte, Leo F. *George Gaylord Simpson: Paleontologist and Evolutionist*. New York: Columbia University Press, 2000.

Lathrop, Ed. "The Jonas Studios." *Prehistoric Times* no. 5 (March/April 1994): 10–11.

Lerangis, Peter. *Time Machine 22: Last of the Dinosaurs*. New York: Byron Preiss, 1988.

Lessem, Don. "Skinning the Dinosaur." *Discover* (March 1989): 39–43.

_____. *Bigger Than T. rex: The Discovery of he Biggest Meat-Eating Dinosaur Ever Found*. New York: Crown Publishers, Inc., 1997.

_____, and Donald F. Glut. *Dinosaur Encyclopedia*. Random House of Canada, Ltd., 1993.

Liston, Jeff. "*Archaeopteryx* and the Evolution of Feathered Flight." *The Quarterly Journal* UK Dinosaur Society) 4, no. 1 (2000): 14.

Long, John A. "Sea Monsters from the Age of Dinosaurs: The Largest Marine Reptiles That Ever Lived." *Dinonews* no. 3 (Western Australian Museum) (September 1991): 6–7.

_____. *Dinosaurs of Australia and New Zealand: And Other Animals of the Mesozoic Era*. Cambridge: Harvard University Press, 1998.

Longman, Heber A. "Some Queensland Fossil Vertebrates." *Memoirs of the Queensland Museum* 8 (1924): 26.

Lucas, Frederic A. "The Restoration of Extinct Animals." *Annual Report of the Board of Regents of the Smithsonian Institution: Showing the Operations, Expenditures and Condition of the Institution for the Year Ending June 30, 1900*. Washington D.C.: Smithsonian (1901): 479–493.

Lull, Richard Swann. "Restoration of the Horned Dinosaur, *Diceratops*." *American Journal of Science* 20, series 4 (1905): 420–422.

Magee, H. W. "The World a Million Years Ago." *Popular Mechanics* 58 (1932): 824–828.

Mantell, Gideon. *Wonders of Geology: Or, a Familiar Exposition of Geological Phenomena. Being the Substance of a Course of Lectures Delivered at Brighton, from Notes Taken by G. F. Richardson, Curator of the Mantellian Museum* (2 volumes). London: Relfe and Fletcher, 1838.

_____. *Petrifactions and Their Teachings: Or, a Hand-Book to the Gallery of Organic Remains of the British Museum*. London: Bohn, 1851.

Marsh, O. C. "The Wings of Pterodactyles." *American Journal of Science* 23, no. 3 (1882): 251–256.

Massare, Judy A. "Ancient Mariners." *Natural History* (September 1992): 49–52.

Mayor, Adrienne. *The First Fossil Hunters: Paleontology in Greek and Roman Times*. Princeton: Princeton University Press, 2000.

McCarthy, Steve. "Owen's Visions: The Crystal Palace Dinosaurs." *Dinosaur World* 1, no. 1 (February 1997): 41.

_____. "Restoring the Vision: Saving the Crystal Palace Dinosaurs." *Dinosaur World* 1, no. 2 (June 1997): 19–22.

_____, and Mick Gilbert. *The Crystal Palace Dinosaurs: The Story of the World's First Prehistoric Sculptures*. London: The Crystal Palace Foundation, 1994.

McGloughlin, John C. *Archosauria: A New Look at the Old Dinosaur*. London: Allen Lane, 1979.

McGowan, Christopher. "The Ichthyosaurian Tailbend: A Verification Problem Facilitated by Computed Tomography." *Paleobiology* 15, no. 4 (1989): 429–436.

_____. *Dinosaurs, Spitfires and Dragons*. Cambridge: Harvard University Press, 1991.

_____. "Giant Ichthyosaurs of the Early Jurassic." *Canadian Journal of Earth Sciences* 33 (1996): 1011–1021.

_____. "Transitional Ichthyosaur Fauna." *Ancient Marine Reptiles*. Jack M. Calloway (ed.), San Diego: Academic Press, 1997.

Merritt, Zella Jonas. "The World of Louis Paul Jonas." *American Artist* (January 1967).

Messmore, Francis B. "The History of the World a Million Years Ago." *Dinosaur World* 1, no. 3 (October 1997): 27–32.

Mitchell, William J. T. *The Last Dinosaur Book: The Life and Times of a Cultural Icon*. Chicago: The University of Chicago Press, 1998.

Muller, Richard. *Nemesis: The Death Star*. New York: Weidenfeld & Nicholson, 1988.

Naish, Darren. "Tree-Climbing Dinosaurs." *Quarterly: The Journal of the Dinosaur Society* 4, no. 1 (2000): 20–24.

Narendra, Barbara. "From the Archives: The Early Years of the Great Hall." *Discovery (Magazine of the Yale Peabody Museum of Natural History)* 22, no. 1 (1990): 18–23.

Nicholls, Robert. "*Diplodocus longus*—The Real Double Beam." *Down to Earth* 31 (May 2000): 5.

Nopsca, Francis Baron. "Notes on British Dinosaurs—Part 1: *Hypsilophodon*." *Geological Magazine* 5 (May 1905): 205.

_____. "Notes on British Dinosaurs—Part 2: *Polacanthus*." *Geological Magazine* 5 (June 1905): 241–250.

Norell, Mark, Luis Chiappe, and James Clark. "New Limb on the Avian Family Tree." *Natural History* (September 1993): 38–42.

Norman, David. *The Illustrated Encyclopedia of Dinosaurs*. New York: Crescent Books, 1985.

_____. *Dinosaur!* New York: Prentice Hall, 1991.

Novas, Fernando E., and Pablo F. Puerta. "New Evidence Concerning Avian Origins from the Late Cretaceous of Patagonia." *Nature* 387 (May 1997): 390–397.

Olshevsky, George. "The Ankylosaur Family Tree." 1979 (self-published wall chart).

_____ "The Origin and Evolution of the *Stegosaurus*." *Kyoryugaku Saizenzen (Frontiers of Dinosaurology—Gakken Mook)* no. 4 (1993): 65–102 (English translation provided by Mr. Olshevsky).

_____. "The Birds First? A Theory to Fit the Facts." *Omni* (July 1994): 34–43, 80–86.

Osborn, Henry Fairfield. "Prehistoric Quadrupeds of the Rockies." *The Century Magazine* LII (September 1896): 705–715.

_____. "A Great Naturalist: Edward Drinker Cope." *The Century Magazine* LV (November 1897): 10–15.

_____. "Fossil Wonders of the West." *The Century Magazine* 68 (1904): 689.

_____. *The Titanotheres of Ancient Wyoming* (Monograph 55) Vols. 1 and 2. Washington D.C. U. S. Government Printing Office: U.S. Geological Survey—Department of the Interior, 1929.

_____. *Cope: Master Naturalist.* Princeton: Princeton University Press, 1931.

Ostrom, John H. "Osteology of *Deinonychus antirrhopus*, an Unusual Theropod from the Lower Cretaceous of Montana." (Bulletin 30) New Haven: Peabody Museum of Natural History—Yale University (July 1969).

_____. "The Supporting Chain." *Discovery* 5, no. 1, Magazine of the Peabody Museum of Natural History (Fall 1969): 10–16.

_____. "*Archaeopteryx* and the Origin of Flight." *Quarterly Review of Biology* 49 (1974): 27–47.

Owen, Richard. "Report on British Fossil Reptiles." *Report of the British Association for the Advancement of Science* II (1842): 60–204.

_____. *A History of Fossil Mammals and Birds.* London: J. Van Voorst, 1846.

Padian, Kevin. "The Case of the Bat-Winged Pterosaur: Typological Taxonomy and the Influence of Pictorial Representation on Scientific Perception." *Dinosaurs Past and Present* (Vol. 2). Eds. Sylvia Czerkas and Everett C. Olson, Seattle: Natural Museum of Los Angeles County in association with University of Washington Press, 1987.

Parrish, Michael. "The Big Guy's Back, Flying High Again and Creating a Flap." *Smithsonian* (March 1986): 73–81.

Paul, Gregory S. "The Science and Art of Restoring the Life Appearance of Dinosaurs and Their Relatives—a Rigorous How-to Guide." *Dinosaurs Past and Present* (Vol. 2). Eds. Sylvia Czerkas and Everett C. Olson, Seattle: Natural Museum of Los Angeles County in association with University of Washington Press, 1987.

_____. *Predatory Dinosaurs of the World: A Complete Illustrated Guide.* New York: Simon & Schuster, 1988.

_____. (Interviewed by Bob Morales—Parts 1 and 2) *Prehistoric Times* nos. 35–36 (April/May 1999): 9–11; and (June/July 1999): 9–11.

_____ (ed.). *The Scientific American Book of Dinosaurs.* New York: St. Martin's Press, 2000.

Pereda-Suberbiola, Javier. "*Hylaeosaurus, Polacanthus*, and the Systematics and Stratigraphy of Wealden Armoured Dinosaurs." *Geological Magazine* 130 (1993): 767–781.

Pettigrew, Neil. *The Stop-Motion Filmography.* Jefferson, NC: McFarland & Company, Inc., Publishers, 1999.

Porter, Elbert H. *Dinosaurs Return: Elbert H. Porter—Dinosaur Sculptor.* Davis printing Co., 1982.

Powell, James Lawrence. *Night Comes to the Cretaceous.* New York: W. H. Freeman and Company, 1998.

Psihoyos, Louis. *Hunting Dinosaurs.* New York: Random House, 1994.

Publications International. *Encyclopedia of Dinosaurs: A Visual Guide to Prehistoric Life with Over 140 Dinosaur Profiles.* New York: Beekman House, 1990.

Rainger, Ronald. *An Agenda for Antiquity: Henry Fairfield Osborn and Vertebrate Paleontology at the American Museum of Natural History.* Tuscaloosa: The University of Alabama Press, 1991.

Reed, Michael, and Jean Sellar. "Ziggy Finds New Home at Field Museum." *Field Museum of Natural History Bulletin* 53, no. 10 (November 1982): 8.

Rovin, Jeff. *From the Land Beyond Beyond: The Making of the Movie Monsters You've Known and Loved.* New York: Berkley Publishing Company Corporation, 1977.

Rudwick, Martin J. S. *The Meaning of Fossils.* Chicago: The University of Chicago Press, 1985.

_____. *Scenes from Deep Time: Early Pictorial Representations of the Prehistoric World.* Chicago: The University of Chicago Press, 1992.

_____. *Georges Cuvier, Fossil Bones and Geological Catastrophes: New Translations and Interpretations of the Primary Texts.* Chicago: The University of Chicago Press, 1997.

Runestead, Todd. "The Making of the Terminator Pig." *Museum Quarterly Notes*, Denver Museum of Natural History (1994).

Rupke, Nicolas, A. "The Picturing of Geological Time." *History of Science* 36, no. 111 (March 1998): 61–90.

Russell, Dale. "Models and Paintings of North American Dinosaurs." *Dinosaurs Past and Present* (Vol. 1). Eds. Sylvia Czerkas and Everett C. Olson, Seattle: Natural Museum of Los Angeles County in association with University of Washington Press, 1987.

_____. *An Odyssey in Time: The Dinosaurs of North America.* Toronto: University of Toronto Press in Association with the National Museum of Canada, 1989.

Russell, Loris. "Body Temperature of Dinosaurs and Its Relationship to Their Extinction." *Journal of Paleontology* 39, no. 3 (May 1965): 497–501.

Ryder, Richard C. "Dusting Off America's First Dinosaur." *American Heritage* 39, no. 2 (March 1988): 68–73.

Sabath, Karol. "Poland's 'Valley of the Dinosaurs.'" *Dinosaur World* no. 6 (Spring/Summer 1999): 33–36.

Sagan, Carl. *The Dragons of Eden.* New York: Ballantine Books, 1977.

_____. *Cosmos.* New York: Ballantine Productions, 1980.

_____, and Ann Druyan. *Comet.* New York: Random House, 1985.

Sander, Martin P. "The Paleobiogeography of *Shastasaurus.*" *Ancient Marine Reptiles.* Jack M. Calloway (ed.), San Diego: Academic Press, 1987.

Savage, R. J. G., and M. R. Long. *Mammal Evolution: An Illustrated Guide.* New York: Facts on File Publications, 1986.

Scheele, William E. *The First Mammals.* New York: The World Publishing Company, 1955.

Schoell, William. "The Making of Jules Verne's *Journey to the Center of the Earth.*" *Filmfax: The Magazine of Unusual Film and Television* no. 55 (March/April 1996): 51–57, 74.

Schuchert, Charles, and Clara LeVene. *O. C. Marsh: Pioneer Paleontologist.* New Haven: Yale University Press, 1940.

Scott, William Berryman. *A History of Land Mammals in the Western Hemisphere.* New York: The Macmillan Company, 1937.

Scully, Vincent, Rudolph Zallinger, and Leo J. Hickey. *The Age of Reptiles: The Great Dinosaur Mural at Yale.* New Haven: Peabody Museum of Natural History, 1990.

Seeley, H. G. *Dragons of the Air.* London: Methuen, 1901.

Semonin, Paul. *American Monster: How the Nation's First Prehistoric Creature Became a Symbol of National Identity.* New York: New York University Press, 2000.

Sereno, Paul. "Early Cretaceous Dinosaurs from the Sahara." *Science* 226 (October 1994): 267–271.

_____. "Predatory Dinosaurs from the Sahara and Late Cretaceous Faunal Differentiation." *Science* 272 (May 1996): 986–991.

_____. "Africa's Dinosaur Castaways." *National Geographic* (June 1996): 106–119.

_____. (Interviewed by Allen A. Debus) *Dinosaur World* no. 4 (Spring 1998): 11–12.

_____. (Interviewed by Tony Campagna) *Prehistoric Times* no. 31 (August/September 1998): 36–38, 48, 52.

_____. "A Long-Snouted Predatory Dinosaur from Africa and the Evolution of Spinosaurids." *Science* 282 (November 1998): 1298–1302.
_____. "The Evolution of Dinosaurs." *Science* 284 (June 1999): 2137–2147.
_____, et al. "Cretaceous Sauropods from the Sahara and the Uneven Rate of Skeletal Evolution Among Dinosaurs." *Science* 286 (November 1999): 1342–1347.
Shay, Don, and Jody Duncan. *The Making of Jurassic Park*. Ballantine Trade, 1993.
Shipman, Pat. *Taking Wing: Archaeopteryx and the Evolution of Bird Flight*. New York: Simon & Schuster, 1998.
Sibbick, John. (Interviewed by Mike Fredericks) *Prehistoric Times* no. 26 (September/October 1997): 22–25.
Simons, Lewis M. "Archaeoraptor Fossil Trail." *National Geographic* 198 (October 2000): 128–133.
Skrepnick, Michael. (Interviewed by Allen A. Debus) *Dinosaur World* no. 8 (Spring 2000): 47–48.
Slater, Barbara A. "A Visit to Munchehagen." *Dinosaur World* 5 (Summer/Fall 1998): 72–73.
Sloan, Christopher. "Feathers for *T. rex*: New Bird-Like Fossils Are Missing Links in Dinosaur Evolution?" *National Geographic* 196 (November 1999): 98–107.
Smith, Dwight, and Gary S. Mangiacopra. "Carl Hagenbeck and the Rhodesian Dinosaurs." *Strange Magazine* no. 5 (1990): 50–52.
Smith, Riff. "A Trip Back in Time: A Journey to Prehistoric Gardens." *Prehistoric Times* no. 33 (December/January 1998/1999): 6–7.
Spinar, Zdenek. *Life Before Man. The Fascinating Story of the Evolution of Living Things on Our Earth*. New York: Crescent Books, 1981.
Staab, Gary. (Interviewed by Allen A. Debus) *Dinosaur World* no. 9 (Winter 2000/2001): 96–98.
Stern, Jane, and Michael Stern. *The Encyclopedia of Bad Taste*. New York: Harper Collins, 1990.
Stevens, Kent A. "I Walked with a Dinosaur!" *Quarterly—The UK Dinosaur Society Quarterly Magazine* 3, no. 4 (2000): 10–11.
_____, and Michael J. Parrish. "Neck Posture and Feeding Habits of Two Jurassic Sauropod Dinosaurs." *Science* 284 (April 1999): 798–799.
Stokstad, Erik. "Fossils Made to Order." *Science* 288 (June 2000): 1729.
_____. "Feathers, or Flight of Fancy?" *Science* 288 (June 2000): 2124–2125.
_____. "Tiny, Feathered Dino Is Most Birdlike Yet." *Science* 290 (December 2000): 1871–1872.
_____. "Doubts Raised About Dinosaur Heart." *Science* 291 (February 2001): 811.
Stout, William. *The Dinosaurs: A Fantastic New View of a Lost Era*. New York: A Byron Preiss Book, 1981.
_____. *The New Dinosaurs*. New York: A Byron Preiss Book, 2000.
_____. (Interviewed by Rick Savage—Parts 1 and 2) *Prehistoric Times* nos. 44–45 (October/November 2000): 34–37; and (December/January 2000/2001): 28–29.
Swinton, William E. "A New Exhibit of *Hypsilophodon*." *Natural History Magazine* 5, London: The Trustees of the British Museum (1936): 331–336.
_____. "Notes on the Osteology of *Hypsilophodon*, and on the Family Hypsilophontidae." *Proceedings of the General Meetings for Scientific Business of the Zoological Society of London* (1936): 555–578.
_____. *Fossil Amphibians and Reptiles*. London: British Museum (Natural History), 1965.
_____. *Dinosaurs*. London: British Museum—Natural History, publication no. 42, 1969.
_____. *The Dinosaurs*. New York: Wiley Interscience, 1970.
Symmes, John Cleves. *Symzonia: Voyage of Discovery*. New York: J. Seymour, 1820 (reprinted by New York: Arno Press, 1975).
Taylor, Michael. "Before the Dinosaur: The Historical Significance of the Fossil Marine Reptiles." *Ancient Marine Reptiles*. San Diego: Academic Press, 1997.
Thomas, David. (Interviewed by Allen A. Debus) *Dinosaur World* no. 5 (Summer/Fall 1998): 49–53.
Trinkaus, Erik, and Pat Shipman. *The Neandertals: Changing the Image of Mankind*. New York: Alfred A. Knopf, 1993.
Tropea, Michael (photographer). *Dinosaur Imagery: The Science of Lost Worlds and Jurassic Art—The Lanzendorf Collection*. San Diego, Academic Press: 2000.
Turner, Susan. "Vertebrate Paleontology in Australia." *Earth Sciences History* 5, no. 1 (1986): 50–65.
Verne, Jules. *Journey to the Center of the Earth: A New Translation by William Butcher*. Oxford: Oxford University Press, 1988.
Vickers-Rich, Patricia, and Thomas Hewitt Rich. "Australia's Polar Dinosaurs." *Scientific American* (July 1993): 50–55.
Walker, Molly. "Celebrating 40 Years of Building a Boyhood Dream: Ernie Nelson's Love of Prehistoric Gardens." *Rogue Coast* (August 25, 1993): 13–18.
Wallace, David Rains. *The Bonehunters' Revenge: Dinosaur Greed and the Greatest Scientific Feud of the Gilded Age*. Boston: Houghton Mifflin Company, 1999.
Wang, Justin. "Scientists Flock to Explore China's 'Site of the Century.'" *Science* 279 (March 1998): 1626–1627.
Ward, Henry A. *Catalogue of Casts of Fossils from the Principal Museums of Europe and America with short Descriptions and Illustrations*. Rochester, New York, 1866.
_____. *The Mammoth: Being Articles Concerning This Celebrated Specimen Which Have Appeared in Various American and European Newspapers to Which Is Added 'The Last of the Mammoths.'* Stevenson, Foster & Co., 1879.
Ward, Roswell. *Henry A. Ward: Museum Builder to America*. Publication XXIV. New York: The Rochester Historical Society, 1948.
Warren, Leonard. *Joseph Leidy: The Last Man Who Knew Everything*. New Haven: Yale University Press, 1998.
Watson, Andrew, and Erik Stokstad. "Dinosaur Docudrama Blends Fact, Fantasy." *Science* 288 (April 2000): 29–31.
Watson, Jane Werner. *The Giant Golden Book of Dinosaurs and Other Prehistoric Animals*. New York: Golden Press, 1965.
Weissman, Paul R. "Are Periodic Bombardments Real?" *Sky and Telescope* (March 1990).
Westphal, Klaus. (Interviewed by Allen A. Debus) *Dinosaur World* no. 8 (Spring/Summer 2000): 84–89.
Williams, Gary. "The Canadian Connection: Canada's 'Hidden' Prehistoric World." *Prehistoric Times* no. 9 (November/December 1995): 18.
_____. "Prehistoric Journeys of Karel Zeman." *Dinosaur World* 1, no. 3 (October 1997): 49–52.
Williston, Samuel W. "Restorations of Some American Permocarboniferous Amphibians and Reptiles." *Journal of Geology* 22, no. 1 (1914): 57–70.

_____. *Water Reptiles of the Past*. Chicago: The University of Chicago Press, 1914.

Winchell, D. Alexander. *Sketches of Creation: A Popular View of Some of the Grand Conclusions of the Sciences in Reference to the History of Matter and of Life Together with a Statement of the Intimation of Science Respecting the Primordial Condition and the Ultimate Destiny of the Earth and the Solar System*. New York: Harper & Brothers, 1874.

Womack, Todd. "Plentifully Charged with Fossils: The 1822 Discovery of the *Eremotherium* at Skidaway." *Fossil News: Journal of Avocational Paleontology* 6, no. 7 (July 2000): 14–16.

Xu, Xing, Xiao-Lin Wang, and Xiao-Chun Wu. "A Dromaeosaurid Dinosaur with a Filamentous Integument from the Yixian of China." *Nature* 401 (September 1999): 262–266.

Xu, Xing, Zhonghe Zhou, and Xiaolin Wang. "The Smallest Known Non-Avian Theropod Dinosaur." *Nature* 408 (December 2000): 705–708.

Zimmer, Carl. "Ruffled Feathers." *Discover* (May 1992): 44–54.

_____ "Dinosaurs in Motion." *Discover* (November 1997): 96–109.

_____. "A Sickle in the Clouds." *Discover* (June 1998): 32.

_____. *At the Water's Edge: Macroevolution and the Transformation of Life*. New York: Simon & Schuster, Inc., 1998.

Index

Numbers in **bold** indicate illustrations.